CAMBRIDGE STUDIES IN ADVANCED MATHEMATICS

AUTOMORPHIC FORMS AND L-FUNCTIONS FOR THE GROUP *GL* (*n*, R)

L-functions associated with automorphic forms encode all classical number theoretic information. They are akin to elementary particles in physics. This book provides an entirely self-contained introduction to the theory of L-functions in a style accessible to graduate students with a basic knowledge of classical analysis, complex variable theory, and algebra. Also within the volume are many new results not yet found in the literature. The exposition provides complete detailed proofs of results in an easy-to-read format using many examples and without the need to know and remember many complex definitions. The main themes of the book are first worked out for GL(2,R) and GL(3,R), and then for the general case of GL(n,R). In an appendix to the book, a set of *Mathematica*® functions is presented, designed to allow the reader to explore the theory from a computational point of view.

CAMBRIDGE STUDIES IN ADVANCED MATHEMATICS

Already published

30 D.J. Benson *Representations and cohomology I*
31 D.J. Benson *Representations and cohomology II*
32 C. Allday & V. Puppe *Cohomological methods in transformation groups*
33 C. Soule *et al. Lectures on Arakelov geometry*
34 A. Ambrosetti & G. Prodi *A primer of nonlinear analysis*
35 J. Palis & F. Takens *Hyperbolicity, stability and chaos at homoclinic bifurcations*
37 Y. Meyer *Wavelets and operators I*
38 C. Weibel *An introduction to homological algebra*
39 W. Bruns & J. Herzog *Cohen-Macaulay rings*
40 V. Snaith *Explicit Brauer induction*
41 G. Laumon *Cohomology of Drinfeld modular varieties I*
42 E.B. Davies *Spectral theory and differential operators*
43 J. Diestel, H. Jarchow, & A. Tonge *Absolutely summing operators*
44 P. Mattila *Geometry of sets and measures in Euclidean spaces*
45 R. Pinsky *Positive harmonic functions and diffusion*
46 G. Tenenbaum *Introduction to analytic and probabilistic number theory*
47 C. Peskine *An algebraic introduction to complex projective geometry*
48 Y. Meyer & R. Coifman *Wavelets*
49 R. Stanley *Enumerative combinatorics I*
50 I. Porteous *Clifford algebras and the classical groups*
51 M. Audin *Spinning tops*
52 V. Jurdjevic *Geometric control theory*
53 H. Volklein *Groups as Galois groups*
54 J. Le Potier *Lectures on vector bundles*
55 D. Bump *Automorphic forms and representations*
56 G. Laumon *Cohomology of Drinfeld modular varieties II*
57 D.M. Clark & B.A. Davey *Natural dualities for the working algebraist*
58 J. McCleary *A user's guide to spectral sequences II*
59 P. Taylor *Practical foundations of mathematics*
60 M.P. Brodmann & R.Y. Sharp *Local cohomology*
61 J.D. Dixon *et al. Analytic pro-P groups*
62 R. Stanley *Enumerative combinatorics II*
63 R.M. Dudley *Uniform central limit theorems*
64 J. Jost & X. Li-Jost *Calculus of variations*
65 A.J. Berrick & M.E. Keating *An introduction to rings and modules*
66 S. Morosawa *Holomorphic dynamics*
67 A.J. Berrick & M.E. Keating *Categories and modules with K-theory in view*
68 K. Sato *Levy processes and infinitely divisible distributions*
69 H. Hida *Modular forms and Galois cohomology*
70 R. Iorio & V. Iorio *Fourier analysis and partial differential equations*
71 R. Blei *Analysis in integer and fractional dimensions*
72 F. Borceaux & G. Janelidze *Galois theories*
73 B. Bollobas *Random graphs*
74 R.M. Dudley *Real analysis and probability*
75 T. Sheil-Small *Complex polynomials*
76 C. Voisin *Hodge theory and complex algebraic geometry I*
77 C. Voisin *Hodge theory and complex algebraic geometry II*
78 V. Paulsen *Completely bounded maps and operator algebras*
79 F. Gesztesy & H. Holden *Soliton Equations and their Algebro-Geometric Solutions Volume 1*
81 Shigeru Mukai *An Introduction to Invariants and Moduli*
82 G. Tourlakis *Lectures in logic and set theory I*
83 G. Tourlakis *Lectures in logic and set theory II*
84 R.A. Bailey *Association Schemes*
85 James Carlson, Stefan Müller-Stach, & Chris Peters *Period Mappings and Period Domains*
86 J.J. Duistermaat & J.A.C. Kolk *Multidimensional Real Analysis I*
87 J.J. Duistermaat & J.A.C. Kolk *Multidimensional Real Analysis II*
89 M. Golumbic & A.N. Trenk *Tolerance Graphs*
90 L.H. Harper *Global Methods for Combinatorial Isoperimetric Problems*
91 I. Moerdijk & J. Mrcun *Introduction to Foliations and Lie Groupoids*
92 János Kollár, Karen E. Smith, & Alessio Corti *Rational and Nearly Rational Varieties*
93 David Applebaum *Lévy Processes and Stochastic Calculus*
95 Martin Schechter *An Introduction to Nonlinear Analysis*

See http:www.cambridge.org for a complete list of books available in this series

Automorphic Forms and L-Functions for the Group *GL* (n, R)

DORIAN GOLDFELD

Columbia University

With an Appendix by Kevin A. Broughan

University of Waikato

CAMBRIDGE
UNIVERSITY PRESS

CAMBRIDGE
UNIVERSITY PRESS

University Printing House, Cambridge CB2 8BS, United Kingdom

One Liberty Plaza, 20th Floor, New York, NY 10006, USA

477 Williamstown Road, Port Melbourne, VIC 3207, Australia

314-321, 3rd Floor, Plot 3, Splendor Forum, Jasola District Centre, New Delhi - 110025, India

103 Penang Road, #05-06/07, Visioncrest Commercial, Singapore 238467

Cambridge University Press is part of the University of Cambridge.

It furthers the University's mission by disseminating knowledge in the pursuit of education, learning and research at the highest international levels of excellence.

www.cambridge.org
Information on this title: www.cambridge.org/9781107565029

First published 2006
First paperback edition 2015

A catalogue record for this publication is available from the British Library

ISBN 978-0-521-83771-2 Hardback
ISBN 978-1-107-56502-9 Paperback

Dedicated to Ada, Dahlia, and Iris

Contents

Introduction

The theory of automorphic forms and L-functions for the group of $n \times n$ invertible real matrices (denoted $GL(n, \mathbb{R})$) with $n \geq 3$ is a relatively new subject. The current literature is rife with 150+ page papers requiring knowledge of a large breadth of modern mathematics making it difficult for a novice to begin working in the subject. The main aim of this book is to provide an essentially self-contained introduction to the subject that can be read by someone with a mathematical background consisting only of classical analysis, complex variable theory, and basic algebra – groups, rings, fields. Preparation in selected topics from advanced linear algebra (such as wedge products) and from the theory of differential forms would be helpful, but is not strictly necessary for a successful reading of the text. Any Lie or representation theory required is developed from first principles.

This is a low definition text which means that it is not necessary for the reader to memorize a large number of definitions. While there are many definitions, they are repeated over and over again; in fact, the book is designed so that a reader can open to almost any page and understand the material at hand without having to backtrack and awkwardly hunt for definitions of symbols and terms.

The philosophy of the exposition is to demonstrate the theory by simple, fully worked out examples. Thus, the book is restricted to the action of the discrete group $SL(n, \mathbb{Z})$ (the group of invertible $n \times n$ matrices with integer coefficients) acting on $GL(n, \mathbb{R})$. The main themes are first developed for $SL(2, \mathbb{Z})$ then repeated again for $SL(3, \mathbb{Z})$, and yet again repeated in the more general case of $SL(n, \mathbb{Z})$ with $n \geq 2$ arbitrary. All of the proofs are carefully worked out over the real numbers $\mathbb{R}$, but the knowledgeable reader will see that the proofs will generalize to any local field. In line with the philosophy of understanding by simple example, we have avoided the use of adeles, and as much as possible the theory of representations of Lie groups. This very explicit language appears particularly useful for analytic number theory where precise growth estimates of L-functions and automorphic forms play a major role.

The theory of L-functions and automorphic forms is an old subject with roots going back to Gauss, Dirichlet, and Riemann. An L-function is a Dirichlet series

$$\sum_{n=1}^{\infty} \frac{a_n}{n^s}$$

where the coefficients a_n, $n = 1, 2, \ldots$, are interesting number theoretic functions. A simple example is where a_n is the number of representations of n as a sum of two squares. If we knew a lot about this series as an analytic function of s then we would obtain deep knowledge about the statistical distribution of the values of a_n. An automorphic form is a function that satisfies a certain differential equation and also satisfies a group of periodicity relations. An example is given by the exponential function $e^{2\pi ix}$ which is periodic (i.e., it has the same value if we transform $x \to x + 1$) and it satisfies the differential equation $\frac{d^2}{dx^2}e^{2\pi ix} = -4\pi^2 e^{2\pi ix}$. In this example the group of periodicity relations is just the infinite additive group of integers, denoted $\mathbb{Z}$. Remarkably, a vast theory has been developed exposing the relationship between L-functions and automorphic forms associated to various infinite dimensional Lie groups such as $GL(n, \mathbb{R})$.

The choice of material covered is very much guided by the beautiful paper (Jacquet, 1981), titled *Dirichlet series for the group* $GL(n)$, a presentation of which I heard in person in Bombay, 1979, where a classical outline of the theory of L-functions for the group $GL(n, \mathbb{R})$ is presented, but without any proofs. Our aim has been to fill in the gaps and to give detailed proofs. Another motivating factor has been the grand vision of Langlands' philosophy wherein L-functions are akin to elementary particles which can be combined in the same way as one combines representations of Lie groups. The entire book builds upon this underlying hidden theme which then explodes in the last chapter.

In the appendix a set of Mathematica functions is presented. These have been designed to assist the reader to explore many of the concepts and results contained in the chapters that go before. The software can be downloaded by going to the website given in the appendix.

This book could not have been written without the help I have received from many people. I am particularly grateful to Qiao Zhang for his painstaking reading of the entire manuscript. Hervé Jacquet, Daniel Bump, and Adrian Diaconu have provided invaluable help to me in clarifying many points in the theory. I would also like to express my deep gratitude to Xiaoqing Li, Elon Lindenstrauss, Meera Thillainatesan, and Akshay Venkatesh for allowing me to include their original material as sections in the text. I would like to especially thank

Dan Bump, Kevin Broughan, Sol Friedberg, Jeff Hoffstein, Alex Kontorovich, Wenzhi Luo, Carlos Moreno, Yannan Qiu, Ian Florian Sprung, C. J. Mozzochi, Peter Sarnak, Freydoon Shahidi, Meera Thillainatesan, Qiao Zhang, Alberto Perelli and Steve Miller, for clarifying and improving various proofs, definitions, and historical remarks in the book. Finally, Kevin Broughan has provided an invaluable service to the mathematical community by creating computer code for many of the functions studied in this book.

Dorian Goldfeld

1
Discrete group actions

The genesis of analytic number theory formally began with the epoch making memoir of Riemann (1859) where he introduced the zeta function,

$$\zeta(s) := \sum_{n=1}^{\infty} n^{-s}, \quad (\Re(s) > 1),$$

and obtained its meromorphic continuation and functional equation

$$\pi^{-s/2}\Gamma\left(\frac{s}{2}\right)\zeta(s) = \pi^{-(1-s)/2}\Gamma\left(\frac{1-s}{2}\right)\zeta(1-s), \qquad \Gamma(s) = \int\limits_0^{\infty} e^{-u}u^s\,\frac{du}{u}.$$

Riemann showed that the Euler product representation

$$\zeta(s) = \prod_p \left(1 - \frac{1}{p^s}\right)^{-1},$$

together with precise knowledge of the analytic behavior of $\zeta(s)$ could be used to obtain deep information on the distribution of prime numbers.

One of Riemann's original proofs of the functional equation is based on the Poisson summation formula

$$\sum_{n\in\mathbb{Z}} f(ny) = y^{-1}\sum_{n\in\mathbb{Z}} \hat{f}(ny^{-1}),$$

where f is a function with rapid decay as $y \to \infty$ and

$$\hat{f}(y) = \int_{-\infty}^{\infty} f(t)e^{-2\pi ity}\,dt,$$

is the Fourier transform of f. This is proved by expanding the periodic function

$$F(x) = \sum_{n\in\mathbb{Z}} f(x+n)$$

in a Fourier series. If f is an even function, the Poisson summation formula may be rewritten as

$$\sum_{n=1}^{\infty} f(ny^{-1}) = y\sum_{n=1}^{\infty} \hat{f}(ny) - \frac{1}{2}(y\hat{f}(0) - f(0)),$$

from which it follows that for $\Re(s) > 1$,

$$\begin{aligned}\zeta(s)\int_0^\infty f(y)y^s\,\frac{dy}{y} &= \int_0^\infty \sum_{n=1}^{\infty} f(ny)y^s\,\frac{dy}{y}\\ &= \int_1^\infty \sum_{n=1}^{\infty}\left(f(ny)y^s + f(ny^{-1})y^{-s}\right)\frac{dy}{y}\\ &= \int_1^\infty \sum_{n=1}^{\infty}\left(f(ny)y^s + \hat{f}(ny)y^{1-s}\right)\frac{dy}{y} - \frac{1}{2}\left(\frac{f(0)}{s} + \frac{\hat{f}(0)}{1-s}\right).\end{aligned}$$

If $f(y)$ and $\hat{f}(y)$ have sufficient decay as $y \to \infty$, then the integral above converges absolutely for all complex s and, therefore, defines an entire function of s. Let

$$\tilde{f}(s) = \int_0^\infty f(y)y^s\,\frac{dy}{y}$$

denote the Mellin transform of f, then we see from the above integral representation and the fact that $\hat{\hat{f}}(y) = f(-y) = f(y)$ (for an even function f) that

$$\zeta(s)\tilde{f}(s) = \zeta(1-s)\tilde{\hat{f}}\,(1-s).$$

Choosing $f(y) = e^{-\pi y^2}$, a function with the property that it is invariant under Fourier transform, we obtain Riemann's original form of the functional equation. This idea of introducing an arbitrary test function f in the proof of the functional equation first appeared in Tate's thesis (Tate, 1950).

A more profound understanding of the above proof did not emerge until much later. If we choose $f(y) = e^{-\pi y^2}$ in the Poisson summation formula, then since $\hat{f}(y) = f(y)$, one observes that for $y > 0$,

$$\sum_{n=-\infty}^{\infty} e^{-\pi n^2 y} = \frac{1}{\sqrt{y}}\sum_{n=-\infty}^{\infty} e^{-\pi n^2/y}.$$

This identity is at the heart of the functional equation of the Riemann zeta function, and is a known transformation formula for Jacobi's theta function

$$\theta(z) = \sum_{n=-\infty}^{\infty} e^{2\pi i n^2 z},$$

where $z = x + iy$ with $x \in \mathbb{R}$ and $y > 0$. If $\begin{pmatrix} a & b \\ c & d \end{pmatrix}$ is a matrix with integer coefficients a, b, c, d satisfiying $ad - bc = 1, c \equiv 0 \pmod 4, c \neq 0$, then the Poisson summation formula can be used to obtain the more general transformation formula (Shimura, 1973)

$$\theta\left(\frac{az+b}{cz+d}\right) = \epsilon_d^{-1}\chi_c(d)(cz+d)^{\frac{1}{2}}\theta(z).$$

Here χ_c is the primitive character of order ≤ 2 corresponding to the field extension $\mathbb{Q}(c^{\frac{1}{2}})/\mathbb{Q}$,

$$\epsilon_d = \begin{cases} 1 & \text{if } d \equiv 1 \pmod 4 \\ i & \text{if } d \equiv -1 \pmod 4, \end{cases}$$

and $(cz+d)^{\frac{1}{2}}$ is the "principal determination" of the square root of $cz + d$, i.e., the one whose real part is > 0.

It is now well understood that underlying the functional equation of the Riemann zeta function are the above transformation formulae for $\theta(z)$. These transformation formulae are induced from the action of a group of matrices $\begin{pmatrix} a & b \\ c & d \end{pmatrix}$ on the upper half-plane $\mathfrak{h} = \{x + iy \mid x \in \mathbb{R}, y > 0\}$ given by

$$z \mapsto \frac{az+b}{cz+d}.$$

The concept of a group acting on a topological space appears to be absolutely fundamental in analytic number theory and should be the starting point for any serious investigations.

1.1 Action of a group on a topological space

Definition 1.1.1 *Given a topological space X and a group G, we say that G* **acts continuously** *on X (on the left) if there exists a map $\circ : G \to \text{Func}(X \to X)$ (functions from X to X), $g \mapsto g\circ$ which satisfies:*

- $x \mapsto g \circ x$ *is a continuous function of x for all $g \in G$;*
- $g \circ (g' \circ x) = (g \cdot g') \circ x$, *for all $g, g' \in G, x \in X$ where $\cdot$ denotes the internal operation in the group G;*
- $e \circ x = x$, *for all $x \in X$ and $e =$ identity element in G.*

Example 1.1.2 Let G denote the additive group of integers $\mathbb{Z}$. Then it is easy to verify that the group $\mathbb{Z}$ acts continuously on the real numbers $\mathbb{R}$ with group action $\circ$ defined by

$$n \circ x := n + x,$$

for all $n \in \mathbb{Z}$, $x \in \mathbb{R}$. In this case $e = 0$.

Example 1.1.3 Let $G = GL(2,\mathbb{R})^+$ denote the group of 2×2 matrices $\begin{pmatrix} a & b \\ c & d \end{pmatrix}$ with $a, b, c, d \in \mathbb{R}$ and determinant $ad - bc > 0$. Let

$$\mathfrak{h} := \{x + iy \mid x \in \mathbb{R},\ y > 0\}$$

denote the upper half-plane. For $g = \begin{pmatrix} a & b \\ c & d \end{pmatrix} \in GL(2,\mathbb{R})^+$ and $z \in \mathfrak{h}$ define:

$$g \circ z := \frac{az + b}{cz + d}.$$

Since

$$\frac{az+b}{cz+d} = \frac{ac|z|^2 + (ad+bc)x + bd}{|cz+d|^2} + i \cdot \frac{(ad-bc)\cdot y}{|cz+d|^2}$$

it immediately follows that $g \circ z \in \mathfrak{h}$. We leave as an exercise to the reader, the verification that $\circ$ satisfies the additional axioms of a continuous action. One usually extends this action to the larger space $\mathfrak{h}^* = \mathfrak{h} \cup \{\infty\}$, by defining

$$\begin{pmatrix} a & b \\ c & d \end{pmatrix} \circ \infty = \begin{cases} a/c & \text{if } c \neq 0, \\ \infty & \text{if } c = 0. \end{cases}$$

Assume that a group G acts continously on a topological space X. Two elements $x_1, x_2 \in X$ are said to be equivalent (mod G) if there exists $g \in G$ such that $x_2 = g \circ x_1$. We define

$$Gx := \{g \circ x \mid g \in G\}$$

to be the equivalence class or orbit of x, and let $G\backslash X$ denote the set of equivalence classes.

Definition 1.1.4 *Let a group G act continuously on a topological space X. We say a subset $\Gamma \subset G$ is* **discrete** *if for any two compact subsets $A, B \subset X$, there are only finitely many $g \in \Gamma$ such that $(g \circ A) \cap B \neq \phi$, where ϕ denotes the empty set.*

Example 1.1.5 The discrete subgroup $SL(2,\mathbb{Z})$. Let

$$\Gamma = SL(2,\mathbb{Z}) := \left\{ \begin{pmatrix} a & b \\ c & d \end{pmatrix} \;\middle|\; a,b,c,d \in \mathbb{Z},\ ad-bc=1 \right\},$$

and let

$$\Gamma_\infty := \left\{ \begin{pmatrix} 1 & m \\ 0 & 1 \end{pmatrix} \;\middle|\; m \in \mathbb{Z} \right\}$$

be the subgroup of Γ which fixes ∞. Note that $\Gamma_\infty\backslash\Gamma$ is just a set of coset representatives of the form $\begin{pmatrix} a & b \\ c & d \end{pmatrix}$ where for each pair of relatively prime integers $(c,d)=1$ we choose a unique a,b satisfying $ad-bc=1$. This follows immediately from the identity

$$\begin{pmatrix} 1 & m \\ 0 & 1 \end{pmatrix} \cdot \begin{pmatrix} a & b \\ c & d \end{pmatrix} = \begin{pmatrix} a+mc & b+md \\ c & d \end{pmatrix}.$$

The fact that $SL(2,\mathbb{Z})$ is discrete will be deduced from the following lemma.

Lemma 1.1.6 *Fix real numbers $0<r$, $0<\delta<1$. Let $R_{r,\delta}$ denote the rectangle*

$$R_{r,\delta} = \left\{ x+iy \mid -r \le x \le r,\ 0<\delta \le y \le \delta^{-1} \right\}.$$

Then for every $\epsilon>0$, and any fixed set $\mathcal{S}$ of coset representatives for $\Gamma_\infty\backslash SL(2,\mathbb{Z})$, there are at most $4+(4(r+1)/\epsilon\delta)$ elements $g\in\mathcal{S}$ such that $\mathrm{Im}(g\circ z)>\epsilon$ holds for some $z\in R_{r,\delta}$.

Proof Let $g=\begin{pmatrix} a & b \\ c & d \end{pmatrix}$. Then for $z\in R_{r,\delta}$,

$$\mathrm{Im}(g\circ z) = \frac{y}{c^2y^2+(cx+d)^2} < \epsilon$$

if $|c|>(y\epsilon)^{-\frac{1}{2}}$. On the other hand, for $|c|\le(y\epsilon)^{-\frac{1}{2}}\le(\delta\epsilon)^{-\frac{1}{2}}$, we have

$$\frac{y}{(cx+d)^2} < \epsilon$$

if the following inequalities hold:

$$|d| > |c|r + (y\epsilon^{-1})^{\frac{1}{2}} \ge |c|r + (\epsilon\delta)^{-\frac{1}{2}}.$$

Consequently, $\mathrm{Im}(g\circ z)>\epsilon$ only if

$$|c| \le (\delta\epsilon)^{-\frac{1}{2}} \quad \text{and} \quad |d| \le (\epsilon\delta)^{-\frac{1}{2}}(r+1),$$

and the total number of such pairs (not counting $(c,d)=(0,\pm1),(\pm1,0)$) is at most $4(\epsilon\delta)^{-1}(r+1)$. □

It follows from Lemma 1.1.6 that $\Gamma = SL(2,\mathbb{Z})$ is a discrete subgroup of $SL(2,\mathbb{R})$. This is because:

(1) it is enough to show that for any compact subset $A \subset \mathfrak{h}$ there are only finitely many $g \in SL(2,\mathbb{Z})$ such that $(g \circ A) \cap A \neq \phi$;
(2) every compact subset of $A \subset \mathfrak{h}$ is contained in a rectangle $R_{r,\delta}$ for some $r > 0$ and $0 < \delta < \delta^{-1}$;
(3) $((\alpha g) \circ R_{r,\delta}) \cap R_{r,\delta} = \phi$, except for finitely many $\alpha \in \Gamma_\infty,\ g \in \Gamma_\infty\backslash\Gamma$.

To prove (3), note that Lemma 1.1.6 implies that $(g \circ R_{r,\delta}) \cap R_{r,\delta} = \phi$ except for finitely many $g \in \Gamma_\infty\backslash\Gamma$. Let $S \subset \Gamma_\infty\backslash\Gamma$ denote this finite set of such elements g. If $g \notin S$, then Lemma 1.1.6 tells us that it is because $\mathrm{Im}(gz) < \delta$ for all $z \in R_{r,\delta}$. Since $\mathrm{Im}(\alpha gz) = \mathrm{Im}(gz)$ for $\alpha \in \Gamma_\infty$, it is enough to show that for each $g \in S$, there are only finitely many $\alpha \in \Gamma_\infty$ such that $((\alpha g) \circ R_{r,\delta}) \cap R_{r,\delta} \neq \phi$. This last statement follows from the fact that $g \circ R_{r,\delta}$ itself lies in some other rectangle $R_{r',\delta'}$, and every $\alpha \in \Gamma_\infty$ is of the form $\alpha = \begin{pmatrix} 1 & m \\ 0 & 1 \end{pmatrix}$ $(m \in \mathbb{Z})$, so that

$$\alpha \circ R_{r',\delta'} = \left\{ x + iy \;\middle|\; -r' + m \le x \le r' + m,\ 0 < \delta' \le \delta'^{-1} \right\},$$

which implies $(\alpha \circ R_{r',\delta'}) \cap R_{r,\delta} = \phi$ for $|m|$ sufficiently large.

Definition 1.1.7 *Suppose the group G acts continuously on a connected topological space X. A* **fundamental domain** *for $G\backslash X$ is a connected region $D \subset X$ such that every $x \in X$ is equivalent* (mod G) *to a point in D and such that no two points in D are equivalent to each other.*

Example 1.1.8 A fundamental domain for the action of $\mathbb{Z}$ on $\mathbb{R}$ of Example 1.1.2 is given by

$$\mathbb{Z}\backslash\mathbb{R} = \{0 \le x < 1 \mid x \in \mathbb{R}\}.$$

The proof of this is left as an easy exercise for the reader.

Example 1.1.9 A fundamental domain for $SL(2,\mathbb{Z})\backslash\mathfrak{h}$ can be given as the region $\mathcal{D} \subset \mathfrak{h}$ where

$$\mathcal{D} = \left\{ z \;\middle|\; -\frac{1}{2} \le \mathrm{Re}(z) \le \frac{1}{2},\ |z| \ge 1 \right\},$$

with congruent boundary points symmetric with respect to the imaginary axis.

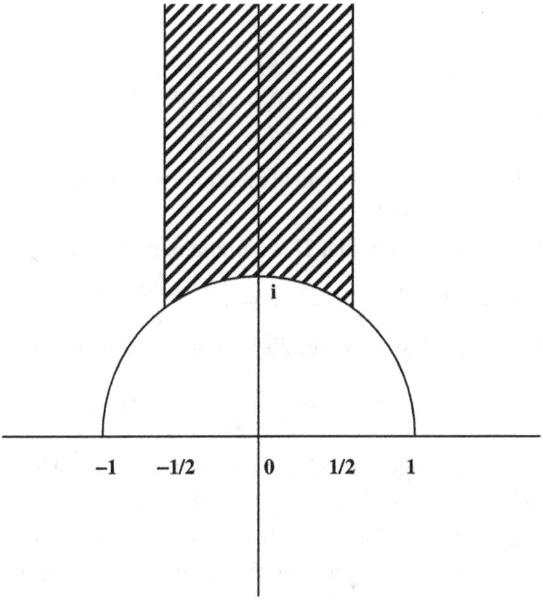

Note that the vertical line $V' := \left\{-\frac{1}{2} + iy \;\middle|\; y \geq \frac{\sqrt{3}}{2}\right\}$ is equivalent to the vertical line $V := \left\{\frac{1}{2} + iy \;\middle|\; y \geq \frac{\sqrt{3}}{2}\right\}$ under the transformation $z \mapsto z + 1$. Furthermore, the arc $A' := \left\{z \;\middle|\; -\frac{1}{2} \leq \text{Re}(z) < 0,\ |z| = 1\right\}$ is equivalent to the reflected arc $A := \left\{z \;\middle|\; 0 < \text{Re}(z) \leq \frac{1}{2},\ |z| = 1\right\}$, under the transformation $z \mapsto -1/z$. To show that $\mathcal{D}$ is a fundamental domain, we must prove:

(1) *For any* $z \in \mathfrak{h}$, *there exists* $g \in SL(2, \mathbb{Z})$ *such that* $g \circ z \in \mathcal{D}$;
(2) *If two distinct points* $z, z' \in \mathcal{D}$ *are congruent* $\pmod{SL(2, \mathbb{Z})}$ *then* $\text{Re}(z) = \pm\frac{1}{2}$ *and* $z' = z \pm 1$, or $|z| = 1$ *and* $z' = -1/z$.

We first prove (1). Fix $z \in \mathfrak{h}$. It follows from Lemma 1.1.6 that for every $\epsilon > 0$, there are at most finitely many $g \in SL(2, \mathbb{Z})$ such that $g \circ z$ lies in the strip

$$D_\epsilon := \left\{ w \;\middle|\; -\frac{1}{2} \leq \text{Re}(w) \leq \frac{1}{2}, \quad \epsilon \leq \text{Im}(w) \right\}.$$

Let B_ϵ denote the finite set of such $g \in SL(2, \mathbb{Z})$. Clearly, for sufficiently small ϵ, the set B_ϵ contains at least one element. We will show that there is at least one $g \in B_\epsilon$ such that $g \circ z \in \mathcal{D}$. Among these finitely many $g \in B_\epsilon$, choose one such that $\text{Im}(g \circ z)$ is maximal in D_ϵ. If $|g \circ z| < 1$, then for $S = \begin{pmatrix} 0 & -1 \\ 1 & 0 \end{pmatrix}$,

$T = \begin{pmatrix} 1 & 1 \\ 0 & 1 \end{pmatrix}$, and any $m \in \mathbb{Z}$,

$$\operatorname{Im}(T^m S g \circ z) = \operatorname{Im}\left(\frac{-1}{g \circ z}\right) = \frac{\operatorname{Im}(g \circ z)}{|g \circ z|^2} > \operatorname{Im}(g \circ z).$$

This is a contradiction because we can always choose m so that $T^m S g \circ z \in D_\epsilon$. So in fact, $g \circ z$ must be in $\mathcal{D}$.

To complete the verification that $\mathcal{D}$ is a fundamental domain, it only remains to prove the assertion (2). Let $z \in \mathcal{D}$, $g = \begin{pmatrix} a & b \\ c & d \end{pmatrix} \in SL(2, \mathbb{Z})$, and assume that $g \circ z \in \mathcal{D}$. Without loss of generality, we may assume that

$$\operatorname{Im}(g \circ z) = \frac{y}{|cz + d|^2} \geq \operatorname{Im}(z),$$

(otherwise just interchange z and $g \circ z$ and use g^{-1}). This implies that $|cz + d| \leq 1$ which implies that $1 \geq |cy| \geq \frac{\sqrt{3}}{2}|c|$. This is clearly impossible if $|c| \geq 2$. So we only have to consider the cases $c = 0, \pm 1$. If $c = 0$ then $d = \pm 1$ and g is a translation by b. Since $-\frac{1}{2} \leq \operatorname{Re}(z), \operatorname{Re}(g \circ z) \leq \frac{1}{2}$, this implies that either $b = 0$ and $z = g \circ z$ or else $b = \pm 1$ and $\operatorname{Re}(z) = \pm\frac{1}{2}$ while $\operatorname{Re}(g \circ z) = \mp\frac{1}{2}$. If $c = 1$, then $|z + d| \leq 1$ implies that $d = 0$ unless $z = e^{2\pi i/3}$ and $d = 0, 1$ or $z = e^{\pi i/3}$ and $d = 0, -1$. The case $d = 0$ implies that $|z| \leq 1$ which implies $|z| = 1$. Also, in this case, $c = 1, d = 0$, we must have $b = -1$ because $ad - bc = 1$. Then $g \circ z = a - \frac{1}{z}$. It follows that $a = 0$. If $z = e^{2\pi i/3}$ and $d = 1$, then we must have $a - b = 1$. It follows that $g \circ e^{2\pi i/3} = a - \frac{1}{1+e^{2\pi i/3}} = a + e^{2\pi i/3}$, which implies that $a = 0$ or 1. A similar argument holds when $z = e^{\pi i/3}$ and $d = -1$. Finally, the case $c = -1$ can be reduced to the previous case $c = 1$ by reversing the signs of a, b, c, d.

1.2 Iwasawa decomposition

This monograph focusses on the general linear group $GL(n, \mathbb{R})$ with $n \geq 2$. This is the multiplicative group of all $n \times n$ matrices with coefficients in $\mathbb{R}$ and non-zero determinant. We will show that every matrix in $GL(n, \mathbb{R})$ can be written as an upper triangular matrix times an orthogonal matrix. This is called the Iwasawa decomposition (Iwasawa, 1949).

The Iwasawa decomposition, in the special case of $GL(2, \mathbb{R})$, states that every $g \in GL(2, \mathbb{R})$ can be written in the form:

$$g = \begin{pmatrix} y & x \\ 0 & 1 \end{pmatrix} \begin{pmatrix} \alpha & \beta \\ \gamma & \delta \end{pmatrix} \begin{pmatrix} d & 0 \\ 0 & d \end{pmatrix} \tag{1.2.1}$$

where $y > 0, x, d \in \mathbb{R}$ with $d \neq 0$ and

$$\begin{pmatrix} \alpha & \beta \\ \gamma & \delta \end{pmatrix} \in O(2, \mathbb{R}),$$

where

$$O(n, \mathbb{R}) = \{g \in GL(n, \mathbb{R}) \mid g \cdot {}^t g = I\}$$

is the orthogonal group. Here I denotes the identity matrix on $GL(n, \mathbb{R})$ and ${}^t g$ denotes the transpose of the matrix g. The matrix $\begin{pmatrix} y & x \\ 0 & 1 \end{pmatrix}$ in the decomposition (1.2.1) is actually uniquely determined. Furthermore, the matrices $\begin{pmatrix} \alpha & \beta \\ \gamma & \delta \end{pmatrix}$ and $\begin{pmatrix} d & 0 \\ 0 & d \end{pmatrix}$ are uniquely determined up to multiplication by $\begin{pmatrix} \pm 1 & 0 \\ 0 & \pm 1 \end{pmatrix}$.

Note that explicitly,

$$O(2, \mathbb{R}) = \left\{ \begin{pmatrix} \pm \cos t & -\sin t \\ \pm \sin t & \cos t \end{pmatrix} \;\middle|\; 0 \leq t \leq 2\pi \right\}.$$

We shall shortly give a detailed proof of (1.2.1) for $GL(n, \mathbb{R})$ with $n \geq 2$.

The decomposition (1.2.1) allows us to realize the upper half-plane

$$\mathfrak{h} = \{x + iy \mid x \in \mathbb{R}, y > 0\}$$

as the set of two by two matrices of type

$$\left\{ \begin{pmatrix} y & x \\ 0 & 1 \end{pmatrix} \;\middle|\; x \in \mathbb{R},\ y > 0 \right\},$$

or by the isomorphism

$$\mathfrak{h} \equiv GL(2, \mathbb{R}) / \langle O(2, \mathbb{R}), Z_2 \rangle, \tag{1.2.2}$$

where

$$Z_n = \left\{ \begin{pmatrix} d & & 0 \\ & \ddots & \\ 0 & & d \end{pmatrix} \;\middle|\; d \in \mathbb{R},\ d \neq 0 \right\}$$

is the center of $GL(n, \mathbb{R})$, and $\langle O(2, \mathbb{R}), Z_2 \rangle$ denotes the group generated by $O(2, \mathbb{R})$ and Z_2.

The isomorphism (1.2.2) is the starting point for generalizing the classical theory of modular forms on $GL(2, \mathbb{R})$ to $GL(n, \mathbb{R})$ with $n > 2$. Accordingly, we define the generalized upper half-plane $\mathfrak{h}^n$ associated to $GL(n, \mathbb{R})$.

Definition 1.2.3 *Let $n \geq 2$. The* **generalized upper half-plane** $\mathfrak{h}^n$ *associated to $GL(n, \mathbb{R})$ is defined to be the set of all $n \times n$ matrices of the form $z = x \cdot y$ where*

$$x = \begin{pmatrix} 1 & x_{1,2} & x_{1,3} & \cdots & & x_{1,n} \\ & 1 & x_{2,3} & \cdots & & x_{2,n} \\ & & \ddots & & & \vdots \\ & & & & 1 & x_{n-1,n} \\ & & & & & 1 \end{pmatrix}, \quad y = \begin{pmatrix} y'_{n-1} & & & & \\ & y'_{n-2} & & & \\ & & \ddots & & \\ & & & y'_1 & \\ & & & & 1 \end{pmatrix},$$

with $x_{i,j} \in \mathbb{R}$ for $1 \leq i < j \leq n$ and $y'_i > 0$ for $1 \leq i \leq n-1$.

To simplify later formulae and notation in this book, we will always express y in the form:

$$y = \begin{pmatrix} y_1 y_2 \cdots y_{n-1} & & & & \\ & y_1 y_2 \cdots y_{n-2} & & & \\ & & \ddots & & \\ & & & y_1 & \\ & & & & 1 \end{pmatrix},$$

with $y_i > 0$ for $1 \leq i \leq n-1$. Note that this can always be done since $y'_i \neq 0$ for $1 \leq i \leq n-1$.

Explicitly, x is an upper triangular matrix with 1s on the diagonal and y is a diagonal matrix beginning with a 1 in the lowest right entry. Note that x is parameterized by $n \cdot (n-1)/2$ real variables $x_{i,j}$ and y is parameterized by $n-1$ positive real variables y_i.

Example 1.2.4 The generalized upper half plane $\mathfrak{h}^3$ is the set of all matrices $z = x \cdot y$ with

$$x = \begin{pmatrix} 1 & x_{1,2} & x_{1,3} \\ 0 & 1 & x_{2,3} \\ 0 & 0 & 1 \end{pmatrix}, \quad y = \begin{pmatrix} y_1 y_2 & 0 & 0 \\ 0 & y_1 & 0 \\ 0 & 0 & 1 \end{pmatrix},$$

where $x_{1,2}, x_{1,3}, x_{2,3} \in \mathbb{R}$, $y_1, y_2 > 0$. Explicitly, every $z \in \mathfrak{h}^3$ can be written in the form

$$z = \begin{pmatrix} y_1 y_2 & x_{1,2} y_1 & x_{1,3} \\ 0 & y_1 & x_{2,3} \\ 0 & 0 & 1 \end{pmatrix}.$$

Remark 1.2.5 The generalized upper half-plane $\mathfrak{h}^3$ does not have a complex structure. Thus $\mathfrak{h}^3$ is quite different from $\mathfrak{h}^2$, which does have a complex structure.

Proposition 1.2.6 *Fix $n \geq 2$. Then we have the Iwasawa decomposition:*

$$GL(n, \mathbb{R}) = \mathfrak{h}^n \cdot O(n, \mathbb{R}) \cdot Z_n,$$

i.e., every $g \in GL(n, \mathbb{R})$ may be expressed in the form

$$g = z \cdot k \cdot d, \qquad (\cdot \text{ denotes matrix multiplication})$$

where $z \in \mathfrak{h}^n$ is uniquely determined, $k \in O(n, \mathbb{R})$, and $d \in Z_n$ is a non-zero diagonal matrix which lies in the center of $GL(n, \mathbb{R})$. Further, k and d are also uniquely determined up to multiplication by $\pm I$ where I is the identity matrix on $GL(n, \mathbb{R})$.

Remark Note that for every $n = 1, 2, 3, \ldots,$ we have $Z_n \cong \mathbb{R}^\times$. We shall, henceforth, write

$$\mathfrak{h}^n \cong GL(n, \mathbb{R})/(O(n, \mathbb{R}) \cdot \mathbb{R}^\times).$$

Proof Let $g \in GL(n, \mathbb{R})$. Then $g \cdot {}^t g$ is a positive definite non–singular matrix. We claim there exists $u, \ell \in GL(n, \mathbb{R})$, where u is upper triangular with 1s on the diagonal and ℓ is lower triangular with 1s on the diagonal, such that

$$u \cdot g \cdot {}^t g = \ell \cdot d \tag{1.2.7}$$

with

$$d = \begin{pmatrix} d_1 & & \\ & \ddots & \\ & & d_n \end{pmatrix}, \qquad d_1, \ldots, d_n > 0.$$

For example, consider $n = 2$, and $g = \begin{pmatrix} a & b \\ c & d \end{pmatrix}$. Then

$$g \cdot {}^t g = \begin{pmatrix} a & b \\ c & d \end{pmatrix} \cdot \begin{pmatrix} a & c \\ b & d \end{pmatrix} = \begin{pmatrix} a^2 + b^2 & ac + bd \\ ac + bd & c^2 + d^2 \end{pmatrix}.$$

If we set $u = \begin{pmatrix} 1 & t \\ 0 & 1 \end{pmatrix}$, then u satisfies (1.2.7) if

$$\begin{pmatrix} 1 & t \\ 0 & 1 \end{pmatrix} \cdot \begin{pmatrix} a^2 + b^2 & ac + bd \\ ac + bd & c^2 + d^2 \end{pmatrix} = \begin{pmatrix} * & 0 \\ * & * \end{pmatrix},$$

so that we may take $t = (-ac - bd)/(c^2 + d^2)$. More generally, the upper triangular matrix u will have $n(n-1)/2$ free variables, and we will have to solve $n(n-1)/2$ equations to satisfy (1.2.7). This system of linear equations has a unique solution because its matrix $g \cdot {}^t g$ is non–singular.

It immediately follows from (1.2.7) that $u^{-1}\ell d = g \cdot {}^t g = d \cdot {}^t \ell ({}^t u)^{-1}$, or equivalently

$$\underbrace{\ell \cdot d \cdot {}^t u}_{\text{lower } \Delta} = \underbrace{u \cdot d \cdot {}^t \ell}_{\text{upper } \Delta} = d.$$

The above follows from the fact that a lower triangular matrix can only equal an upper triangular matrix if it is diagonal, and that this diagonal matrix must be d by comparing diagonal entries. The entries $d_i > 0$ because $g \cdot {}^t g$ is positive definite.

Consequently $\ell d = d({}^t u)^{-1}$. Substituting this into (1.2.7) gives

$$u \cdot g \cdot {}^t g \cdot {}^t u = d = a^{-1} \cdot ({}^t a)^{-1}$$

for

$$a = \begin{pmatrix} d_1^{-\frac{1}{2}} & & \\ & \ddots & \\ & & d_n^{-\frac{1}{2}} \end{pmatrix}.$$

Hence $aug \cdot ({}^t g \cdot {}^t u \cdot {}^t a) = I$ so that $aug \in O(n, \mathbb{R})$. Thus, we have expressed g in the form

$$g = (au)^{-1} \cdot (aug),$$

from which the Iwasawa decomposition immediately follows after dividing and multiplying by the scalar $d_n^{-\frac{1}{2}}$ to arrange the bottom right entry of $(au)^{-1}$ to be 1.

It only remains to show the uniqueness of the Iwasawa decomposition. Suppose that $zkd = z'k'd'$ with $z, z' \in \mathfrak{h}^n,\ k, k' \in O(n, \mathbb{R}),\ d, d' \in Z_n$. Then, since the only matrices in $\mathfrak{h}^n$ and $O(n, \mathbb{R})$ which lie in Z_n are $\pm I$ where I is the identity matrix, it follows that $d' = \pm d$. Further, the only matrix in $\mathfrak{h}^n \cap O(n, \mathbb{R})$ is I. Consequently $z = z'$ and $k = \pm k'$. □

We shall now work out some important instances of the Iwasawa decomposition which will be useful later.

Proposition 1.2.8 *Let I denote the identity matrix on $GL(n,\mathbb{R})$, and for every $1 \le j < i \le n$, let $E_{i,j}$ denote the matrix with a 1 at the $\{i, j\}$th position and zeros elsewhere. Then, for an arbitrary real number t, we have*

$$I + tE_{i,j} = \begin{pmatrix} 1 & & & & & \\ & \ddots & & & & \\ & & \frac{1}{(t^2+1)^{\frac{1}{2}}} & \cdots & \frac{t}{(t^2+1)^{\frac{1}{2}}} & & \\ & & & \ddots & \vdots & & \\ & & & & (t^2+1)^{\frac{1}{2}} & & \\ & & & & & \ddots & \\ & & & & & & 1 \end{pmatrix} \big(\text{mod}\,(O(n,\mathbb{R})\cdot\mathbb{R}^\times)\big),$$

where, in the above matrix, $\frac{1}{(t^2+1)^{\frac{1}{2}}}$ occurs at position $\{j, j\}$, $(t^2+1)^{\frac{1}{2}}$ occurs at position $\{i, i\}$, all other diagonal entries are ones, $\frac{t}{(t^2+1)^{\frac{1}{2}}}$ occurs at position $\{j, i\}$, and, otherwise, all other entries are zero.

Proof Let $g = I + tE_{i,j}$. Then

$$g \cdot {}^t g = (I + tE_{i,j}) \cdot (I + tE_{j,i}) = I + tE_{i,j} + tE_{j,i} + t^2 E_{i,i}.$$

If we define a matrix $u = I - (t/(t^2+1))E_{j,i}$, then $u \cdot g \cdot {}^t g \cdot {}^t u$ must be a diagonal matrix d. Setting $d = a^{-1} \cdot ({}^t a)^{-1}$, we may directly compute:

$$\begin{aligned} u \cdot g \cdot {}^t g \cdot {}^t u &= I + t^2 E_{i,i} - \frac{t^2}{t^2+1} E_{j,j}, \\ u^{-1} &= I + \frac{t}{t^2+1} E_{j,i}, \\ a^{-1} &= I + \left(\frac{1}{\sqrt{t^2+1}} - 1 \right) E_{j,j} + \left(\sqrt{t^2+1} - 1\right) E_{i,i}. \end{aligned}$$

Therefore,

$$u^{-1} a^{-1} = I + \left(\frac{1}{\sqrt{t^2+1}} - 1 \right) E_{j,j} + \left(\sqrt{t^2+1} - 1\right) E_{i,i} + \frac{t}{\sqrt{t^2+1}} E_{j,i}.$$

As in the proof of Proposition 1.2.6, we have $g = u^{-1} \cdot a^{-1}$ $(\text{mod}\,(O(n,\mathbb{R}), \mathbb{R}^\times))$. □

Proposition 1.2.9 *Let $n \geq 2$, and let $z = xy \in \mathfrak{h}^n$ have the form given in Definition 1.2.3. For $i = 1, 2, \ldots, n-1$, define*

$$\omega_i = \begin{pmatrix} 1 & & & & & \\ & \ddots & & & & \\ & & 0 & 1 & & \\ & & 1 & 0 & & \\ & & & & \ddots & \\ & & & & & 1 \end{pmatrix},$$

to be the $n \times n$ identity matrix except for the ith *and* $(i+1)$th *rows where we have $\begin{pmatrix} 0 & 1 \\ 1 & 0 \end{pmatrix}$ on the diagonal. Then*

$$\omega_i z \equiv \begin{pmatrix} 1 & x'_{1,2} & x'_{1,3} & \cdots & x'_{1,n} \\ & 1 & x'_{2,3} & \cdots & x'_{2,n} \\ & & \ddots & & \vdots \\ & & & 1 & x'_{n-1,n} \\ & & & & 1 \end{pmatrix} \cdot \begin{pmatrix} y'_1 y'_2 \cdots y'_{n-1} & & & & \\ & y'_1 y'_2 \cdots y'_{n-2} & & & \\ & & \ddots & & \\ & & & y'_1 & \\ & & & & 1 \end{pmatrix}$$

$\big(\mathrm{mod}\, (O(n,\mathbb{R}) \cdot \mathbb{R}^\times)\big)$, *where $y'_k = y_k$ except for $k = n-i+1, n-i, n-i-1$, in which case*

$$y'_{n-i} = \frac{y_{n-i}}{x_{i,i+1}^2 + y_{n-i}^2}, \qquad y'_{n-i\pm 1} = y_{n-i\pm 1} \cdot \sqrt{x_{i,i+1}^2 + y_{n-i}^2},$$

and $x_{k,\ell} = x'_{k,\ell}$ except for $\ell = i, i+1$, in which case

$$x'_{i-j,i} = x_{i-j,i+1} - x_{i-j,i}x_{i,i+1}, \qquad x'_{i-j,i+1} = \frac{x_{i-j,i}y_{n-i}^2 + x_{i-j,i+1}x_{i,i+1}}{x_{i,i+1}^2 + y_{n-i}^2},$$

for $j = 1, 2, \ldots, i-2$.

Proof Brute force computation which is omitted. □

Proposition 1.2.10 *The group $GL(n,\mathbb{Z})$ acts on $\mathfrak{h}^n$.*

Proof Recall the definition of a group acting on a topological space given in Definition 1.1.1. The fact that $GL(n,\mathbb{Z})$ acts on $GL(n,\mathbb{R})$ follows immediately from the fact that $GL(n,\mathbb{Z})$ acts on the left on $GL(n,\mathbb{R})$ by matrix multiplication and that we have the realization $\mathfrak{h}^n = GL(n,\mathbb{R})/(O(n,\mathbb{R}) \cdot \mathbb{R}^\times)$, as a set of cosets, by the Iwasawa decomposition given in Proposition 1.2.6. □

1.3 Siegel sets

We would like to show that $\Gamma^n = GL(n, \mathbb{Z})$ acts discretely on the generalized upper half-plane $\mathfrak{h}^n$ defined in Definition 1.2.3. This was already proved for $n = 2$ in Lemma 1.1.6, but the generalization to $n > 2$ requires more subtle arguments. In order to find an approximation to a fundamental domain for $GL(n, \mathbb{Z})\backslash\mathfrak{h}^n$, we shall introduce for every $t, u \geq 0$ the Siegel set $\Sigma_{t,u}$.

Definition 1.3.1 *Let $a, b \geq 0$ be fixed. We define the Siegel set $\Sigma_{a,b} \subset \mathfrak{h}^n$ to be the set of all*

$$\begin{pmatrix} 1 & x_{1,2} & x_{1,3} & \cdots & & x_{1,n} \\ & 1 & x_{2,3} & \cdots & & x_{2,n} \\ & & \ddots & & & \vdots \\ & & & & 1 & x_{n-1,n} \\ & & & & & 1 \end{pmatrix} \cdot \begin{pmatrix} y_1 y_2 \cdots y_{n-1} & & & & \\ & y_1 y_2 \cdots y_{n-2} & & & \\ & & \ddots & & \\ & & & y_1 & \\ & & & & 1 \end{pmatrix}$$

with $|x_{i,j}| \leq b$ for $1 \leq i < j \leq n$, and $y_i > a$ for $1 \leq i \leq n-1$.

Let $\Gamma^n = GL(n, \mathbb{Z})$ and $\Gamma^n_\infty \subset \Gamma^n$ denote the subgroup of upper triangular matrices with 1s on the diagonal. We have shown in Proposition 1.2.10 that Γ^n acts on $\mathfrak{h}^n$. For $g \in \Gamma^n$ and $z \in \mathfrak{h}^n$, we shall denote this action by $g \circ z$. The following proposition proves that the action is discrete and that $\Sigma_{\frac{\sqrt{3}}{2},\frac{1}{2}}$ is a good approximation to a fundamental domain.

Proposition 1.3.2 *Fix an integer $n \geq 2$. For any $z \in \mathfrak{h}^n$ there are only finitely many $g \in \Gamma^n$ such that $g \circ z \in \Sigma_{\frac{\sqrt{3}}{2},\frac{1}{2}}$. Furthermore,*

$$GL(n, \mathbb{R}) = \bigcup_{g \in \Gamma^n} g \circ \Sigma_{\frac{\sqrt{3}}{2},\frac{1}{2}}. \tag{1.3.3}$$

Remarks The bound $\frac{\sqrt{3}}{2}$ is implicit in the work of Hermite, and a proof can be found in (Korkine and Zolotareff, 1873). The first part of Proposition 1.3.2 is a well known theorem due to Siegel (1939). For the proof, we follow the exposition of Borel and Harish-Chandra (1962).

Proof of Proposition 1.3.2 In order to prove (1.3.3), it is enough to show that

$$SL(n, \mathbb{R}) = \bigcup_{g \in SL(n,\mathbb{Z})} g \circ \Sigma^*_{\frac{\sqrt{3}}{2},\frac{1}{2}}, \tag{1.3.4}$$

where $\Sigma^*_{t,u}$ denotes the subset of matrices $\Sigma_{t,u} \cdot Z_n$ which have determinant 1 and $\circ$ denotes the action of $SL(n, \mathbb{Z})$ on $\Sigma^*_{0,\infty}$. Note that every element in $\Sigma^*_{a,b}$

is of the form

$$\begin{pmatrix} 1 & x_{1,2} & x_{1,3} & \cdots & x_{1,n} \\ & 1 & x_{2,3} & \cdots & x_{2,n} \\ & & \ddots & & \vdots \\ & & & 1 & x_{n-1,n} \\ & & & & 1 \end{pmatrix} \cdot \begin{pmatrix} dy_1y_2\cdots y_{n-1} & & & & \\ & dy_1y_2\cdots y_{n-2} & & & \\ & & \ddots & & \\ & & & dy_1 & \\ & & & & d \end{pmatrix} \tag{1.3.5}$$

where the determinant

$$\mathrm{Det}\begin{pmatrix} dy_1y_2\cdots y_{n-1} & & & & \\ & dy_1y_2\cdots y_{n-2} & & & \\ & & \ddots & & \\ & & & dy_1 & \\ & & & & d \end{pmatrix} = 1,$$

so that

$$d = \left(\prod_{i=1}^{n-1} y_i^{n-i}\right)^{-1/n}.$$

In view of the Iwasawa decomposition of Proposition 1.2.6, we may identify $\Sigma^*_{0,\infty}$ as the set of coset representatives $SL(n,\mathbb{R})/SO(n,\mathbb{R})$, where $SO(n,\mathbb{R})$ denotes the subgroup $O(n,\mathbb{R}) \cap SL(n,\mathbb{R})$. □

In order to prove (1.3.4), we first introduce some basic notation. Let

$$e_1 = (1,0,\ldots,0), \quad e_2 = (0,1,\ldots,0), \quad \ldots, \quad e_n = (0,0,\ldots,1),$$

denote the canonical basis for $\mathbb{R}^n$. For $1 \le i \le n$ and any matrix $g \in GL(n,\mathbb{R})$, let $e_i \cdot g$ denote the usual multiplication of a $1 \times n$ matrix with an $n \times n$ matrix. For an arbitrary $v = (v_1, v_2, \ldots, v_n) \in \mathbb{R}^n$, define the norm: $||v|| := \sqrt{v_1^2 + v_2^2 + \cdots + v_n^2}$. We now introduce a function

$$\phi : SL(n,\mathbb{R}) \to \mathbb{R}^{>0}$$

from $SL(n,\mathbb{R})$ to the positive real numbers. For all $g = (g_{i,j})_{1\le i,j\le n}$ in $SL(n,\mathbb{R})$ we define

$$\phi(g) := ||e_n \cdot g|| = \sqrt{g_{n,1}^2 + g_{n,2}^2 + \cdots + g_{n,n}^2}.$$

Claim *The function ϕ is well defined on the quotient space $SL(n,\mathbb{R})/SO(n,\mathbb{R})$.*

To verify the claim, note that for $k \in SO(n, \mathbb{R})$, and $v \in \mathbb{R}^n$, we have

$$||v \cdot k|| = \sqrt{(v \cdot k) \cdot {}^t(v \cdot k)} = \sqrt{v \cdot k \cdot {}^tk \cdot {}^tv} = \sqrt{v \cdot {}^tv} = ||v||.$$

This immediately implies that $\phi(gk) = \phi(g)$, i.e., the claim is true.

Note that if $z \in \Sigma^*_{0,\infty}$ is of the form (1.3.5), then

$$\phi(z) = d = \left(\prod_{i=1}^{n-1} y_i^{(n-i)}\right)^{-1/n}. \tag{1.3.6}$$

Now, if $z \in \Sigma^*_{0,\infty}$ is fixed, then

$$e_n \cdot SL(n, \mathbb{Z}) \cdot z \subset \big(\mathbb{Z}e_1 + \cdots + \mathbb{Z}e_n - \{(0, 0, \ldots, 0)\}\big) \cdot z, \tag{1.3.7}$$

where $\cdot$ denotes matrix multiplication. The right-hand side of (1.3.7) consists of non–zero points of a lattice in $\mathbb{R}^n$. This implies that ϕ achieves a **positive minimum** on the coset $SL(n, \mathbb{Z}) \cdot z$. The key to the proof of Proposition 1.3.2 will be the following lemma from which Proposition 1.3.2 follows immediately.

Lemma 1.3.8 *Let $z \in \Sigma^*_{0,\infty}$. Then the minimum of ϕ on $SL(n, \mathbb{Z}) \circ z$ is achieved at a point of $\Sigma^*_{\frac{\sqrt{3}}{2},\frac{1}{2}}$.*

Proof It is enough to prove that the minimum of ϕ is achieved at a point of $\Sigma^*_{\frac{\sqrt{3}}{2},\infty}$ because we can always translate by an upper triangular matrix

$$u = \begin{pmatrix} 1 & u_{1,2} & u_{1,3} & \cdots & & u_{1,n} \\ & 1 & u_{2,3} & \cdots & & u_{2,n} \\ & & \ddots & & & \vdots \\ & & & & 1 & u_{n-1,n} \\ & & & & & 1 \end{pmatrix} \in SL(n, \mathbb{Z})$$

to arrange that the minimum of ϕ lies in $\Sigma^*_{\frac{\sqrt{3}}{2},\frac{1}{2}}$. This does not change the value of ϕ because of the identity $\phi(u \cdot z) = ||e_n \cdot u \cdot z|| = ||e_n \cdot z||$. We shall use induction on n. We have already proved a stronger statement for $n = 2$ in Example 1.1.9. Fix $\gamma \in SL(n, \mathbb{Z})$ such that $\phi(\gamma \circ z)$ is minimized. We set $\gamma \circ z = x \cdot y$ with

$$x = \begin{pmatrix} 1 & x_{1,2} & x_{1,3} & \cdots & & x_{1,n} \\ & 1 & x_{2,3} & \cdots & & x_{2,n} \\ & & \ddots & & & \vdots \\ & & & & 1 & x_{n-1,n} \\ & & & & & 1 \end{pmatrix}, \quad y = \begin{pmatrix} dy_1y_2\cdots y_{n-1} & & & & \\ & dy_1y_2\cdots y_{n-2} & & & \\ & & \ddots & & \\ & & & dy_1 & \\ & & & & d \end{pmatrix},$$

with $d = \left(\prod_{i=1}^{n-1} y_i^{n-i}\right)^{-1/n}$ as before. We must show $y_i \geq \frac{\sqrt{3}}{2}$ for $i = 1, 2, \ldots, n-1$. The proof proceeds in 3 steps.

Step 1 $y_1 \geq \frac{\sqrt{3}}{2}$.

This follows from the action of $\alpha := \begin{pmatrix} I_{n-2} & & \\ & 0 & -1 \\ & 1 & 0 \end{pmatrix}$ on $\gamma \circ z$. Here I_{n-2} denotes the identity $(n-2) \times (n-2)$–matrix. First of all

$$\begin{aligned}\phi(\alpha \circ \gamma \circ z) &= ||e_n \cdot \alpha \circ \gamma \circ z|| = ||e_{n-1} \cdot x \cdot y|| = ||(e_{n-1} + x_{n-1,n} e_n) \cdot y|| \\ &= d\sqrt{y_1^2 + x_{n-1,n}^2}.\end{aligned}$$

Since $|x_{n-1,n}| \leq \frac{1}{2}$ we see that $\phi(\alpha\gamma z)^2 \leq d^2(y_1^2 + \frac{1}{4})$. On the other hand, the assumption of minimality forces $\phi(\gamma z)^2 = d^2 \leq d^2\left(y_1^2 + \frac{1}{4}\right)$. This implies that $y_1 \geq \frac{\sqrt{3}}{2}$.

Step 2 Let $g' \in SL(n-1, \mathbb{Z})$, $g = \begin{pmatrix} g' & 1 \\ 0 & 1 \end{pmatrix}$. Then $\phi(g\gamma z) = \phi(\gamma z)$.
This follows immediately from the fact that $e_n \cdot g = e_n$.

Step 3 $y_i \geq \frac{\sqrt{3}}{2}$ for $i = 2, 3, \ldots, n-1$.

Let us write $\gamma \circ z = \begin{pmatrix} z' \cdot d' & * \\ & d \end{pmatrix}$ with $z' \in SL(n-1, \mathbb{R})$ and $d' \in Z_{n-1}$ a suitable diagonal matrix. By induction, there exists $g' \in SL(n-1, \mathbb{Z})$ such that $g' \circ z' = x' \cdot y' \in \Sigma^*_{\frac{\sqrt{3}}{2}, \frac{1}{2}} \subset \mathfrak{h}^{n-1}$, the Siegel set for $GL(n-1, \mathbb{R})$. This is equivalent to the fact that

$$y' = \begin{pmatrix} a_{n-1} & & & \\ & a_{n-2} & & \\ & & \ddots & \\ & & & a_1 \end{pmatrix}$$

and

$$\frac{a_{j+1}}{a_j} \geq \frac{\sqrt{3}}{2} \quad \text{for } j = 1, 2, \ldots, n-2. \tag{1.3.9}$$

Define $g := \begin{pmatrix} g' & 0 \\ 0 & 1 \end{pmatrix} \in SL(n, \mathbb{Z})$. Then

$$g \circ \gamma \circ z = \begin{pmatrix} g' & 0 \\ 0 & 1 \end{pmatrix} \circ \begin{pmatrix} z' \cdot d' & * \\ & d \end{pmatrix} = \begin{pmatrix} g' \circ z' \cdot d' & * \\ 0 & d \end{pmatrix} = x'' \cdot y'',$$

where $y'' = \begin{pmatrix} y'd' & 0 \\ 0 & d \end{pmatrix}$, $x'' = \begin{pmatrix} x' & * \\ 0 & 1 \end{pmatrix}$. The inequalities (1.3.9) applied to

$$y'' = \begin{pmatrix} y'd' & 0 \\ 0 & d \end{pmatrix} = \begin{pmatrix} y_1 y_2 \cdots y_{n-1} d & & & \\ & \ddots & & \\ & & y_1 d & \\ & & & d \end{pmatrix},$$

imply that $y_i \geq \frac{\sqrt{3}}{2}$ for $i = 2, 3, \ldots, n-1$. Step 2 insures that multiplying by g on the left does not change the value of $\phi(\gamma z)$. Step 1 gives $y_1 \geq \frac{\sqrt{3}}{2}$. □

1.4 Haar measure

Let $n \geq 2$. The discrete subgroup $SL(n, \mathbb{Z})$ acts on $SL(n, \mathbb{R})$ by left multiplication. The quotient space $SL(n, \mathbb{Z}) \backslash SL(n, \mathbb{R})$ turns out to be of fundamental importance in number theory. Now, we turn our attention to a theory of integration on this quotient space.

We briefly review the theory of Haar measure and integration on locally compact Hausdorff topological groups. Good references for this material are (Halmos, 1974), (Lang, 1969), (Hewitt and Ross, 1979). Excellent introductary books on matrix groups and elementary Lie theory are (Curtis, 1984), (Baker, 2002), (Lang, 2002).

Recall that a topological group G is a topological space G where G is also a group and the map

$$(g, h) \mapsto g \cdot h^{-1}$$

of $G \times G$ onto G is continuous in both variables. Here $\cdot$ again denotes the internal group operation and h^{-1} denotes the inverse of the element h. The assumption that G is locally compact means that every point has a compact neighborhood. Recall that G is termed Hausdorff provided every pair of distinct elements in G lie in disjoint open sets.

Example 1.4.1 The general linear group $GL(n, \mathbb{R})$ is a locally compact Hausdorff topological group.

Let $\mathfrak{gl}(n, \mathbb{R})$ denote the Lie algebra of $GL(n, \mathbb{R})$. Viewed as a set, $\mathfrak{gl}(n, \mathbb{R})$ is just the set of all $n \times n$ matrices with coefficients in $\mathbb{R}$. We assign a topology to $\mathfrak{gl}(n, \mathbb{R})$ by identifying every matrix

$$g = \begin{pmatrix} g_{1,1} & g_{1,2} & \cdots & g_{1,n} \\ g_{2,1} & g_{2,2} & \cdots & g_{2,n} \\ \vdots & & \cdots & \vdots \\ g_{n,1} & g_{n,2} & \cdots & g_{n,n} \end{pmatrix}$$

with a point

$$(g_{1,1}, g_{1,2}, \ldots, g_{1,n}, g_{2,1}, g_{2,2}, \ldots, g_{2,n}, \ldots, g_{n,n}) \in \mathbb{R}^{n^2}.$$

This identification is a one–to–one correspondence. One checks that $\mathfrak{gl}(n, \mathbb{R})$ is a locally compact Hausdorff topological space under the usual Euclidean topology on $\mathbb{R}^{n^2}$. The determinant function $\text{Det} : \mathfrak{gl}(n, \mathbb{R}) \to \mathbb{R}$ is clearly continuous. It follows that

$$GL(n, \mathbb{R}) = \mathfrak{gl}(n, \mathbb{R}) - \text{Det}^{-1}(0)$$

must be an open set since $\{0\}$ is closed. Also, the operations of addition and multiplication of matrices in $\mathfrak{gl}(n, \mathbb{R})$ are continuous maps from

$$\mathfrak{gl}(n, \mathbb{R}) \times \mathfrak{gl}(n, \mathbb{R}) \to \mathfrak{gl}(n, \mathbb{R}).$$

The inverse map

$$\text{Inv} : GL(n, \mathbb{R}) \to GL(n, \mathbb{R}),$$

given by $\text{Inv}(g) = g^{-1}$ for all $g \in GL(n, \mathbb{R})$, is also continuous since each entry of g^{-1} is a polynomial in the entries of g divided by $\text{Det}(g)$. Thus, $GL(n, \mathbb{R})$ is a topological subspace of $\mathfrak{gl}(n, \mathbb{R})$ and we may view $GL(n, \mathbb{R}) \times GL(n, \mathbb{R})$ as the product space. Since the multiplication and inversion maps: $GL(n, \mathbb{R}) \times GL(n, \mathbb{R}) \to GL(n, \mathbb{R})$ are continuous, it follows that $GL(n, \mathbb{R})$ is a topological group.

By a **left Haar measure** on a locally compact Hausdorff topological group G, we mean a positive Borel measure (Halmos, 1974)

$$\mu : \{\text{measurable subsets of } G\} \to \mathbb{R}^+,$$

which is left invariant under the action of G on G via left multiplication. This means that for every measurable set $E \subset G$ and every $g \in G$, we have

$$\mu(gE) = \mu(E).$$

In a similar manner, one may define a **right Haar measure.** If every left invariant Haar measure on G is also a right invariant Haar measure, then we say that G is **unimodular**.

Given a left invariant Haar measure μ on G, one may define (in the usual manner) a differential one-form $d\mu(g)$, and for compactly supported functions $f : G \to \mathbb{C}$ an integral

$$\int_G f(g)\, d\mu(g),$$

which is characterized by the fact that

$$\int_E d\mu(g) = \mu(E)$$

for every measurable set E. We shall also refer to $d\mu(g)$ as a Haar measure. The fundamental theorem in the subject is due to Haar.

Theorem 1.4.2 (Haar) *Let G be a locally compact Hausdorff topological group. Then there exists a left Haar measure on G. Further, any two such Haar measures must be positive real multiples of each other.*

We shall not need this general existence theorem, because in the situations we are interested in, we can explicitly construct the Haar measure and Haar integral. For unimodular groups, the uniqueness of Haar measure follows easily from Fubini's theorem. The proof goes as follows. Assume we have two Haar measures μ, ν on G, which are both left and right invariant. Let $h : G \to \mathbb{C}$ be a compactly supported function satisfying

$$\int_G h(g)\, d\mu(g) = 1.$$

Then for an arbitrary compactly supported function $f : G \to \mathbb{C}$,

$$\begin{aligned}
\int_G f(g) d\nu(g) &= \int_G h(g') d\mu(g') \int_G f(g) d\nu(g) \\
&= \int_G \int_G h(g') f(g) d\nu(g) d\mu(g') \\
&= \int_G \int_G h(g') f(g \cdot g')\, d\nu(g)\, d\mu(g') \\
&= \int_G \int_G h(g') f(g \cdot g')\, d\mu(g')\, d\nu(g) \\
&= \int_G \int_G h(g^{-1} \cdot g') f(g')\, d\mu(g')\, d\nu(g) \\
&= \int_G \int_G h(g^{-1} \cdot g') f(g')\, d\nu(g)\, d\mu(g') \\
&= c \cdot \int_G f(g')\, d\mu(g')
\end{aligned}$$

where $c = \int_G h(g^{-1}) d\nu(g)$.

Proposition 1.4.3 *For $n = 1, 2, \ldots$, let*

$$g = \begin{pmatrix} g_{1,1} & g_{1,2} & \cdots & g_{1,n} \\ g_{2,1} & g_{2,2} & \cdots & g_{2,n} \\ \vdots & & \cdots & \vdots \\ g_{n,1} & g_{n,2} & \cdots & g_{n,n} \end{pmatrix} \in GL(n, \mathbb{R}),$$

where $g_{1,1}, g_{1,2}, \ldots, g_{1,n}, g_{2,1}, \ldots, g_{n,n}$ are n^2 real variables. Define

$$d\mu(g) := \frac{\prod\limits_{1 \le i,j \le n} dg_{i,j}}{\mathrm{Det}(g)^n}, \qquad \text{(wedge product of differential one-forms)}$$

where $dg_{i,j}$ denotes the usual differential one–form on $\mathbb{R}$ and $\mathrm{Det}(g)$ denotes the determinant of the matrix g. Then $d\mu(g)$ is the unique left–right invariant Haar measure on $GL(n, \mathbb{R})$.

Proof Every matrix in $GL(n, \mathbb{R})$ may be expressed as a product of a diagonal matrix in Z_n and matrices of the form $\tilde{x}_{r,s}$ (with $1 \le r, s \le n$) where $\tilde{x}_{r,s}$ denotes the matrix with the real number $x_{r,s}$ at position r, s, and, otherwise, has 1s on the diagonal and zeros off the diagonal. It is easy to see that

$$d\mu(g) = d\mu(ag)$$

for $a \in Z_n$. To complete the proof, it is, therefore, enough to check that

$$d\mu(\tilde{x}_{r,s} \cdot g) = d\mu(g \cdot \tilde{x}_{r,s}) = d\mu(g),$$

for all $1 \le r, s \le n$. We check the left invariance and leave the right invariance to the reader.

It follows from the definition that in the case $r \ne s$,

$$d\mu(\tilde{x}_{r,s} \cdot g) = \frac{\Bigg(\prod\limits_{\substack{1 \le i,j \le n \\ i \ne r}} dg_{i,j}\Bigg)\Bigg(\prod\limits_{1 \le j \le n} d(g_{r,j} + g_{s,j}x_{r,s})\Bigg)}{\mathrm{Det}(\tilde{x}_{r,s} \cdot g)^n}.$$

First of all,

$$\mathrm{Det}(\tilde{x}_{r,s} \cdot g) = \mathrm{Det}(\tilde{x}_{r,s}) \cdot \mathrm{Det}(g) = \mathrm{Det}(g)$$

because $\mathrm{Det}(\tilde{x}_{r,s}) = 1$.

Second, for any $1 \le j \le n$,

$$\Bigg(\prod\limits_{\substack{1 \le i,j \le n \\ i \ne r}} dg_{i,j}\Bigg) \wedge dg_{s,j} = 0$$

because $g_{s,j}$ also occurs in the product $\left(\prod_{\substack{1\le i<j\le n\\ i\ne r}} dg_{i,j}\right)$ and $dg_{s,j} \wedge dg_{s,j} = 0$.

Consequently, the measure is invariant under left multiplication by $\tilde{x}_{r,s}$. □

On the other hand, if $r = s$, then

$$\begin{aligned} d\mu(\tilde{x}_{r,s}\cdot g) &= \frac{\left(\prod\limits_{\substack{1\le i,j\le n\\ i\ne r}} dg_{i,j}\right)\left(\prod\limits_{1\le j\le n}(x_{r,s}\cdot dg_{r,j})\right)}{\mathrm{Det}(\tilde{x}_{r,s}\cdot g)^n} \\ &= d\mu(g)\cdot\frac{x_{r,s}^n}{\mathrm{Det}(\tilde{x}_{r,s})^n} \\ &= d\mu(g). \end{aligned}$$

1.5 Invariant measure on coset spaces

This monograph focusses on the coset space

$$GL(n,\mathbb{R})/(O(n,\mathbb{R})\cdot\mathbb{R}^\times).$$

We need to establish explicit invariant measures on this space. The basic principle which allows us to define invariant measures on coset spaces, in general, is given in the following theorem.

Theorem 1.5.1 *Let G be a locally compact Hausdorff topological group, and let H be a compact closed subgroup of G. Let μ be a Haar measure on G, and let ν be a Haar measure on H, normalized so that $\int_H d\nu(h) = 1$. Then there exists a unique (up to scalar multiple) quotient measure $\tilde{\mu}$ on G/H. Furthermore*

$$\int_G f(g)\,d\mu(g) = \int_{G/H}\left(\int_H f(gh)\,d\nu(h)\right)d\tilde{\mu}(gH),$$

for all integrable functions $f : G \to \mathbb{C}$.

Proof For a proof see (Halmos, 1974). We indicate, however, why the formula in Theorem 1.5.1 holds. First of all note that if $f : G \to \mathbb{C}$, is an integrable function on G, and if we define a new function, $f^H : G \to \mathbb{C}$, by the recipe

$$f^H(g) := \int_H f(gh)\,d\nu(h),$$

then $f^H(gh) = f^H(g)$ for all $h \in H$. Thus, f^H is well defined on the coset space G/H. We write $f^H(g) = f^H(gH)$, to stress that f^H is a function on the coset space. For any measurable subset $E \subset G/H$, we may easily choose a measurable function $\delta_E : G \to \mathbb{C}$ so that

$$\delta_E(g) = \delta_E^H(gH) = \begin{cases} 1 & \text{if } gH \in E, \\ 0 & \text{if } gH \notin E. \end{cases}$$

We may then define an H–invariant quotient measure $\tilde{\mu}$ satisfying:

$$\tilde{\mu}(E) = \int_G \delta_E(g)\, d\mu(g) = \int_{G/H} \delta_E^H(gH)\, d\tilde{\mu}(gH),$$

and

$$\int_G f(g)\, d\mu(g) = \int_{G/H} f^H(gH)\, d\tilde{\mu}(gH),$$

for all integrable functions $f : G \to \mathbb{C}$. □

Remarks There is an analogous version of Theorem 1.5.1 for left coset spaces $H\backslash G$. Note that we are not assuming that H is a normal subgroup of G. Thus G/H (respectively $H\backslash G$) may not be a group.

Example 1.5.2 (Left invariant measure on $GL(n,\mathbb{R})/(O(n,\mathbb{R})\cdot\mathbb{R}^\times)$)

For $n \geq 2$, we now explicitly construct a left invariant measure on the generalized upper half-plane $\mathfrak{h}^n = GL(n,\mathbb{R})/(O(n,\mathbb{R})\cdot\mathbb{R}^\times)$. Returning to the Iwasawa decomposition (Proposition 1.2.6), every $z \in \mathfrak{h}^n$ has a representation in the form $z = xy$ with

$$x = \begin{pmatrix} 1 & x_{1,2} & x_{1,3} & \cdots & x_{1,n} \\ & 1 & x_{2,3} & \cdots & x_{2,n} \\ & & \ddots & & \vdots \\ & & & 1 & x_{n-1,n} \\ & & & & 1 \end{pmatrix}, \quad y = \begin{pmatrix} y_1 y_2 \cdots y_{n-1} & & & & \\ & y_1 y_2 \cdots y_{n-2} & & & \\ & & \ddots & & \\ & & & y_1 & \\ & & & & 1 \end{pmatrix},$$

with $x_{i,j} \in \mathbb{R}$ for $1 \leq i < j \leq n$ and $y_i > 0$ for $1 \leq i \leq n-1$. Let d^*z denote the left invariant measure on $\mathfrak{h}^n$. Then d^*z has the property that

$$d^*(gz) = d^*z$$

for all $g \in GL(n,\mathbb{R})$.

Proposition 1.5.3 *The left invariant $GL(n, \mathbb{R})$–measure d^*z on $\mathfrak{h}^n$ can be given explicitly by the formula*

$$d^*z = d^*x\, d^*y$$

where

$$d^*x = \prod_{1\le i<j\le n} dx_{i,j}, \quad d^*y = \prod_{k=1}^{n-1} y_k^{-k(n-k)-1}\, dy_k. \tag{1.5.4}$$

For example, for $n = 2$, with $z = \begin{pmatrix} y & x \\ 0 & 1 \end{pmatrix}$, we have $d^*z = \frac{dxdy}{y^2}$, while for $n = 3$ with

$$z = \begin{pmatrix} y_1y_2 & x_{1,2}y_1 & x_{1,3} \\ 0 & y_1 & x_{2,3} \\ 0 & 0 & 1 \end{pmatrix},$$

we have

$$d^*z = dx_{1,2}dx_{1,3}dx_{2,3}\,\frac{dy_1dy_2}{(y_1y_2)^3}.$$

Proof We sketch the proof. The group $GL(n, \mathbb{R})$ is generated by diagonal matrices, upper triangular matrices with 1s on the diagonal, and the Weyl group W_n which consists of all $n \times n$ matrices with exactly one 1 in each row and column and zeros everywhere else. For example,

$$W_2 = \left\{ \begin{pmatrix} 1 & 0 \\ 0 & 1 \end{pmatrix}, \begin{pmatrix} 0 & 1 \\ 1 & 0 \end{pmatrix} \right\},$$

$$W_3 = \left\{ \begin{pmatrix} 1 & 0 & 0 \\ 0 & 1 & 0 \\ 0 & 0 & 1 \end{pmatrix}, \begin{pmatrix} 1 & 0 & 0 \\ 0 & 0 & 1 \\ 0 & 1 & 0 \end{pmatrix}, \begin{pmatrix} 0 & 1 & 0 \\ 1 & 0 & 0 \\ 0 & 0 & 1 \end{pmatrix}, \right.$$
$$\left. \begin{pmatrix} 0 & 1 & 0 \\ 0 & 0 & 1 \\ 1 & 0 & 0 \end{pmatrix}, \begin{pmatrix} 0 & 0 & 1 \\ 1 & 0 & 0 \\ 0 & 1 & 0 \end{pmatrix}, \begin{pmatrix} 0 & 0 & 1 \\ 0 & 1 & 0 \\ 1 & 0 & 0 \end{pmatrix} \right\}.$$

Note that the Weyl group W_n has order $n!$ and is simply the symmetric group on n symbols. It is clear that $d^*(gz) = d^*z$ if g is an upper triangular matrix with 1s on the diagonal. This is because the measures $dx_{i,j}$ (with $1 \le i < j \le n$) are all invariant under translation. It is clear that the differential d^*z is Z_n-invariant where $Z_n \cong \mathbb{R}^\times$ denotes the center of $GL(n, \mathbb{R})$. So, without loss of generality, we may define a diagonal matrix a with its lower-right entry to be one:

$$a = \begin{pmatrix} a_1a_2\cdots a_{n-1} & & & & \\ & a_1a_2\cdots a_{n-2} & & & \\ & & \ddots & & \\ & & & a_1 & \\ & & & & 1 \end{pmatrix}.$$

Then

$$\begin{aligned} az = axy &= (axa^{-1})\cdot ay \\ &= \begin{pmatrix} 1 & a_{n-1}x_{1,2} & a_{n-1}a_{n-2}x_{1,3} & \cdots & a_{n-1}\cdots a_1\, x_{1,n} \\ & 1 & a_{n-2}x_{2,3} & \cdots & a_{n-2}\cdots a_1\, x_{2,n} \\ & & \ddots & & \vdots \\ & & & 1 & a_1\, x_{n-1,n} \\ & & & & 1 \end{pmatrix} \\ &\quad \times \begin{pmatrix} a_1y_1\cdots a_{n-1}y_{n-1} & & & \\ & \ddots & & \\ & & a_1y_1 & \\ & & & 1 \end{pmatrix}. \end{aligned}$$

Thus $d^*(axa^{-1}) = \left(\prod_{k=1}^{n-1} a_k^{k(n-k)}\right) d^*x$. It easily follows that

$$d^*(az) = d^*(axa^{-1}\cdot ay) = d^*z.$$

It remains to check the invariance of d^*z under the Weyl group W_n. Now, if $w \in W_n$ and

$$d = \begin{pmatrix} d_n & & & \\ & d_{n-1} & & \\ & & \ddots & \\ & & & d_1 \end{pmatrix} \in GL(n,\mathbb{R})$$

is a diagonal matrix, then wdw^{-1} is again a diagonal matrix whose diagonal entries are a permutation of $\{d_1, d_2, \ldots, d_n\}$. The Weyl group is generated by the transpositions ω_i $(i = 1, 2, \ldots n-1)$ given in Proposition 1.2.9 which interchange (transpose) d_i and d_{i+1} when d is conjugated by ω_i. After a tedious calculation using Proposition 1.2.9 one checks that $d^*(\omega_i z) = d^*z$. □

1.6 Volume of $SL(n,\mathbb{Z})\backslash SL(n,\mathbb{R})/SO(n,\mathbb{R})$

Following earlier work of Minkowski, Siegel (1936) showed that the volume of

$$\begin{aligned} SL(n,\mathbb{Z})\backslash SL(n,\mathbb{R})/SO(n,\mathbb{R}) &\cong SL(n,\mathbb{Z})\backslash GL(n,\mathbb{R})/(O(n,\mathbb{R})\cdot\mathbb{R}^\times) \\ &\cong SL(n,\mathbb{Z})\backslash\mathfrak{h}^n, \end{aligned}$$

can be given in terms of

$$\zeta(2)\cdot\zeta(3)\cdots\zeta(n)$$

where $\zeta(s)$ is the Riemann zeta function. The fact that the special values (taken at integral points) of the Riemann zeta function appear in the formula for the volume is remarkable. Later, Weil (1946) found another method to prove such results based on a direct application of the Poisson summation formula. A vast generalization of Siegel's computation of fundamental domains for the case of arithmetic subgroups acting on Chevalley groups was obtained by Langlands (1966). See also (Terras, 1988) for interesting discussions on the history of this subject.

The main aim of this section is to explicitly compute the volume

$$\int\limits_{SL(n,\mathbb{Z})\backslash SL(n,\mathbb{R})/SO(n,\mathbb{R})} d^*z,$$

where d^*z is the left–invariant measure given in Proposition 1.5.3. We follow the exposition of Garret (2002).

Theorem 1.6.1 *Let $n\geq 2$. As in Proposition 1.5.3, fix*

$$d^*z=\prod_{1\leq i<j\leq n} dx_{i,j}\prod_{k=1}^{n-1} y_k^{-k(n-k)-1}\,dy_k$$

to be the left $SL(n,\mathbb{R})$–invariant measure on $\mathfrak{h}^n = SL(n,\mathbb{R})/SO(n,\mathbb{R})$. Then

$$\int_{SL(n,\mathbb{Z})\backslash\mathfrak{h}^n} d^*z = n\,2^{n-1}\cdot\prod_{\ell=2}^{n}\frac{\zeta(\ell)}{\mathrm{Vol}(S^{\ell-1})},$$

where

$$\mathrm{Vol}(S^{\ell-1})=\frac{2(\sqrt{\pi})^\ell}{\Gamma(\ell/2)}$$

denotes the volume of the $(\ell-1)$–dimensional sphere $S^{\ell-1}$ and $\zeta(\ell)=\sum\limits_{n=1}^{\infty} n^{-\ell}$ denotes the Riemann zeta function.

Proof for the case of $SL(2,\mathbb{R})$ We first prove the theorem for $SL(2,\mathbb{R})$. The more general result will follow by induction. Let $K = SO(2,\mathbb{R})$ denote the maximal compact subgroup of $SL(2,\mathbb{R})$. We use the Iwasawa decomposition which says that

$$SL(2,\mathbb{R})/K \cong \left\{ z = \begin{pmatrix} 1 & x \\ 0 & 1 \end{pmatrix} \begin{pmatrix} y^{\frac{1}{2}} & 0 \\ 0 & y^{-\frac{1}{2}} \end{pmatrix} \;\middle|\; x \in \mathbb{R},\ y > 0 \right\}.$$

Let $f : \mathbb{R}^2/K \to \mathbb{C}$ be an arbitrary smooth compactly supported function. Then, by definition, $f((u,v)\cdot k) = f((u,v))$ for all $(u,v) \in \mathbb{R}^2$ and all $k \in K$. We can define a function $F : SL(2,\mathbb{R})/K \to \mathbb{C}$ by letting

$$F(z) := \sum_{(m,n)\in\mathbb{Z}^2} f((m,n)\cdot z).$$

If $\gamma = \begin{pmatrix} a & b \\ c & d \end{pmatrix} \in SL\,2(\mathbb{Z})$, then

$$\begin{aligned} F(\gamma z) &= \sum_{(m,n)\in\mathbb{Z}^2} f\left((m,n)\cdot \begin{pmatrix} a & b \\ c & d \end{pmatrix} \cdot z \right) \\ &= \sum_{(m,n)\in\mathbb{Z}^2} f\big((ma+nc,\, mb+nd)\cdot z\big) \\ &= F(z). \end{aligned}$$

Thus, $F(z)$ is $SL(2,\mathbb{Z})$–invariant.

Note that we may express

$$\{(m,n)\in\mathbb{Z}^2\} = (0,0) \cup \left\{ \ell\cdot(0,1)\cdot\gamma \;\middle|\; 0<\ell\in\mathbb{Z},\ \gamma\in\Gamma_\infty\backslash SL(2,\mathbb{Z}) \right\}, \tag{1.6.2}$$

where

$$\Gamma_\infty = \left\{ \begin{pmatrix} 1 & r \\ 0 & 1 \end{pmatrix} \;\middle|\; r\in\mathbb{Z} \right\}.$$

We now integrate F over $\Gamma\backslash\mathfrak{h}^2$, where $\mathfrak{h}^2 = SL(2,\mathbb{R})/K$, $\Gamma = SL(2,\mathbb{Z})$, and $dxdy/y^2$ is the invariant measure on $\mathfrak{h}^2$ given in Proposition 1.5.3. It immediately follows from (1.6.2) that

$$\begin{aligned}\int\limits_{\Gamma\backslash\mathfrak{h}^2} F(z)\,\frac{dxdy}{y^2} &= f((0,0))\cdot \mathrm{Vol}(\Gamma\backslash\mathfrak{h}^2)\\ &\quad + \sum_{\ell>0}\sum_{\gamma\in\Gamma_\infty\backslash\Gamma}\int\limits_{\Gamma\backslash\mathfrak{h}^2} f\big(\ell(0,1)\cdot\gamma\cdot z\big)\,\frac{dxdy}{y^2}\\ &= f((0,0))\cdot\mathrm{Vol}(\Gamma\backslash\mathfrak{h}^2) + 2\sum_{\ell>0}\int\limits_{\Gamma_\infty\backslash\mathfrak{h}^2} f\big(\ell(0,1)\cdot z\big)\,\frac{dxdy}{y^2}.\end{aligned}$$

The factor 2 occurs because $\begin{pmatrix}-1 & \\ & -1\end{pmatrix}$ acts trivially on $\mathfrak{h}^2$. We easily observe that

$$f\big(\ell(0,1)\cdot z\big) = f\left(\ell(0,1)\cdot\begin{pmatrix} y^{\frac{1}{2}} & 0\\ 0 & y^{-\frac{1}{2}}\end{pmatrix}\right) = f\left(\left(0,\ell y^{-\frac{1}{2}}\right)\right).$$

It follows, after making the elementary transformations

$$y\mapsto \ell^2 y,\qquad y\mapsto y^{-2}$$

that

$$\int\limits_{\Gamma\backslash\mathfrak{h}^2} F(z)\,\frac{dxdy}{y^2} = f((0,0))\cdot\mathrm{Vol}(\Gamma\backslash\mathfrak{h}^2) + 2^2\zeta(2)\int\limits_0^\infty f((0,y))\,y\,dy. \tag{1.6.3}$$

Now, the function $f((u,v))$ is invariant under multiplication by $k\in K$ on the right. Since $\begin{pmatrix}\sin\theta & -\cos\theta\\ \cos\theta & \sin\theta\end{pmatrix}\in K$, we see that

$$f((0,y)) = f((y\cos\theta,\ y\sin\theta))$$

for any $0\le\theta\le 2\pi$. Consequently

$$\begin{aligned}\int_0^\infty f((0,y))\,y\,dy &= \frac{1}{2\pi}\int_0^{2\pi}\int_0^\infty f((y\cos\theta,\ y\sin\theta))\,d\theta\,y\,dy\\ &= \frac{1}{2\pi}\int\limits_{\mathbb{R}^2} f((u,v))\,dudv\\ &= \frac{1}{2\pi}\hat{f}((0,0)).\end{aligned} \tag{1.6.4}$$

Here $\hat{f}$ denotes the Fourier transform of f in $\mathbb{R}^2$. If we now combine (1.6.3) and (1.6.4), we obtain

$$\int_{\Gamma\backslash\mathfrak{h}^2} F(z)\frac{dxdy}{y^2} = f((0,0))\cdot \mathrm{Vol}(\Gamma\backslash\mathfrak{h}^2) + \frac{2\zeta(2)}{\pi}\hat{f}((0,0)). \tag{1.6.5}$$

To complete the proof, we make use of the Poisson summation formula (see appendix) which states that for any $z \in GL(2,\mathbb{R})$

$$\begin{aligned} F(z) &= \sum_{(m,n)\in\mathbb{Z}^2} f((m,n)z) = \frac{1}{|\mathrm{Det}(z)|}\sum_{(m,n)\in\mathbb{Z}^2} \hat{f}\big((m,n)\cdot({}^tz)^{-1}\big) \\ &= \sum_{(m,n)\in\mathbb{Z}^2} \hat{f}\big((m,n)\cdot({}^tz)^{-1}\big), \end{aligned}$$

since $z = \begin{pmatrix} y^{\frac{1}{2}} & y^{-\frac{1}{2}}x \\ 0 & y^{-\frac{1}{2}} \end{pmatrix}$ and $\mathrm{Det}(z) = 1$. We now repeat all our computations with the roles of f and $\hat{f}$ reversed. Since the group Γ is stable under transpose–inverse, one easily sees (from the Poisson summation formula above), by letting $z \mapsto ({}^tz)^{-1}$, that the integral

$$\int_{\Gamma\backslash\mathfrak{h}^2} F(z)\frac{dxdy}{y^2}$$

is unchanged if we replace f by $\hat{f}$.

Also, since $\hat{\hat{f}}(x) = f(-x)$, the formula (1.6.5) now becomes

$$\int_{\Gamma\backslash\mathfrak{h}^2} F(z)\frac{dxdy}{y^2} = \hat{f}((0,0))\cdot \mathrm{Vol}(\Gamma\backslash\mathfrak{h}^2) + \frac{2\zeta(2)}{\pi}f((0,0)). \tag{1.6.6}$$

If we combine (1.6.5) and (1.6.6) and solve for the volume, we obtain

$$\big(f((0,0)) - \hat{f}((0,0))\big)\cdot \mathrm{vol}(\Gamma\backslash\mathfrak{h}^2) = \big(f((0,0)) - \hat{f}((0,0))\big)\cdot\frac{2\zeta(2)}{\pi}.$$

Since f is arbitrary, we can choose f so that $f((0,0)) - \hat{f}((0,0)) \neq 0$. It follows that

$$\mathrm{Vol}(\Gamma\backslash\mathfrak{h}^2) = \frac{2\zeta(2)}{\pi} = \frac{\pi}{3}. \qquad \square$$

Proof for the case of $SL(n,\mathbb{R})$ We shall now complete the proof of Theorem 1.6.1 using induction on n. □

The proof of Theorem 1.6.1 requires two preliminary lemmas which we straightaway state and prove. For $n > 2$, let $U_n(\mathbb{R})$ (respectively $U_n(\mathbb{Z})$) denote the group of all matrices of the form

$$\begin{pmatrix} 1 & & & u_1 \\ & \ddots & & \vdots \\ & & 1 & u_{n-1} \\ & & & 1 \end{pmatrix}$$

with $u_i \in \mathbb{R}$ (respectively, $u_i \in \mathbb{Z}$), for $i = 1, 2, \ldots, n-1$.

Lemma 1.6.7 *Let $n > 2$ and fix an element $\gamma \in SL(n-1,\mathbb{Z})$. Consider the action of $U_n(\mathbb{Z})$ on $\mathbb{R}^{n-1}$ given by left matrix multiplication of $U_n(\mathbb{Z})$ on $\begin{pmatrix} \gamma & 0 \\ 0 & 1 \end{pmatrix} \cdot U_n(\mathbb{R})$. Then a fundamental domain for this action is given by the set of all matrices*

$$\begin{pmatrix} \gamma & 0 \\ 0 & 1 \end{pmatrix} \cdot \begin{pmatrix} 1 & & & u_1 \\ & \ddots & & \vdots \\ & & 1 & u_{n-1} \\ & & & 1 \end{pmatrix}$$

with $0 \le u_i < 1$ for $1 \le i \le n-1$. In particular,

$$U_n(\mathbb{Z}) \Big\backslash \begin{pmatrix} \gamma & 0 \\ 0 & 1 \end{pmatrix} \cdot U_n(\mathbb{R}) \cong (\mathbb{Z}\backslash\mathbb{R})^{n-1}.$$

Proof of Lemma 1.6.7 Let m be a column vector with $(m_1, m_2, \ldots, m_{n-1})$ as entries. Then one easily checks that

$$\begin{pmatrix} I_{n-1} & m \\ & 1 \end{pmatrix} \cdot \begin{pmatrix} \gamma & \\ & 1 \end{pmatrix} = \begin{pmatrix} \gamma & \\ & 1 \end{pmatrix} \cdot \begin{pmatrix} I_{n-1} & \gamma^{-1}m \\ & 1 \end{pmatrix},$$

where I_{n-1} denotes the $(n-1)\times(n-1)$ identity matrix. It follows that

$$\begin{aligned}
&\bigcup_{m\in\mathbb{Z}^{n-1}} \begin{pmatrix} I_{n-1} & m \\ & 1 \end{pmatrix} \cdot \begin{pmatrix} \gamma & \\ & 1 \end{pmatrix} \cdot \begin{pmatrix} I_{n-1} & (\mathbb{Z}\backslash\mathbb{R})^{n-1} \\ & 1 \end{pmatrix} \\
&\quad = \bigcup_{m\in\mathbb{Z}^{n-1}} \begin{pmatrix} \gamma & \\ & 1 \end{pmatrix} \cdot \begin{pmatrix} I_{n-1} & \gamma^{-1}m \\ & 1 \end{pmatrix} \cdot \begin{pmatrix} I_{n-1} & (\mathbb{Z}\backslash\mathbb{R})^{n-1} \\ & 1 \end{pmatrix} \\
&\quad = \begin{pmatrix} \gamma & \\ & 1 \end{pmatrix} \cdot \bigcup_{m\in\mathbb{Z}^{n-1}} \begin{pmatrix} I_{n-1} & (\mathbb{Z}\backslash\mathbb{R})^{n-1} + \gamma^{-1}m \\ & 1 \end{pmatrix} \\
&\quad = \begin{pmatrix} \gamma & \\ & 1 \end{pmatrix} \cdot U_n(\mathbb{R}).
\end{aligned}$$

It is also clear that the above union is over non-overlapping sets. This is because $\gamma^{-1}\mathbb{Z}^{n-1} = \mathbb{Z}^{n-1}$ for $\gamma \in SL(n-1,\mathbb{Z})$. □

The second lemma we need is a generalization of the identity (1.6.4).

Lemma 1.6.8 *Let $n > 2$ and let $f : \mathbb{R}^n \to \mathbb{C}$ be a smooth function, with sufficient decay at ∞, which satisfies $f(u_1, \ldots, u_n) = f(v_1, \ldots, v_n)$ whenever $u_1^2 + \cdots + u_n^2 = v_1^2 + \cdots + v_n^2$. Then*

$$\int_0^\infty f(0, \ldots, 0, t)\, t^{n-1}\, dt = \frac{1}{\mathrm{Vol}(S^{n-1})} \int_{\mathbb{R}^n} f(x_1, \ldots, x_n)\, dx_1 \cdots dx_n$$
$$= \frac{\hat{f}(0)}{\mathrm{Vol}(S^{n-1})},$$

where

$$\mathrm{Vol}(S^{n-1}) = \frac{2(\sqrt{\pi})^n}{\Gamma(n/2)}$$

denotes the volume of the $(n-1)$–dimensional sphere S^{n-1}.

Proof of Lemma 1.6.8 For $n \geq 2$ consider the spherical coordinates:

$$\begin{aligned} x_1 &= t \cdot \sin\theta_{n-1} \cdots \sin\theta_2 \sin\theta_1, \\ x_2 &= t \cdot \sin\theta_{n-1} \cdots \sin\theta_2 \cos\theta_1, \\ x_3 &= t \cdot \sin\theta_{n-1} \cdots \sin\theta_3 \cos\theta_2, \\ &\ \vdots \\ x_{n-1} &= t \cdot \sin\theta_{n-1} \cos\theta_{n-2}, \\ x_n &= t \cdot \cos\theta_{n-1}, \end{aligned} \tag{1.6.9}$$

with

$$0 < t < \infty, \qquad 0 \leq \theta_1 < 2\pi, \qquad 0 \leq \theta_j < \pi, \ (1 < j < n).$$

Clearly $x_1^2 + \cdots + x_n^2 = t^2$. One may also show that the invariant measure on the sphere S^{n-1} is given by

$$d\mu(\theta) = \prod_{1 \leq j < n} (\sin\theta_j)^{j-1}\, d\theta_j,$$

and that $dx_1 dx_2 \cdots dx_n = t^{n-1} dt\ d\mu(\theta)$. Then the volume of the unit sphere, $\mathrm{Vol}(S^{n-1})$, is given by

$$\mathrm{Vol}(S^{n-1}) = \int_{S^{n-1}} d\mu(\theta) = \frac{2(\sqrt{\pi})^n}{\Gamma(n/2)}.$$

Since f is a rotationally invariant function, it follows that

$$f(0, \ldots, 0, t) = \frac{1}{\mathrm{Vol}(S^{n-1})} \int_{S^{n-1}} f(x_1, \ldots, x_n)\, d\mu(\theta)$$

with $x_1, \ldots, x_n$ given by (1.6.9). Consequently

$$\int_0^\infty f(0, \ldots, 0, t)t^{n-1}\, dt = \frac{1}{\mathrm{Vol}(S^{n-1})} \int_0^\infty \int_{S^{n-1}} f(x_1, \ldots, x_n)t^{n-1} d\mu(\theta) dt$$
$$= \frac{1}{\mathrm{Vol}(S^{n-1})} \int_{\mathbb{R}^n} f(x_1, \ldots, x_n)\, dx_1 \cdots dx_n.$$

□

We now return to the proof of Theorem 1.6.1. Let $K_n = SO(n, \mathbb{R})$ denote the maximal compact subgroup of $SL(n, \mathbb{R})$. In this case, the Iwasawa decomposition (Proposition 1.2.6) says that every $z \in SL(n, \mathbb{R})/K_n$ is of the form $z = xy$ with

$$x = \begin{pmatrix} 1 & x_{1,2} & x_{1,3} & \cdots & & x_{1,n} \\ & 1 & x_{2,3} & \cdots & & x_{2,n} \\ & & \ddots & & & \vdots \\ & & & & 1 & x_{n-1,n} \\ & & & & & 1 \end{pmatrix},$$
$$y = \begin{pmatrix} y_1 y_2 \cdots y_{n-1} t & & & & \\ & y_1 y_2 \cdots y_{n-2} t & & & \\ & & \ddots & & \\ & & & y_1 t & \\ & & & & t \end{pmatrix}, \qquad (1.6.10)$$

with $t = \mathrm{Det}(y)^{-1/n} = \left(\prod_{i=1}^{n-1} y_i^{n-i} \right)^{-1/n}$.

In analogy to the previous proof for $SL(2, \mathbb{R})$ we let $f : \mathbb{R}^n / K_n \to \mathbb{C}$ be an arbitrary smooth compactly supported function. We shall also define a function $F : SL(n, \mathbb{R})/K_n \to \mathbb{C}$ by letting

$$F(z) := \sum_{m \in \mathbb{Z}^n} f(m \cdot z).$$

As before, the function $F(z)$ will be invariant under left multiplication by $SL(n, \mathbb{Z})$.

Let

$$P_n = \begin{pmatrix} & & * & & \\ 0 & 0 & \cdots & 0 & 1 \end{pmatrix} \in SL(n, \mathbb{Z})$$

denote the set of all $n \times n$ matrices in $SL(n, \mathbb{Z})$ with last row $(0, 0, \ldots, 0, 1)$. Let $\Gamma_n = SL(n, \mathbb{Z})$. Then we have as before:

$$F(z) = f(0) + \sum_{0<\ell\in\mathbb{Z}} \sum_{\gamma\in P_n\backslash\Gamma_n} f(\ell e_n \cdot \gamma \cdot z),$$

where $f(0)$ denotes $f((0, 0, \ldots, 0))$ and $e_n = (0, 0, \ldots, 0, 1)$.

We now integrate $F(z)$ over a fundamental domain for $\Gamma_n\backslash\mathfrak{h}^n$. It follows that

$$\int_{\Gamma_n\backslash\mathfrak{h}^n} F(z)\, d^*z = f(0) \cdot \mathrm{Vol}(\Gamma_n\backslash\mathfrak{h}^n) + 2\sum_{\ell>0} \int_{P_n\backslash\mathfrak{h}^n} f(\ell e_n \cdot z)\, d^*z. \qquad (1.6.11)$$

The factor 2 occurs because $-I_n$ ($I_n = n \times n$ identity matrix) acts trivially on $\mathfrak{h}^n$. The computation of the integral above requires some preparations.

We may express $z \in \mathfrak{h}^n$ in the form

$$z = x \cdot \begin{pmatrix} y_1 y_2 \cdots y_{n-1} t & & & & \\ & y_1 y_2 \cdots y_{n-2} t & & & \\ & & \ddots & & \\ & & & y_1 t & \\ & & & & t \end{pmatrix} \cdot \begin{pmatrix} t^{\frac{1}{n-1}} \cdot I_{n-1} & \\ & t^{-1} \end{pmatrix} \cdot \begin{pmatrix} t^{-\frac{1}{n-1}} \cdot I_{n-1} & \\ & t \end{pmatrix},$$

where x and t are given by (1.6.10). It follows that

$$z = \begin{pmatrix} 1 & & & & x_{1,n} \\ & 1 & & & x_{2,n} \\ & & \ddots & & \vdots \\ & & & 1 & x_{n-1,n} \\ & & & & 1 \end{pmatrix} \cdot \begin{pmatrix} 1 & x_{1,2} & x_{1,3} & \cdots & x_{1,n-1} & 0 \\ & 1 & x_{2,3} & \cdots & x_{2,n-1} & 0 \\ & & \ddots & & \vdots & \vdots \\ & & & & 1 & 0 \\ & & & & & 1 \end{pmatrix}$$

$$\times \begin{pmatrix} y_1 y_2 \cdots y_{n-1} \cdot t^{\frac{n}{n-1}} & & & & \\ & y_1 y_2 \cdots y_{n-2} \cdot t^{\frac{n}{n-1}} & & & \\ & & \ddots & & \\ & & & y_1 \cdot t^{\frac{n}{n-1}} & \\ & & & & 1 \end{pmatrix} \cdot \begin{pmatrix} t^{-\frac{1}{n-1}} \cdot I_{n-1} & \\ & t \end{pmatrix}$$

$$= \begin{pmatrix} 1 & & & & x_{1,n} \\ & 1 & & & x_{2,n} \\ & & \ddots & & \vdots \\ & & & 1 & x_{n-1,n} \\ & & & & 1 \end{pmatrix} \cdot \begin{pmatrix} z' & \\ & 1 \end{pmatrix} \cdot \begin{pmatrix} t^{-\frac{1}{n-1}} \cdot I_{n-1} & \\ & t \end{pmatrix},$$

(1.6.12)

where

$$z' = \begin{pmatrix} 1 & x_{1,2} & x_{1,3} & \cdots & x_{1,n-1} \\ & 1 & x_{2,3} & \cdots & x_{2,n-1} \\ & & \ddots & & \vdots \\ & & & 1 & x_{n-2,n-1} \\ & & & & 1 \end{pmatrix}$$

$$\times \begin{pmatrix} y_1 y_2 \cdots y_{n-1} \cdot t^{\frac{n}{n-1}} & & & \\ & y_1 y_2 \cdots y_{n-2} \cdot t^{\frac{n}{n-1}} & & \\ & & \ddots & \\ & & & y_1 \cdot t^{\frac{n}{n-1}} \end{pmatrix}.$$

Now z' represents the Iwasawa coordinate for $SL(n-1,\mathbb{R})/SO(n-1,\mathbb{R}) = \mathfrak{h}^{n-1}$, and the Haar measure d^*z' can be computed using Proposition 1.5.3 and is given by

$$d^*z' = \prod_{1\le i<j\le n-1} dx_{i,j} \prod_{k=1}^{n-2} y_{k+1}^{-k(n-1-k)-1} dy_{k+1}.$$

If we compare this with

$$\begin{aligned} d^*z &= \prod_{1\le i<j\le n} dx_{i,j} \prod_{k=1}^{n-1} y_k^{-k(n-k)-1} dy_k \\ &= \prod_{1\le i<j\le n} dx_{i,j} \prod_{k=0}^{n-2} y_{k+1}^{-(k+1)(n-1-k)-1} dy_k, \end{aligned}$$

we see that

$$d^*z = d^*z' \prod_{j=1}^{n-1} dx_{j,n}\, t^n \frac{dy_1}{y_1}. \tag{1.6.13}$$

Here, the product of differentials is understood as a wedge product satisfying the usual rule: $du \wedge du = 0$, given by the theory of differential forms. Since

$$t = y_1^{-(n-1)/n} \prod_{i=2}^{n-1} y_i^{-(n-i)/n},$$

we see that

$$\frac{dt}{t} = -\frac{n-1}{n}\frac{dy_1}{y_1} + \Omega,$$

where Ω is a differential form involving dy_j for each $j = 2, 3, \ldots, n-1$, but not involving dy_1. It follows from (1.6.13) that

$$d^*z = -\frac{n}{n-1} d^*z' \prod_{j=1}^{n-1} dx_{j,n}\, t^n \,\frac{dt}{t}. \tag{1.6.14}$$

We also note that, by (1.6.12), we have

$$f(\ell e_n \cdot z) = f\left(\ell e_n \cdot \begin{pmatrix} t^{-\frac{1}{n-1}} \cdot I_{n-1} & \\ & t \end{pmatrix}\right) = f(\ell t\, e_n). \tag{1.6.15}$$

The last thing we need to do is to construct a fundamental domain for the action of P_n on $\mathfrak{h}^n$. Every $p \in P_n$ can be written in the form

$$p = \begin{pmatrix} \gamma & b \\ & 1 \end{pmatrix}, \qquad (\text{with } \gamma \in SL(n-1,\mathbb{Z}),\ b \in \mathbb{Z}^{n-1}).$$

By (1.6.12), we may express $z \in \mathfrak{h}^n$ in the form

$$z = \begin{pmatrix} z' & u \\ & 1 \end{pmatrix} \cdot \begin{pmatrix} t^{-\frac{1}{n-1}} \cdot I_{n-1} & \\ & t \end{pmatrix}, \qquad u = \begin{pmatrix} x_{1,n} \\ x_{2,n} \\ \vdots \\ x_{n-1,n} \end{pmatrix} \in \mathbb{R}^{n-1}.$$

It follows that

$$p \cdot z = \begin{pmatrix} \gamma z' & \gamma \cdot u + b \\ & 1 \end{pmatrix} \cdot \begin{pmatrix} t^{-\frac{1}{n-1}} \cdot I_{n-1} & \\ & t \end{pmatrix},$$

from which one deduces from Lemma 1.6.7 that

$$P_n \backslash \mathfrak{h}^n \cong SL(n-1,\mathbb{Z}) \backslash \mathfrak{h}^{n-1} \times (\mathbb{R}/\mathbb{Z})^{n-1} \times (0,\infty),$$

With these preliminaries, we can now continue the calculation of (1.6.11). It follows from (1.6.14), (1.6.15), and Lemma 1.6.8 that

$$\begin{aligned} 2\sum_{\ell>0} \int\limits_{P_n\backslash\mathfrak{h}^n} & f(\ell e_n \cdot z)\, d^*z \\ &= \frac{2n}{n-1} \sum_{\ell>0} \left(\int\limits_{\Gamma_{n-1}\backslash\mathfrak{h}^{n-1}} d^*z' \right) \left(\int\limits_{(\mathbb{R}/\mathbb{Z})^{n-1}} \prod_{j=1}^{n-1} dx_{j,n} \right) \int_0^\infty f(\ell t\, e_n)\, t^n \,\frac{dt}{t} \\ &= \frac{2n}{n-1} \zeta(n)\, \mathrm{Vol}\big(\Gamma_{n-1}\backslash\mathfrak{h}^{n-1}\big) \cdot \int_{t=0}^{\infty} f(t\, e_n)\, t^n \,\frac{dt}{t} \\ &= \frac{2n}{n-1} \zeta(n)\, \mathrm{Vol}\big(\Gamma_{n-1}\backslash\mathfrak{h}^{n-1}\big) \cdot \frac{\hat{f}(0)}{\mathrm{Vol}(S^{n-1})}. \end{aligned} \tag{1.6.16}$$

Here, we used the facts that

$$\int\limits_{\Gamma_{n-1}\backslash\mathfrak{h}^{n-1}} d^*z' = \mathrm{Vol}\big(\Gamma_{n-1}\backslash\mathfrak{h}^{n-1}\big), \qquad \int\limits_{(\mathbb{R}/\mathbb{Z})^{n-1}} \prod_{j=1}^{n-1} dx_{j,n} = 1.$$

Combining (1.6.11) and (1.6.16) gives

$$\int\limits_{\Gamma_n\backslash\mathfrak{h}^n} F(z)\,d^*z = f(0)\cdot \mathrm{Vol}(\Gamma_n\backslash\mathfrak{h}^n) + \hat{f}(0)\cdot \frac{2n\,\zeta(n)\,\mathrm{Vol}\big(\Gamma_{n-1}\backslash\mathfrak{h}^{n-1}\big)}{(n-1)\mathrm{Vol}(S^{n-1})}. \tag{1.6.17}$$

As before, we make use of the Poisson summation formula

$$F(z) = \sum_{m\in\mathbb{Z}^n} f(m\cdot z) = \sum_{m\in\mathbb{Z}^n} \hat{f}(m\cdot({}^tz)^{-1}),$$

which holds for $\mathrm{Det}(z)=1$. Since the group Γ_n is stable under transpose–inverse, we can repeat all our computations with the roles of f and $\hat{f}$ reversed, and the integral

$$\int\limits_{\Gamma_n\backslash\mathfrak{h}^n} F(z)\,d^*z$$

again remains unchanged. The formula (1.6.17) now becomes

$$f(0)\cdot \mathrm{Vol}(\Gamma_n\backslash\mathfrak{h}^n) + \hat{f}(0)\cdot \frac{2n\,\zeta(n)\,\mathrm{Vol}\big(\Gamma_{n-1}\backslash\mathfrak{h}^{n-1}\big)}{(n-1)\mathrm{Vol}(S^{n-1})}$$
$$= \hat{f}(0)\cdot \mathrm{Vol}(\Gamma_n\backslash\mathfrak{h}^n) + f(0)\cdot \frac{2n\,\zeta(n)\,\mathrm{Vol}\big(\Gamma_{n-1}\backslash\mathfrak{h}^{n-1}\big)}{(n-1)\mathrm{Vol}(S^{n-1})}.$$

Taking f so that $f(0)\neq\hat{f}(0)$, we obtain

$$\mathrm{Vol}(\Gamma_n\backslash\mathfrak{h}^n) = \frac{2n\,\zeta(n)\,\mathrm{Vol}\big(\Gamma_{n-1}\backslash\mathfrak{h}^{n-1}\big)}{(n-1)\mathrm{Vol}(S^{n-1})}.$$

Theorem 1.6.1 immediately follows from this by induction. □

GL(n)pack functions The following **GL(n)pack** functions, described in the appendix, relate to the material in this chapter:

IwasawaForm	IwasawaXMatrix	IwasawaXVariables
IwasawaYMatrix	IwasawaYVariables	IwasawaQ
MakeXMatrix	MakeXVariables	MakeYMatrix
MakeYVariables	MakeZMatrix	MakeZVariables
VolumeFormDiagonal	VolumeFormHn	VolumeFormUnimodular
VolumeHn	Wedge	d.

2
Invariant differential operators

It has been shown in the previous chapter that discrete group actions can give rise to functional equations associated to important number theoretic objects such as the Riemann zeta function. Thus, there is great motivation for studying discrete group actions from all points of view. Let us explore this situation in one of the simplest cases. Consider the additive group of integers $\mathbb{Z}$ acting on the real line $\mathbb{R}$ by translation as in Example 1.1.2. The quotient space $\mathbb{Z}\backslash\mathbb{R}$ is just the circle S^1. One may study S^1 by considering the space of all possible smooth functions $f: S^1 \to \mathbb{C}$. These are the periodic functions that arise in classical Fourier theory. The Fourier theorem says that every smooth periodic function $f: S^1 \to \mathbb{C}$ can be written as a linear combination

$$f(x) = \sum_{n\in\mathbb{Z}} a_n e^{2\pi i n x}$$

where

$$a_n = \int_0^1 f(x) e^{-2\pi i n x}\, dx,$$

for all $n \in \mathbb{Z}$. In other words, the basic periodic functions, $e^{2\pi i n x}$ with $n \in \mathbb{Z}$, form a basis for the space $\mathcal{L}^2(\mathbb{Z}\backslash\mathbb{R})$. It is clear that a deeper understanding of this space is an important question in number theory. We shall approach this question from the viewpoint of differential operators and obtain a fresh and illuminating perspective thereby.

A basis for the space $\mathcal{L}^2(\mathbb{Z}\backslash\mathbb{R})$ may be easily described by using the Laplace operator $\frac{d^2}{dx^2}$. One sees that the basic periodic functions $e^{2\pi i n x}$ (with $n \in \mathbb{Z}$) are all eigenfunctions of this operator with eigenvalue $-4\pi^2 n^2$, i.e.,

$$\frac{d^2}{dx^2} e^{2\pi i n x} = -4\pi^2 n^2 \cdot e^{2\pi i n x}.$$

Consequently, the space $\mathcal{L}^2(\mathbb{Z}\backslash\mathbb{R})$ can be realized as the space generated by the eigenfunctions of the Laplacian. What we have pointed out here is a simple example of spectral theory. Good references for spectral theory are (Aupetit, 1991), (Arveson, 2002).

In the higher-dimensional setting, we shall investigate smooth functions invariant under discrete group actions by studying invariant differential operators. These are operators that do not change under discrete group actions. For example, the classical Laplace operator $\frac{d^2}{dx^2}$ does not change under the action $x \mapsto x + n$ for any fixed integer n. It is not so obvious that the operator

$$-y^2\left(\frac{\partial^2}{\partial x^2} + \frac{\partial^2}{\partial y^2}\right)$$

is invariant under the discrete group actions

$$z \mapsto \frac{az+b}{cz+d},$$

with $\begin{pmatrix} a & b \\ c & d \end{pmatrix} \in GL(2, \mathbb{R})$. The main aim of this chapter is to develop a general theory of invariant differential operators and the best framework to do this in is the setting of elementary Lie theory. For introductary texts on differential operators, see (Boothby, 1986), (Munkres, 1991). The classic reference on Lie groups and Lie algebras is (Bourbaki, 1998b), but see also (Bump, 2004). In order to give a self-contained exposition, we begin with some basic definitions.

2.1 Lie algebras

Definition 2.1.1 *An* **associative algebra** *A over a field K is a vector space A over K with an associative product $\circ$, which satisfies for all $a, b, c \in A$, the following conditions:*

- *$a \circ b$ is uniquely defined and $a \circ b \in A$,*
- *$a \circ (b \circ c) = (a \circ b) \circ c$, (associative law)*
- *$(a + b) \circ c = a \circ c + b \circ c$,*
  $c \circ (a + b) = c \circ a + c \circ b$, (distributive law).

Note that in a vector space we can either add vectors or multiply them by scalars (elements of A). The associative product gives a way of multiplying vectors themselves.

Example Let $\mathbb{R}$ be the field of real numbers and let

$$A = M(n, \mathbb{R})$$

denote the set of $n \times n$ matrices with coefficients in $\mathbb{R}$. Then $A = M(n, \mathbb{R})$ is an associative algebra over $\mathbb{R}$ where $\circ$ denotes matrix multiplication. A basis for the vector space A over $\mathbb{R}$ is given by the set of n^2 vectors $E_{i,j}$ where $E_{i,j}$ denotes the matrix with a 1 at position $\{i, j\}$ and zeros everywhere else.

Definition 2.1.2 *A* **Lie algebra L** *over a field K is a vector space* **L** *over K together with a bilinear map* $[,]$ *(pronounced bracket), of* **L** *into itself, which satisfies for all $a, b, c \in \mathbf{L}$:*

- $[a, b]$ *is uniquely defined and* $[a, b] \in \mathbf{L}$;
- $[a,\ \beta b + \gamma c] = \beta[a, b] + \gamma[a, c], \qquad \forall \beta, \gamma \in K;$
- $[a, a] = 0;$
- $[a, b] = -[b, a],$ *(skew symmetry);*
- $[a, [b, c]] + [b, [c, a]] + [c, [a, b]] = 0,$ *(Jacobi identity).*

Example Every associative algebra A (with associative product $\circ$) can be made into a Lie algebra (denoted Lie(A)) by defining a bracket on A:

$$[a, b] = a \circ b - b \circ a, \qquad \forall\, a, b \in A.$$

This is easily proved since $[a, b]$ is clearly bilinear, $[a, a] = 0$ and

$$\begin{aligned}
&[[a, b], c] + [[b, c], a] + [[c, a], b] \\
&= (a \circ b - b \circ a) \circ c - c \circ (a \circ b - b \circ a) \\
&\quad + (b \circ c - c \circ b) \circ a - a \circ (b \circ c - c \circ b) \\
&\quad + (c \circ a - a \circ c) \circ b - b \circ (c \circ a - a \circ c).
\end{aligned}$$

We now show that it is also possible to go in the other direction. Namely, given a Lie algebra **L**, we show how to construct an associative algebra $U(\mathbf{L})$ (called the universal enveloping algebra) where

$$\mathbf{L} \subseteq \text{Lie}(U(\mathbf{L})).$$

In order to construct the universal enveloping algebra, we remind the reader of some basic concepts and notation in the theory of vector spaces. Let V denote a vector space over a field K with basis vectors $\mathbf{v}_1, \mathbf{v}_2, \ldots$ Then we may write

$$V = \oplus_i \ K\mathbf{v}_i.$$

Similarly, if W is another vector space with basis vectors $\mathbf{w}_1, \mathbf{w}_2, \ldots$ such that $\mathbf{w}_i \notin V$ for $i = 1, 2, \ldots,$ then we may form the vector space

$$V \oplus W$$

(defined over K) with basis vectors $\mathbf{v}_1, \mathbf{w}_1, \mathbf{v}_2, \mathbf{w}_2, \ldots$ Similarly, we can also define higher direct sums

$$\bigoplus_{\ell} V_\ell$$

of a set of linearly independent vector spaces V_ℓ. We shall also consider the tensor product $V \otimes W$ which is the vector space with basis vectors

$$\mathbf{v}_i \otimes \mathbf{w}_j, \qquad (i = 1, 2, \ldots, j = 1, 2, \ldots)$$

and the higher tensor products $\otimes^k V$ (for $k = 0, 1, 2, 3, \ldots$) where $\otimes^0 V = K$ and for $k \geq 1$, $\otimes^k V$ denotes the vector space with basis vectors

$$\mathbf{v}_{i_1} \otimes \mathbf{v}_{i_2} \otimes \cdots \otimes \mathbf{v}_{i_k}$$

where $i_j = 1, 2, \ldots$ for $1 \leq j \leq k$. If $\mathbf{L}$ is a Lie algebra, then when we take direct sums or tensor products of $\mathbf{L}$ the convention is to forget the Lie bracket and simply consider $\mathbf{L}$ as a vector space.

Definition 2.1.3 *Let $\mathbf{L}$ be a Lie algebra with bracket $[,]$. Define*

$$T(\mathbf{L}) = \bigoplus_{k=0}^{\infty} \otimes^k \mathbf{L},$$

and define $I(\mathbf{L})$ to be the two–sided ideal of $T(\mathbf{L})$ generated by all the tensors (linear combinations of tensor products), $X \otimes Y - Y \otimes X - [X, Y]$, with $X, Y \in \mathbf{L}$. The **universal enveloping algebra** *$U(\mathbf{L})$ of $\mathbf{L}$ is defined to be*

$$U(\mathbf{L}) = T(\mathbf{L})/I(\mathbf{L})$$

with an associative multiplication $\circ$ given by

$$\eta \circ \xi = \eta \otimes \xi \pmod{I(\mathbf{L})}.$$

Example 2.1.4 Let $\mathbf{L} = M(3, \mathbb{R})$, the Lie algebra of 3×3 matrices with coefficients in $\mathbb{R}$ and with Lie bracket

$$[X, Y] = X \cdot Y - Y \cdot X$$

where $\cdot$ denotes matrix multiplication. We shall now exhibit two examples of multiplication $\circ$ in $U(\mathbf{L})$. First:

$$\begin{pmatrix} 1 & 0 & 0 \\ 0 & 0 & 0 \\ 0 & 0 & 0 \end{pmatrix} \circ \begin{pmatrix} 0 & 1 & 0 \\ 0 & 0 & 0 \\ 0 & 0 & 0 \end{pmatrix} = \begin{pmatrix} 1 & 0 & 0 \\ 0 & 0 & 0 \\ 0 & 0 & 0 \end{pmatrix} \otimes \begin{pmatrix} 0 & 1 & 0 \\ 0 & 0 & 0 \\ 0 & 0 & 0 \end{pmatrix} \pmod{I(\mathbf{L})}.$$

The second example is:

$$\begin{pmatrix}1&0&0\\0&0&0\\0&0&0\end{pmatrix}\circ\begin{pmatrix}0&1&0\\0&0&0\\0&0&0\end{pmatrix}-\begin{pmatrix}0&1&0\\0&0&0\\0&0&0\end{pmatrix}\circ\begin{pmatrix}1&0&0\\0&0&0\\0&0&0\end{pmatrix}$$
$$=\begin{pmatrix}1&0&0\\0&0&0\\0&0&0\end{pmatrix}\otimes\begin{pmatrix}0&1&0\\0&0&0\\0&0&0\end{pmatrix}-\begin{pmatrix}0&1&0\\0&0&0\\0&0&0\end{pmatrix}\otimes\begin{pmatrix}1&0&0\\0&0&0\\0&0&0\end{pmatrix}\pmod{I(\mathbf{L})}$$
$$=\begin{pmatrix}1&0&0\\0&0&0\\0&0&0\end{pmatrix}\cdot\begin{pmatrix}0&1&0\\0&0&0\\0&0&0\end{pmatrix}-\begin{pmatrix}0&1&0\\0&0&0\\0&0&0\end{pmatrix}\cdot\begin{pmatrix}1&0&0\\0&0&0\\0&0&0\end{pmatrix}\pmod{I(\mathbf{L})}$$
$$=\begin{pmatrix}0&1&0\\0&0&0\\0&0&0\end{pmatrix}\pmod{I(\mathbf{L})}.$$

The latter example shows that in general

$$X\circ Y-Y\circ X=X\cdot Y-Y\cdot X\pmod{I(\mathbf{L})},$$

which is easily proved.

2.2 Universal enveloping algebra of $\mathfrak{gl}(n,\mathbb{R})$

The group $GL(n,\mathbb{R})$ is a Lie group and there is a standard procedure to pass from a Lie group to a Lie algebra. We shall not need this construction because the Lie algebra of $GL(n,\mathbb{R})$ is very simply described. Let $\mathfrak{gl}(n,\mathbb{R})$ be the Lie algebra of $GL(n,\mathbb{R})$. Then $\mathfrak{gl}(n,\mathbb{R})$ is the additive vector space (over $\mathbb{R}$) of all $n\times n$ matrices with coefficients in $\mathbb{R}$ with Lie bracket given by

$$[\alpha,\beta]=\alpha\cdot\beta-\beta\cdot\alpha$$

for all $\alpha,\beta\in\mathfrak{gl}(n,\mathbb{R})$, and where $\cdot$ denotes matrix multiplication. We shall find an explicit realization of the universal enveloping algebra of $\mathfrak{gl}(n,\mathbb{R})$ as an algebra of differential operators. We shall consider the space $\mathcal{S}$ consisting of smooth (infinitely differentiable) functions $F:GL(n,\mathbb{R})\to\mathbb{C}$.

Definition 2.2.1 *Let $\alpha\in\mathfrak{gl}(n,\mathbb{R})$ and $F\in\mathcal{S}$. Then we define a differential operator D_α acting on F by the rule:*

$$D_\alpha F(g):=\frac{\partial}{\partial t}F(g\cdot\exp(t\alpha))\Big|_{t=0}=\frac{\partial}{\partial t}F(g+t(g\cdot\alpha))\Big|_{t=0}.$$

Remark Recall that $\exp(t\alpha) = I + \sum_{k=1}^{\infty} (t\alpha)^k/k!$, where I denotes the identity matrix on $\mathfrak{gl}(n, \mathbb{R})$. Since we are differentiating with respect to t and then setting $t = 0$, only the first two terms in the Taylor series for $\exp(t\alpha)$ matter.

The differential operator D_α satisfies the usual properties of a derivation:

$$D_\alpha\left(F(g)\cdot G(g)\right) = D_\alpha F(g)\cdot G(g) + F(g)\cdot D_\alpha G(g), \qquad \text{(product rule)},$$
$$D_\alpha F((G(g)) = (D_\alpha F)(G(g))\cdot D_\alpha G(g), \qquad \text{(chain rule)},$$

for all $F, G \in \mathcal{S}$, and $g \in GL(n, \mathbb{R})$.

Example 2.2.2 Let $g = \begin{pmatrix} a & b \\ c & d \end{pmatrix}$, $F(g) := 2a + a^2 + b + d + d^3$, $\alpha = \begin{pmatrix} 0 & 1 \\ 0 & 0 \end{pmatrix}$. Then we have

$$\begin{aligned} D_\alpha F(g) &= \frac{\partial}{\partial t} F\left(\begin{pmatrix} a & b \\ c & d \end{pmatrix}\begin{pmatrix} 1 & t \\ 0 & 1 \end{pmatrix}\right)\Bigg|_{t=0} = \frac{\partial}{\partial t} F\left(\begin{pmatrix} a & at+b \\ c & ct+d \end{pmatrix}\right)\Bigg|_{t=0} \\ &= \frac{\partial}{\partial t}\left(2a + a^2 + at + b + ct + d + (ct+d)^3\right)\Big|_{t=0} \\ &= a + c + 3cd^2. \end{aligned}$$

The differential operators D_α with $\alpha \in \mathfrak{gl}(n, \mathbb{R})$ generate an associative algebra $\mathcal{D}^n$ defined over $\mathbb{R}$. Then every element of $\mathcal{D}^n$ is a linear combination (with coefficients in $\mathbb{R}$) of differential operators $D_{\alpha_1} \circ D_{\alpha_2} \circ \cdots \circ D_{\alpha_k}$ with $\alpha_1, \alpha_2, \ldots, \alpha_k \in \mathfrak{gl}(n, \mathbb{R})$, where $\circ$ denotes multiplication in $\mathcal{D}^n$, which is explicitly given by composition (repeated iteration) of differential operators.

Proposition 2.2.3 *Fix $n \geq 2$. Let $D_\alpha, D_\beta \in \mathcal{D}^n$ with $\alpha, \beta \in \mathfrak{gl}(n, \mathbb{R})$. Then*

$$\begin{aligned} D_{\alpha+\beta} &= D_\alpha + D_\beta, \\ D_\alpha \circ D_\beta - D_\beta \circ D_\alpha &= D_{[\alpha,\beta]}, \end{aligned}$$

where $[\alpha, \beta] = \alpha\cdot\beta - \beta\cdot\alpha$, denotes the Lie bracket in $\mathfrak{gl}(n, \mathbb{R})$, i.e., $\cdot$ denotes matrix multiplication.

Proof Let $g = (g_{i,j})_{1\leq i\leq n, 1\leq j\leq n} \in GL(n, \mathbb{R})$. A smooth complex–valued function $F(g)$, defined on $GL(n, \mathbb{R})$, can be thought of as a function of n^2 real variables $g_{i,j}$ with $1 \leq i \leq n$, $1 \leq j \leq n$. It immediately follows from the chain rule for functions of several real variables and Definition 2.2.1 that

$$D_\alpha F(g) = \sum_{i,j=1}^{n} (g\cdot\alpha)_{i,j} \cdot \frac{\partial}{\partial g_{i,j}} F(g). \tag{2.2.4}$$

Here, $(g \cdot \alpha)_{i,j}$ denotes the i, j entry of the matrix $g \cdot \alpha$. It immediately follows from (2.2.4) that $D_{\alpha+\beta} = D_\alpha + D_\beta$. If we now apply D_β to the above expression (2.2.4), we see that

$$\begin{aligned} D_\beta \circ D_\alpha\, F(g) &= \sum_{i,j=1}^{n} \frac{\partial}{\partial t}\left[((g + t(g\cdot\beta))\cdot\alpha)_{i,j} \cdot \frac{\partial}{\partial g_{i,j}} F(g + t(g\cdot\beta))\right]\Bigg|_{t=0} \\ &= \sum_{i,j=1}^{n} (g\cdot\beta\cdot\alpha)_{i,j} \cdot \frac{\partial F}{\partial g_{i,j}} \\ &\quad + \sum_{i',j'=1}^{n} (g\cdot\beta)_{i',j'}(g\cdot\alpha)_{i,j} \cdot \frac{\partial^2 F}{\partial g_{i,j}\partial g_{i',j'}}. \end{aligned}$$

Consequently,

$$\begin{aligned} (D_\beta \circ D_\alpha - D_\alpha \circ D_\beta)F(g) &= \sum_{i,j=1}^{n} (g\cdot(\beta\cdot\alpha - \alpha\cdot\beta))_{i,j} \cdot \frac{\partial F}{\partial g_{i,j}} \\ &= D_{[\beta,\alpha]}F(g), \end{aligned}$$

which completes the proof of the proposition. □

Proposition 2.2.3 shows that the ring of differential operators $\mathcal{D}^n$ is a realization of the universal enveloping algebra of the Lie algebra $\mathfrak{gl}(n, \mathbb{R})$.

Corollary 2.2.5 *If* $D_\alpha \circ D_\beta = D_\beta \circ D_\alpha$ *for* $\alpha, \beta \in \mathfrak{gl}(n, \mathbb{R})$, *then* $D_{\alpha\cdot\beta} = D_{\beta\cdot\alpha}$.

Proof It follows from Proposition 2.2.3 that

$$0 = D_\alpha \circ D_\beta - D_\beta \circ D_\alpha = D_{\alpha\cdot\beta-\beta\cdot\alpha} = D_{\alpha\cdot\beta} - D_{\beta\cdot\alpha}. \qquad \square$$

Define

$$\delta_1 := \begin{pmatrix} -1 & & & \\ & 1 & & \\ & & \ddots & \\ & & & 1 \end{pmatrix}.$$

Proposition 2.2.6 *For* $n \geq 2$, *let* $f : GL(n, \mathbb{R}) \to \mathbb{C}$ *be a smooth function which is left invariant by* $GL(n, \mathbb{Z})$, *and right invariant by the center* Z_n. *Then for all* $D \in \mathcal{D}^n$, Df *is also left invariant by* $GL(n, \mathbb{Z})$, *right invariant by* Z_n, *and right invariant by the element* δ_1.

Proof It is enough to consider the case when

$$D = D_{\alpha_1} \circ D_{\alpha_2} \circ \cdots \circ D_{\alpha_m}$$

with $m \geq 1$, and $\alpha_i \in \mathfrak{gl}(n, \mathbb{R})$ for $i = 1, 2, \ldots, m$. Since f is left invariant by $GL(n, \mathbb{Z})$, i.e., $f(\gamma \cdot g) = f(g)$ for all $g \in GL(n, \mathbb{R})$, it immediately follows from Definition 2.2.1 that

$$\begin{aligned} Df(\gamma g) &= \frac{\partial}{\partial t_1}\frac{\partial}{\partial t_2}\cdots\frac{\partial}{\partial t_m} f\left(\gamma g e^{t_1\alpha_1+\cdots+t_m\alpha_m}\right)\Big|_{t_1=0,\ldots,t_m=0} \\ &= \frac{\partial}{\partial t_1}\frac{\partial}{\partial t_2}\cdots\frac{\partial}{\partial t_m} f\left(g e^{t_1\alpha_1+\cdots+t_m\alpha_m}\right)\Big|_{t_1=0,\ldots,t_m=0} \\ &= Df(g). \end{aligned}$$

Similarly, let $\delta \in Z_n$. Then $\delta g = g\delta$ for all $g \in GL(n, \mathbb{R})$. It follows, as above, that

$$\begin{aligned} Df(g\delta) &= \frac{\partial}{\partial t_1}\frac{\partial}{\partial t_2}\cdots\frac{\partial}{\partial t_m} f\left(g\delta e^{t_1\alpha_1+\cdots+t_m\alpha_m}\right)\Big|_{t_1=0,\ldots,t_m=0} \\ &= \frac{\partial}{\partial t_1}\frac{\partial}{\partial t_2}\cdots\frac{\partial}{\partial t_m} f\left(g e^{t_1\alpha_1+\cdots+t_m\alpha_m}\delta\right)\Big|_{t_1=0,\ldots,t_m=0} \\ &= \frac{\partial}{\partial t_1}\frac{\partial}{\partial t_2}\cdots\frac{\partial}{\partial t_m} f\left(g e^{t_1\alpha_1+\cdots+t_m\alpha_m}\right)\Big|_{t_1=0,\ldots,t_m=0} \\ &= Df(g). \end{aligned}$$

Finally, we must show that Df is right invariant by δ_1. Note that for all $g \in GL(n, \mathbb{R})$,

$$\begin{aligned} (Df)((\delta_1 \cdot g \cdot \delta_1) \cdot \delta_1) &= (Df)(\delta_1 \cdot g) = (Df)(g) = D(f(g)) \\ &= D(f(\delta_1 \cdot g \cdot \delta_1)), \end{aligned}$$

because $\delta_1 \in O(n, \mathbb{R}) \cap GL(n, \mathbb{Z})$. Thus, $(Df)(g_1 \cdot \delta_1) = D(f(g_1))$ with $g_1 = \delta_1 \cdot g \cdot \delta_1$. The proposition follows because the map $g \to g_1$ is an isomorphism of $\mathfrak{h}^n$. □

The associative algebra $\mathcal{D}^n$ can also be made into a Lie algebra by defining a bracket $[D, D'] = D \circ D' - D' \circ D$ for all $D, D' \in \mathcal{D}^n$. There is a useful identity given in the next proposition.

Proposition 2.2.7 *For $n \geq 2$, let $\alpha, \beta \in \mathfrak{gl}(n, \mathbb{R})$ and $D \in \mathcal{D}^n$. Then*

$$[D_\alpha, D_\beta \circ D] = [D_\alpha, D_\beta] \circ D + D_\beta \circ [D_\alpha, D].$$

Proof We have $[D_\alpha,\ D_\beta \circ D] = D_\alpha \circ D_\beta \circ D - D_\beta \circ D \circ D_\alpha$. On the other hand, $[D_\alpha, D_\beta] \circ D + D_\beta \circ [D_\alpha, D] = (D_\alpha \circ D_\beta - D_\beta \circ D_\alpha) \circ D + D_\beta \circ (D_\alpha \circ D - D \circ D_\alpha)$. It is obvious that these expressions are the same. □

2.3 The center of the universal enveloping algebra of $\mathfrak{gl}(n, \mathbb{R})$

Let $n \geq 2$. We now consider the center $\mathfrak{D}^n$ of $\mathcal{D}^n$. Every $D \in \mathfrak{D}^n$ satisfies $D \circ D' = D' \circ D$ for all $D' \in \mathcal{D}^n$.

Proposition 2.3.1 *Let $n \geq 2$ and let $D \in \mathfrak{D}^n$ lie in the center of $\mathcal{D}^n$. Then D is well defined on the space of smooth functions*

$$f : GL(n, \mathbb{Z}) \backslash GL(n, \mathbb{R}) / (O(n, \mathbb{R}) Z_n) \to \mathbb{C},$$

i.e.,

$$(Df)(\gamma \cdot g \cdot k \cdot \delta) = Df(g),$$

for all $g \in GL(n, \mathbb{R})$, $\gamma \in GL(n, \mathbb{Z})$, $\delta \in Z_n$, and $k \in O(n, \mathbb{R})$.

Proof Proposition 2.2.6 proves Proposition 2.3.1 for the left action of $GL(n, \mathbb{Z})$ and the right action by the center Z_n. It only remains to show that $(Df)(g \cdot k) = Df(g)$ for $k \in O(n, \mathbb{R})$ and $g \in GL(n, \mathbb{R})$.

Fix the function f, the differential operator $D \in \mathfrak{D}^n$,the matrix $g \in GL(n, \mathbb{R})$, and, in addition, fix a matrix $h \in \mathfrak{gl}(n, \mathbb{R})$ which satisfies $h + {}^t h = 0$. Given f, D, g, h, we define a function $\{\phi_{f,D,g,h} = \phi\} : \mathbb{R} \to \mathbb{C}$ as follows:

$$\phi(u) := D\big(f(g \cdot \exp(uh))\big) - (Df)\big(g \cdot \exp(uh)\big).$$

Clearly $\phi(0) = 0$. We will now show that $d\phi/du = 0$, which implies by elementary calculus that $\phi(u)$ is identically zero. The proof that $\phi'(u) = 0$ goes as follows:

$$\begin{aligned}
\phi'(u) &= \frac{\partial}{\partial t} \phi(u + t) \Big|_{t=0} \\
&= \frac{\partial}{\partial t} \Big(D\big(f(g \cdot \exp((u + t) \cdot h)\big) - (Df)\big(g \cdot \exp((u + t) \cdot h)\big) \Big) \Big|_{t=0} \\
&= \frac{\partial}{\partial t} \Big(D\big(f(g \cdot \exp(uh) \cdot \exp(th)\big) - (Df)\big(g \cdot \exp(uh) \cdot \exp(th)\big) \Big) \Big|_{t=0} \\
&= (D \circ D_h)(f(g \cdot \exp(uh))) - ((D_h \circ D) f)(g \cdot \exp(uh)) \\
&= 0,
\end{aligned}$$

because $D \circ D_h = D_h \circ D$.

It follows, as explained before, that $\phi(u) = 0$. Now, the elements $\exp(uh)$ with $h + {}^t h = 0$ generate $O(n, \mathbb{R})^+$ (the elements of the orthogonal group with positive determinant). This is because $\text{Det}(\exp(uh)) > 0$ and

$$\exp(uh) \cdot {}^t(\exp(uh)) = \exp(u(h + {}^t h)) = I,$$

the identity matrix. Consequently,

$$\begin{aligned} 0 = \phi(u) &= D\big(f(g \cdot \exp(uh))\big) - (Df)\big(g \cdot \exp(uh)\big) \\ &= D\big(f(g)\big) - (Df)\big(g \cdot \exp(uh)\big). \end{aligned}$$

Thus, Df is invariant on the right by $O(n, \mathbb{R})^+$. On the other hand, we already know by Proposition 2.2.6 that Df is invariant on the right by the special element δ_1 of determinant -1. It follows that Df must be right–invariant by the entire orthogonal group $O(n, \mathbb{R})$. This proves the proposition. □

We now show how to explicitly construct certain differential operators (called **Casimir operators**) that lie in $\mathfrak{D}^n$, the center of the universal enveloping algebra of $\mathfrak{gl}(n, \mathbb{R})$. For $1 \le i \le n$, $1 \le j \le n$ let $E_{i,j} \in \mathfrak{gl}(n, \mathbb{R})$ denote the matrix with a 1 at the i, jth component and zeros elsewhere. Then, computing the bracket of two such elements, we easily see that

$$\begin{aligned} [E_{i,j}, E_{i',j'}] &= E_{i,j} \cdot E_{i',j'} - E_{i',j'} \cdot E_{i,j} \\ &= \delta_{i',j}\, E_{i,j'} - \delta_{i,j'}\, E_{i',j}, \end{aligned} \tag{2.3.2}$$

where $\delta_{i,j} = \begin{cases} 1 & \text{if } i = j \\ 0 & \text{otherwise,} \end{cases}$ is Kronecker's delta function.

Proposition 2.3.3 *Let $n \ge 2$ and $E_{i,j}$ (with $1 \le i, j \le n$) be as above. Define $D_{i,j} = D_{E_{i,j}}$ with $D_{E_{i,j}}$ given by Definition 2.2.1. Then for $2 \le m \le n$, the differential operator (Casimir operator)*

$$\sum_{i_1=1}^{n} \sum_{i_2=1}^{n} \cdots \sum_{i_m=1}^{n} D_{i_1,i_2} \circ D_{i_2,i_3} \circ \quad \cdots \quad \circ D_{i_m,i_1}$$

lies in $\mathfrak{D}^n$, the center of the universal enveloping algebra of $\mathfrak{gl}(n, \mathbb{R})$.

Proof Let

$$D = \sum_{i_1=1}^{n} \sum_{i_2=1}^{n} \cdots \sum_{i_m=1}^{n} D_{i_1,i_2} \circ D_{i_2,i_3} \circ \quad \cdots \quad \circ D_{i_m,i_1}.$$

It is enough to show that $[D_{r,s}, D] = 0$ for all integers $1 \le r \le n$, $1 \le s \le n$. We shall give the proof for $m = 2$. The case of general m follows by induction.

It follows from Proposition 2.2.7, (2.3.2) and Proposition 2.2.3 that

$$\begin{aligned}
[D_{r,s}, D] &= \sum_{i_1=1}^{n}\sum_{i_2=1}^{n} \left([D_{r,s}, D_{i_1,i_2}] \circ D_{i_2,i_1} + D_{i_1,i_2} \circ [D_{r,s}, D_{i_2,i_1}]\right)\\
&= \sum_{i_1=1}^{n}\sum_{i_2=1}^{n} \Big(\left(\delta_{i_1,s} D_{r,i_2} - \delta_{r,i_2} D_{i_1,s}\right) \circ D_{i_2,i_1} + D_{i_1,i_2} \circ \left(\delta_{i_2,s} D_{r,i_1} - \delta_{r,i_1} D_{i_2,s}\right)\Big)\\
&= \sum_{i_2=1}^{n} D_{r,i_2} \circ D_{i_2,s} - \sum_{i_1=1}^{n} D_{i_1,s} \circ D_{r,i_1} + \sum_{i_1=1}^{n} D_{i_1,s} \circ D_{r,i_1} - \sum_{i_2=1}^{n} D_{r,i_2} \circ D_{i_2,s}\\
&= 0.
\end{aligned}$$

□

Example 2.3.4 (The Casimir operator for $\mathfrak{gl}(2, \mathbb{R})$) We use the notation of Proposition 2.3.3 in the case $n = 2$, and let $z = \begin{pmatrix} y & x \\ 0 & 1 \end{pmatrix} \in GL(2, \mathbb{R})$. By Definition 2.2.1 we have the following explicit differential operators acting on smooth functions $f : \mathfrak{h}^2 \to \mathbb{C}$.

$$\begin{aligned}
D_{1,1} f(z) &:= \frac{\partial}{\partial t} f\left(\begin{pmatrix} y & x \\ 0 & 1 \end{pmatrix} + t \begin{pmatrix} y & x \\ 0 & 1 \end{pmatrix}\begin{pmatrix} 1 & 0 \\ 0 & 0 \end{pmatrix}\right)\Bigg|_{t=0}\\
&= \frac{\partial}{\partial t} f\left(\begin{pmatrix} y(1+t) & x \\ 0 & 1 \end{pmatrix}\right)\Bigg|_{t=0}\\
&= y \frac{\partial}{\partial y} f(z).\\
D_{1,2} f(z) &:= \frac{\partial}{\partial t} f\left(\begin{pmatrix} y & x \\ 0 & 1 \end{pmatrix} + t \begin{pmatrix} y & x \\ 0 & 1 \end{pmatrix}\begin{pmatrix} 0 & 1 \\ 0 & 0 \end{pmatrix}\right)\Bigg|_{t=0}\\
&= \frac{\partial}{\partial t} f\left(\begin{pmatrix} y & x+ty \\ 0 & 1 \end{pmatrix}\right)\Bigg|_{t=0}\\
&= y \frac{\partial}{\partial x} f(z).\\
D_{2,1} f(z) &:= \frac{\partial}{\partial t} f\left(\begin{pmatrix} y & x \\ 0 & 1 \end{pmatrix} + t \begin{pmatrix} y & x \\ 0 & 1 \end{pmatrix}\begin{pmatrix} 0 & 0 \\ 1 & 0 \end{pmatrix}\right)\Bigg|_{t=0}\\
&= \frac{\partial}{\partial t} f\left(\begin{pmatrix} y+tx & x \\ t & 1 \end{pmatrix}\right)\Bigg|_{t=0}\\
&= \frac{\partial}{\partial t} f\left(\begin{pmatrix} \frac{y}{t^2+1} & \frac{xt^2+yt+x}{t^2+1} \\ 0 & 1 \end{pmatrix}\right)\Bigg|_{t=0}\\
&= y \frac{\partial}{\partial x} f(z)
\end{aligned}$$

$$
\begin{aligned}
D_{2,2}f(z) &:= \frac{\partial}{\partial t} f\left(\begin{pmatrix} y & x \\ 0 & 1\end{pmatrix} + t\begin{pmatrix} y & x \\ 0 & 1\end{pmatrix}\begin{pmatrix} 0 & 0 \\ 0 & 1\end{pmatrix}\right)\Bigg|_{t=0}\\
&= \frac{\partial}{\partial t} f\left(\begin{pmatrix} y & x+tx \\ 0 & 1+t\end{pmatrix}\right)\Bigg|_{t=0}\\
&= \frac{\partial}{\partial t} f\left(\begin{pmatrix} \frac{y}{1+t} & x \\ 0 & 1\end{pmatrix}\right)\Bigg|_{t=0}\\
&= -y\frac{\partial}{\partial y} f(z).
\end{aligned}
$$

Warning Although $f:\mathfrak{h}^2 \to \mathbb{C}$ satisfies $f(z\cdot k) = f(z)$ for all $k \in O(2,\mathbb{R})\cdot Z_2$, it is not the case that $D_{i,j}f$ is well defined on $\mathfrak{h}^2$ for $1 \le i, j \le 2$. For example, although $D_{1,2} = D_{2,1} = y\partial/\partial x$, it is not true that $D_{2,1}\circ D_{1,2} = (y\partial/\partial x)^2$.

We compute the second order differential operators as follows.

$$
\begin{aligned}
D_{1,1}\circ D_{1,1}f(z) &= \frac{\partial}{\partial t_1}\frac{\partial}{\partial t_2} f\left(\begin{pmatrix} y & x \\ 0 & 1\end{pmatrix}\cdot\begin{pmatrix} 1+t_1 & 0 \\ 0 & 1\end{pmatrix}\cdot\begin{pmatrix} 1+t_2 & 0 \\ 0 & 1\end{pmatrix}\right)\Bigg|_{t_1=0,\,t_2=0}\\
&= \frac{\partial}{\partial t_1}\frac{\partial}{\partial t_2} f\left(\begin{pmatrix} y & x \\ 0 & 1\end{pmatrix}\begin{pmatrix} 1+t_1+t_2+t_1t_2 & 0 \\ 0 & 1\end{pmatrix}\right)\Bigg|_{t_1=0,\,t_2=0}\\
&= \frac{\partial}{\partial t_1}\frac{\partial}{\partial t_2} f\left(\begin{pmatrix} y(1+t_1+t_2+t_1t_2) & x \\ 0 & 1\end{pmatrix}\right)\Bigg|_{t_1=0,\,t_2=0}\\
&= \left(y\frac{\partial}{\partial y} + y^2\frac{\partial^2}{\partial y^2}\right) f(z).
\end{aligned}
$$

$$
\begin{aligned}
D_{1,2}\circ D_{2,1}f(z) &= \frac{\partial}{\partial t_1}\frac{\partial}{\partial t_2} f\left(\begin{pmatrix} y & x \\ 0 & 1\end{pmatrix}\cdot\begin{pmatrix} 1 & 0 \\ t_1 & 1\end{pmatrix}\cdot\begin{pmatrix} 1 & t_2 \\ 0 & 1\end{pmatrix}\right)\Bigg|_{t_1=0,\,t_2=0}\\
&= \frac{\partial}{\partial t_1}\frac{\partial}{\partial t_2} f\left(\begin{pmatrix} y & x \\ 0 & 1\end{pmatrix}\cdot\begin{pmatrix} 1 & t_2 \\ t_1 & 1+t_1t_2\end{pmatrix}\right)\Bigg|_{t_1=0,\,t_2=0}\\
&= \frac{\partial}{\partial t_1}\frac{\partial}{\partial t_2} f\left(\begin{pmatrix} y+xt_1 & yt_2+x(1+t_1t_2) \\ t_1 & 1+t_1t_2\end{pmatrix}\right)\Bigg|_{t_1=0,\,t_2=0}\\
&= \frac{\partial}{\partial t_1}\frac{\partial}{\partial t_2} f\left(\begin{pmatrix} \frac{y}{t_1(t_1t_2^2+2t_2+t_1)+1} & x+\frac{(t_1t_2^2+t_2+t_1)y}{t_1(t_1t_2^2+2t_2+t_1)+1} \\ 0 & 1\end{pmatrix}\right)\Bigg|_{t_1=0,\,t_2=0}\\
&= \left(-2y\frac{\partial}{\partial y} + y^2\frac{\partial^2}{\partial x^2}\right) f(z).
\end{aligned}
$$

$$D_{2,1} \circ D_{1,2} f(z) = \frac{\partial}{\partial t_1}\frac{\partial}{\partial t_2} f\left(\begin{pmatrix} y & x \\ 0 & 1 \end{pmatrix} \cdot \begin{pmatrix} 1 & t_1 \\ 0 & 1 \end{pmatrix} \cdot \begin{pmatrix} 1 & 0 \\ t_2 & 1 \end{pmatrix}\right)\Bigg|_{t_1=0,\, t_2=0}$$

$$= \frac{\partial}{\partial t_1}\frac{\partial}{\partial t_2} f\left(\begin{pmatrix} t_2 x + y(1+t_1 t_2) & x + t_1 y \\ t_2 & 1 \end{pmatrix}\right)\Bigg|_{t_1=0,\, t_2=0}$$

$$= \frac{\partial}{\partial t_1}\frac{\partial}{\partial t_2} f\left(\begin{pmatrix} \frac{y}{t_2^2+1} & x + \frac{(t_1 t_2^2 + t_2 + t_1)y}{t_2^2+1} \\ 0 & 1 \end{pmatrix}\right)\Bigg|_{t_1=0,\, t_2=0}$$

$$= y^2 \frac{\partial^2}{\partial x^2} f(z).$$

$$D_{2,2} \circ D_{2,2} f(z) = \left(y\frac{\partial}{\partial y} + y^2 \frac{\partial^2}{\partial y^2}\right) f(z).$$

The Casimir operator is then given by

$$D_{1,1} \circ D_{1,1} + D_{1,2} \circ D_{2,1} + D_{2,1} \circ D_{1,2} + D_{2,2} \circ D_{2,2} = 2y^2 \frac{\partial^2}{\partial y^2} + 2y^2 \frac{\partial^2}{\partial x^2}.$$

The following proposition is the basic result in the subject. As pointed out in (Borel, 2001), it was first proved by Capelli (1890).

Proposition 2.3.5 *Every differential operator which lies in $\mathfrak{D}^n$ (the center of the universal enveloping algebra of $\mathfrak{gl}(n, \mathbb{R})$) can be expressed as a polynomial (with coefficients in $\mathbb{R}$) in the Casimir operators defined in Example 2.3.4. Furthermore, $\mathfrak{D}^n$ is a polynomial algebra of rank $n-1$.*

2.4 Eigenfunctions of invariant differential operators

We would like to construct an eigenfunction of all differential operators $D \in \mathfrak{D}^n$, where $\mathfrak{D}^n$ denotes the center of the universal enveloping algebra of $\mathfrak{gl}(n, \mathbb{R})$. Here $\mathfrak{gl}(n, \mathbb{R})$ is the Lie algebra of $GL(n, \mathbb{R})$, i.e., the vector space of all $n \times n$ matrices with coefficients in $\mathbb{R}$. We would like the eigenfunction f to be a smooth function $f : \mathfrak{h}^n \to \mathbb{C}$, where $\mathfrak{h}^n = GL(n, \mathbb{R})/(O(n, \mathbb{R}) \cdot \mathbb{R}^\times)$. Then, we say f is an eigenfunction of $D \in \mathcal{D}^n$, if there exists a complex number λ_D such that

$$Df(z) = \lambda_D f(z)$$

for all $z \in \mathfrak{h}^n$. Let $z = x \cdot y \in \mathfrak{h}^n$ where

$$x = \begin{pmatrix} 1 & x_{1,2} & x_{1,3} & \cdots & & x_{1,n} \\ & 1 & x_{2,3} & \cdots & & x_{2,n} \\ & & \ddots & & & \vdots \\ & & & & 1 & x_{n-1,n} \\ & & & & & 1 \end{pmatrix}, \quad y = \begin{pmatrix} y_1 y_2 \cdots y_{n-1} & & & & \\ & y_1 y_2 \cdots y_{n-2} & & & \\ & & \ddots & & \\ & & & y_1 & \\ & & & & 1 \end{pmatrix},$$

with $x_{i,j} \in \mathbb{R}$ for $1 \le i < j \le n$ and $y_i > 0$ for $1 \le i \le n-1$.

We shall now define the important I_s–function, which is a generalization of the imaginary part function (raised to a complex power s) on the classical upper half-plane. It will be shown that the I_s–function is an eigenfunction of $\mathfrak{D}^n$.

Definition 2.4.1 *For $n \ge 2$, $s = (s_1, s_2, \ldots, s_{n-1}) \in \mathbb{C}^{n-1}$, and $z = x \cdot y \in \mathfrak{h}^n$, as above, we define the function, $I_s : \mathfrak{h}^n \to \mathbb{C}$, by the condition:*

$$I_s(z) := \prod_{i=1}^{n-1} \prod_{j=1}^{n-1} y_i^{b_{i,j} s_j},$$

where

$$b_{i,j} = \begin{cases} ij & \text{if } i + j \le n, \\ (n-i)(n-j) & \text{if } i + j \ge n. \end{cases}$$

The coefficients $b_{i,j}$ are incorporated into the definition because they make later formulae simpler. Note that since $I_s(z)$ is defined on the generalized upper half-plane, $\mathfrak{h}^n$, it must satisfy

$$I_s(z \cdot k \cdot a) = I_s(z)$$

for all $k \in O(n, \mathbb{R})$, $a \in \mathbb{R}^\times$.

Example 2.4.2 (Eigenfunction for $GL(2, \mathbb{R})$) In this example, we may take $z = \begin{pmatrix} y & x \\ 0 & 1 \end{pmatrix}$, $s \in \mathbb{C}$, and $I_s(z) = y^s$. Then, we have shown in Example 2.3.4 that $\Delta = y^2 \left(\frac{\partial^2}{\partial x^2} + \frac{\partial^2}{\partial y^2} \right)$ is a generator of $\mathfrak{D}^2$. Clearly,

$$\Delta I_s(z) = s(s-1) I_s(z).$$

Proposition 2.4.3 *Let $n \geq 2$, and let $I_s(z)$ be as given in Definition 2.4.1. Define $D_{i,j} = D_{E_{i,j}}$ where $E_{i,j} \in \mathfrak{gl}(n, \mathbb{R})$ is the matrix with a 1 at the i, jth component and zeros elsewhere. Then for $1 \leq i, j \leq n$, and $k = 1, 2, \ldots,$ we have*

$$D_{i,j}^k \, I_s(z) = \begin{cases} s_{n-i}^k \cdot I_s(z) & \text{if } i = j \\ 0 & \text{otherwise,} \end{cases}$$

where $D_{i,j}^k = D_{i,j} \circ \cdots \circ D_{i,j}$ denotes composition of differential operators iterated k times.

Proof Note that the function $I_s(z)$ satisfies $I_s(x \cdot y) = I_s(y)$ for all x, y of the form

$$x = \begin{pmatrix} 1 & x_{1,2} & x_{1,3} & \cdots & x_{1,n} \\ & 1 & x_{2,3} & \cdots & x_{2,n} \\ & & \ddots & & \vdots \\ & & & 1 & x_{n-1,n} \\ & & & & 1 \end{pmatrix}, \quad y = \begin{pmatrix} y_1 y_2 \cdots y_{n-1} & & & & \\ & y_1 y_2 \cdots y_{n-2} & & & \\ & & \ddots & & \\ & & & y_1 & \\ & & & & 1 \end{pmatrix},$$

with $x_{i,j} \in \mathbb{R}$ for $1 \leq i < j \leq n$ and $y_i > 0$ for $1 \leq i \leq n-1$. It easily follows that $D_{i,j} I_s(x \cdot y) = D_{i,j} I_s(y)$. If $i < j$, then by Definition 2.2.1, we have

$$D_{i,j} I_s(y) = \frac{\partial}{\partial t} I_s(y + ty \cdot E_{i,j})\Big|_{t=0} = y_1 y_2 \cdots y_{n-i} \frac{\partial}{\partial x_{i,j}} I_s(y) = 0.$$

If $i = j$, then

$$\begin{aligned} D_{i,i} I_s(y) &= \frac{\partial}{\partial t} I_s(y + ty \cdot E_{i,i})\Big|_{t=0} \\ &= \left(y_{n-i} \frac{\partial}{\partial y_{n-i}} - \sum_{\ell=n-i+1}^{n-1} y_\ell \frac{\partial}{\partial y_\ell} \right) I_s(y) \\ &= s_{n-i} \cdot I_s(y). \end{aligned}$$

In a similar manner,

$$D_{i,i}^k I_s(y) = \left(\frac{\partial}{\partial t} \right)^k I_s\left(y \cdot e^{tE_{i,i}} \right) \Big|_{t=0} = s_{n-i}^k I_s(y).$$

If $i > j$, then the argument is more complicated. We make use of Proposition 1.2.8. It follows as before that

$$D_{i,j} I_s(y) = \frac{\partial}{\partial t} I_s\big(y \cdot (I + tE_{i,j})\big) \Big|_{t=0},$$

where I is the identity matrix. By Proposition 1.2.8, $I + tE_{i,j}$ is a matrix with either 1, $(t^2+1)^{\frac{1}{2}}$, or $(t^2+1)^{-\frac{1}{2}}$ on the diagonal. When you take the derivative of any of these with respect to t and then set $t = 0$ you must get zero as an answer. So the only contribution comes from the off diagonal entry $t/(t^2+1)^{\frac{1}{2}}$. Consequently

$$D_{i,j} I_s(y) = y_1 y_2 \cdots y_{n-j} \frac{\partial}{\partial x_{j,i}} I_s(y) = 0. \qquad \square$$

GL(n)pack functions The following **GL(n)pack** functions, described in the appendix, relate to the material in this chapter:

ApplyCasimirOperator GetCasimirOperator IFun.

3
Automorphic forms and L–functions for $SL(2, \mathbb{Z})$

The spectral theory of non-holomorphic automorphic forms formally began with Maass (1949). His book (Maass, 1964) has been a source of inspiration to many. Some other references for this material are (Hejhal, 1976), (Venkov, 1981), (Sarnak, 1990), (Terras, 1985), (Iwaniec-Kowalski, 2004).

Maass gave examples of non-holomorphic forms for congruence subgroups of $SL(2, \mathbb{Z})$ and took the very modern viewpoint, originally due to Hecke (1936), that automorphicity should be equivalent to the existence of functional equations for the associated L-functions. This is the famous converse theorem given in Section 3.15, and is a central theme of this entire book. The first converse theorem was proved by Hamburger (1921) and states that any Dirichlet series satisfying the functional equation of the Riemann zeta function $\zeta(s)$ (and suitable regularity criteria) must actually be a multiple of $\zeta(s)$.

Hyperbolic Fourier expansions of automorphic forms were first introduced in (Neunhöffer, 1973). In (Siegel, 1980), the hyperbolic Fourier expansion of $GL(2)$ Eisenstein series is used to obtain the functional equation of certain Hecke L-functions of real quadratic fields with Grössencharakter (Hecke, 1920). When this is combined with the converse theorem, it gives explicit examples of Maass forms. These ideas are worked out in Sections 3.2 and 3.15.

Another important theme of this chapter is the theory of Hecke operators (Hecke, 1937a,b). We follow the beautiful exposition of Shimura (1971), but reduce the key computations to the Hermite and Smith normal forms (Cohen, 1993), a method which easily generalizes to $SL(n, \mathbb{Z})$ with $n \geq 2$. The Hecke operators map automorphic forms to automorphic forms. Hecke proved the remarkable theorem that if an automorphic form is an eigenfunction of all the Hecke operators then its associated L-function has an Euler product expansion.

Finally, the chapter concludes with the Selberg spectral decomposition (Selberg, 1956) which has played such a pivotal role in modern number theory.

3.1 Eisenstein series

Let

$$\mathfrak{h}^2 = \left\{ \begin{pmatrix} y & x \\ 0 & 1 \end{pmatrix} \,\middle|\, x \in \mathbb{R},\ y > 0 \right\}$$

be the upper half-plane associated to $GL(2,\mathbb{R})$, i.e., $\mathfrak{h}^2 = GL(2,\mathbb{R})/(O(2,\mathbb{R})\cdot \mathbb{R}^\times)$. It is clear that for $s \in \mathbb{C}$ and $z \in \mathfrak{h}$, the function $I_s(z) = y^s$ is an eigenfunction of the hyperbolic Laplacian $\Delta = -y^2\left(\frac{\partial^2}{\partial x^2} + \frac{\partial^2}{\partial y^2}\right)$, with eigenvalue $s(1-s)$. If $\Re(s) \geq \frac{1}{2}$, the function $I_s(z)$ is neither automorphic for $SL(2,\mathbb{Z})$ nor is it square integrable with respect to the $GL^+(2,\mathbb{R})$ invariant measure $dxdy/y^2$ (over the standard fundamental domain $SL(2,\mathbb{Z})\backslash\mathfrak{h}$ given in Example 1.1.9). The fact that Δ is an invariant differential operator does imply that

$$\Delta I_s(gz) = s(1-s) I_s(gz) \tag{3.1.1}$$

for any $g \in GL^+(2,\mathbb{R})$. An automorphic function for $SL(2,\mathbb{Z})$ is a smooth function $f : SL(2,\mathbb{Z})\mathfrak{h}^2 \to \mathbb{C}$. One way to construct an automorphic function for $SL(2,\mathbb{Z})$ which is also an eigenfunction of the Laplacian Δ is to average over the group. Since $I_s(\alpha z) = I_s(z)$ for any

$$\alpha \in \Gamma_\infty := \left\{ \begin{pmatrix} 1 & m \\ 0 & 1 \end{pmatrix} \,\middle|\, m \in \mathbb{Z} \right\},$$

and Γ_∞ is an infinite group, it is necessary to factor out by this subgroup. The cosets $\Gamma_\infty\backslash SL(2,\mathbb{Z})$ are determined by the bottom row of a representative

$$\Gamma_\infty \begin{pmatrix} a & b \\ c & d \end{pmatrix} = \left\{ \begin{pmatrix} * & * \\ c & d \end{pmatrix} \right\} = \left\{ \begin{pmatrix} u & -v \\ c & d \end{pmatrix} \,\middle|\, du + cv = 1 \right\}.$$

Each relatively prime pair (c, d) determines a coset.

Definition 3.1.2 *Let $z \in \mathfrak{h}^2$, $\Re(s) > 1$. We define the Eisenstein series:*

$$E(z,s) := \sum_{\gamma \in \Gamma_\infty\backslash SL(2,\mathbb{Z})} \frac{I_s(\gamma z)}{2} = \frac{1}{2} \sum_{\substack{c,d\in\mathbb{Z}\\ (c,d)=1}} \frac{y^s}{|cz+d|^{2s}}.$$

Proposition 3.1.3 *The Eisenstein series $E(z,s)$ converges absolutely and uniformly on compact sets for $z \in \mathfrak{h}^2$ and $\Re(s) > 1$. It is real analytic in z and complex analytic in s.*

In addition, we have the following:

(1) *Let $\epsilon > 0$. For $\sigma = \Re(s) \geq 1 + \epsilon > 1$, there exists a constant $c(\epsilon)$ such that*

$$|E(z,s) - y^s| \leq c(\epsilon) y^{-\epsilon}, \qquad \textit{for } y \geq 1.$$

(2) $E\left(\frac{az+b}{cz+d}, s\right) = E(z,s)$ *for all* $\begin{pmatrix} a & b \\ c & d \end{pmatrix} \in SL(2,\mathbb{Z})$.

(3) $\Delta E(z,s) = s(1-s)E(z,s)$.

Proof First of all, for $y \geq 1$, we have

$$\begin{aligned} |E(z,s) - y^s| &\leq \sum_{\substack{(c,d)=1 \\ c>0}} \frac{1}{c^{2\sigma}} \cdot \frac{y^\sigma}{\left|z + \frac{d}{c}\right|^{2\sigma}} \\ &= y^\sigma \sum_{c\geq 1} \sum_{\substack{r=1 \\ (r,c)=1}}^{c} \frac{1}{c^{2\sigma}} \sum_{m\in\mathbb{Z}} \frac{1}{\left|z + \frac{r}{c} + m\right|^{2\sigma}}. \end{aligned}$$

Since the set $\{|z + (r/c) + m| \mid m \in \mathbb{Z}, 1 \leq r \leq c, \ (r,c) = 1\}$ forms a set of points spaced by $(1/c)$, we may majorize each term so that

$$\ell \leq \left|x + \frac{r}{c} + m\right| < \ell + 1$$

for some integer ℓ. There are at most $\phi(c)$ such terms for each ℓ. It follows that

$$\begin{aligned} |E(z,s) - y^s| &\leq y^\sigma \sum_{c=1}^{\infty} \frac{\phi(c)}{c^{2\sigma}} \sum_{\ell\in\mathbb{Z}} \frac{1}{(\ell^2 + y^2)^\sigma} \\ &\leq 2y^\sigma \frac{\zeta(2\sigma - 1)}{\zeta(2\sigma)} \sum_{\ell=0}^{\infty} \frac{1}{(\ell^2 + y^2)^\sigma} \\ &\leq 2y^\sigma \frac{\zeta(2\sigma - 1)}{\zeta(2\sigma)} \left(y^{-2\sigma} + \int_0^\infty \frac{du}{(u^2 + y^2)^\sigma} \right) \\ &\ll y^{1-\sigma}. \end{aligned}$$

The second statement of Proposition 3.1.3 follows easily from the fact that for every $\gamma \in SL(2,\mathbb{Z})$, we have $\gamma(\Gamma_\infty\backslash SL(2,\mathbb{Z})) = (\Gamma_\infty\backslash SL(2,\mathbb{Z}))$. The third statement is an easy consequence of (3.1.1) and the definition of $E(z,s)$. □

We now determine the Fourier expansion of the Eisenstein series. This requires some computations involving Ramanujan sums whose definitions and theory we briefly review.

Definition 3.1.4 *For fixed integers n, c with $c \geq 1$, the* **Ramanujan sum** *is the exponential sum*

$$S(n;c) = \sum_{\substack{r=1 \\ (r,c)=1}}^{c} e^{2\pi i n \frac{r}{c}}.$$

The Ramanujan sum can be explicitly evaluated using the **Moebius function** $\mu(n)$ which is defined by the conditions:

$$\mu(1) = 1, \tag{3.1.5}$$

$$\sum_{d|n} \mu(d) = \begin{cases} 1 & \text{if } n = 1 \\ 0 & \text{if } n > 1, \end{cases} \tag{3.1.6}$$

or equivalently by the identity

$$\frac{1}{\zeta(s)} = \sum_{n=1}^{\infty} \frac{\mu(n)}{n^s} = \prod_p (1 - p^{-s}).$$

Proposition 3.1.7 *We have*

$$S(n; c) = \sum_{\ell|n, \ell|c} \ell\, \mu\left(\frac{c}{\ell}\right),$$

and for $\Re(s) > 1$,

$$\sum_{c=1}^{\infty} \frac{S(n; c)}{c^s} = \frac{\sigma_{1-s}(n)}{\zeta(s)},$$

where $\sigma_z(n) = \sum_{d|n} d^z$ *is the divisor function.*

Proof Using the properties (3.1.5), (3.1.6) of the Moebius function, we sift out those integers r relatively prime to c as follows:

$$\begin{aligned} S(n; c) &= \sum_{r=1}^{c} e^{2\pi i n \frac{r}{c}} \sum_{d|c, d|r} \mu(d) \\ &= \sum_{d|c} \mu(d) \sum_{\substack{r=1 \\ r \equiv 0 \,(\mathrm{mod}\, d)}}^{c} e^{2\pi i n \frac{r}{c}} \\ &= \sum_{d|c} \mu(d) \sum_{m=1}^{\frac{c}{d}} e^{2\pi i n \frac{md}{c}} \\ &= \sum_{d|c,\, \frac{c}{d}|_n} \frac{c}{d} \mu(d) = \sum_{\ell|n, \ell|c} \ell\, \mu\left(\frac{c}{\ell}\right). \end{aligned}$$

Here, we have use the fact that the sum $\sum\limits_{m=1}^{q} e^{2\pi i \frac{nm}{q}}$ is zero unless $q|n$, in which case it is q.

For the second part of the proposition, we calculate

$$\sum_{c=1}^{\infty} \frac{S(n; c)}{c^s} = \sum_{c=1}^{\infty} c^{-s} \sum_{\ell|n, \ell|c} \ell\, \mu\left(\frac{c}{\ell}\right) = \sum_{\ell|n} \ell \sum_{m=1}^{\infty} \frac{\mu(m)}{(m\ell)^s} = \sum_{\ell|n} \ell^{1-s} \cdot \zeta(s)^{-1}.$$

It is easily verified that all the above sums converge absolutely for $\Re(s) > 1$. □

Theorem 3.1.8 *Let* $\Re(s) > 1$ *and* $z = \begin{pmatrix} y & x \\ 0 & 1 \end{pmatrix} \in \mathfrak{h}^2$. *The Eisenstein series* $E(z,s)$ *has the Fourier expansion*

$$E(z,s) = y^s + \phi(s)y^{1-s} + \frac{2\pi^s\sqrt{y}}{\Gamma(s)\zeta(2s)}\sum_{n\neq 0}\sigma_{1-2s}(n)|n|^{s-\frac{1}{2}}K_{s-\frac{1}{2}}(2\pi|n|y)e^{2\pi inx}$$

where

$$\phi(s) = \sqrt{\pi}\frac{\Gamma\left(s-\frac{1}{2}\right)}{\Gamma(s)}\frac{\zeta(2s-1)}{\zeta(2s)},$$
$$\sigma_s(n) = \sum_{\substack{d|n\\ d>0}} d^s,$$

and

$$K_s(y) = \frac{1}{2}\int_0^\infty e^{-\frac{1}{2}y\left(u+\frac{1}{u}\right)}u^s\frac{du}{u}.$$

Proof First note that

$$\zeta(2s)E(z,s) = \zeta(2s)y^s + \sum_{c>0}\sum_{d\in\mathbb{Z}}\frac{y^s}{|cz+d|^{2s}}.$$

If we let $\delta_{n,0} = \begin{cases} 1 & n=0 \\ 0 & n\neq 0, \end{cases}$ and $d = mc + r$, it follows that

$$\zeta(2s)\int_0^1 E(z,s)e^{-2\pi inx}\,dx$$
$$= \zeta(2s)y^s\delta_{n,0} + \sum_{c=1}^\infty c^{-2s}\sum_{r=1}^{c}\sum_{m\in\mathbb{Z}}\int_0^1\frac{y^se^{-2\pi inx}}{\left|z+m+\frac{r}{c}\right|^{2s}}\,dx$$
$$= \zeta(2s)y^s\delta_{n,0} + \sum_{c=1}^\infty c^{-2s}\sum_{r=1}^{c}\sum_{m\in\mathbb{Z}}\int_{m+\frac{r}{c}}^{1+m+\frac{r}{c}}\frac{y^se^{-2\pi in\left(x-\frac{r}{c}\right)}}{|z|^{2s}}\,dx$$
$$= \zeta(2s)y^s\delta_{n,0} + \sum_{c=1}^\infty c^{-2s}\sum_{r=1}^{c}e^{\frac{2\pi inr}{c}}\int_{-\infty}^{\infty}\frac{y^se^{-2\pi inx}}{(x^2+y^2)^s}\,dx.$$

Since

$$\sum_{r=1}^{c}e^{\frac{2\pi inr}{c}} = \begin{cases} c & c|n \\ 0 & c\not|n, \end{cases}$$

it is clear that

$$\zeta(2s)\int_0^1 E(z,s)e^{-2\pi inx}\,dx = \zeta(2s)y^s\delta_{n,0} + \sigma_{1-2s}(n)y^{1-s}\int_{-\infty}^{\infty}\frac{e^{-2\pi inxy}}{(x^2+1)^s}\,dx,$$

with the understanding that $\sigma_{1-2s}(0) = \zeta(1-2s)$. The proof of the theorem now immediately follows from the well-known Fourier transform:

$$\int_{-\infty}^{\infty}\frac{e^{-2\pi ixy}}{(x^2+1)^s}\,dx = \begin{cases} \sqrt{\pi}\frac{\Gamma(s-\frac{1}{2})}{\Gamma(s)} & \text{if } y=0, \\ \frac{2\pi^s|y|^{s-\frac{1}{2}}}{\Gamma(s)}K_{s-\frac{1}{2}}(2\pi|y|) & \text{if } y\neq 0. \end{cases} \tag{3.1.9}$$

We may easily prove (3.1.9) as follows.

$$\begin{aligned}\Gamma(s)\int_{-\infty}^{\infty}\frac{e^{-2\pi ixy}}{(x^2+1)^s}\,dx &= \int_0^{\infty}\int_{-\infty}^{\infty} e^{-u-2\pi ixy}\left(\frac{u}{1+x^2}\right)^s dx\,\frac{du}{u} \\ &= \int_0^{\infty} e^{-u}u^s\int_{-\infty}^{\infty} e^{-ux^2}e^{-2\pi ixy}\,dx\,\frac{du}{u} \\ &= \sqrt{\pi}\int_0^{\infty} e^{-u-\frac{\pi^2y^2}{u}}u^{s-\frac{1}{2}}\,\frac{du}{u},\end{aligned}$$

since $e^{-\pi x^2}$ is its own Fourier transform. □

Theorem 3.1.10 *Let $z\in\mathfrak{h}^2$ and $s\in\mathbb{C}$ with $\Re(s)>1$. The Eisenstein series $E(z,s)$ and the function $\phi(s)$ appearing in the constant term of the Fourier expansion of $E(z,s)$ can be continued to meromorphic functions on $\mathbb{C}$ satisfying the functional equations:*

(1) $\phi(s)\phi(1-s) = 1$;
(2) $E(z,s) = \phi(s)E(z,1-s)$.

The modified function $E^*(z,s) = \pi^{-s}\Gamma(s)\zeta(2s)E(z,s)$ is regular except for simple poles at $s=0,1$ and satisfies the functional equation $E^*(z,s) = E^*(z,1-s)$. Furthermore the residue of the pole at $s=1$ is given by

$$\operatorname*{Res}_{s=1} E(z,s) = \frac{3}{\pi}$$

for all $z\in\mathfrak{h}^2$.

3.2 Hyperbolic Fourier expansion of Eisenstein series

In this section we use the classical upper half-plane model for $\mathfrak{h}^2$. Thus $z = x+iy$ with $y>0$, $x\in\mathbb{R}$, and the action of $\begin{pmatrix} a & b \\ c & d\end{pmatrix}\in SL(2,\mathbb{R})$ on z is given by $(az+b)/(cz+d)$.

Let $\rho = \begin{pmatrix} \alpha & \beta \\ \gamma & \delta \end{pmatrix} \in SL(2,\mathbb{Z})$ with trace $|\alpha+\delta| > 2$ and $\gamma > 0$. Such a matrix is termed hyperbolic. Set $D = (\alpha+\delta)^2 - 4$. Then, a point $w \in \mathbb{C}$ is termed fixed under ρ if $(\alpha w + \beta)/(\gamma w + \delta) = w$. It is easily seen that ρ has exactly two real fixed points

$$\omega = \frac{\alpha - \delta + \sqrt{D}}{2\gamma}, \qquad \omega' = \frac{\alpha - \delta - \sqrt{D}}{2\gamma}.$$

Now, define $\kappa = \begin{pmatrix} 1 & -\omega \\ 1 & -\omega' \end{pmatrix}$. Then $\kappa\rho\kappa^{-1} = \begin{pmatrix} \epsilon & \\ & \epsilon^{-1} \end{pmatrix}$ is a diagonal matrix with action on $z \in \mathfrak{h}^2$ given by $\epsilon^2 z$. Since conjugation preserves the trace, we see that $\epsilon + \epsilon^{-1} = \alpha + \delta$. Consequently, $\epsilon = \gamma\omega' + \delta = (\alpha + \delta - \sqrt{D})/2$ is a unit in the quadratic field $\mathbb{Q}(\sqrt{D})$. We shall assume that it is a fundamental unit, i.e., every unit in $\mathbb{Q}(\sqrt{D})$ is, up to ± 1, an integral power of $\epsilon > 0$. The Eisenstein series $E(\kappa^{-1}z, s)$ is invariant under $z \to \epsilon^2 z$. This is because

$$E(\kappa^{-1}z, s) = E(\rho\kappa^{-1}z, s) = E(\kappa^{-1}(\epsilon^2 z), s).$$

Therefore, on the positive imaginary axis (i.e., choosing $z = iv$), the Eisenstein series $\zeta(2s)\cdot E(\kappa^{-1}z, s)$ (for $\Re(s) > 1$) has a Fourier expansion

$$\zeta(2s)\cdot E(\kappa^{-1}(iv), s) = \sum_{n\in\mathbb{Z}} b_n(s)\, v^{\frac{\pi i n}{\log\epsilon}}, \tag{3.2.1}$$

with

$$b_n(s) = \frac{1}{2\log\epsilon}\int_1^{\epsilon^2} \zeta(2s)\cdot E(\kappa^{-1}(iv), s)\, v^{-\frac{\pi i n}{\log\epsilon}}\,\frac{dv}{v}.$$

A direct computation shows that

$$\zeta(2s)\cdot E(\kappa^{-1}(iv), s) = \sum_{\substack{c,d\in\mathbb{Z}\\ \{c,d\}\neq\{0,0\}}} \frac{v^s\cdot(\omega-\omega')^s}{\left((c\omega'+d)^2v^2 + (c\omega+d)^2\right)^s}.$$

The reason for multiplying by $\zeta(2s)$ on the left is to have the sum go over all $c, d \in \mathbb{Z}$ $(\{c,d\} \neq \{0,0\})$ and not just coprime pairs of c, d. Thus,

$$b_n(s) = \frac{(\omega-\omega')^s}{2\log\epsilon}\sum_{\beta\neq 0} N(\beta)^{-s}\left|\frac{\beta}{\beta'}\right|^{-\frac{\pi i n}{\log\epsilon}} \int_{\frac{\beta'}{\beta}}^{\epsilon^2\cdot\frac{\beta'}{\beta}} \left(\frac{v^2}{v^2+1}\right)^s v^{-\frac{\pi i n}{\log\epsilon}}\,\frac{dv}{v},$$

where the sum goes over all non-zero $\beta = c\omega + d$, $\beta' = c\omega' + d$ with $c, d \in \mathbb{Z}$ and $\{c,d\} \neq \{0,0\}$. These elements β lie in an ideal $\mathfrak{b}$ where $\omega - \omega' = N(\mathfrak{b})\sqrt{D}$, and where N denotes the norm from $\mathbb{Q}(\sqrt{D})$ to $\mathbb{Q}$. The above integral can be further simplified by using an idea of Hecke. For an algebraic integer $\beta \in \mathbb{Q}(\sqrt{D})$, let (β) denote the principal ideal generated by β. Two integers $\beta_1, \beta_2 \in \mathbb{Q}(\sqrt{D})$ satisfy $(\beta_1) = (\beta_2)$ if and only if $\beta_1 = \epsilon^m \beta_2$ for some integer m. Consequently

$$\begin{aligned}
b_n(s) &= \frac{(N(\mathfrak{b})\sqrt{D})^s}{2\log\epsilon} \sum_{\mathfrak{b}|(\beta)\neq 0} N(\beta)^{-s} \left|\frac{\beta}{\beta'}\right|^{-\frac{\pi i n}{\log\epsilon}} \\
&\qquad \times \sum_{m\in\mathbb{Z}} \int_{\frac{\beta'\epsilon'^m}{\beta\epsilon^m}}^{\epsilon^2\cdot\frac{\beta'\epsilon'^m}{\beta\epsilon^m}} \left(\frac{v^2}{v^2+1}\right)^s v^{-\frac{\pi i n}{\log\epsilon}} \frac{dv}{v} \\
&= \frac{(N(\mathfrak{b})\sqrt{D})^s}{2\log\epsilon} \sum_{\mathfrak{b}|(\beta)\neq 0} N(\beta)^{-s} \left|\frac{\beta}{\beta'}\right|^{-\frac{\pi i n}{\log\epsilon}} \\
&\qquad \times \sum_{m\in\mathbb{Z}} \int_{\epsilon^{-2m}\cdot\frac{\beta'}{\beta}}^{\epsilon^{-2m+2}\cdot\frac{\beta'}{\beta}} \left(\frac{v^2}{v^2+1}\right)^s v^{-\frac{\pi i n}{\log\epsilon}} \frac{dv}{v} \\
&= \frac{(N(\mathfrak{b})\sqrt{D})^s}{2\log\epsilon} \sum_{\mathfrak{b}|(\beta)\neq 0} N(\beta)^{-s} \left|\frac{\beta}{\beta'}\right|^{-\frac{\pi i n}{\log\epsilon}} \int_0^\infty \left(\frac{v^2}{v^2+1}\right)^s v^{-\frac{\pi i n}{\log\epsilon}} \frac{dv}{v} \\
&= \frac{\Gamma\left(\frac{s-\frac{\pi i n}{\log\epsilon}}{2}\right)\Gamma\left(\frac{s+\frac{\pi i n}{\log\epsilon}}{2}\right)}{\Gamma(s)} \cdot \frac{(N(\mathfrak{b})\sqrt{D})^s}{2\log\epsilon} \\
&\qquad \times \sum_{\mathfrak{b}|(\beta)\neq 0} \left|\frac{\beta}{\beta'}\right|^{-\frac{\pi i n}{\log\epsilon}} N(\beta)^{-s}.
\end{aligned}$$

For a principal ideal (β) of $\mathbb{Q}(\sqrt{D})$, we define the Hecke grössencharakter

$$\psi((\beta)) := \left|\frac{\beta}{\beta'}\right|^{-\frac{\pi i}{\log\epsilon}},$$

and for an ideal $\mathfrak{b}$, we define the the Hecke L–function

$$L_{\mathfrak{b}}(s, \psi^n) = \sum_{\mathfrak{b}|(\beta)\neq 0} \left|\frac{\beta}{\beta'}\right|^{-\frac{\pi i n}{\log\epsilon}} N(\beta)^{-s}. \tag{3.2.2}$$

It now immediately follows from (3.2.1) that

$$\begin{aligned}
E^*(\kappa^{-1}(iv), s) = \frac{\left(N(\mathfrak{b})\sqrt{D}\right)^s}{2\pi^s\log\epsilon} \sum_{n\in\mathbb{Z}} \Gamma\left(\frac{s-\frac{\pi i n}{\log\epsilon}}{2}\right)\Gamma\left(\frac{s+\frac{\pi i n}{\log\epsilon}}{2}\right) \\
\times L_{\mathfrak{b}}(s, \psi^n)\cdot v^{\frac{\pi i n}{\log\epsilon}}, \quad (3.2.3)
\end{aligned}$$

where $E^*(z, s) = \pi^{-s}\Gamma(s)\zeta(2s)E(z, s) = E^*(z, 1-s)$. The expansion (3.2.3) is termed the hyperbolic Fourier expansion of the Eisenstein series. An immediate consequence of this expansion is the following proposition.

Proposition 3.2.4 *Let $K = \mathbb{Q}(\sqrt{D})$ be a real quadratic field with ring of integers $O(K)$. For an ideal $\mathfrak{b}$ in $O(K)$, let $L_{\mathfrak{b}}(s, \psi^n)$ denote the Hecke L–function with grössencharakter given in (3.2.2). Then $L_{\mathfrak{b}}(s, \psi^n)$ has a meromorphic continuation to all s with at most a simple pole at $s = 1$, and satisfies the functional equation*

$$\Lambda_{\mathfrak{b}}^n(s) := \left(\frac{N(\mathfrak{b})\sqrt{D}}{\pi}\right)^s \Gamma\left(\frac{s - \frac{\pi i n}{\log \epsilon}}{2}\right)\Gamma\left(\frac{s + \frac{\pi i n}{\log \epsilon}}{2}\right) L_{\mathfrak{b}}(s, \psi^n)$$
$$= \Lambda_{\mathfrak{b}}^n(1-s).$$

3.3 Maass forms

We shall study the vector space $\mathcal{L}^2(SL(2, \mathbb{Z})\backslash\mathfrak{h}^2)$ (defined over $\mathbb{C}$) which is the completion of the subspace consisting of all smooth functions $f : SL(2, \mathbb{Z})\backslash\mathfrak{h}^2 \to \mathbb{C}$ satisfying the $\mathcal{L}^2$ condition

$$\iint_{SL(2,\mathbb{Z})\backslash\mathfrak{h}^2} |f(z)|^2 \frac{dxdy}{y^2} < \infty.$$

The space $\mathcal{L}^2(SL(2, \mathbb{Z})\backslash\mathfrak{h}^2)$ is actually a Hilbert space with inner product given by

$$\langle f, g\rangle := \iint_{SL(2,\mathbb{Z})\backslash\mathfrak{h}^2} f(z)\overline{g(z)} \frac{dxdy}{y^2}$$

for all $f, g \in \mathcal{L}^2(SL(2, \mathbb{Z})\backslash\mathfrak{h}^2)$. This inner product was first introduced by Petersson.

Definition 3.3.1 *Let $\nu \in \mathbb{C}$. A* **Maass form of type** *ν for $SL(2, \mathbb{Z})$ is a non–zero function $f \in \mathcal{L}^2(SL(2, \mathbb{Z})\backslash\mathfrak{h}^2)$ which satisfies:*

- $f(\gamma z) = f(z)$, *for all* $\gamma \in SL(2, \mathbb{Z})$, $z = \begin{pmatrix} y & x \\ 0 & 1 \end{pmatrix} \in \mathfrak{h}^2$;
- $\Delta f = \nu(1-\nu)f$;
- $\int_0^1 f(z)dx = 0$.

Proposition 3.3.2 *Let f be a Maass form of type ν for $SL(2, \mathbb{Z})$. Then $\nu(1-\nu)$ is real and ≥ 0.*

Proof The proof is based on the fact that the eigenvalues of a symmetric operator on a Hilbert space are real. We have by Green's theorem (integration by parts) that

$$\begin{aligned}
\nu(1-\nu)\langle f, f\rangle &= \langle \Delta f, f\rangle \\
&= \iint\limits_{SL(2,\mathbb{Z})\backslash\mathfrak{h}^2} -\left(\left(\frac{\partial^2}{\partial x^2}+\frac{\partial^2}{\partial y^2}\right) f(z)\right)\cdot \overline{f(z)}\, dxdy \\
&= \iint\limits_{SL(2,\mathbb{Z})\backslash\mathfrak{h}^2} \left(\left|\frac{\partial f}{\partial x}\right|^2+\left|\frac{\partial f}{\partial y}\right|^2\right) dxdy \\
&= \langle f, \Delta f\rangle \;=\; \overline{\nu(1-\nu)}\,\langle f, f\rangle.
\end{aligned}$$

The positivity of $\langle f, f\rangle$ and the inner integral above implies that $\nu(1-\nu)$ is real and non–negative. □

Proposition 3.3.3 *A Maass form of type* 0 *or* 1 *for* $SL(2,\mathbb{Z})$ *must be a constant function.*

Proof Let f be a Maass form of type 0 or 1. Then $f(z)$ is a harmonic function because $\Delta f = 0$. Furthermore, since f is a Maass form it is bounded as $\Im(z)\to\infty$. The only harmonic functions on $SL(2,\mathbb{Z})\backslash\mathfrak{h}^2$ which are bounded at infinity are the constant functions. □

3.4 Whittaker expansions and multiplicity one for $GL(2,\mathbb{R})$

Let f be a Maass form of type ν for $SL(2,\mathbb{Z})$, as in Definition 3.3.1. Since the element $\begin{pmatrix}1 & 1\\ 0 & 1\end{pmatrix}$ is in $SL(2,\mathbb{Z})$ it follows that a Maass form $f(z)$ satisfies

$$f\left(\begin{pmatrix}y & x\\ 0 & 1\end{pmatrix}\right) = f\left(\begin{pmatrix}1 & 1\\ 0 & 1\end{pmatrix}\begin{pmatrix}y & x\\ 0 & 1\end{pmatrix}\right) = f\left(\begin{pmatrix}y & x+1\\ 0 & 1\end{pmatrix}\right).$$

Thus $f(z)$ is a periodic function of x and must have a Fourier expansion of type

$$f(z) = \sum_{m\in\mathbb{Z}} A_m(y)e^{2\pi i m x}. \tag{3.4.1}$$

Define $W_m(z) = A_m(y)e^{2\pi i m x}$. Then $W_m(z)$ satisfies the following two conditions:

$$\begin{aligned}
\Delta W_m(z) &= \nu(1-\nu)W_m(z),\\
W_m\left(\begin{pmatrix}1 & u\\ 0 & 1\end{pmatrix}\cdot z\right) &= W_m(z)e^{2\pi i m u}.
\end{aligned}$$

We call such a function a Whittaker function of type ν associated to the additive character $e^{2\pi i m x}$. Recall that an additive character $\psi:\mathbb{R}\to U$, where U denotes the unit circle, is characterized by the fact that $\psi(x+x') = \psi(x)\psi(x')$, for all $x, x' \in \mathbb{R}$. Formally, we have the following definition.

Definition 3.4.2 *A* **Whittaker function of type** ν *associated to an additive character* $\psi : \mathbb{R} \to U$ *is a smooth non–zero function* $W : \mathfrak{h}^2 \to \mathbb{C}$ *which satisfies the following two conditions:*

$$\Delta W(z) = \nu(1-\nu)W(z),$$
$$W\left(\begin{pmatrix} 1 & u \\ 0 & 1 \end{pmatrix} \cdot z\right) = W(z)\psi(u).$$

Remark 3.4.3 A Whittaker function $W(z)$, of type ν and character ψ, can always be written in the form

$$W(z) = A_\psi(y) \cdot \psi(x)$$

where $A_\psi(y)$ is a function of y only. This is because the function $W(z)/\psi(x)$ is invariant under translations $x \to x+u$ for any $u \in \mathbb{R}$ and, hence, must be the constant function for any fixed y.

Whittaker functions can be constructed explicitly. We know that the function $I_\nu(z) = y^\nu$ satisfies $\Delta I_\nu(z) = \nu(1-\nu)I_\nu(z)$, which is the first condition a Whittaker function must satisfy. In order to impose the second condition, we need the following simple lemma.

Lemma 3.4.4 *Let* $h : \mathbb{R} \to \mathbb{C}$ *be a smooth* $\mathcal{L}^1$ *function. Let* ψ *be an additive character of* $\mathbb{R}$. *Then the function* $H(x) := \int_{-\infty}^{\infty} h(u_1 + x)\psi(-u_1)\,du_1$ *satisfies*

$$H(u+x) = \psi(u)H(x)$$

for all $u \in \mathbb{R}$.

Proof Just make the change of variables $u_1 + u \to u_1$ in the integral for $H(u+x)$. □

Now, $\Delta I_\nu(\gamma z) = \nu(1-\nu)I_\nu(\gamma z)$ for any $\gamma \in GL(2,\mathbb{R})$ because Δ is an invariant differential operator. It follows from this and Lemma 3.4.4 that the function

$$\begin{aligned} W(z,\nu,\psi) &:= \int_{-\infty}^{\infty} I_\nu\left(\begin{pmatrix} 0 & -1 \\ 1 & 0 \end{pmatrix} \cdot \begin{pmatrix} 1 & u \\ 0 & 1 \end{pmatrix} \cdot z\right)\psi(-u)\,du \\ &= \int_{-\infty}^{\infty} \left(\frac{y}{(u+x)^2+y^2}\right)^\nu \psi(-u)\,du \\ &= \psi(x)\int_{-\infty}^{\infty} \left(\frac{y}{u^2+y^2}\right)^\nu \psi(-u)\,du \end{aligned} \tag{3.4.5}$$

must be a Whittaker function of type ν associated to ψ. Actually, we may use any matrix in $GL^+(2,\mathbb{R})$ instead of $\begin{pmatrix} 0 & -1 \\ 1 & 0 \end{pmatrix}$ in the integral on the right-hand side of (3.4.5). All that is required is that the integral converges absolutely, which happens in our case provided $\Re(\nu) > \frac{1}{2}$.

The $GL(2)$ theory of Whittaker functions is considerably simplified because the one Whittaker function which we can construct, $W(z,\nu,\psi)$, can be evaluated exactly in terms of classical Bessel functions. Unfortunately, we are not able to obtain such explicit realizations for Whittaker functions on $GL(n)$ if $n > 3$, and this will lead to considerable complications in the development of the theory in the higher-rank case.

Proposition 3.4.6 *Let $\psi_m(u) = e^{2\pi i m u}$, and let $W(z,\nu,\psi_m)$ be the Whittaker function (3.4.5). Then we have*

$$W(z,\nu,\psi_m) = \sqrt{2}\frac{(\pi|m|)^{\nu-\frac{1}{2}}}{\Gamma(\nu)}\sqrt{2\pi y}\; K_{\nu-\frac{1}{2}}(2\pi|m|y)\cdot e^{2\pi i m x},$$

where

$$K_\nu(y) = \frac{1}{2}\int_0^\infty e^{-\frac{1}{2}y(u+\frac{1}{u})}u^\nu\,\frac{du}{u}$$

is the classical K-Bessel function.

Proof It follows from Remark 3.4.3 and (3.4.5) that

$$W(z,\nu,\psi_m) = W(y,\nu,\psi_m)\cdot e^{2\pi i m x}$$

where

$$\begin{aligned} W(y,\nu,\psi_m) &= \int_{-\infty}^{\infty}\left(\frac{y}{u^2+y^2}\right)^{\nu} e^{-2\pi i u m}\,du \\ &= y^{1-\nu}\int_{-\infty}^{\infty}\frac{e^{-2\pi i u y m}}{(u^2+1)^\nu}\,du. \end{aligned}$$

Note that we made the transformation $u \mapsto y\cdot u$ to identify the above integrals. The result now follows from (3.1.9). □

Definition 3.4.7 *Let $f : \mathfrak{h}^2 \to \mathbb{C}$ be a smooth function. We say that f is of* **polynomial growth** *at ∞ if for fixed $x \in \mathbb{R}$ and $z = \begin{pmatrix} y & x \\ 0 & 1 \end{pmatrix} \in \mathfrak{h}^2$, we have $f(z)$ is bounded by a fixed polynomial in y as $y \to \infty$. We say f is of* **rapid decay** *if for any fixed $N > 1$, $|y^N f(z)| \to 0$ as $y \to \infty$. Similarly, we say f is of* **rapid growth** *if for any fixed $N > 1$, $|y^{-N} f(z)| \to \infty$ as $y \to \infty$.*

We will now state and prove the multiplicity one theorem for Whittaker functions on $GL(2,\mathbb{R})$. This theorem is a cornerstone of the entire theory and provides the basis for the Fourier–Whittaker expansions given in the next section.

Theorem 3.4.8 (Multiplicity one) *Let $\Psi(z)$ be an $SL(2,\mathbb{Z})$–Whittaker function of type $\nu \neq 0, 1$, associated to an additive character ψ, which has rapid decay at ∞. Then*

$$\Psi(z) = aW(z, \nu, \psi)$$

for some $a \in \mathbb{C}$ with $W(z, \nu, \psi)$ given by (3.4.5). If $\psi \equiv 1$ is trivial, then $a = 0$.

Proof Let $\Psi(z) = \Psi(y)\psi(x)$ be a Whittaker function of type ν associated to ψ. We may assume that $\psi(x) = e^{2\pi i m x}$ for some $m \in \mathbb{Z}$, because every additive character is of this form. It follows from Definition 3.4.2 that the differential equation

$$\Delta(\Psi(y)e^{2\pi i m x}) = -y^2\left(\frac{\partial^2}{\partial x^2} + \frac{\partial^2}{\partial y^2}\right)\left(\Psi(y)e^{2\pi i m x}\right) = \nu(1-\nu)\Psi(y)e^{2\pi i m x}$$

implies that $\Psi(y)$ satisfies the differential equation

$$\Psi''(y) - \left(4\pi^2 m^2 - \frac{\nu(1-\nu)}{y^2}\right)\Psi(y) = 0. \tag{3.4.9}$$

By the classical theory of differential equations, (3.4.9) will have exactly two linearly independent solutions over $\mathbb{C}$.

If $\psi \equiv 1$ is trivial, then $m = 0$. Assume $\nu \neq \frac{1}{2}$. Then there are precisely two solutions, to the above differential equation, namely: $y^\nu, y^{1-\nu}$. Thus

$$\Psi(y) = ay^\nu + by^{1-\nu}$$

for certain complex constants a, b. The assumption that $\Psi(y)$ is of rapid decay imples that $a = b = 0$. Similarly, if $m = 0$ and $\nu = \frac{1}{2}$, then $\Psi(y) = ay^{\frac{1}{2}} + by^{\frac{1}{2}}\log y$ and $a = b = 0$ as before.

If $\nu(1-\nu) = 0$, then the equation becomes $\Psi''(y) = 4\pi^2 m^2 \Psi(y)$, which has the general solution $\Psi(y) = ae^{-2\pi m y} + be^{2\pi m y}$ for complex constants $a, b \in \mathbb{C}$, but this case does not come up in our theorem.

For $\nu(1-\nu) \neq 0$, $m \neq 0$, the differential equation (3.4.9) has precisely two smooth solutions (see (Whittaker and Watson, 1935)), namely: $\sqrt{2\pi|m|y} \cdot K_{\nu-\frac{1}{2}}(2\pi|m|y)$, $\sqrt{2\pi|m|y} \cdot I_{\nu-\frac{1}{2}}(2\pi|m|y)$, where

$$I_\nu(y) = \sum_{k=0}^{\infty} \frac{\left(\frac{1}{2}y\right)^{\nu+2k}}{k!\,\Gamma(k+\nu+1)},$$

and

$$K_\nu(y) = \frac{\pi}{2} \cdot \frac{I_{-\nu}(y) - I_\nu(y)}{\sin \pi\nu} = \frac{1}{2}\int_0^\infty e^{-\frac{1}{2}y(t+\frac{1}{t})}\, t^\nu \frac{dt}{t}$$

are classical Bessel functions which have the following asymptotic behavior:

$$\lim_{y\to\infty} \sqrt{y}\, I_\nu(y) e^{-y} = \frac{1}{\sqrt{2\pi}},$$

$$\lim_{y\to\infty} \sqrt{y}\, K_\nu(y) e^{y} = \sqrt{\frac{\pi}{2}}.$$

The assumption that $\Psi(y)$ has polynomial growth at ∞ forces

$$\Psi(y) = a\sqrt{2\pi|m|y} \cdot K_{\nu-\frac{1}{2}}(2\pi|m|y), \qquad (a \in \mathbb{C})$$

which gives us multiplicity one. □

3.5 Fourier–Whittaker expansions on $GL(2, \mathbb{R})$

We shall now show that every non–constant Maass form for $SL(2, \mathbb{Z})$ can be expressed as an infinite sum of Whittaker functions of type (3.4.5).

Proposition 3.5.1 *Let f be a non–constant Maass form of type ν for $SL(2, \mathbb{Z})$. Then for $z \in \mathfrak{h}^2$ we have the Whittaker expansion*

$$f(z) = \sum_{n\neq 0} a_n \sqrt{2\pi y} \cdot K_{\nu-\frac{1}{2}}(2\pi|n|y) \cdot e^{2\pi i n x}$$

for complex coefficients a_n ($n \in \mathbb{Z}$).

Proof Recall (3.4.1) which says that the fact that $f(z)$ is periodic in x implies that f has a Fourier expansion of type

$$f(z) = \sum_{n\in\mathbb{Z}} A_n(y) e^{2\pi i n x}.$$

Since $\Delta f = \nu(1-\nu) f$ it follows that

$$\Delta\left(A_n(y)e^{2\pi i n x}\right) = \nu(1-\nu) A_n(y) e^{2\pi i n x}.$$

Consequently, $A_n(y)e^{2\pi inx}$ must be a Whittaker function of type ν associated to $e^{2\pi inx}$. The assumption that f is not the constant function implies, by Proposition 3.3.3, that $\nu \neq 0, 1$. Since a Maass form f is an $\mathcal{L}^2$ function, i.e.,

$$\iint\limits_{SL(2,\mathbb{Z})\backslash\mathfrak{h}^2} |f(z)|^2 \frac{dxdy}{y^2} < \infty,$$

it easily follows that $A_n(y)e^{2\pi inx}$ must have polynomial growth at ∞. The proof of Proposition 3.5.1 is now an immediate consequence of the multiplicity one Theorem 3.4.8. □

3.6 Ramanujan–Petersson conjecture

Ramanujan had great interest in the Fourier coefficients of the Δ function, which is defined by the product

$$\Delta(z) = e^{2\pi iz}\prod_{n=1}^{\infty}(1 - e^{2\pi inz})^{24} = \sum_{n=1}^{\infty}\tau(n)e^{2\pi inz}$$

for $z = x + iy$ with $x \in \mathbb{R},\ y > 0$. One may easily compute that

$$\tau(1) = 1,\ \tau(2) = -24,\ \tau(3) = 252,\ \tau(4) = -1472,\ \tau(5) = 4830,$$
$$\tau(6) = -6048,\ \tau(7) = -16744,\ \tau(8) = 84480, \ldots$$

and Ramanujan conjectured that

$$\tau(n) \leq n^{\frac{11}{2}}\, d(n)$$

for all $n = 1, 2, 3, \ldots$ The Δ function is a cusp form of weight 12 for $SL(2, \mathbb{Z})$. This means that

$$\Delta\left(\frac{az+b}{cz+d}\right) = (cz+d)^{12}\Delta(z)$$

for all $\begin{pmatrix} a & b \\ c & d \end{pmatrix} \in SL(2, \mathbb{Z})$. Petersson generalized Ramanujan's conjecture to holomorphic cusp forms f of weight k which satisfy

$$f\left(\frac{az+b}{cz+d}\right) = (cz+d)^k f(z)$$

for all $\begin{pmatrix} a & b \\ c & d \end{pmatrix} \in SL(2, \mathbb{Z})$. Petersson conjectured that the nth Fourier coefficient of a weight k cusp form is bounded by $\mathcal{O}\left(n^{(k-1)/2}d(n)\right)$. This explains the $\frac{11}{2}$ in Ramanujan's conjecture. We remark that Petersson's conjecture has been proved by Deligne for all holomorphic cusp forms (of even integral weight) associated to congruence subgroups of $SL(2,\mathbb{Z})$. Deligne's proof is based on the very deep fact that the coefficients of holomorphic cusp forms can be expressed in terms of the number of points on certain varieties defined over certain finite fields, and that optimal error terms for the number of points on a variety over a finite field are a consequence of the Riemann hypothesis (proved by Deligne) for such varieties.

Non–constant Maass forms for $SL(2,\mathbb{Z})$ can be thought of as non–holomorphic automorphic functions of weight zero. One might be tempted, by analogy with the classical theory of holomorphic modular forms, to make a Petersson type conjecture about the growth of the Fourier coefficients in the Fourier–Whittaker expansion in Proposition 3.5.1. Remarkably, all evidence points to the truth of such a conjecture.

We have shown in Proposition 3.5.1 that every non–constant $SL(2,\mathbb{Z})$–Maass form of type ν has a Fourier expansion of type

$$f(z) = \sum_{n\neq 0} a_n\sqrt{2\pi y}\cdot K_{\nu-\frac{1}{2}}(2\pi|n|y)\cdot e^{2\pi i n x}. \tag{3.6.1}$$

We now state the famous Ramanujan–Petersson conjecture for Maass forms.

Conjecture 3.6.2 (Ramanujan–Petersson) *The Fourier–Whittaker coefficients a_n ($n \in \mathbb{Z}$, $n \neq 0$), occurring in the expansion (3.6.1), satisfy the growth condition*

$$|a_n| = \mathcal{O}(d(n))$$

where $d(n) = \sum_{d|n} 1$ denotes the number of divisors of n, and the $\mathcal{O}$–constant depends only on the Petersson norm of f.

It is not hard to show that the nth Fourier–Whittaker coefficient of a non–constant Maass form for $SL(2,\mathbb{Z})$ is bounded by $\sqrt{|n|}$.

Proposition 3.6.3 *Let $f(z)$ be a non–constant Maass form of type ν for $SL(2,\mathbb{Z})$, normalized to have Petersson norm equal to 1, i.e.,*

$$\langle f, f\rangle = \iint_{SL(2,\mathbb{Z})\backslash\mathfrak{h}^2} |f(z)|^2\,\frac{dx\,dy}{y^2} = 1.$$

Then the Fourier–Whittaker coefficients a_n ($n \in \mathbb{Z}$, $n \neq 0$), occurring in the expansion (3.6.1), satisfy the growth condition

$$|a_n| = \mathcal{O}_\nu\big(\sqrt{|n|}\big).$$

Proof It follows from Proposition 3.5.1 that for any fixed $Y > 0$, and any fixed $n \neq 0$,

$$\int_Y^\infty \int_0^1 |f(z)|^2 \frac{dxdy}{y^2} = 2\pi \sum_{m\neq 0} |a_m|^2 \cdot \int_{2\pi|m|Y}^\infty |K_{\nu-\frac{1}{2}}(y)|^2 \frac{dy}{y}$$
$$\gg |a_n|^2 \cdot \int_{2\pi|n|Y}^\infty |K_{\nu-\frac{1}{2}}(y)|^2 \frac{dy}{y}. \tag{3.6.4}$$

Now, it is a simple consequence of Lemma 1.1.6 that

$$\int_Y^\infty \int_0^1 |f(z)|^2 \frac{dxdy}{y^2} = \mathcal{O}\left(Y^{-1} \cdot \iint\limits_{SL(2,\mathbb{Z})\backslash\mathfrak{h}^2} |f(z)|^2 \frac{dxdy}{y^2} \right) = \mathcal{O}(Y^{-1}).$$

If we now choose $Y = |n|^{-1}$ and combine this estimate with (3.6.4) the proposition immediately follows. □

3.7 Selberg eigenvalue conjecture

A non–constant Maass form f of type ν for $SL(2,\mathbb{Z})$ satisfies the partial differential equation $\Delta f(z) = \nu(1-\nu) f(z)$. It is, therefore, an eigenfunction of the Laplace operator Δ with eigenvalue $\lambda = \nu(1-\nu)$. It follows from Propositions 3.3.2, 3.3.3 that $\lambda > 0$. The question arises as to how small λ can be? Selberg proved that $\lambda \geq \frac{1}{4}$ for $SL(2,\mathbb{Z})$, and conjectured that the smallest eigenvalue is greater or equal to $\frac{1}{4}$ for Maass forms associated to any congruence subgroup of $SL(2,\mathbb{Z})$. Recall that a congruence subgroup Γ of $SL(2,\mathbb{Z})$ is a subgroup which contains the so–called principal congruence subgroup of level N:

$$\Gamma(N) = \left\{ \begin{pmatrix} a & b \\ c & d \end{pmatrix} \in SL(2,\mathbb{Z}) \;\middle|\; \begin{pmatrix} a & b \\ c & d \end{pmatrix} \equiv \begin{pmatrix} 1 & 0 \\ 0 & 1 \end{pmatrix} \pmod{N} \right\},$$

for some integer $N \geq 1$. A Maass form of type ν for Γ is a smooth non–zero function $f : \mathfrak{h}^2 \to \mathbb{C}$ which is automorphic for Γ, i.e., $f(\gamma z) = f(z)$ for all $\gamma \in \Gamma$, $z \in \mathfrak{h}^2$, is square integrable on a fundamental domain $\Gamma\backslash\mathfrak{h}^2$, all constant terms in Fourier expansions at cusps (which are rational numbers or ∞) vanish, and satisfies $\Delta f = \nu(1-\nu) f$.

Conjecture 3.7.1 (Selberg) *Let f be a Maass form of type ν for a congruence subgroup $\Gamma \subset SL(2,\mathbb{Z})$. Then $\nu(1-\nu) \geq \frac{1}{4}$, or equivalently, $\Re(\nu) = \frac{1}{2}$.*

We shall now prove this conjecture for $SL(2,\mathbb{Z})$ with a much better lower bound than $\frac{1}{4}$. It should be remarked that the eigenvalue $\frac{1}{4}$ can occur for congruence subgroups (see (3.15.1)).

Theorem 3.7.2 (M–F Vigneras) *Let f be a Maass form of type ν for $SL(2,\mathbb{Z})$. Then $\nu(1-\nu) \geq 3\pi^2/2$.*

Proof Let $D = SL(2,\mathbb{Z})\backslash\mathfrak{h}^2$ be the standard fundamental domain (see Example 1.1.9) for the action of $SL(2,\mathbb{Z})$ on $\mathfrak{h}^2$. We also let D^* denote the transform by $z \to -1/z$ of D. Now, f is a Maass form of type ν with Fourier–Whittaker expansion given by (3.6.1):

$$f(z) = \sum_{n\neq 0} a_n \sqrt{2\pi y}\cdot K_{\nu-\frac{1}{2}}(2\pi|n|y)\cdot e^{2\pi i n x}.$$

Set $\lambda = \nu(1-\nu)$. Then we calculate

$$\begin{aligned}
2\lambda\langle f, f\rangle &= \iint_{D\cup D^*} \Delta f(z)\cdot\overline{f(z)}\,\frac{dxdy}{y^2} = \iint_{D\cup D^*} \left(\left|\frac{\partial f}{\partial x}\right|^2 + \left|\frac{\partial f}{\partial y}\right|^2\right) dxdy\\
&\geq \iint_{D\cup D^*} \left|\frac{\partial f}{\partial x}\right|^2 dxdy \geq \int_{\frac{\sqrt{3}}{2}}^{\infty}\int_{-\frac{1}{2}}^{\frac{1}{2}} \left|\frac{\partial f}{\partial x}\right|^2 dxdy\\
&= \int_{\frac{\sqrt{3}}{2}}^{\infty} \sum_{n\neq 0} |a_n|^2\cdot 4\pi^2 n^2\cdot\left|\sqrt{2\pi y}\cdot K_{\nu-\frac{1}{2}}(2\pi|n|y)\right|^2 dy\\
&\geq \frac{3}{4}\cdot 4\pi^2 \int_{\frac{\sqrt{3}}{2}}^{\infty}\int_{-\frac{1}{2}}^{\frac{1}{2}} |f(z)|^2\,\frac{dxdy}{y^2}\\
&\geq 3\pi^2\langle f, f\rangle.
\end{aligned}$$

Hence, it follows that $\lambda \geq 3\pi^2/2$. □

3.8 Finite dimensionality of the eigenspaces

Let $\mathfrak{S}_\lambda$ denote subspace of all $f \in \mathcal{L}^2(SL(2,\mathbb{Z})\backslash\mathfrak{h}^2)$ which are Maass forms of type ν with $\lambda = \nu(1-\nu)$.

Theorem 3.8.1 (Maass) *For any $\lambda \geq 0$, the space $\mathfrak{S}_\lambda$ is finite dimensional.*

Remark We already know that the space $\mathfrak{S}_0$ is one-dimensional and just contains the constant functions.

Proof By Proposition 3.5.1, every $f \in \mathfrak{S}_\lambda$ has a Fourier–Whittaker expansion of type

$$f(z) = \sum_{n\neq 0} a_n \sqrt{2\pi y} \cdot K_{\nu-\frac{1}{2}}(2\pi|n|y) \cdot e^{2\pi i n x} \tag{3.8.2}$$

with $\lambda = \nu(1-\nu)$. If the dimension of $\mathfrak{S}_\lambda$ were infinite, it would be possible, for every integer $n_0 > 1$, to construct a non–zero finite linear combination of Maass forms of type ν which had a Fourier–Whittaker expansion of the form (3.8.2) where $a_n = 0$ for all $|n| < n_0$. To complete the proof of Theorem 3.8.1 it is enough to prove the following lemma. □

Lemma 3.8.3 *Assume that the function*

$$f(z) = \sum_{|n|>n_0} a_n \sqrt{2\pi y} \cdot K_{\nu-\frac{1}{2}}(2\pi|n|y) e^{2\pi i n x}$$

is a Maass form of type ν for $SL(2,\mathbb{Z})$. Then, if n_0 is sufficiently large, it follows that $f(z) = 0$ for all $z \in \mathfrak{h}^2$.

Proof Without loss of generality, we may assume that the Petersson norm $\langle f, f\rangle = 1$. We have

$$\begin{aligned}
1 &= \iint\limits_{SL(2,\mathbb{Z})\backslash\mathfrak{h}^2} |f(z)|^2 \frac{dxdy}{y^2} \\
&\le \int\limits_{\frac{\sqrt{3}}{2}}^{\infty}\int\limits_{-\frac{1}{2}}^{\frac{1}{2}} \left| \sum_{|n|>n_0} a_n \sqrt{2\pi y} \cdot K_{\nu-\frac{1}{2}}(2\pi|n|y) e^{2\pi i n x}\right|^2 \frac{dxdy}{y^2} \\
&= \sum_{|n|>n_0} \int\limits_{\frac{\sqrt{3}}{2}}^{\infty} |a_n|^2 \cdot \left|\sqrt{2\pi y} \cdot K_{\nu-\frac{1}{2}}(2\pi|n|y)\right|^2 \frac{dy}{y^2} \\
&\ll \sum_{|n|>n_0} \int\limits_{\frac{\sqrt{3}}{2}}^{\infty} |a_n|^2 \cdot \left|K_{\nu-\frac{1}{2}}(2\pi|n|y)\right|^2 \frac{dy}{y}.
\end{aligned} \tag{3.8.4}$$

We now make use of Proposition 3.6.3 which says that a_n is bounded by $\sqrt{|n|}$. We also make use of the well-known asymptotic formula

$$\lim_{y\to\infty} \sqrt{y}\, K_\nu(y) e^y = \sqrt{\tfrac{\pi}{2}}$$

for the K–Bessel function. It immediately follows from these remarks and

(3.8.4) that

$$1 \ll \sum_{|n|>n_0} \int_{\frac{\sqrt{3}}{2}}^{\infty} |n| \cdot e^{-4\pi|n|y} \frac{dy}{y^2} \ll \sum_{|n|>n_0} |n| e^{-2\sqrt{3}\pi|n|} \ll e^{-2n_0},$$

which is a contradition for n_0 sufficiently large. □

3.9 Even and odd Maass forms

We introduce the operator T_{-1} which maps Maass forms to Maass forms. It is defined as follows. Let f be a Maass form of type ν for $SL(2,\mathbb{Z})$. Then we define

$$T_{-1} f\left(\begin{pmatrix} y & x \\ 0 & 1 \end{pmatrix}\right) := f\left(\begin{pmatrix} y & -x \\ 0 & 1 \end{pmatrix}\right).$$

The notation T_{-1} is used at this stage to conform to standard notation for Hecke operators which will be defined later in this book. Note that $(T_{-1})^2$ is the identity transformation, so that the eigenvalues of T_{-1} can only be ± 1.

Since the Laplace operator $\Delta = -y^2((\partial^2/\partial x^2) + (\partial^2/\partial y^2))$ is invariant under the transformation $x \mapsto -x$, it is easy to see that T_{-1} maps Maass forms of type ν to Maass forms of type ν.

Definition 3.9.1 *A Maass form f of type ν for $SL(2,\mathbb{Z})$ is said to be* **even** *if $T_{-1} f = f$. It is said to be* **odd** *if $T_{-1} f = -f$.*

Proposition 3.9.2 *Let f be a Maass form of type ν for $SL(2,\mathbb{Z})$ with Fourier–Whittaker expansion*

$$f(z) = \sum_{n\neq 0} a(n)\sqrt{2\pi y} \cdot K_{\nu-\frac{1}{2}}(2\pi|n|y) \cdot e^{2\pi i n x},$$

as in Proposition 3.5.1. Then $a(n) = a(-n)$ if f is an even Maass form and $a(n) = -a(-n)$ if f is an odd Maass form.

Proof We have

$$a(n)\sqrt{2\pi y} \cdot K_{\nu-\frac{1}{2}}(2\pi|n|y) = \int_0^1 f(z)e^{-2\pi i n x}\, dx$$

$$a(-n)\sqrt{2\pi y} \cdot K_{\nu-\frac{1}{2}}(2\pi|n|y) = \int_0^1 f(z)e^{2\pi i n x}\, dx$$

$$= \int_0^1 (T_{-1} f(z))e^{-2\pi i n x}\, dx,$$

after making the transformation $x \to -x$. The result immediately follows. □

Finally, we remark that if f is an arbitrary Maass form of type ν for $SL(2,\mathbb{Z})$, then

$$f(z) = \frac{1}{2}\big[f(z) + T_{-1}f(z)\big] \; + \; \frac{1}{2}\big[f(z) - T_{-1}f(z)\big]$$

where $\frac{1}{2}\left[f(z) + T_{-1}f(z)\right]$ is an even Maass form and $\frac{1}{2}\left[f(z) - T_{-1}f(z)\right]$ is an odd Maass form.

3.10 Hecke operators

We shall first define Hecke operators in a quite general setting, and then return to the specific case of $\mathcal{L}^2\left(SL(2,\mathbb{Z})\backslash\mathfrak{h}^2\right)$.

Let G be a group which acts continuously on a topological space X as in Definition 1.1.1. Let $\Gamma \subset G$ be a discrete subgroup of G as in Definition 1.1.4. Assume the quotient space $\Gamma\backslash X$ has a left Γ-invariant measure dx and define the $\mathbb{C}$-vector space:

$$\mathcal{L}^2(\Gamma\backslash X) = \left\{ f : \Gamma\backslash X \to \mathbb{C} \;\middle|\; \int_{\Gamma\backslash X} |f(x)|^2\, dx < \infty \right\}. \tag{3.10.1}$$

The commensurator of Γ, denoted $C_G(\Gamma)$, defined by

$$C_G(\Gamma) := \left\{ g \in G \;\middle|\; (g^{-1}\Gamma g) \cap \Gamma \text{ has finite index in both } \Gamma \text{ and } g^{-1}\Gamma g \right\}, \tag{3.10.2}$$

is of fundamental importance in the theory of Hecke operators. Here, for example, $g^{-1}\Gamma g$ denotes the set of all elements of the form $g^{-1}\gamma g$ with $\gamma \in \Gamma$ and we have identities of type $\Gamma \cdot \Gamma = \Gamma$. This notation and its obvious generalizations are used in all that follows.

For every fixed $g \in C_G(\Gamma)$, the group Γ can be expressed as a disjoint union of right cosets

$$\Gamma = \bigcup_{i=1}^{d} \left((g^{-1}\Gamma g) \cap \Gamma\right) \delta_i, \tag{3.10.3}$$

where d is the index of $(g^{-1}\Gamma g) \cap \Gamma$ in Γ. Note also that (3.10.3) may be rewritten as $g^{-1}\Gamma g\Gamma = \bigcup_i g^{-1}\Gamma g\delta_i$ which is equivalent to

$$\Gamma g\Gamma = \bigcup_{i=1}^{d} \Gamma g\delta_i. \tag{3.10.4}$$

Definition 3.10.5 (Hecke operators) *Let a group G act continuously on a topological space X and let Γ be a discrete subgroup of G as in Definition 1.1.4. For each $g \in C_G(\Gamma)$, with $C_G(\Gamma)$ given by (3.10.2), we define a* **Hecke operator**

$$T_g : \mathcal{L}^2(\Gamma\backslash X) \to \mathcal{L}^2(\Gamma\backslash X)$$

by the formula

$$T_g(f(x)) = \sum_{i=1}^{d} f(g\delta_i x),$$

for all $f \in \mathcal{L}^2(\Gamma\backslash X)$, $x \in X$ and δ_i, $(i = 1, 2, \ldots, d)$ given by (3.10.4).

In order for T_g to be well defined, we just need to check that $T_g(f) \in \mathcal{L}^2(\Gamma\backslash X)$. Now, for every $\gamma \in \Gamma$,

$$T_g(f(\gamma x)) = \sum_{i=1}^{d} f(g\delta_i\gamma x).$$

But $\delta_i\gamma = \delta_i'\delta_{\sigma(i)}$ for some permutation σ of $\{1, 2, \ldots, d\}$ and some $\delta_i' \in (g^{-1}\Gamma g) \cap \Gamma$. Also, note that $g\delta_i\gamma = g\delta_i'\delta_{\sigma(i)} = \delta_i'' g\delta_{\sigma(i)}$ for some other $\delta_i'' \in \Gamma$. It follows that

$$T_g\left(f(\gamma x)\right) = \sum_{i=1}^{d} f\left(\delta_i'' g\delta_{\sigma(i)}x\right) = \sum_{i=1}^{d} f(g\delta_i x) = T_g(f(x)),$$

so T_g is well defined.

We also consider for any integer m, the multiple mT_g, which acts on $f \in \mathcal{L}^2(\Gamma\backslash X)$ by the canonical formula $(mT_g)(f(x)) = m \cdot T_g(f(x))$. In this manner, one constructs the Hecke ring of all formal sums $\sum\limits_{k\in\mathbb{Z}} m_k T_{g_k}$ with $m_k \in \mathbb{Z}$.

We now define a way of multiplying Hecke operators so that the product of two Hecke operators is a sum of other Hecke operators. For $g, h \in C_G(\Gamma)$, consider the coset decompositions

$$\Gamma g\Gamma = \bigcup_i \Gamma\alpha_i, \qquad \Gamma h\Gamma = \bigcup_j \Gamma\beta_j, \tag{3.10.6}$$

as in (3.10.4). Then

$$(\Gamma g\Gamma)\cdot(\Gamma h\Gamma) = \bigcup_j \Gamma g\Gamma\beta_j = \bigcup_{i,j}\Gamma\alpha_i\beta_j = \bigcup_{\Gamma w\subset\Gamma g\Gamma h\Gamma}\Gamma w = \bigcup_{\Gamma w\Gamma\subset\Gamma g\Gamma h\Gamma}\Gamma w\Gamma.$$

One may then define, for $g, h \in C_G(\Gamma)$, the product of the Hecke operators, $T_g T_h$, by the formula

$$T_g T_h = \sum_{\Gamma w \Gamma \subset \Gamma g \Gamma h \Gamma} m(g, h, w) T_w, \tag{3.10.7}$$

where $m(g, h, w)$ denotes the number of i, j such that $\Gamma \alpha_i \beta_j = \Gamma w$ with α_i, β_j given by (3.10.6). One checks that this multiplication law is associative.

Definition 3.10.8 (The Hecke ring) *Let a group G act continuously on a topological space X and let Γ be a discrete subgroup of G as in Definition 1.1.4. Fix any semigroup Δ such that $\Gamma \subset \Delta \subset C_G(\Gamma)$. The* **Hecke ring** $\mathcal{R}_{\Gamma,\Delta}$ *is defined to be the set of all formal sums*

$$\sum_k c_k T_{g_k}$$

with $c_k \in \mathbb{Z}, g_k \in \Delta$. The multiplication law in this ring is induced from (3.10.7).

Definition 3.10.9 (Antiautomorphism) *By an* **antiautomorphism** *of a group G we mean a map $g \to g^*$ (for $g \in G$) satisfying $(gh)^* = h^* g^*$ for all $g, h \in G$.*

The following theorem is of supreme importance in the theory of automorphic forms. It is the key to understanding Euler products in the theory of L-functions.

Theorem 3.10.10 (Commutativity of the Hecke ring) *Let $\mathcal{R}_{\Gamma,\Delta}$ be the Hecke ring as in Definition 3.10.8. If there exists an antiautomorphism $g \to g^*$ of $C_G(\Gamma)$ such that $\Gamma^* = \Gamma$ and $(\Gamma g \Gamma)^* = \Gamma g \Gamma$ for every $g \in \Delta$, then $\mathcal{R}_{\Gamma,\Delta}$ is a commutative ring.*

Proof Let $g \in \Delta$. Since $(\Gamma g \Gamma)^* = \Gamma g \Gamma$, it immediately follows that if we decompose $\Gamma g \Gamma$ into either left or right cosets of Γ, then the number of left cosets must be the same as the number of right cosets. Let $\Gamma \alpha \subset \Gamma g \Gamma$ be a right coset and let $\beta \Gamma \subset \Gamma g \Gamma$ be a left coset. Then $\alpha \in \Gamma g \Gamma = \Gamma \beta \Gamma$. Consequently $\alpha = \gamma \beta \gamma'$ with $\gamma, \gamma' \in \Gamma$. Then $\gamma^{-1}\alpha = \beta\gamma'$ must be both a left and right coset representative. It easily follows that there exists a common set of representatives $\{\alpha_i\}$ such that

$$\Gamma g \Gamma = \bigcup_i \Gamma \alpha_i = \bigcup_i \alpha_i \Gamma.$$

Similarly, for any other $h \in \Delta$, we have

$$\Gamma h \Gamma = \bigcup_j \Gamma \beta_j = \bigcup_j \beta_j \Gamma$$

for some set of common representatives $\{\beta_j\}$.

Now, by our assumptions about the antiautomorphism $*$ we immediately see that $(\Gamma g\Gamma)^* = \Gamma g^*\Gamma$ and $(\Gamma h\Gamma)^* = \Gamma h^*\Gamma$. We also know that $\Gamma g\Gamma h\Gamma = \bigcup_w \Gamma w\Gamma$ which implies that $(\Gamma g\Gamma h\Gamma)^* = \Gamma g\Gamma h\Gamma$. It follows that

$$(\Gamma h\Gamma)\cdot(\Gamma g\Gamma) = (\Gamma h^*\Gamma)\cdot(\Gamma g^*\Gamma) = (\Gamma g\Gamma h\Gamma)^* = \bigcup_w \Gamma w\Gamma.$$

This tells us that

$$T_g \cdot T_h = \sum_w m_w \; \Gamma w\Gamma,$$
$$T_h \cdot T_g = \sum_w m'_w \; \Gamma w\Gamma,$$

with the same components w but with possibly different integers m_w, m'_w. To complete the proof, we must prove that $m_w = m'_w$. But

$$m_w = \#\left\{i, j \mid \Gamma\alpha_i\beta_j = \Gamma w\right\} = \frac{\#\left\{i, j \mid \Gamma\alpha_i\beta_j\Gamma = \Gamma w\Gamma\right\}}{\#\left\{u \in \Gamma w\Gamma/\Gamma \mid \Gamma u \subset \Gamma w\Gamma\right\}}.$$

Similarly,

$$m'_w = \frac{\#\left\{i, j \mid \Gamma\beta_j\alpha_i\Gamma = \Gamma w\Gamma\right\}}{\#\left\{u \in \Gamma w\Gamma/\Gamma \mid \Gamma u \subset \Gamma w\Gamma\right\}}.$$

It only remains to show that

$$\#\left\{i, j \mid \Gamma\alpha_i\beta_j\Gamma = \Gamma w\Gamma\right\} = \#\left\{i, j \mid \Gamma\beta_j\alpha_i\Gamma = \Gamma w\Gamma\right\}. \tag{3.10.11}$$

Now $\Gamma g\Gamma = (\Gamma g\Gamma)^* = \bigcup \Gamma\alpha_i^*$ and $\Gamma h\Gamma = (\Gamma h\Gamma)^* = \bigcup \Gamma\beta_j^* = \bigcup \beta_j^*\Gamma$. Therefore, we have

$$(\Gamma g\Gamma)\cdot(\Gamma h\Gamma) = \bigcup_{i,j} \Gamma\alpha_i^*\beta_j^*.$$

So (3.10.11) will follow if

$$\#\left\{i, j \mid \Gamma\alpha_i^*\beta_j^*\Gamma = \Gamma w\Gamma\right\} = \#\left\{i, j \mid \Gamma\beta_j\alpha_i\Gamma = \Gamma w\Gamma\right\}. \tag{3.10.12}$$

Clearly, (3.10.12) holds, as one easily sees by applying the antiautomorphism $*$. □

3.11 Hermite and Smith normal forms

We have seen in the previous Section 3.10 that Hecke operators are defined by expressing a double coset as a union of right cosets. In the classical literature, if two matrices are in the same right coset then they are said to be right equivalent, while if they are in the same double coset, then they are said to be equivalent.

Hermite found a canonical or normal form for right or left equivalent matrices while Smith found a normal form for equivalent matrices. The Hermite and Smith normal forms play such an important role in Hecke theory that we have decided to give a self-contained exposition at this point.

Theorem 3.11.1 (Hermite normal form) *For $n \geq 2$, let A be an $n \times n$ matrix with integer coefficients and* $\mathrm{Det}(A) > 0$. *Then there exists a unique upper triangular matrix $B \in GL(n, \mathbb{Z})$ which is left equivalent to A, (i.e., $B = \gamma A$ with $\gamma \in SL(n, \mathbb{Z})$) such that the diagonal entries of B are positive and each element above the main diagonal lies in a prescribed complete set of residues modulo the diagonal element below it.*

Note that B takes the form:

$$B = \begin{pmatrix} d_1 & \alpha_{2,1} & \alpha_{3,1} & \cdots & \alpha_{n-1,1} & \alpha_{n,1} \\ & d_2 & \alpha_{3,2} & \cdots & \alpha_{n-1,2} & \alpha_{n,2} \\ & & d_3 & \cdots & \alpha_{n-1,3} & \alpha_{n,3} \\ & & & \ddots & \vdots & \vdots \\ & & & & d_{n-1} & \alpha_{n,n-1} \\ & & & & & d_n \end{pmatrix}$$

where each $\alpha_{k,j}$ satisfies $0 \leq \alpha_{k,j} < d_k$.

Proof We first prove that every matrix in $GL(n, \mathbb{Z})$, with $n \geq 1$, is left equivalent to an upper triangular matrix. The result is obvious if $n = 1$. We proceed by induction on n. Let ${}^t(a_1, a_2, \ldots, a_n)$ denote the first column of A. Either every $a_i = 0$, $(i = 1, 2, \ldots, n)$ or some element is non-zero. Suppose the latter and set $\delta \neq 0$ to be the greatest common divisor of the elements a_i, $(i = 1, 2, \ldots, n)$. Then there exist coprime integers $\gamma_1, \gamma_2, \ldots, \gamma_n$ such that $\sum_{i=1}^{n} a_i \gamma_i = \delta$. Thus, there exists a matrix $\gamma' \in SL(n, \mathbb{Z})$ with first row $(\gamma_1, \gamma_2, \ldots, \gamma_n)$ such that the matrix $\gamma' A$ has δ in the $(1, 1)$ position and all the remaining elements of the first column are multiples of δ. By successively multiplying $\gamma' A$ on the left by matrices which have 1s on the diagonal, zeros everywhere else except for one element in the first row one may easily show that there exists $\gamma \in SL(n, \mathbb{Z})$ such that γA takes the form

$$\gamma A = \begin{pmatrix} \delta & * & * & \cdots & * \\ 0 & & & & \\ 0 & & A' & & \\ \vdots & & & & \\ 0 & & & & \end{pmatrix}$$

where A' is an $(n-1)\times(n-1)$ matrix. By induction, we may bring A' to upper triangular form by left multiplication by $\begin{pmatrix}1 & \\ & M\end{pmatrix}$, with $M \in SL(n-1,\mathbb{Z})$. This proves that A is left equivalent to an upper triangular matrix T. It is clear that we can make the diagonal elements of T positive by multiplying on the left by a suitable diagonal matrix with entries ± 1. Let T' denote this upper triangular matrix with positive diagonal entries.

In order to make each element of T' above the main diagonal lie in a prescribed complete set of residues modulo the diagonal element which lies below it, we repeatedly multiply T' on the left by matrices of the form $I + m_{i,j}E_{i,j}$ (with suitable integers $m_{i,j}$) where I is the $n\times n$ identity matrix and $E_{i,j}$ is the $n\times n$ matrix with zeros everywhere except at the (i,j) position where there is a 1. For example, if $t_{i,j}$ is the (j,i) entry of T' and d_i is the ith diagonal entry of T', then $(I + m_{i,j}E_{i,j})T'$ has a new (j,i) entry which is $t_{i,j} + d_i \cdot m_{i,j}$.

We leave the proof of the uniqueness of the Hermite normal form to the reader. □

Theorem 3.11.2 (Smith normal form) *For $n \geq 2$, let A be an $n\times n$ integer matrix with* $\mathrm{Det}(A) > 0$. *Then there exists a unique diagonal matrix D of the form*

$$D = \begin{pmatrix} d_n & & & \\ & \ddots & & \\ & & d_2 & \\ & & & d_1 \end{pmatrix}$$

where $0 < d_1 | d_2,\ d_2 | d_3, \ldots, d_{n-1} | d_n$, and $A = \gamma_1 D \gamma_2$ for $\gamma_1, \gamma_2 \in SL(n,\mathbb{Z})$.

Proof We may assume that A contains a non-zero element which may be brought to the (n,n) position by suitable row and column interchanges. As in the proof of Theorem 3.11.1, this element may be replaced by the greatest common divisor of the last column and the last row, and will divide every element of the last row and column. By further elementary row and column operations, we may obtain a new matrix B where every element of B (except at the (n,n) position) in the last row and column is zero. Thus

$$B = \begin{pmatrix} & & & & 0 \\ & B' & & & 0 \\ & & & & \vdots \\ & & & & 0 \\ 0 & 0 & \cdots & 0 & b_{n,n} \end{pmatrix} \tag{3.11.3}$$

where $b_{n,n} \neq 0$ and B' is an $(n-1)\times(n-1)$ matrix.

Suppose that the submatrix B' contains an element $b_{i,j}$ which is not divisible by $b_{n,n}$. If we add column j to column n, then the new column n will be of the form

$$^t(b_{1,j}, b_{2,j}, \ldots, b_{i,j}, \ldots, b_{n-1,j}, b_{n,n}).$$

We may then repeat the previous process and replace $b_{n,n}$ with a proper divisor of itself. Continuing in this manner, we obtain a matrix B of the form (3.11.3) which is equivalent to A and where $b_{n,n}$ divides every element of the matrix B'.

The entire previous process can then be repeated on B'. Continuing inductively, we prove our theorem. Again, we leave the proof of uniqueness to the reader. □

3.12 Hecke operators for $\mathcal{L}^2(SL(2,\mathbb{Z})\backslash\mathfrak{h}^2)$

We shall now work out the theory of Hecke operators for $\mathcal{L}^2\,(SL(2,\mathbb{Z}))\,\backslash\mathfrak{h}^2$. In this case let

$$G = GL(2,\mathbb{R}), \quad \Gamma = SL(2,\mathbb{Z}), \quad X = GL(2,\mathbb{R})/(O(2,\mathbb{R})\cdot\mathbb{R}^\times) = \mathfrak{h}^2.$$

For integers $n_0, n_1 \geq 1$, it is easily seen that the matrix $\begin{pmatrix} n_0n_1 & 0 \\ 0 & n_0 \end{pmatrix} \in C_G(\Gamma)$.

Let Δ denote the semigroup generated by the matrices $\begin{pmatrix} n_0n_1 & 0 \\ 0 & n_0 \end{pmatrix}$, $(n_0, n_1 \geq 1)$ and the modular group Γ. In this situation, we have the antiautomorphism

$$g \to {}^t g, \qquad g \in \Delta,$$

where ${}^t g$ denotes the transpose of the matrix g. Since diagonal matrices are always invariant under transposition, one immediately sees that the conditions of Theorem 3.10.10 are satisfied so that the Hecke ring $\mathcal{R}_{\Gamma,\Delta}$ is commutative.

Lemma 3.12.1 *Fix $n \geq 1$. Define the set,*

$$S_n := \left\{ \begin{pmatrix} a & b \\ 0 & d \end{pmatrix} \;\middle|\; ad = n,\ 0 \leq b < d \right\}.$$

Then one has the disjoint partition

$$\bigcup_{m_0^2 m_1 = n} \Gamma \begin{pmatrix} m_0m_1 & 0 \\ 0 & m_0 \end{pmatrix} \Gamma = \bigcup_{\alpha \in S_n} \Gamma\alpha. \tag{3.12.2}$$

Proof It is easy to see that the decomposition is disjoint, because if

$$\begin{pmatrix} r & s \\ t & u \end{pmatrix}\begin{pmatrix} a & b \\ 0 & d \end{pmatrix} = \begin{pmatrix} a' & b' \\ 0 & d' \end{pmatrix}$$

then we must have $t = 0$, and, therefore, $r = u = 1, s = 0$.

Next, by Theorem 3.11.2, every element on the right-hand side of (3.12.2) can be put into Smith normal form, so must occur as an element on the left-hand side of (3.12.2). Similarly, by Theorem 3.11.1, every element on the left-hand side of (3.12.2) can be put into Hermite normal form, so must occur as an element on the right-hand side of (3.12.2). This proves the equality of the two sides of (3.12.2). □

Now, two double cosets are either the same or totally disjoint and different. Thus a union of double cosets can be viewed as an element in the Hecke ring as in Definition 3.10.8. It follows that for each integer $n \geq 1$, we have a Hecke operator T_n acting on the space of square integrable automorphic forms $f(z)$ with $z \in \mathfrak{h}^2$. The action is given by the formula

$$T_n f(z) = \frac{1}{\sqrt{n}} \sum_{\substack{ad=n \\ 0\leq b<d}} f\left(\frac{az+b}{d}\right). \tag{3.12.3}$$

Clearly, T_1 is just the identity operator. Note that we have introduced the normalizing factor of $1/\sqrt{n}$ to simplify later formulae.

The $\mathbb{C}$-vector space $\mathcal{L}^2(\Gamma\backslash\mathfrak{h}^2)$ has a natural inner product (called the Petersson inner product), denoted $\langle\ ,\ \rangle$, and defined by

$$\langle f, g\rangle = \iint\limits_{\Gamma\backslash\mathfrak{h}^2} f(z)\overline{g(z)}\,\frac{dxdy}{y^2},$$

for all $f, g \in \mathcal{L}^2(\Gamma\backslash\mathfrak{h}^2)$, where $\frac{dxdy}{y^2}$ denotes the left invariant measure as given in Proposition 1.5.3.

Theorem 3.12.4 (Hecke operators are self-adjoint) *The Hecke operators T_n, $(n = 1, 2, \ldots)$ defined in (3.12.3) satisfy*

$$\langle T_n f, g\rangle = \langle f, T_n g\rangle,$$

for all $f, g \in \mathcal{L}^2(\Gamma\backslash\mathfrak{h}^2)$.

Proof Lemma 3.12.1 says that

$$\bigcup_{m_0^2 m_1 = n} \Gamma\begin{pmatrix} m_0 m_1 & 0 \\ 0 & m_0 \end{pmatrix}\Gamma = \bigcup_{\alpha\in S_n} \Gamma\alpha.$$

Since the left union above is invariant on the right by any $\sigma \in \Gamma$, it follows that for any $\sigma_1, \sigma_2 \in \Gamma$:

$$\bigcup_{m_0^2 m_1 = n} \Gamma \begin{pmatrix} m_0 m_1 & 0 \\ 0 & m_0 \end{pmatrix} \Gamma = \bigcup_{\alpha \in S_n} \Gamma \sigma_1 \alpha \sigma_2.$$

Further, since diagonal matrices are invariant under transposition, we see that

$$\bigcup_{m_0^2 m_1 = n} \Gamma \begin{pmatrix} m_0 m_1 & 0 \\ 0 & m_0 \end{pmatrix} \Gamma = \bigcup_{\alpha \in S_n} \Gamma\, \sigma_1 \cdot {}^t\alpha \cdot \sigma_2.$$

But the action of the Hecke operator is independent of the choice of right coset decomposition. Consequently

$$\begin{aligned} \langle T_n f, g\rangle &= \frac{1}{\sqrt{n}} \iint\limits_{\Gamma\backslash\mathfrak{h}^2} \sum_{\alpha \in S_n} f(\sigma_1 \cdot {}^t\alpha \cdot \sigma_2 z)\overline{g(z)}\, \frac{dxdy}{y^2} \\ &= \frac{1}{\sqrt{n}} \iint\limits_{\Gamma\backslash\mathfrak{h}^2} f(z) \sum_{\alpha \in S_n} \overline{g\left(\sigma_2^{-1} \cdot {}^t\alpha^{-1} \cdot \sigma_1^{-1} z\right)}\, \frac{dxdy}{y^2}, \end{aligned} \tag{3.12.5}$$

after making the change of variables $z \to \sigma_2^{-1} \cdot {}^t\alpha^{-1} \cdot \sigma_1^{-1} z$.

We shall now choose $\sigma_1 = \begin{pmatrix} 0 & -1 \\ 1 & 0 \end{pmatrix}$ and $\sigma_2 = \sigma_1^{-1} = \begin{pmatrix} 0 & 1 \\ -1 & 0 \end{pmatrix}$. A simple computation shows that for $\alpha = \begin{pmatrix} a & b \\ 0 & d \end{pmatrix}$,

$$\sigma_2^{-1} \cdot {}^t\alpha^{-1} \cdot \sigma_1^{-1} = \frac{1}{ad} \begin{pmatrix} a & b \\ 0 & d \end{pmatrix}.$$

Since the action of $\begin{pmatrix} a & b \\ 0 & d \end{pmatrix}$ on $z \in \mathfrak{h}^2$ is the same as the action of $(1/ad)\begin{pmatrix} a & b \\ 0 & d \end{pmatrix}$, we immediately get that

$$\frac{1}{\sqrt{n}} \sum_{\alpha \in S_n} \overline{g\left(\sigma_2^{-1} \cdot {}^t\alpha^{-1} \cdot \sigma_1^{-1} z\right)} = \overline{T_n g(z)}.$$

Plugging this into (3.12.5) completes the proof of the theorem. □

Theorem 3.12.6 *The Hecke operators $T_n, (n = 1, 2, \ldots)$ as defined in (3.12.3) commute with each other, commute with the operator T_{-1} in Definition 3.9.1, and commute with the Laplacian $\Delta = -y^2\left(\frac{\partial^2}{\partial x^2} + \frac{\partial^2}{\partial y^2}\right)$. Furthermore T_{-1} commutes with Δ.*

Proof We have already pointed out at the very beginning of this section that matrix transposition satisfies the conditions of Theorem 3.10.10 which implies that the Hecke operators commute with each other, i.e.,

$$T_m T_n = T_n T_m, \qquad \forall m, n \geq 1.$$

That the Hecke operators also commute with T_{-1} and Δ, and T_{-1} and Δ commute with each other is a fairly straightforward calculation which we leave to the reader. □

By standard methods in functional analysis, it follows from Theorems 3.12.4 and 3.12.6 that the space $\mathcal{L}^2(\Gamma\backslash\mathfrak{h}^2)$ may be simultaneously diagonaliazed by the set of operators $\mathcal{T} := \{T_n \mid n = -1,\ n = 1, 2, \ldots\} \bigcup \{\Delta\}$. We may, therefore, consider Maass forms f which are simultaneous eigenfunctions of $\mathcal{T}$. If $\Delta f = \nu(1-\nu)f$, then f will be a Maass form of type ν as in Definition 3.3.1. It will be even or odd depending on whether $T_{-1}f = f$ or $T_{-1}f = -f$ as in Definition 3.9.1. Further, there will exist real numbers λ_n such that

$$T_n f = \lambda_n f, \tag{3.12.7}$$

for all $n = 1, 2, \ldots$

Theorem 3.12.8 (Muliplicativity of the Fourier coefficients) *Consider*

$$f(z) = \sum_{n\neq 0} a(n)\sqrt{2\pi y}\cdot K_{\nu-\frac{1}{2}}(2\pi|n|y)\cdot e^{2\pi i n x},$$

a Maass form of type ν, as in Proposition 3.5.1, which is an eigenfunction of all the Hecke operators, i.e., (3.12.7) holds. If $a(1) = 0$, then f vanishes identically. Assume $f \neq 0$, and it is normalized so that $a(1) = 1$. Then

$$T_n f = a(n)\cdot f, \qquad \forall n = 1, 2, \ldots$$

Furthermore, we have the following multiplicativity relations:

$$a(m)a(n) = a(mn), \qquad \text{if } (m,n) = 1,$$
$$a(m)a(n) = \sum_{d|(m,n)} a\left(\frac{mn}{d^2}\right),$$
$$a(p^{r+1}) = a(p)a(p^r) - a(p^{r-1}),$$

for all primes p, and all integers $r \geq 1$.

Proof It follows from Proposition 3.5.1 that

$$f(z) = \sum_{M\neq 0} a(M)\sqrt{2\pi y}\cdot K_{\nu-\frac{1}{2}}(2\pi|M|y)\cdot e^{2\pi i M x}.$$

It is convenient to rewrite (3.12.3) in the form

$$T_n f(z) = \frac{1}{\sqrt{n}} \sum_{\substack{bd=n\\0\le a<b}} f\left(\frac{dz+a}{b}\right).$$

It follows that

$$T_n f(z) = \sum_{M\neq 0} a(M) \sum_{bd=n}\sum_{0\le a<b} \sqrt{\frac{2\pi d y}{nb}}\cdot K_{\nu-\frac12}\left(2\pi|M|\frac{dy}{b}\right)\cdot e^{2\pi i M\frac{dx+a}{b}}.$$

The sum over $0 \le a < b$ is zero unless $b|M$, in which case the sum is b. Consequently, if we let $M = bm'$, and afterwards $m = dm'$, then we see that

$$\begin{aligned}
T_n f(z) &= \sum_{m'\neq 0} a(bm') \sum_{bd=n} \sqrt{\frac{2\pi d b y}{n}}\cdot K_{\nu-\frac12}(2\pi|m'|dy)e^{2\pi i d m' x}\\
&= \sum_{m\neq 0}\sum_{\substack{bd=n\\d|m}} a\left(\frac{mb}{d}\right)\cdot\sqrt{2\pi y}\cdot K_{\nu-\frac12}(2\pi|m|y)\,e^{2\pi i m x}\\
&= \sum_{m\neq 0}\sum_{d|m,\,d|n} a\left(\frac{mn}{d^2}\right)\cdot\sqrt{2\pi y}\cdot K_{\nu-\frac12}(2\pi|m|y)\,e^{2\pi i m x}.
\end{aligned}$$

Since $T_n f = \lambda_n f$, it immediately follows that

$$\lambda_n a(m) = \sum_{d|(m,n)} a\left(\frac{mn}{d^2}\right). \tag{3.12.9}$$

For $m = 1$, the identity (3.12.9) gives

$$a(n) = \lambda_n a(1) \tag{3.12.10}$$

for all values of n. Consequently, if $a(1) = 0$ the Maass form f would have to vanish identically. Thus, we may assume that if $f \neq 0$, then it is normalized so that $a(1) = 1$. In this case, the identity (3.12.10) shows that $\lambda_n = a(n)$. The other claims of Theorem 3.12.8 follow easily. □

3.13 L–functions associated to Maass forms

Let

$$f(z) = \sum_{n\neq 0} a(n)\sqrt{2\pi y}\cdot K_{\nu-\frac12}(2\pi|n|y)\cdot e^{2\pi i n x} \tag{3.13.1}$$

be a non-zero Maass form of type ν for $SL(2,\mathbb{Z})$ which is an eigenfunction of all the Hecke operators T_n, $(n = 1, 2, \dots)$ given by (3.12.3), and in addition, is also an eigenfunction of T_{-1} given in Section 3.9. Assume f is normalized so that $a(1) = 1$.

We have shown in Theorem 3.12.8 that

$$T_n f = a(n) \cdot f$$

and that the eigenvalues $a(n)$ satisfy multiplicative relations. It follows from the bound $a(n) = \mathcal{O}(\sqrt{n})$ given in Proposition 3.6.3 that for $\Re(s) > \frac{3}{2}$, the series $\sum_{n=1}^{\infty} a(n)n^{-s}$ converges absolutely. Theorem 3.12.8 tells us that the Fourier coefficients $a(n)$ are multiplicative and satisfy $a(m)a(n) = a(mn)$ if $(m, n) = 1$. Consequently, we may write

$$\sum_{n=1}^{\infty} a(n)n^{-s} = \prod_p \left(\sum_{\ell=0}^{\infty} \frac{a(p^\ell)}{p^{\ell s}} \right).$$

Let us evaluate a typical factor

$$\phi_p(s) := \sum_{\ell=0}^{\infty} \frac{a(p^\ell)}{p^{\ell s}}.$$

We compute (with the convention that $a(p^{-1}) = 0$)

$$\begin{aligned} \frac{a(p)}{p^s}\phi_p(s) &= \sum_{\ell=0}^{\infty} \frac{a(p^\ell)a(p)}{p^{(\ell+1)s}} = \sum_{\ell=0}^{\infty} \frac{a(p^{\ell+1}) + a(p^{\ell-1})}{p^{(\ell+1)s}} \\ &= \phi_p(s) - 1 + p^{-2s}\phi_p(s), \end{aligned}$$

and, therefore,

$$\phi_p(s) = (1 - a(p)p^{-s} + p^{-2s})^{-1}.$$

It immediately follows that

$$\begin{aligned} \sum_{n=1}^{\infty} a(n)n^{-s} &= \prod_p (1 - \alpha_p p^{-s})^{-1}(1 - \beta_p p^{-s})^{-1} \qquad (3.13.2) \\ &= \prod_p (1 - a(p)p^{-s} + p^{-2s})^{-1}, \end{aligned}$$

where for each prime p, we have $\alpha_p \cdot \beta_p = 1$, $\alpha_p + \beta_p = a(p)$.

Thus, the Dirichlet series (3.13.2) has an Euler product which is very similar to the Euler product

$$\sum_{n=1}^{\infty} n^{-s} = \prod_p (1 - p^{-s})^{-1}$$

of the Riemann zeta function, the major difference being that there are two Euler factors for each prime instead of one. We shall show that (3.13.2) also satisfies a functional equation $s \to 1-s$. It is natural, therefore, to make the following definition.

Definition 3.13.3 *Let $s \in \mathbb{C}$ with $\Re(s) > \frac{3}{2}$, and let $f(z)$ be a Maass form given by (3.13.1). We define the* **L–function** *$L_f(s)$ (termed the L–function associated to f) by the absolutely convergent series*

$$L_f(s) = \sum_{n=1}^{\infty} a(n) n^{-s}. \tag{3.13.4}$$

We now show that the L–function, $L_f(s)$, satisfies a functional equation $s \to 1-s$. The only additional complication is the fact that there are two distinct functional equations depending on whether f is an even or odd Maass form as defined in Section 3.9.

Proposition 3.13.5 *Let f be a Maass form of type ν for $SL(2,\mathbb{Z})$. Then the L–function, $L_f(s)$, (given in Definition 3.13.3) has a holomorphic continuation to all $s \in \mathbb{C}$ and satisfies the functional equation*

$$\begin{aligned}\Lambda_f(s) &:= \pi^{-s}\Gamma\left(\frac{s+\epsilon-\frac{1}{2}+\nu}{2}\right)\Gamma\left(\frac{s+\epsilon+\frac{1}{2}-\nu}{2}\right)L_f(s)\\ &= (-1)^{\epsilon}\cdot\Lambda_f(1-s),\end{aligned}$$

where $\epsilon = 0$ if f is even ($T_{-1}f = f$) and $\epsilon = 1$ if f is odd ($T_{-1}f = -f$).

Proof We follow the line of reasoning in Riemann's original proof of the functional equation for $\zeta(s)$, which is to set $x = 0$ and take the Mellin transform in y of $f(z)$. We shall need the well-known transform

$$\int_0^{\infty} K_{\nu-\frac{1}{2}}(y)\, y^{s+\frac{1}{2}} \frac{dy}{y} = 2^{-\frac{3}{2}+s}\Gamma\left(\frac{1+s-\nu}{2}\right)\Gamma\left(\frac{s+\nu}{2}\right),$$

which is valid for $\Re(s-\nu) > -1$.

If f is an even Maass form, the computation goes as follows. For $\Re(s) > 1$, the following integrals converge absolutely.

$$\begin{aligned}\int_0^{\infty} f\left(\begin{pmatrix} y & 0\\ 0 & 1\end{pmatrix}\right) y^s \frac{dy}{y} &= 2\int_0^{\infty}\sum_{n=1}^{\infty} a_n\sqrt{2\pi y}\cdot K_{\nu-\frac{1}{2}}(2\pi|n|y)\, y^s\frac{dy}{y}\\ &= 2(2\pi)^{-s}L_f\left(s+\tfrac{1}{2}\right)\int_0^{\infty} K_{\nu-\frac{1}{2}}(y)\, y^{s+\frac{1}{2}}\frac{dy}{y}\\ &= 2^{-\frac{1}{2}}\pi^{-s}\Gamma\left(\frac{s+\nu}{2}\right)\Gamma\left(\frac{1+s-\nu}{2}\right)L_f\left(s+\tfrac{1}{2}\right).\end{aligned} \tag{3.13.6}$$

On the other hand, since $f\left(\begin{pmatrix} y & 0 \\ 0 & 1 \end{pmatrix}\right)$ is invariant under $y \mapsto y^{-1}$, it follows that

$$\begin{aligned}
&\int_0^\infty f\left(\begin{pmatrix} y & 0 \\ 0 & 1 \end{pmatrix}\right) y^s \frac{dy}{y} \\
&\quad = \int_0^1 f\left(\begin{pmatrix} y^{-1} & 0 \\ 0 & 1 \end{pmatrix}\right) y^s \frac{dy}{y} + \int_1^\infty f\left(\begin{pmatrix} y & 0 \\ 0 & 1 \end{pmatrix}\right) y^s \frac{dy}{y} \\
&\quad = \int_1^\infty f\left(\begin{pmatrix} y & 0 \\ 0 & 1 \end{pmatrix}\right) \left(y^s + y^{-s}\right) \frac{dy}{y}.
\end{aligned} \tag{3.13.7}$$

Since $f\left(\begin{pmatrix} y & 0 \\ 0 & 1 \end{pmatrix}\right)$ has exponential decay in y as $y \to \infty$, the above integral is easily seen to converge for all $s \in \mathbb{C}$, and thus defines an entire function. It is also invariant under the transformation $s \mapsto -s$. This gives the functional equation for even Maass forms.

We now consider the case when f is an odd Maass form. The above argument does not work because by Proposition 3.9.2, you would have $a_n = -a_{-n}$, and, therefore, $\sum\limits_{n \neq 0} a_n \cdot |n|^{-s} = 0$ which implies by the calculation (3.13.6) that

$$\int_0^\infty f\left(\begin{pmatrix} y & 0 \\ 0 & 1 \end{pmatrix}\right) y^s \frac{dy}{y} = 0.$$

To get around this difficulty, we consider

$$\begin{aligned}
\frac{\partial}{\partial x} f\left(\begin{pmatrix} y & x \\ 0 & 1 \end{pmatrix}\right) &= \frac{\partial}{\partial x}\left(\sum_{n \neq 0} a_n \sqrt{2\pi y} \cdot K_{\nu-\frac{1}{2}}(2\pi |n| y) \cdot e^{2\pi i n x}\right) \\
&= 2\pi i \sum_{n \neq 0} a_n \cdot n \cdot \sqrt{2\pi y} \cdot K_{\nu-\frac{1}{2}}(2\pi |n| y) \cdot e^{2\pi i n x}.
\end{aligned}$$

Hence,

$$\begin{aligned}
&\frac{\partial}{\partial x}\left(\sum_{n \neq 0} a_n \sqrt{2\pi y} \cdot K_{\nu-\frac{1}{2}}(2\pi |n| y) \cdot e^{2\pi i n x}\right)\Bigg|_{x=0} \\
&\quad = 2\pi i \sum_{n \neq 0} a_n \cdot n \cdot \sqrt{2\pi y} \cdot K_{\nu-\frac{1}{2}}(2\pi |n| y).
\end{aligned}$$

It follows as in (3.13.6), that

$$\int_0^\infty \frac{\partial}{\partial x}\Bigg(\sum_{n\neq 0} a_n\sqrt{2\pi y}\cdot K_{\nu-\frac{1}{2}}(2\pi|n|y)e^{2\pi i n x}\Bigg)\Bigg|_{x=0} y^s\,\frac{dy}{y}$$
$$= \sqrt{2}i\cdot\pi^{1-s}\,\Gamma\left(\frac{s+\nu}{2}\right)\Gamma\left(\frac{1+s-\nu}{2}\right)L_f(s-\tfrac{1}{2}). \tag{3.13.8}$$

Now, because f is automorphic, we have:

$$f\left(\begin{pmatrix} y & x\\ 0 & 1\end{pmatrix}\right) = f\left(\begin{pmatrix} 0 & -1\\ 1 & 0\end{pmatrix}\begin{pmatrix} y & x\\ 0 & 1\end{pmatrix}\right)$$
$$= f\left(\begin{pmatrix} \frac{y}{x^2+y^2} & \frac{-x}{x^2+y^2}\\ 0 & 1\end{pmatrix}\right).$$

Consequently

$$\frac{\partial}{\partial x} f\left(\begin{pmatrix} y & x\\ 0 & 1\end{pmatrix}\right)\Bigg|_{x=0} = \frac{\partial}{\partial x} f\left(\begin{pmatrix} \frac{y}{x^2+y^2} & \frac{-x}{x^2+y^2}\\ 0 & 1\end{pmatrix}\right)\Bigg|_{x=0}$$
$$= -\frac{1}{y^2}\cdot\frac{\partial}{\partial x} f\left(\begin{pmatrix} y^{-1} & x\\ 0 & 1\end{pmatrix}\right)\Bigg|_{x=0}$$
$$= -\frac{1}{y^2}\,\frac{\partial f}{\partial x}\left(\begin{pmatrix} y^{-1} & 0\\ 0 & 1\end{pmatrix}\right).$$

Here, if we let $f\left(\begin{pmatrix} y & x\\ z & w\end{pmatrix}\right) = f(y,x,z,w)$, then $\frac{d}{dx}f = f^{(0,1,0,0)}$. It then follows as in (3.13.7) that

$$\int_0^\infty \frac{\partial f}{\partial x}\left(\begin{pmatrix} y & 0\\ 0 & 1\end{pmatrix}\right) y^s\,\frac{dy}{y}$$
$$= -\int_0^1 \frac{\partial f}{\partial x}\left(\begin{pmatrix} y^{-1} & 0\\ 0 & 1\end{pmatrix}\right) y^{s-2}\,\frac{dy}{y} + \int_1^\infty \frac{\partial f}{\partial x}\left(\begin{pmatrix} y & 0\\ 0 & 1\end{pmatrix}\right) y^s\,\frac{dy}{y}$$
$$= \int_1^\infty \frac{\partial f}{\partial x}\left(\begin{pmatrix} y & 0\\ 0 & 1\end{pmatrix}\right)\left[y^s - y^{2-s}\right]\frac{dy}{y}. \tag{3.13.9}$$

The functional equation for odd Maass forms is an immediate consequence of (3.13.8) and (3.13.9). □

3.14 L-functions associated to Eisenstein series

Let $w \in \mathbb{C}$ with $\Re(w) > 1$. The Eisenstein series

$$E(z, w) = \frac{1}{2} \sum_{\substack{c,d \in \mathbb{Z} \\ (c,d)=1}} \frac{y^w}{|cz + d|^{2w}},$$

defined in Definition 3.1.2 has the Fourier–Whittaker expansion

$$\begin{aligned} E(z, w) &= y^w + \phi(w) y^{1-w} \\ &\quad + \frac{2^{\frac{1}{2}} \pi^{w-\frac{1}{2}}}{\Gamma(w)\zeta(2w)} \sum_{n \neq 0} \sigma_{1-2w}(n) |n|^{w-1} \sqrt{2\pi |n| y} \cdot K_{w-\frac{1}{2}}(2\pi |n| y) e^{2\pi i n x}, \end{aligned}$$

given in Theorem 3.1.8. Then $E(z, w)$ is an even automorphic form, i.e., it is invariant under the transformation $x \mapsto -x$. In an analogous manner to Definition 3.13.3, we may define the L–function associated to an Eisenstein series.

Definition 3.14.1 *We define the* **L–function associated to the Eisenstein series** $E(z, w)$ *to be*

$$L_{E(*,w)}(s) = \sum_{n=1}^{\infty} \sigma_{1-2w}(n) \cdot n^{w-\frac{1}{2}-s}.$$

The following elementary computation shows that $L_{E(*,w)}(s)$ is just a product of two Riemann zeta functions at shifted arguments.

$$\begin{aligned} L_{E(*,w)}(s) &= \sum_{n=1}^{\infty} \sigma_{1-2w}(n) \cdot n^{w-\frac{1}{2}-s} \\ &= \sum_{n=1}^{\infty} n^{w-s-\frac{1}{2}} \sum_{d|n} d^{1-2w} = \sum_{d=1}^{\infty} d^{1-2w} \sum_{m=1}^{\infty} (md)^{w-s-\frac{1}{2}} \\ &= \zeta\left(s + w - \tfrac{1}{2}\right) \zeta\left(s - w + \tfrac{1}{2}\right). \end{aligned}$$

Consequently, if we define

$$\begin{aligned} \Lambda_{E(*,w)}(s) &:= \pi^{-\frac{s+w-\frac{1}{2}}{2}} \Gamma\left(\frac{s + w - \frac{1}{2}}{2}\right) \zeta\left(s + w - \tfrac{1}{2}\right) \\ &\quad \times \pi^{-\frac{s-w+\frac{1}{2}}{2}} \Gamma\left(\frac{s - w + \frac{1}{2}}{2}\right) \zeta\left(s - w + \tfrac{1}{2}\right) \\ &= \pi^{-s} \Gamma\left(\frac{s + w - \frac{1}{2}}{2}\right) \Gamma\left(\frac{s - w + \frac{1}{2}}{2}\right) \zeta\left(s + w - \tfrac{1}{2}\right) \zeta\left(s - w + \tfrac{1}{2}\right), \end{aligned}$$

then the functional equation,

$$\pi^{-s/2}\Gamma\left(\frac{s}{2}\right)\zeta(s) = \pi^{-(1-s)/2}\Gamma\left(\frac{1-s}{2}\right)\zeta(1-s),$$

of the Riemann zeta function immediately implies that

$$\Lambda_{E(*,w)}(s) = \Lambda_{E(*,w)}(1-s).$$

Note that this matches perfectly the functional equation of an even Maass form of type w (as in Proposition 3.13.5) as it should be.

We shall now show directly that the Eisenstein series is an eigenfunction of all the Hecke operators. This explains why the L-function associated to the Eisenstein series has an Euler product.

Proposition 3.14.2 *The Eisenstein series $E(z,s)$ is an eigenfunction of all the Hecke operators. For $n \geq 1$, let T_n denote the Hecke operator (3.12.3). Then*

$$T_n E(z,s) \;=\; n^{s-\frac{1}{2}}\sigma_{1-2s}(n)\cdot E(z,s).$$

Proof For $n \geq 1$, let

$$\Gamma_n = \left\{ \begin{pmatrix} a & b \\ c & d \end{pmatrix} \;\middle|\; ad - bc = n \right\}.$$

Then the set S_n given in Lemma 3.12.1 is just a set of coset representatives for $\Gamma_1\backslash\Gamma_n$. If R is any set of coset representatives for $\Gamma_\infty\backslash\Gamma_1$ then naturally RS_n is a set of coset representatives for $\Gamma_\infty\backslash\Gamma_n$. On the other hand, S_nR is also a set of coset representatives for $\Gamma_\infty\backslash\Gamma_n$. It follows that

$$\begin{aligned} T_nE(z,s) &= \frac{1}{\sqrt{n}} \sum_{\alpha\in\Gamma_\infty\backslash\Gamma_n} E(\alpha z, s) \\ &= \frac{1}{\sqrt{n}} \sum_{\gamma\in\Gamma_\infty\backslash\Gamma_1}\ \sum_{\alpha\in\Gamma_\infty\backslash\Gamma_n} \Im(\gamma\alpha z)^s \\ &= \frac{1}{\sqrt{n}} \sum_{\gamma\in\Gamma_\infty\backslash\Gamma_1}\ \sum_{\alpha\in\Gamma_\infty\backslash\Gamma_n} \Im(\alpha\gamma z)^s \\ &= \frac{1}{\sqrt{n}} \sum_{d\mid n} d^{1-s}\left(\frac{n}{d}\right)^s E(z,s) \\ &= n^{s-\frac{1}{2}}\sigma_{1-2s}(n)E(z,s). \end{aligned}$$

□

3.15 Converse theorems for $SL(2, \mathbb{Z})$

We have shown that the L–function associated to a Maass form for $SL(2, \mathbb{Z})$ is an entire function which satisfies a simple functional equation (see Proposition 3.13.5). The converse theorem of Maass–Hecke states that if a Dirichlet series is entire and bounded in vertical strips and satisfies the same functional equation as the L-function of a Maass form on $SL(2, \mathbb{Z})$, then it must be an L–function coming from a Maass form for $SL(2, \mathbb{Z})$.

Now, L–functions have been studied by number theorists for a long time, and many different L–functions satisfying all sorts of functional equations have been discovered. Surprisingly, not a single L–function has been found which satisfies exactly the right functional equation associated to a Maass form on $SL(2, \mathbb{Z})$. The closest examples known are the Hecke L–functions with grössencharakter of real quadratic fields discussed in Section 3.2. If we compare the functional equation of the L–function of an even Maass form given in Proposition 3.13.5:

$$\Lambda_f(s) := \pi^{-s}\Gamma\left(\frac{s-\frac{1}{2}+\nu}{2}\right)\Gamma\left(\frac{s+\frac{1}{2}-\nu}{2}\right)L_f(s) = \Lambda_f(1-s)$$

with the functional equation of the Hecke L–function with grössencharakter given in Proposition 3.2.4:

$$\begin{aligned}\Lambda_{\mathfrak{b}}^{n}(s) &:= \left(\frac{A}{\pi}\right)^{s}\Gamma\left(\frac{s-\frac{\pi i n}{\log\epsilon}}{2}\right)\Gamma\left(\frac{s+\frac{\pi i n}{\log\epsilon}}{2}\right)L_{\mathfrak{b}}(s,\psi^{n})\\ &= \Lambda_{\mathfrak{b}}^{n}(1-s). \end{aligned} \tag{3.15.1}$$

with $A = N(\mathfrak{b})\sqrt{D}$, we see that the two functional equations would match up if $\nu = \frac{1}{2} + \frac{\pi i n}{\log\epsilon}$ and $A = 1$. Unfortunately, A can never equal 1. It turns out, however, that the functional equation (3.15.1) does match the functional equation of a Maass form $\phi_{\mathfrak{b}}$ for a congruence subgroup of $SL(2, \mathbb{Z})$ and $\Delta\phi_{\mathfrak{b}} = \left(\frac{1}{4} + \frac{\pi^2 n^2}{(\log\epsilon)^2}\right)\cdot\phi_{\mathfrak{b}}$. The converse theorem for congruence subgroups of $SL(2, \mathbb{Z})$ was first discovered by A. Weil. It requires a family of functional equations (for twists of the original L–function by Dirichlet characters) instead of just one functional equation. This is because a congruence subgroup of $SL(2, \mathbb{Z})$, considered as a finitely generated group will, in general, be generated by several (more than 2) matrices. Weil's converse theorem can be used to prove that the Hecke L–function $L_{\mathfrak{b}}(s, \psi^n)$ is, in fact, associated to the Maass form $\phi_{\mathfrak{b}}$ of type $\frac{1}{2} + \frac{\pi i n}{\log\epsilon}$.

We now state and prove the converse theorem for Maass forms for $SL(2, \mathbb{Z})$. In order to simplify the exposition, we define what it means for a function of a complex variable to be EBV (entire and bounded in vertical strips).

Definition 3.15.2 *A function $f:\mathbb{C}\to\mathbb{C}$ is said to be* **EBV** *(entire and bounded on vertical strips) if*

- *$f(s)$ is holomorphic for all $s\in\mathbb{C}$.*
- *For fixed $A<B$, there exists $c>0$ such that $|f(s)|<c$ for $A\le\Re(s)\le B$.*

Theorem 3.15.3 (Hecke–Maass converse theorem) *Let $L(s)=\sum\limits_{n=1}^{\infty}(a(n)/n^s)$ (with $a(n)\in\mathbb{C}$) be a given Dirichlet series which converges absolutely for $\Re(s)$ sufficiently large. Assume that for fixed $\nu\in\mathbb{C}$, $L(s)$ satisfies the functional equation*

$$\Lambda^{\nu}(s):=\pi^{-s}\Gamma\left(\frac{s+\epsilon-\frac{1}{2}+\nu}{2}\right)\Gamma\left(\frac{s+\epsilon+\frac{1}{2}-\nu}{2}\right)L(s)=(-1)^{\epsilon}\Lambda^{\nu}(1-s),$$

with $\epsilon=0$ (respectively $\epsilon=1$), where $\Lambda^{\nu}(s)$ is EBV. Then

$$\sum_{n\neq 0}a(n)\sqrt{2\pi y}\cdot K_{\nu-\frac{1}{2}}(2\pi|n|y)e^{2\pi inx}$$

must be an even (respectively odd) Maass form of type ν for $SL(2,\mathbb{Z})$, where we have defined $a(n)=(-1)^{\epsilon}\cdot a(-n)$ for $n<0$.

Proof For $z\in\mathfrak{h}^2$, define $f(z)=\sum\limits_{n\neq 0}a(n)\sqrt{2\pi y}\cdot K_{\nu-\frac{1}{2}}(2\pi|n|y)e^{2\pi inx}$, which by our assumptions is an absolutely convergent series. Then clearly

$$\Delta f(z)=\nu(1-\nu)f(z),$$

because the Whittaker function $\sqrt{2\pi y}K_{\nu-\frac{1}{2}}(2\pi y)e^{2\pi ix}$ is an eigenfunction of Δ with eigenvalue $\nu(1-\nu)$. We also get for free the fact that $f(z)$ is periodic in x. This implies that

$$f\left(\begin{pmatrix}1&1\\0&1\end{pmatrix}z\right)=f(z).$$

Since $SL(2,\mathbb{Z})$ is generated by the two matrices

$$\begin{pmatrix}1&1\\0&1\end{pmatrix},\qquad\begin{pmatrix}0&-1\\1&0\end{pmatrix}$$

it follows that all that is left to be done to prove the converse theorem is to check that

$$\begin{aligned}f\left(\begin{pmatrix}y&x\\0&1\end{pmatrix}\right)&=f\left(\begin{pmatrix}0&-1\\1&0\end{pmatrix}\begin{pmatrix}y&x\\0&1\end{pmatrix}\right)\\&=f\left(\begin{pmatrix}\frac{y}{x^2+y^2}&\frac{-x}{x^2+y^2}\\0&1\end{pmatrix}\right).\end{aligned}\qquad(3.15.4)$$

In order to accomplish this, we need a lemma. □

Lemma 3.15.5 *Let $F : \mathfrak{h}^2 \to \mathbb{C}$ be a smooth eigenfunction of the Laplacian Δ with eigenvalue λ, i.e., $\Delta F = \lambda F$. Assume that*

$$F\left(\begin{pmatrix} y & 0 \\ 0 & 1 \end{pmatrix}\right) = \frac{\partial F}{\partial x}\left(\begin{pmatrix} y & x \\ 0 & 1 \end{pmatrix}\right)\Bigg|_{x=0} = 0. \tag{3.15.6}$$

Then $F(z)$ is identically zero on $\mathfrak{h}^2$.

Proof Since $\Delta F = \lambda F$ it follows that F is real analytic. Thus, it has a power series expansion in x of the form $F(z) = \sum_{n=0}^{\infty} b_n(y)x^n$. The eigenfunction equation implies that

$$\begin{aligned} 0 &= \Delta F(z) - \lambda F(z) \\ &= \sum_{n=0}^{\infty} \big[-(n+2)(n+1)y^2 b_{n+2}(y) \; - \; y^2 b_n''(y) \; - \; \lambda b_n(y)\big] x^n. \end{aligned}$$

Now, the initial conditions (3.15.6) tell us that $b_0(y) = b_1(y) = 0$, and it immediately follows from the recurrence relation

$$b_{n+2}(y) = -\frac{y^2 b_n''(y) + \lambda b_n(y)}{(n+1)(n+2)y^2}$$

that $b_n(y) = 0$ for all integers $n \geq 0$.

Lemma 3.15.5 implies that we can prove (3.15.4) if the function

$$F(z) = f\left(\begin{pmatrix} y & x \\ 0 & 1 \end{pmatrix}\right) - f\left(\begin{pmatrix} \frac{y}{x^2+y^2} & \frac{-x}{x^2+y^2} \\ 0 & 1 \end{pmatrix}\right)$$

satisfies the initial conditions (3.15.6). These can be written

$$f\left(\begin{pmatrix} y & 0 \\ 0 & 1 \end{pmatrix}\right) - f\left(\begin{pmatrix} y^{-1} & 0 \\ 0 & 1 \end{pmatrix}\right) = 0, \tag{3.15.7}$$

$$\frac{\partial f}{\partial x}\left(\begin{pmatrix} y & 0 \\ 0 & 1 \end{pmatrix}\right) + y^{-2}\frac{\partial f}{\partial x}\left(\begin{pmatrix} y^{-1} & 0 \\ 0 & 1 \end{pmatrix}\right) = 0. \tag{3.15.8}$$

First note that if f is even (i.e., $a(n) = a(-n)$), then it is enough to show that (3.15.7) holds. This is due to the fact that

$$\frac{\partial f}{\partial x}\left(\begin{pmatrix} y & 0 \\ 0 & 1 \end{pmatrix}\right) = \sum_{n \neq 0} 2\pi i n\, a(n)\sqrt{2\pi y} \cdot K_{\nu-\frac{1}{2}}(2\pi |n| y) = 0,$$

and similarly

$$y^{-2}\frac{\partial f}{\partial x}\left(\begin{pmatrix} y^{-1} & 0 \\ 0 & 1 \end{pmatrix}\right) = 0.$$

In an analogous manner, if f is odd, it is enough to prove (3.15.8).

We shall prove (3.15.7), (3.15.8) for even and odd f, respectively, using the Mellin inversion formulae

$$\tilde{h}(s) = \int_0^\infty h(y) y^s \frac{dy}{y} \tag{3.15.9}$$

$$h(y) = \frac{1}{2\pi i} \int_{\sigma - i\infty}^{\sigma + i\infty} \tilde{h}(s) y^{-s}\, ds. \tag{3.15.10}$$

These formulae hold for any smooth function $h : \mathbb{R}^+ \to \mathbb{C}$ and any fixed real σ provided $\tilde{h}(s)$ is EBV.

We apply (3.15.7), (3.15.8) with $\tilde{h}(s) = \Lambda^\nu(s)$. It follows from our previous calculations that we have the Mellin transform pair:

$$\Lambda^\nu(s) = \int_0^\infty \left(\frac{\partial}{\partial x}\right)^\epsilon f(z)\Big]_{x=0} y^{s-\frac{1}{2}} \frac{dy}{y},$$

$$\left(\frac{\partial}{\partial x}\right)^\epsilon f(z)\Big]_{x=0} = \frac{1}{2\pi i} \int_{\sigma - i\infty}^{\sigma + i\infty} \Lambda^\nu(s)\, y^{\frac{1}{2}-s} \frac{dy}{y}. \tag{3.15.11}$$

The functional equation,

$$\Lambda^\nu(s) = (-1)^\epsilon \Lambda^\nu(1-s),$$

combined with (3.15.11) immediately prove (3.15.7) and (3.15.8) for even and odd f, respectively. This completes the proof of the converse Theorem 3.15.3. □

3.16 The Selberg spectral decomposition

Our main goal of this section is the Selberg spectral decomposition for $SL(2,\mathbb{Z})$ which states that

$$\mathcal{L}^2\left(SL(2,\mathbb{Z})\backslash\mathfrak{h}^2\right) = \mathbb{C} \oplus \mathcal{L}^2_{\text{cusp}}\left(SL(2,\mathbb{Z})\backslash\mathfrak{h}^2\right) \oplus \mathcal{L}^2_{\text{cont}}\left(SL(2,\mathbb{Z})\backslash\mathfrak{h}^2\right),$$

where $\mathbb{C}$ is the one–dimensional space of constant functions, $L^2_{\text{cusp}}\left(SL(2,\mathbb{Z})\backslash\mathfrak{h}^2\right)$ represents the Hilbert space of square integrable functions on $\mathfrak{h}^2$ whose constant term is zero, and $L^2_{\text{cont}}\left(SL(2,\mathbb{Z})\backslash\mathfrak{h}^2\right)$ represents all square integrable functions on $\mathfrak{h}^2$ which are representable as integrals of the Eisenstein series. The reason for the terminology $\mathcal{L}^2_{\text{cusp}}$, $\mathcal{L}^2_{\text{cont}}$ is because the classical definition of cusp form, introduced by Hecke, requires that the constant term in the Fourier expansion around any cusp (a real number equivalent to ∞ under the discrete group) be zero, and also because the Eisenstein series is in the continuous spectrum of the Laplace operator. The latter means that $\Delta E(z,s) = s(1-s)E(z,s)$, or that $s(1-s)$ is an eigenvalue of Δ for any complex number s.

Let $\eta_j(z), (j = 1, 2, \dots)$ be an orthonormal basis of Maass forms for $SL(2,\mathbb{Z})$. We may assume as in Theorem 3.12.8 that each η_j is an eigenfunction of all the Hecke operators, so that its L-function has an Euler product. We shall also adopt the convention that

$$\eta_0(z) = \sqrt{\frac{3}{\pi}},$$

is the constant function of norm 1. The Selberg spectral decomposition is given in the following theorem.

Theorem 3.16.1 (Selberg spectral decomposition) *Let $f \in \mathcal{L}^2(SL(2,\mathbb{Z})\backslash\mathfrak{h}^2)$. Then we have*

$$f(z) = \sum_{j=0}^{\infty} \langle f, \eta_j\rangle\, \eta_j(z) \;+\; \frac{1}{4\pi i} \int_{\frac12 - i\infty}^{\frac12 + i\infty} \langle f, E(*,s)\rangle\, E(z,s)\, ds,$$

where

$$\langle f, g\rangle = \iint_{SL(2,\mathbb{Z})\backslash\mathfrak{h}^2} f(z)\overline{g(z)}\, \frac{dxdy}{y^2}$$

denotes the Petersson inner product on $\mathcal{L}^2\left(SL(2,\mathbb{Z})\backslash\mathfrak{h}^2\right)$.

We shall not give a complete proof of this theorem, but will only sketch one of the key ideas of the proof which is contained in the following proposition.

Proposition 3.16.2 *Let $f(z) \in \mathcal{L}^2\left(SL(2,\mathbb{Z})\backslash\mathfrak{h}^2\right)$ be orthogonal to the constant function, i.e.,*

$$\langle f, 1\rangle \;=\; \iint_{SL(2,\mathbb{Z})\backslash\mathfrak{h}^2} f(z)\, \frac{dxdy}{y^2} = 0.$$

Assume that f is of sufficiently rapid decay so that the inner product

$$\left\langle f, E(*,\bar{s})\right\rangle = \iint_{SL(2,\mathbb{Z})\backslash\mathfrak{h}^2} f(z) E(z,s)\, \frac{dxdy}{y^2}$$

converges absolutely for $\Re(s) > 1$. *Then*

$$f(z) = f_0(z) + \frac{1}{4\pi i} \int_{\frac{1}{2}-i\infty}^{\frac{1}{2}+i\infty} \langle f, E(*,s)\rangle E(z,s)\, ds,$$

where $f_0(z)$ *is automorphic for* $SL(2,\mathbb{Z})$ *with constant term in its Fourier expansion equal to zero, i.e.,* $\int_0^1 f_0(z)\, dx = 0$ *or* $f_0 \in \mathcal{L}^2_{\text{cusp}}$.

Proof The main idea of the proof is based on Mellin inversion. Recall that if $h(y)$ is a smooth complex valued function for $y \geq 0$ then the Mellin transform of h is

$$\tilde{h}(s) = \int_0^\infty h(y) y^s \frac{dy}{y}.$$

The transform is well defined provided there exists $c \in \mathbb{R}$ such that the integral converges absolutely for $\Re(s) \geq c$, and in this case, $\tilde{h}(s)$ is analytic for $\Re(s) \geq c$. The inverse transform is given by

$$h(y) = \frac{1}{2\pi i} \int_{c-i\infty}^{c+i\infty} \tilde{h}(s) y^{-s}\, ds.$$

The proof of Proposition 3.16.2 consists of two steps. In the first step it is shown that the inner product $\langle f, E(*,s)\rangle$ is the Mellin transform of the constant term of $f(z)$. In the second step, it is shown that the constant term of

$$\frac{1}{4\pi i} \int_{\frac{1}{2}-i\infty}^{\frac{1}{2}+i\infty} \langle f, E(*,s)\rangle E(z,s)\, ds$$

is the inverse Mellin transform of $\langle f, E(*,s)\rangle$ which brings you back precisely to the constant term of $f(z)$. Thus,

$$f(z) - \frac{1}{4\pi i} \int_{\frac{1}{2}-i\infty}^{\frac{1}{2}+i\infty} \langle f, E(*,s)\rangle E(z,s)\, ds$$

is automorphic with constant term equal to zero.

Step 1 Let

$$f(z) = \sum_{n=-\infty}^{\infty} A_n(y) e^{2\pi i n x}$$

denote the Fourier expansion of f. Recall that for $z \in \mathfrak{h}^2$, $I_s(z) = y^s$, and $\circ$ denotes the action of $SL(2,\mathbb{Z})$ on $\mathfrak{h}^2$. Since the function f and the measure $dxdy/y^2$ are invariant under the action $\circ$, it follows that for $\Re(s) > 1$,

$$\begin{aligned}\langle f, E(*,\bar{s})\rangle &= \iint\limits_{SL(2,\mathbb{Z})\backslash\mathfrak{h}^2} f(z) \sum_{\gamma\in\Gamma_\infty\backslash SL(2,\mathbb{Z})} \frac{1}{2} I_s(\gamma\circ z)\,\frac{dxdy}{y^2} \qquad (3.16.3)\\ &= \frac{1}{2}\sum_{\gamma\in\Gamma_\infty\backslash SL(2,\mathbb{Z})} \iint\limits_{\gamma\circ\left(SL(2,\mathbb{Z})\backslash\mathfrak{h}^2\right)} f(z) I_s(z)\,\frac{dxdy}{y^2}\\ &= \int_0^\infty\int_0^1 f(z)y^s\,\frac{dxdy}{y^2}\\ &= \int_0^\infty A_0(y)y^s\,\frac{dy}{y^2}\\ &= \tilde{A}_0(s-1).\end{aligned}$$

The assumption that f is orthogonal to the constant function implies that the residue at $s = 1$ of $\langle f, E(*,\bar{s})\rangle$ is zero. Further, $E(z,s) = E^*(z,s)/\zeta(2s)$ which implies (by the fact that $\zeta(1+it) \neq 0$ for real t) that $\tilde{A}_0(s-1)$ is holomorphic for $\Re(s) \geq \frac{1}{2}$. The functional equation (Theorem 3.1.10) of the Eisenstein series tells us that

$$\langle f, E(*,\bar{s})\rangle = \phi(s)\langle f, E(*, 1-\bar{s})\rangle,$$

or equivalently that

$$\tilde{A}_0(s-1) = \phi(s)\tilde{A}_0(-s).$$

$$\tilde{A}_0(-s) = \phi(1-s)\tilde{A}_0(s-1). \qquad (3.16.4)$$

Step 2 By Mellin inversion it follows that for $c > 1$,

$$A_0(y) = \frac{1}{2\pi i}\int_{c-i\infty}^{c+i\infty} \tilde{A}_0(s-1)\,y^{1-s}\,ds.$$

Since $\tilde{A}_0(s-1)$ is holomorphic for $\Re(s) \geq \frac{1}{2}$, this implies that we may shift the above line of integration to $\Re(s) = \frac{1}{2}$. It follows from the transformation $s \to 1-s$ that

$$A_0(y) = \frac{1}{2\pi i}\int_{\frac{1}{2}-i\infty}^{\frac{1}{2}+i\infty} \tilde{A}_0(s-1)\,y^{1-s}\,ds = \frac{1}{2\pi i}\int_{\frac{1}{2}-i\infty}^{\frac{1}{2}+i\infty} \tilde{A}_0(-s)\,y^{s}\,ds.$$

If we now make use of the functional equation (3.16.4), we easily obtain

$$A_0(y) = \frac{1}{4\pi i}\int\limits_{\frac{1}{2}-i\infty}^{\frac{1}{2}+i\infty} \tilde{A}_0(-s)(y^s + \phi(s)y^{1-s})\,ds. \tag{3.16.5}$$

But for $\Re(s) = \frac{1}{2}$ we have $\tilde{A}_0(-s) = \tilde{A}_0(\bar{s}-1)$, and by (3.16.3), we have

$$\tilde{A}_0(\bar{s}-1) = \langle f, E(*, s)\rangle.$$

Plugging this into (3.16.5) completes the proof of the theorem. □

GL(n)pack functions The following **GL(n)pack** functions, described in the appendix, relate to the material in this chapter:

EisensteinFourierCoefficient	EisensteinSeriesTerm	HeckeCoefficientSum
HeckeEigenvalues	HeckeMultiplicativeSplit	HeckeOperator
HermiteFormLower	HermiteFormUpper	SmithForm
SmithElementaryDivisors	SmithInvariantFactors	Whittaker
WhittakerStar.		

4
Existence of Maass forms

Maass forms for $SL(2, \mathbb{Z})$ were introduced in Section 3.3. An important objective of this book is to generalize these functions to the higher-rank group $SL(n, \mathbb{Z})$ with $n \geq 3$. It is a highly non-trivial problem to show that infinitely many even Maass forms for $SL(2, \mathbb{Z})$ exist. The first proof was given by Selberg (1956) where he introduced the trace formula as a tool to obtain Weyl's law, which in this context gives an asymptotic count (as $x \to \infty$) for the number of Maass forms of type ν with $|\nu| \leq x$. Selberg's methods were extended by Miller (2001), who obtain Weyl's law for Maass forms on $SL(3, \mathbb{Z})$ and Müller (2004), who obtained Weyl's law for Maass forms on $SL(n, \mathbb{Z})$.

A rather startling revelation was made by Phillips and Sarnak (1985) where it was conjectured that Maass forms should not exist for generic non-congruence subgroups of $SL(2, \mathbb{Z})$, except for certain situations where their existence is ensured by symmetry considerations, see Section 4.1. Up to now no one has found a single example of a Maass form for $SL(2, \mathbb{Z})$, although Maass (1949) discovered some examples for congruence subgroups (see Section 3.15). So it seemed as if Maass forms for $SL(2, \mathbb{Z})$ were elusive mysterious objects and the non-constructive proof of their existence (Selberg, 1956) suggested that they may be unconstructible.

Recently, Lindenstrauss and Venkatesh (to appear) found a new, short, and essentially elementary proof which shows the existence of infinitely many Maass forms on $G(\mathbb{Z})\backslash G(\mathbb{R})/K_\infty$ where G is a split semisimple group over $\mathbb{Z}$ and K_∞ is the maximal compact subgroup. Lindenstrauss and Venkatesh were also able to obtain Weyl's law in a very broad context. Their method works whenever one has Hecke operators. Although in the case of $SL(2, \mathbb{Z})$, the proof is perhaps not much simpler than the trace formula, it has the advantage of being much more explicit (it allows one to *write down* an even cuspidal function). However, a much bigger advantage is that it generalizes in a relatively straightforward way to higher rank (unlike the trace formula, where one encounters formidable technical obstacles). I would very much like to thank Elon and Akshay for preparing and allowing me to incorporate a preliminary manuscript which formed the basis of this chapter.

4.1 The infinitude of odd Maass forms for $SL(2, \mathbb{Z})$

Let $\Delta = -y^2\left(\frac{\partial^2}{\partial x^2} + \frac{\partial^2}{\partial y^2}\right)$ be the hyperbolic Laplacian on $\mathfrak{h}^2$. It was shown in Theorem 3.16.1 that

$$\mathcal{L}^2\left(SL(2,\mathbb{Z})\backslash\mathfrak{h}^2\right) = \mathbb{C} \oplus \mathcal{L}^2_{\text{cusp}}\left(SL(2,\mathbb{Z})\backslash\mathfrak{h}^2\right) \oplus \mathcal{L}^2_{\text{cont}}\left(SL(2,\mathbb{Z})\backslash\mathfrak{h}^2\right),$$

where $\mathcal{L}^2_{\text{cont}}(SL(2,\mathbb{Z})\backslash\mathfrak{h}^2)$ is spanned by the continuous spectrum of Δ, explicitly given by Eisenstein series $E(z, \frac{1}{2} + ir)$ with $r \in \mathbb{R}$, and $\mathcal{L}^2_{\text{cusp}}\left(SL(2,\mathbb{Z})\backslash\mathfrak{h}^2\right)$ is spanned by Maass forms.

Recall from Section 3.9 that a Maass form

$$f(z) = \sum_{n\neq 0} a_n \sqrt{2\pi y} K_{\nu-\frac{1}{2}}(2\pi|n|y)e^{2\pi i n x}$$

of type ν for $SL(2,\mathbb{Z})$ (as in Proposition 3.5.1) is even or odd according to whether

$$T_{-1}f(z) = T_{-1}f\left(\begin{pmatrix} y & x \\ 0 & 1 \end{pmatrix}\right) := f\left(\begin{pmatrix} y & -x \\ 0 & 1 \end{pmatrix}\right)$$

is equal to $f(z)$ or $-f(z)$, respectively.

The following proposition is almost obvious.

Proposition 4.1.1 *There are infinitely many odd Maass forms for $SL(2,\mathbb{Z})$.*

Proof The image of the endomorphism

$$J : \mathcal{L}^2\left(SL(2,\mathbb{Z})\backslash\mathfrak{h}^2\right) \to \mathcal{L}^2\left(SL(2,\mathbb{Z})\backslash\mathfrak{h}^2\right),$$

defined by

$$Jf(z) := f(z) - T_{-1}f(z),$$

is purely cuspidal. This is due to the fact that the constant term of $f - Jf$, given by $\int_0^1 \left(f(x+iy) - f(-x+iy)\right)dx = 0$ for all $y > 0$. We leave it to the reader to show that the image of J is non-trivial. □

The rest of this chapter will be devoted to showing that the space of even Maass forms for $SL(2,\mathbb{Z})$ is also infinite dimensional. The only other known proof of this fact uses the trace formula (Selberg, 1956), see also (Hejhal, 1976), as we already mentioned earlier.

4.2 Integral operators

In this section we shall adopt the classical model of the upper half-plane $\mathfrak{h}^2$ and consider $z \in \mathfrak{h}^2$ in the form $z = x + iy$ with $x \in \mathbb{R}$, $y > 0$. Note that every matrix $\begin{pmatrix} y & x \\ 0 & 1 \end{pmatrix}$ can be put in this form by simply letting it act on $i = \sqrt{-1}$.

Let $d(z, z')$ denote the hyperbolic distance between two points $z, z' \in \mathfrak{h}^2$, which is characterized by the property that $d(\alpha z, \alpha z') = d(z, z')$ for all $\alpha \in SL(2, \mathbb{R})$ and $z, z' \in \mathfrak{h}^2$. It is easy to check that

$$u(z, z') = \frac{|z - z'|^2}{4\Im(z)\Im(z')} \tag{4.2.1}$$

satisfies this property, and we have the relation

$$u(z, z') = \sinh^2\left(\frac{d(z, z')}{2}\right),$$

because the hyperbolic distance between the points i and iy_0, with $y_0 > 1$, is given by

$$\int_1^{y_0} \frac{dy}{y} = \log(y_0).$$

Definition 4.2.2 (The Abel transform) *The Abel transform $F(x)$ of an integrable function $f(x)$ on $[0, +\infty)$, is given by*

$$F(x) = \int_{-\infty}^{\infty} f\left(x + \frac{\xi^2}{2}\right) d\xi = \sqrt{2} \int_x^{\infty} \frac{f(v)dv}{\sqrt{v - x}}.$$

Proposition 4.2.3 *Let $f(x)$ be a continuously differentiable function on $[0, +\infty)$. If $F(x)$ denotes the Abel transform in Definition 4.2.2, then*

$$f(x) = -\frac{1}{2\pi} \int_{-\infty}^{\infty} F'\left(x + \frac{\eta^2}{2}\right) d\eta$$

is the inverse Abel transform.

Proof First differentiate under the integral sign, and then convert to polar coordinates, to obtain

$$\begin{aligned}\int_{-\infty}^{\infty} F'\left(x+\frac{\eta^2}{2}\right)\,d\eta &= \int_{-\infty}^{\infty}\int_{-\infty}^{\infty} f'\left(x+\frac{\xi^2}{2}+\frac{\eta^2}{2}\right)\,d\xi\,d\eta\\ &= \int_0^{2\pi}\int_0^{\infty} f'\left(x+\frac{r^2}{2}\right) r\,dr\,d\theta\\ &= 2\pi\int_0^{\infty} f'(x+w)\,dw,\end{aligned}$$

where $w = r^2/2$ and $dw = r\,dr$. This equals $-2\pi f(x+w)|_{w=0}$ or $-2\pi f(x)$ which proves the proposition. □

Let $g(x)$ be an even smooth function of compact support on the real line with Fourier transform

$$h(t) = \int_{-\infty}^{\infty} g(x)e^{itx}\,dx.$$

Following Selberg (1956), we define a variation of the Abel transform, denoted k, as follows.

$$k(u) = -\frac{1}{\pi}\int_u^{\infty} (v-u)^{-\frac{1}{2}}\,dq(v), \tag{4.2.4}$$

where

$$q(v) := \frac{1}{2}\,g\big(2\log\left(\sqrt{v+1}+\sqrt{v}\right)\big).$$

The relation between k and h is called the Selberg transform. It is clear from the definitions that k is compactly supported and continuous. Indeed, if g is supported in $[-M, M]$, then $k(u)$ vanishes for $u > \sinh^2(M/2)$.

For $z, w \in \mathfrak{h}^2$, let $u(z, w)$ be given by (4.2.1). Then $u(z, w)$ is real valued and positive. Define

$$u := u(z, w). \tag{4.2.5}$$

Definition 4.2.6 (Point pair invariant) *Let $g : \mathbb{R} \to \mathbb{C}$ be an even smooth function of compact support. The point pair invariant $K : \mathfrak{h}^2 \times \mathfrak{h}^2 \to \mathbb{C}$ associated to g is the function defined by*

$$K(z, w) = k(u(z, w)) = k(u), \qquad (\textit{for all } z, w \in \mathfrak{h}^2),$$

where k is given by (4.2.4) and u is given by (4.2.5).

It is clear that K is a continuous function on $\mathfrak{h}^2 \times \mathfrak{h}^2$. Moreover, it follows from the fact that k is compactly supported, that $K(z, w)$ is supported in $d(z, w) \le R$ for some R that depends only on the support of g. Since the point pair invariant K is a function of the hyperbolic distance, it is plain that $K(\alpha z, \alpha w) = K(z, w)$ for all $\alpha \in SL(2, \mathbb{R})$, and $z, w \in \mathfrak{h}^2$.

The point pair invariant K can be used to define an integral operator which acts on functions in $\mathcal{L}^1(SL(2, \mathbb{Z})\backslash\mathfrak{h}^2)$.

Definition 4.2.7 (Integral operator) *For any function $f \in \mathcal{L}^1\left(SL(2, \mathbb{Z})\backslash\mathfrak{h}^2\right)$ and any point pair invariant $K(z, w)$, define the integral operator*

$$K * f(z) := \int\limits_{\mathfrak{h}^2} K(z, w) f(w)\, d^*w$$

$$= \int\limits_0^\infty \int\limits_{-\infty}^\infty K(z, \mu + i\nu) f(\mu + i\nu) \frac{d\mu d\nu}{\nu^2},$$

where for $w = \mu + i\nu \in \mathfrak{h}^2$, the invariant measure $d^*w = d\mu d\nu/\nu^2$, as in Proposition 1.5.3.

It is easy to check that $K * f$ is also invariant by $SL(2, \mathbb{Z})$, and, indeed, that $f \mapsto K * f$ defines a self-adjoint continuous endomorphism of $\mathcal{L}^2(SL(2, \mathbb{Z})\backslash\mathfrak{h}^2)$. The key property of K that we will need is given in the following lemma.

Lemma 4.2.8 *We have $\int\limits_{-\infty}^{\infty} K(i, t + ie^x)\, dt = e^{x/2} g(x)$.*

Proof It follows from (4.2.1) that

$$u(i,\ t + ie^x) = \frac{(e^x - 1)^2 + t^2}{4e^x},$$

so that the integral is exactly equal to

$$\int\limits_{-\infty}^{\infty} k\left(\sinh^2\left(\frac{x}{2}\right) + \frac{t^2}{4e^x}\right) dt.$$

But, by (Iwaniec, 1995), we have

$$g(x) = 2 \int\limits_{\sinh^2(x/2)}^{\infty} k(u)(u - \sinh^2(x/2))^{-\frac{1}{2}}\, du,$$

from which the lemma follows. □

Remark *It is a basic fact (although we will not need it for the proof, it provides valuable intuition) that, if ϕ_λ is an eigenfunction of the hyperbolic Laplacian Δ with eigenvalue $\lambda = \frac{1}{4} + r^2$, then*

$$K * \phi_\lambda = h(r)\phi_\lambda. \tag{4.2.9}$$

This can be readily established by computing $K * y^{\frac{1}{2}+ir}$.

Now, suppose we take $g = g^{(j)}$ to be a Dirac δ−sequence with $j \to \infty$. By this we mean that

$$\lim_{j\to\infty} \int_{\mathbb{R}} g^{(j)}(x)\, dx = 1,$$

and the support of $g^{(j)}$ shrinks to $\{0\}$ as $j \to \infty$. In this situation, the Fourier transform $h^{(j)}$ approaches the constant function, so we expect, in view of (4.2.9), that the associated point pair operators

$$f \to K^{(j)} * f$$

will approach the identity endomorphism. In fact, we have already seen that as the support of $g^{(j)}$ shrinks to zero, then the support of $K^{(j)}$ shrinks to zero also. Moreover, it follows from Lemma 4.2.8, after making the substitution $v = e^x$, that

$$\begin{aligned}
\int_{\mathfrak{h}^2} K^{(j)}(i, w)\, d^*w &= \int_0^\infty \int_{-\infty}^\infty K^{(j)}(i, \mu + iv)\, \frac{d\mu dv}{v^2} \\
&= \int_0^\infty \int_{-\infty}^\infty K^{(j)}(i, \mu + ie^x) \cdot e^{-x}\, d\mu\, dx \\
&= \int_{-\infty}^\infty e^{-x/2} g^{(j)}(x)\, dx \longrightarrow 1.
\end{aligned}$$

On the other hand, by the point pair property, and the fact that d^*w is an $SL(2, \mathbb{R})$-invariant measure, we have for any $\alpha \in SL(2, \mathbb{R})$ that

$$\begin{aligned}
\int_{\mathfrak{h}^2} K^{(j)}(i, w)\, d^*w &= \int_{\mathfrak{h}^2} K^{(j)}(\alpha i, \alpha w)\, d^*w \\
&= \int_{\mathfrak{h}^2} K^{(j)}(\alpha i, \alpha w)\, d^*(\alpha w) \\
&= \int_{\mathfrak{h}^2} K^{(j)}(\alpha i, w)\, d^*w.
\end{aligned}$$

Now, for any $z \in \mathfrak{h}^2$, we may choose $\alpha \in SL(2, \mathbb{R})$ such that $z = \alpha i$. It follows that

$$\int_{\mathfrak{h}^2} K^{(j)}(z, w)\, d^*w \longrightarrow 1$$

for any $z \in \mathfrak{h}^2$, as $j \to \infty$.

Consequently, for any continuous function f, we have

$$\begin{aligned} K^{(j)} * f(z) &= \int_{\mathfrak{h}^2} K^{(j)}(z, w) f(w)\, d^*w \\ &= \int_{\mathfrak{h}^2} K^{(j)}(z, w) f(z)\, d^*w + \int_{\mathfrak{h}^2} K^{(j)}(z, w)(f(w) - f(z))\, d^*w \\ &\longrightarrow f(z), \end{aligned} \tag{4.2.10}$$

as $j \to \infty$. So the corresponding operators $K^{(j)}$ will satisfy

$$K^{(j)} * f(z) \to f(z)$$

for any continuous funtion f and all $z \in \mathfrak{h}^2$.

4.3 The endomorphism $\heartsuit$

In Section 4.1, we showed that there are infinitely many odd Maass forms for $SL(2, \mathbb{Z})$ by showing that the endomorphism J, given in Proposition 4.1.1, has a purely cuspidal image. The key idea of the present approach is to construct an explicit endomorphism $\heartsuit$ of $\mathcal{L}^2_{\text{cusp}} = \mathcal{L}^2_{\text{cusp}}(SL(2, \mathbb{Z}\backslash\mathfrak{h}^2)$ whose image is purely cuspidal. The endomorphism $\heartsuit$, however, will use the arithmetic structure of $SL(2, \mathbb{Z})$ in a much more essential way than J did.

Recall that for any rational prime p, we have the Hecke operator T_p, given in (3.12.3), which acts on functions f on $SL(2, \mathbb{Z})\backslash\mathfrak{h}^2$ via the rule

$$T_p f(z) = \frac{1}{\sqrt{p}} \left(f(pz) + \sum_{k=0}^{p-1} f\left(\frac{z+k}{p}\right)\right). \tag{4.3.1}$$

The Hecke operators T_p commute with Δ, and we showed in Propositions 3.1.3 and 3.14.2 that the Eisenstein series $E(z, \frac{1}{2} + ir)$, defined in Definition 3.1.2, satisfies

$$\Delta E\left(z, \frac{1}{2}+ir\right) = \left(\frac{1}{4}+r^2\right) \cdot E\left(z, \frac{1}{2}+ir\right) \tag{4.3.2}$$

$$T_p E\left(z, \frac{1}{2}+ir\right) = (p^{ir}+p^{-ir}) \cdot E\left(z, \frac{1}{2}+ir\right). \tag{4.3.3}$$

We proceed formally for now. From (4.3.2) and (4.3.3), the operator

$$\heartsuit := T_p - p^{\sqrt{\frac{1}{4}-\Delta}} - p^{-\sqrt{\frac{1}{4}-\Delta}}$$

annihilates $E(z, \frac{1}{2}+ir)$. Similarly (again at the formal level) $\heartsuit$ also annihilates the constant function. The operator $\heartsuit$ may be given a rigorous interpretation either in terms of the wave equation or using convolution operators. (In fact, for technical simplicity in our treatment, we will use not $\heartsuit$ but a certain smoothed version.) For the time being, let us accept – as is indeed the case – that $\heartsuit$ may be given a rigorous interpretation and that it is a self-adjoint endomorphism of the space, $\mathcal{L}^2_+$, of even square integrable automorphic functions.

Since $\heartsuit$ kills the continuous spectrum and is self-adjoint, it has cuspidal image. To show that there exist even Maass forms for $SL(2,\mathbb{Z})$, it suffices to find a *single non-constant function* in $\mathcal{L}^2_+$ not annihilated by $\heartsuit$; this we do by choosing an appropriate test function supported high in the cusp.

Although we have appealed to the theory of Eisenstein series, this is not really necessary: it is possible to prove that the image of $\heartsuit$ is cuspidal directly from the definition, and in fact the proof we present will be completely independent of any knowledge of Eisenstein series.

4.4 How to interpret $\heartsuit$: an explicit operator with purely cuspidal image

Let h be the Fourier transform of g. Proceeding formally for a moment, let us also note that if g were the sum of a δ-mass at $x = \log(p)$ and at $x = -\log(p)$, then $h(r) = p^{ir} + p^{-ir}$, and so in this case (4.2.9) says – if we can make sense of it – that $f \mapsto K * f$ has the properties we would expect of the operator $p^{\sqrt{\Delta-\frac{1}{4}}} + p^{-\sqrt{\Delta-\frac{1}{4}}}$. In this section we will mildly modify this construction (because we prefer not to deal with the technicalities that arise by taking g to be a distribution).

Now, let g_0 be an even smooth function of compact support on $\mathbb{R}$, and for $p \geq 1$ put $g_p(x) := g_0(x+\log(p)) + g_0(x-\log(p))$. Starting from g_p for $p \geq 0$, we define $k_p(u)$ and $K_p(z,w)$ as in Definition 4.2.6.

Let us pause to explain the connection of this with the vague idea that we described at the start of this section. Work formally for a moment and suppose that g_0 were the "delta-function at 0." Then, formally speaking, the Fourier transform h_0 is the constant function 1, whereas the Fourier transform $h_p(r) = p^{ir} + p^{-ir}$. Using (4.2.9) we see (formally speaking – this is not intended to be a rigorous proof!) that the map $f \mapsto K_0 * f$ is just the identity endomorphism and that the operator $f \mapsto K_p * f$ is essentially the operator $p^{\sqrt{\Delta - \frac{1}{4}}} + p^{-\sqrt{\Delta - \frac{1}{4}}}$. So $f \mapsto K_p * f - T_p(K_0 * f)$, formally speaking, gives an interpretation to the operator ♡ that we discussed earlier. In practice, to avoid certain technical complications, we just take g_0 to be smooth rather than the δ function, and K_0 (resp. K_p) approximates the identity endomorphism (resp. $p^{\sqrt{\Delta - \frac{1}{4}}} + p^{-\sqrt{\Delta - \frac{1}{4}}}$); in fact, we rigorously proved at the end of Section 4.2 that, as g_0 varies through a δ sequence, the operator K_0 approaches the identity endomorphism in an appropriate sense.

Lemma 4.4.1 *Let $f \in \mathcal{L}^1\left(SL(2,\mathbb{Z})\backslash\mathfrak{h}^2\right)$. Then $K_p * f - T_p(K_0 * f)$ defines a cuspidal function on $SL(2,\mathbb{Z})\backslash\mathfrak{h}^2$.*

Proof For any function F on $SL(2,\mathbb{Z})\backslash\mathfrak{h}^2$, we define for $y > 0$, the *constant term* (denoted F_{CT}):

$$F_{CT}(y) = \int\limits_{\mathbb{R}/\mathbb{Z}} F(x+iy)\,dx.$$

Using the explicit definition of T_p given in (4.3.1), we see that

$$(T_pF)_{CT}(y) = p^{-\frac{1}{2}}F_{CT}(py) + p^{\frac{1}{2}}F_{CT}(p^{-1}y) \tag{4.4.2}$$

holds for any F on $SL(2,\mathbb{Z})\backslash\mathfrak{h}^2$. To prove Lemma 4.4.1, we need to check that, for any $y > 0$, $(K_p * f)_{CT} = (T_p(K_0 * f))_{CT}$. So we just need to check that

$$(K_p * f)_{CT}(y) = p^{-1/2}(K_0 * f)_{CT}(py) + p^{1/2}(K_0 * f)_{CT}(p^{-1}y). \tag{4.4.3}$$

Now, for any $p \geq 0$, we have

$$(K_p * f)_{CT}(y) = \int\limits_{x\in\mathbb{R}/Z}\ \int\limits_{w\in\mathfrak{h}^2} K_p(x+iy,w)f(w)\,d^*w\,dx \tag{4.4.4}$$

$$= \int\limits_{x\in\mathbb{R}}\ \int\limits_{\substack{w\in\mathfrak{h}^2\\ 0\leq\Re(w)\leq 1}} K_p(x+iy,w)f(w)\,d^*w\,dx,$$

where we have used the fact that $f(w) = f(w+1)$ to *unfold* the integral over $x \in \mathbb{R}/\mathbb{Z}$, at the cost of restricting the w-integration from $\mathfrak{h}^2$ to $\mathfrak{h}^2/\{w \mapsto w+1\}$.

Now, using the fact that $K(gz, gw) = K(z, w)$ for $g \in SL(2, \mathbb{R})$, we see that that if $w = x_w + iy_w$, then

$$\int_{x\in\mathbb{R}} K_p(x+iy, w)\,dx = \int_{x\in\mathbb{R}} K_p(x+iy,\ iy_w)\,dx \tag{4.4.5}$$

$$= y_w \int_{x\in\mathbb{R}} K_p\big(x+iyy_w^{-1},\ i\big)\,dx$$

$$= (yy_w)^{1/2}\, g_p\big(\log\big(yy_w^{-1}\big)\big).$$

So we get

$$(K_p * f)_{CT}(y_0) = \int_0^\infty y_0^{1/2} y^{-1/2}\, g_p(\log(y_0 y^{-1})) f_{CT}(y)\,\frac{dy}{y}.$$

From this and the fact that $g_p(x) := g_0(x + \log(p)) + g_0(x - \log(p))$, (4.4.3) follows by a simple computation. □

4.5 There exist infinitely many even cusp forms for $SL(2, \mathbb{Z})$

Let notations be as in Section 4.4 and let ♡ be the self-adjoint endomorphism of $\mathcal{L}^2(SL(2,\mathbb{Z})\backslash\mathfrak{h}^2)$ defined by $f \mapsto K_p * f - T_p(K_0 * f)$. It is easy to check that ♡ preserves $\mathcal{L}^2_{\text{cusp},+}$ (in fact, all operators in sight do). To show that there exist even cusp forms, we must show that ♡ $\neq 0$ on $\mathcal{L}^2_{\text{cusp},+}$. The idea, in words, is the following. Let Γ_∞ be the stabilizer of the cusp at ∞ in $SL(2, \mathbb{Z})$, that is to say, the group generated by $z \mapsto z + 1$.

High in the cusp, $SL(2,\mathbb{Z})\backslash\mathfrak{h}^2$ looks like the cylinder $\Gamma_\infty\backslash\mathfrak{h}^2$. This cylinder has rotational symmetry, i.e., it admits the action $z \mapsto z + t$ of the group $\mathbb{R}/\mathbb{Z}$. It turns out that the maps $f \mapsto K_p * f$ and $f \mapsto T_p(K_0 * f)$ behave totally differently with respect to this action; so this incompatibility forces ♡ to be non-zero.

Let $T \geq 1$ and let

$$\mathfrak{S}(T) = \Sigma_{T,\frac{1}{2}} = \Gamma_\infty\backslash \left\{ z \in \mathfrak{h}^2 \,\middle|\, \Im(z) > T \right\},$$

be the Siegel set as in Definition 1.3.1. Then the natural projection: $\mathfrak{S}(T)$ to $SL(2,\mathbb{Z})\backslash\mathfrak{h}^2$ is a homeomorphism onto an open subset. We can, therefore, regard $C_c^\infty(\mathfrak{S}(T))$, the space of smooth compactly supported functions on $\mathfrak{S}(T)$, as a subset of $C_c^\infty(SL(2,\mathbb{Z})\backslash\mathfrak{h}^2)$; similarly $\mathcal{L}^2(\mathfrak{S}(T))$ is a subset of

$\mathcal{L}^2(SL(2,\mathbb{Z})\backslash\mathfrak{h}^2)$. We will make these identifications throughout the rest of this argument.

If $f \in C^\infty(\mathfrak{S}(T))$, we define for $n \in \mathbb{Z}$, the nth Fourier coefficient $a_{n,f}(y)$ to be the function on (T,∞) defined by the rule $a_{n,f}(y) = \int_0^1 f(x+iy)e^{-2\pi inx}dx$.

Now, let R be so large that $k_0(z,w)$ and $k_p(z,w)$ are supported in $d(z,w) \le R$, and let $Y \ge pe^R$. Then, one sees from (4.3.2) and Definition 4.2.7 that $\heartsuit$ maps $C^\infty(\mathfrak{S}(Y))$ into $C^\infty(\mathfrak{S}(1))$. Indeed, it is enough to check that this is true for $f \mapsto K_p * f$ and $f \mapsto T_p(K_0 * f)$; we deal with the first and leave the second to the reader. It is clear from Definition 4.2.7 that $K_p * f$ is supported in an R-neighborhood of the support of f. But an R-neighborhood of $\mathfrak{S}(Y)$ is contained in $\mathfrak{S}(Y/R)$, thus the claim.

Moreover, if the nth Fourier coefficient $a_{n,f}(y)$ vanishes identically, then so does $a_{n,K_p*f}(y)$. This follows from (4.4.4):

$$\begin{aligned} a_{n,K_p*f}(y) &:= \int\limits_{x\in\mathbb{Z}\backslash\mathbb{R}} e^{-2\pi inx} \int\limits_{w\in\mathfrak{h}^2} K_p(x+iy,w)f(w)\,d^*w\,dx \\ &= \int\limits_{x\in\mathbb{Z}\backslash\mathbb{R}} e^{-2\pi inx} \int\limits_{w\in\mathfrak{h}^2} K_p(iy,w-x)f(w)\,d^*w\,dx \\ &= \int\limits_{w\in\mathfrak{h}^2}\int\limits_{x\in\mathbb{Z}\backslash\mathbb{R}} K_p(iy,w)f(w+x)e^{-2\pi inx}\,dx\,d^*w, \end{aligned}$$

and the final integral clearly vanishes if $a_{n,f}$ vanishes identically.

Fix an arbitrary integer $N \not\equiv 0 \pmod p$. Let

$$f \in C_+^\infty(\mathfrak{S}(Y)) := C_c^\infty(\mathfrak{S}(Y)) \cap \mathcal{L}^2_{\text{cusp},+}$$

be a non-zero even function so that $a_{n,f}$ vanishes identically for all $n \ne \pm N$. Then a_{n,K_p*f} vanishes identically for $n \ne \pm N$. On the other hand, we see from Lemma 4.4.1 that, for $z \in \mathfrak{h}^2$, we have

$$T_p(K_0 * f)(z) = K_0 * f(pz),$$

so (by the same argument as before), $a_{n,T_p(K_0*f)}$ vanishes identically for $n \ne \pm pN$.

It follows that $\heartsuit f$ is a non-zero even cuspidal function, *as long as* $K_0 * f \ne 0$. But we are still free to choose the function g_0 that entered in the definition of K_0, and it is clear from the discussion of Section 4.2 that, as we let g_0 approximate the δ function, $K_0 * f$ will approach f pointwise; in particular, it will be non-zero.

We have therefore shown that – for appropriate choice of g_0 – the function $F = \heartsuit f$ is non-zero, cuspidal, belongs to $C_+^\infty(\mathfrak{S}(1))$, and, moreover, has the property (as is clear from our discussion above) that $a_{n,F}(y)$ vanishes unless $n \in \{N, -N, pN, -pN\}$. Since N could be any integer not divisible by p, we conclude from this that $\mathcal{L}^2_{\text{cusp},+}$ is infinite dimensional, so there are infinitely many even cusp forms.

4.6 A weak Weyl law

The proof given above shows there exist *infinitely many* cusp forms. It is easy to make this quantitative. Here we will just explain how to prove a weak version of the Weyl law and we will say a few words about how to prove the full Weyl law in Section 4.8. We first recall the following "variational principle":

Lemma 4.6.1 *Suppose H is a Hilbert space and A a non-negative self-adjoint (possibly unbounded) operator on H with discrete spectrum $\lambda_1 \le \lambda_2 \le \dots$ Suppose $V \subset H$ is a finite-dimensional subspace and Λ is such that $\|Av\| \le \Lambda \|v\|$ whenever $v \in V$. Then $\#\{\lambda_i \le \Lambda\} \ge \dim(V)$.*

Proof Let v_i be the eigenvector corresponding to the eigenvalue λ_i. Let W be the space spanned by all the v_i's with $\lambda_i \le \Lambda$. If the claim is false, then $\dim(W) < \dim(V)$, so there is a vector in V perpendicular to W. Such a vector must have the form $v = \sum_j c_j v_j$, where the sum is taken only over eigenvectors v_j with eigenvalue $> \Lambda$. But it is clear that such a vector cannot satisfy $\|Av\| \le \Lambda \|v\|$, contradiction. □

Proposition 4.6.2 *Let $N(\Lambda)$ be the number of eigenfunctions of the Laplacian in $\mathcal{L}^2_{cusp,+}$ with eigenvalue $\le \Lambda$. Then there exists $c > 0$ such that $N(\Lambda) \ge c\Lambda$ for all $\Lambda \ge 1$.*

Proof (Sketch only) We follow the notations of the previous section. Fix a non-zero smooth function h on the real line, supported in $(0, 1)$. Fix an integer $N \ge 1$. For each pair of positive integers j, k satisfying $1 \le j, k \le N$ and so that p does not divide j, we put $f_{jk}(x + iy) = h\left((N(y - Y)/Y) - k\right)\cos(2\pi jx)$, regarded as an element of $C^\infty(\mathfrak{S}(Y))_+ \subset \mathcal{L}^2_{\text{cusp},+}$. Let W be the span of f_{jk}, so an $N^2 - N\,[N/p]$ dimensional subspace of $\mathcal{L}^2_{\text{cusp},+}$. Also, let $V = \heartsuit(W)$, where we take the function g_0 entering in the definition of $\heartsuit$ to be an approximation to a δ function. Now apply the previous lemma to V. □

4.7 Interpretation via wave equation and the role of finite propagation speed

We now comment how ♡ may be understood in terms of the wave equation. Equations (4.3.2) and (4.3.3) admit a nice interpretation in terms of the automorphic wave equation

$$u_{tt} = -\Delta u + \frac{u}{4}. \tag{4.7.1}$$

A solution $u = u(x + iy, t)$ to (4.7.1) may be regarded as describing the amplitude of a wave propagating in the hyperbolic plane. The low order term of $u/4$ is natural for the hyperbolic Laplacian (see (Lax and Phillips, 1976)).

For every $t \in \mathbb{R}$ we can define a linear endomorphism U_t of

$$\mathcal{L}^2\left(SL(2,\mathbb{Z})\backslash\mathfrak{h}^2\right) \cap C^\infty\left(SL(2,\mathbb{Z})\backslash\mathfrak{h}^2\right)$$

to itself, taking a function $f(x + iy)$ to $2u(x + iy, t)$, where u is the solution to (4.7.1) with $u|_{t=0} = f$, $u_t|_{t=0} = 0$. One may show that this operator is well defined in a standard way; moreover, it is self-adjoint ("time reversal symmetry"). Formally speaking, one may write $U_t = p^{t\sqrt{\frac{1}{4}-\Delta}} + p^{-t\sqrt{\frac{1}{4}-\Delta}}$; in fact, U_t gives a rigorous meaning to the right-hand side.

4.8 Interpretation via wave equation: higher rank case

In this section, we briefly detail how the operator $p^{\sqrt{\Delta-\frac{1}{4}}} + p^{-\sqrt{\Delta-\frac{1}{4}}}$ may be viewed in terms of the wave equation. We then conclude by discussing how the considerations of this section generalize. For further details, we refer the reader to (Lindenstrauss and Venkatesh, to appear).

The *automorphic wave equation*

$$u_{tt} = -\Delta u + \frac{u}{4} \tag{4.8.1}$$

describes the propagation of waves on the hyperbolic plane. A solution

$$u = u(x + iy, t)$$

to (4.8.1) may be regarded as describing the amplitude (at time t and position $x + iy$) of a wave propagating in the hyperbolic plane. The low order term of $u/4$ is natural for the hyperbolic Laplacian (see (Lax and Phillips, 1976)).

For every $t \in \mathbb{R}$ we can define a linear endomorphism U_t of

$$\mathcal{L}^2\left(SL(2,\mathbb{Z})\backslash\mathfrak{h}^2\right) \cap \mathbb{C}^\infty\left(SL(2,\mathbb{Z})\backslash\mathfrak{h}^2\right)$$

to itself, taking a function $f(x+iy)$ to $2u(x+iy,t)$, where u is the solution to (4.8.1) with $u|_{t=0} = f$, $u_t|_{t=0} = 0$. One may show that this operator is well defined in a standard way; moreover, it is self-adjoint ("time reversal symmetry"). Formally speaking, one may write $U_t = e^{t\sqrt{\Delta-\frac{1}{4}}} + e^{-t\sqrt{\Delta-\frac{1}{4}}}$; in fact, U_t gives a rigorous meaning to the right-hand side.

Moreover, for any function $f \in \mathbb{C}^\infty(\mathfrak{h}^2)$, the value of $U_t f$ at $z \in \mathfrak{h}^2$ depends only on the values of f at points w with $d(z,w) \leq t$; this fact expresses the *finite propagation speed* of waves in the hyperbolic plane, and corresponds to the fact that the point pair invariants K we used earlier were supported within $d(z,w) \leq R$ for some R.

In this fashion the operator $p^{\sqrt{\Delta-\frac{1}{4}}} + p^{-\sqrt{\Delta-\frac{1}{4}}}$ can be regarded as the operator on functions that corresponds to propagating a wave for a time $\log p$. One can thereby rephrase our previous arguments using the wave equation.

Finally, we note that the methods of this section can be extended to the higher-rank case, e.g. existence of cusp forms on $SL(n,\mathbb{Z})\backslash SL(n,\mathbb{R})$; and moreover, a more careful analysis gives not merely the existence of cusp forms but the full Weyl law, that is to say, the correct asymptotic for the number of cusp forms of eigenvalue $\leq \Lambda$. The idea is that, again, one may construct operators like "♡" by combining Hecke operators and integral convolution operators.

To write such an operator down explicitly for $SL(3,\mathbb{Z})\backslash SL(3,\mathbb{R})$ would be rather a painful process! However, it is not too difficult to convince yourself that they do exist: in the $SL(2,\mathbb{Z})$ case, the crucial point was that both the Hecke eigenvalue $p^{it} + p^{-it}$ and the Laplacian eigenvalue $1/4 + t^2$ of the Eisenstein series $E(z, \frac{1}{2} + it)$ were controlled by just one parameter $t \in \mathbb{R}$, and so it is not too surprising that we can concoct a combination of these parameters that always vanishes. A similar phenomenon occurs for higher-rank: the Eisenstein series is controlled by too few parameters for the archimedean and Hecke eigenvalues to be completely independent. This is perhaps a bit surprising since, on $SL(3,\mathbb{Z})\backslash SL(3,\mathbb{R})$, there exist Eisenstein series indexed not merely by a complex parameter t but also by Maass forms (see Section 10.5) on $SL(2,\mathbb{Z})\backslash SL(2,\mathbb{R})$!

In the higher-rank case one uses a slightly different approach to see that the equivalent of "♡" is non-zero. Let us describe this approach in the $SL(2,\mathbb{Z})\backslash\mathfrak{h}^2$ case; the higher-rank case proceeds analogously but using higher-rank Whittaker functions. The idea is again to explicitly write down a function which ♡ does not annihilate; but we will use instead a somewhat more complicated function than before. The payoff will be that ♡ will act on it in a very simple way.

Choose $\theta, r \in \mathbb{R}$ and define for $k \geq 0$ the complex numbers

$$a_k := e^{-ik\theta} + e^{-i(k-2)\theta} + \cdots + e^{ik\theta}.$$

Put $f(z) = \sum_{k \geq 0} a_k \sqrt{y} K_{ir}(p^k y) \cos(2\pi p^k x)$. Recall that for $T \geq 1$ we defined

$$\mathfrak{S}(T) = \Sigma_{T, \frac{1}{2}} = \Gamma_\infty \backslash \left\{ z \in \mathfrak{h}^2 \,\middle|\, \Im(z) > T \right\}$$

to be the Siegel set as in Definition 1.3.1. If we restrict f to $\mathfrak{S}(T)$, we may, thereby, regard it as belonging to $\mathcal{L}^2(SL(2,\mathbb{Z})\backslash\mathfrak{h}^2)$. One verifies that, if $T' \gg T$ is sufficiently large, then for any $w \in \mathfrak{S}(T')$, we have $\heartsuit f(w) = h_0(ir)(p^{ir} + p^{-ir} - e^{i\theta} - e^{-i\theta}) f(w)$; here h_0 is the Fourier transform of the basic function g_0 that was chosen at the start of Section 4.4. In other words, for this particular function f, "high in the cusp," $\heartsuit$ actually acts on f by a *scalar*, namely $h_0(ir)(p^{ir} + p^{-ir} - e^{i\theta} - e^{-i\theta})$. In particular, we can choose r and θ so that $\heartsuit f \neq 0$.

It is possible to reproduce this behavior in any rank using higher-rank Whittaker functions, and this allows one to show that the relevant convolution operators are non-zero.

5

Maass forms and Whittaker functions for $SL(n, \mathbb{Z})$

5.1 Maass forms

Maass forms for $SL(2, \mathbb{Z})$ were introduced in Section 3.3. We want to generalize the theory to $SL(n, \mathbb{Z})$ with $n > 2$. Accordingly, we will define, for $n \geq 2$, a Maass form as a smooth complex valued cuspidal function on

$$\mathfrak{h}^n = GL(n, \mathbb{R})/(O(n, \mathbb{R}) \cdot \mathbb{R}^\times)$$

which is invariant under the discrete group $SL(n, \mathbb{Z})$ and which is also an eigenfunction of every invariant differential operator in $\mathfrak{D}^n$, the center of the universal enveloping algebra as defined in Section 2.3. A cuspidal function (or cuspform) on $\mathfrak{h}^2$ was defined by the condition that the constant term in its Fourier expansion vanishes which in turn is equivalent to the condition that $\phi(z)$ has exponential decay as $y \to \infty$. These notions are generalized in the formal Definition 5.1.3.

Harish-Chandra was the first to systematically study spaces of automorphic forms in a much more general situation than $GL(n)$. He proved (Harish-Chandra, 1959, 1966, 1968) that the space of automorphic functions of a certain type (characterized by a cuspidality condition, eigenfunction condition, and good growth) is finite dimensional. Godement (1966) explains why Maass forms on $GL(n)$ are rapidly decreasing. It was not at all clear at that time if a theory of L-functions, analogous to the $GL(2)$ theory could be developed for $GL(n)$ with $n > 2$. The first important breakthrough came in (Piatetski-Shapiro, 1975), and independently in (Shalika, 1973, 1974), where the Fourier expansion of a Maass form for $SL(n, \mathbb{Z})$ was obtained for the first time. The Fourier expansion involved Whittaker functions. In his thesis, Jacquet introduced and obtained the meromorphic continuation and functional equations of Whittaker functions on an arbitrary Chevalley group (see (Jacquet, 1967)). These papers provided the cornerstone for an arithmetic theory of L-functions in the higher-rank situation.

Recall that for $n \ge 2$, an element $z \in \mathfrak{h}^n$ takes the form $z = x \cdot y$ where

$$x = \begin{pmatrix} 1 & x_{1,2} & x_{1,3} & \cdots & & x_{1,n} \\ & 1 & x_{2,3} & \cdots & & x_{2,n} \\ & & \ddots & & & \vdots \\ & & & & 1 & x_{n-1,n} \\ & & & & & 1 \end{pmatrix}, \quad y = \begin{pmatrix} y_1 y_2 \cdots y_{n-1} & & & & \\ & y_1 y_2 \cdots y_{n-2} & & & \\ & & \ddots & & \\ & & & y_1 & \\ & & & & 1 \end{pmatrix},$$

with $x_{i,j} \in \mathbb{R}$ for $1 \le i < j \le n$ and $y_i > 0$ for $1 \le i \le n-1$.

Let $\nu = (\nu_1, \nu_2, \ldots \nu_{n-1}) \in \mathbb{C}^{n-1}$. We have shown in Section 2.4 that the function

$$I_\nu(z) = \prod_{i=1}^{n-1} \prod_{j=1}^{n-1} y_i^{b_{i,j}\nu_j} \tag{5.1.1}$$

with

$$b_{i,j} = \begin{cases} ij & \text{if } i + j \le n, \\ (n-i)(n-j) & \text{if } i + j \ge n, \end{cases}$$

is an eigenfunction of every $D \in \mathfrak{D}^n$. Let us write

$$DI_\nu(z) = \lambda_D \cdot I_\nu(z) \qquad \text{for every} D \in \mathfrak{D}^n. \tag{5.1.2}$$

The function λ_D (viewed as a function of D) is a character of $\mathfrak{D}^n$ because it satisfies

$$\lambda_{D_1 \cdot D_2} = \lambda_{D_1} \cdot \lambda_{D_2}$$

for all $D_1, D_2 \in \mathfrak{D}^n$. It is sometimes called the Harish–Chandra character.

Definition 5.1.3 *Let $n \ge 2$, and let $\nu = (\nu_1, \nu_2, \ldots \nu_{n-1}) \in \mathbb{C}^{n-1}$. A* **Maass form** *for $SL(n, \mathbb{Z})$ of type ν is a smooth function $f \in \mathcal{L}^2(SL(n, \mathbb{Z})\backslash\mathfrak{h}^n)$ which satisfies*

(1) $f(\gamma z) = f(z)$, *for all* $\gamma \in SL(n, \mathbb{Z})$, $z \in \mathfrak{h}^n$,
(2) $Df(z) = \lambda_D f(z)$, *for all* $D \in \mathfrak{D}^n$, *with* λ_D *given by (5.1.2),*
(3) $\displaystyle\int_{(SL(n,\mathbb{Z})\cap U)\backslash U} f(uz)\, du = 0,$

for all upper triangular groups U of the form

$$U = \left\{ \begin{pmatrix} I_{r_1} & & & \\ & I_{r_2} & & * \\ & & \ddots & \\ & & & I_{r_b} \end{pmatrix} \right\},$$

with $r_1 + r_2 + \cdots + r_b = n$. *Here* I_r *denotes the* $r \times r$ *identity matrix, and* $*$ *denotes arbitrary real entries.*

5.2 Whittaker functions associated to Maass forms

For $n \geq 2$, let $U_n(\mathbb{R})$ denote the group of upper triangular matrices with *1s* on the diagonal and real entries above the diagonal. Then every $u \in U_n(\mathbb{R})$ is of the form

$$u = \begin{pmatrix} 1 & u_{1,2} & u_{1,3} & \cdots & u_{1,n} \\ & 1 & u_{2,3} & \cdots & u_{2,n} \\ & & \ddots & & \vdots \\ & & & 1 & u_{n-1,n} \\ & & & & 1 \end{pmatrix}, \tag{5.2.1}$$

with $u_{i,j} \in \mathbb{R}$ for $1 \leq i < j \leq n$. Similarly, we define $U_n(\mathbb{Z})$ with entries $u_{i,j} \in \mathbb{Z}$ for $1 \leq i < j \leq n$.

If $m = (m_1, m_2, \ldots, m_{n-1}) \in \mathbb{Z}^{n-1}$, then the function $\psi_m : U_n(\mathbb{R}) \to \mathbb{C}^\times$ defined by

$$\psi_m(u) = e^{2\pi i (m_1 u_{1,2} + m_2 u_{2,3} + \cdots + m_{n-1} u_{n-1,n})}, \qquad \text{(with } u \in U_n(\mathbb{R})\text{)}$$

is a character of $U_n(\mathbb{R})$. This means that

$$\psi_m(u \cdot v) = \psi_m(u)\psi_m(v) \tag{5.2.2}$$

for all $u, v \in U_n(\mathbb{R})$. This can be quickly verified in the case $n = 3$ because

$$\begin{pmatrix} 1 & u_{1,2} & u_{1,3} \\ 0 & 1 & u_{2,3} \\ 0 & 0 & 1 \end{pmatrix} \cdot \begin{pmatrix} 1 & v_{1,2} & v_{1,3} \\ 0 & 1 & v_{2,3} \\ 0 & 0 & 1 \end{pmatrix} = \begin{pmatrix} 1 & u_{1,2} + v_{1,2} & * \\ 0 & 1 & u_{2,3} + v_{2,3} \\ 0 & 0 & 1 \end{pmatrix},$$

and it is easy to see that (5.2.2) holds in general.

For $n \geq 2$, let ϕ be a Maass form for $SL(n, \mathbb{Z})$ of type $\nu = (\nu_1, \ldots, \nu_{n-1}) \in \mathbb{C}^{n-1}$. By analogy with the Fourier expansion techniques introduced in Section 3.5, it is natural to introduce the function $\tilde{\phi}_m(z)$ defined as follows:

$$\tilde{\phi}_m(z) := \int_0^1 \cdots \int_0^1 \phi(u \cdot z)\, \overline{\psi_m(u)} \prod_{1 \leq i < j \leq n} du_{i,j}. \tag{5.2.3}$$

One might reasonably expect that $\tilde{\phi}_m(z)$ is a Fourier coefficient of ϕ and that ϕ might be recoverable as a sum of such Fourier coefficients. Unfortunately, the fact that $U_n(\mathbb{R})$ is a non–abelian group (for $n > 2$) complicates the issue enormously, and it is necessary to go through various contortions in order to obtain a useful Fourier theory. We shall study this issue carefully in the next section. For the moment, we focus on the integral (5.2.3).

Proposition 5.2.4 *For $n \geq 2$ and $\nu = (\nu_1, \nu_2, \ldots, \nu_{n-1}) \in \mathbb{C}^{n-1}$, let ϕ be a Maass form of type ν for $SL(n, \mathbb{Z})$. Let $m = (m_1, \ldots, m_{n-1}) \in \mathbb{Z}^{n-1}$, and let ψ_m be an additive character as in (5.2.2). Then the function $\tilde{\phi}_m(z)$ defined in (5.2.3) satisfies the following conditions:*

(1) $\tilde{\phi}_m(u \cdot z) = \psi_m(u) \cdot \tilde{\phi}_m(z)$ *(for all $u \in U_n(\mathbb{R})$),*
(2) $D\tilde{\phi}_m = \lambda_D \tilde{\phi}_m$, *(for all $D \in \mathfrak{D}^n$),*
(3) $\displaystyle\int\limits_{\Sigma_{\frac{\sqrt{3}}{2}, \frac{1}{2}}} |\tilde{\phi}_m(z)|^2 d^*z < \infty,$

where $\Sigma_{\frac{\sqrt{3}}{2}, \frac{1}{2}}$ denotes the Siegel set as in Definition 1.3.1, and d^*z is the left invariant measure given in Proposition 1.5.3.

Remark Any smooth function: $\mathfrak{h}^n \to \mathbb{C}$ which satisfies conditions (1), (2), (3) of Proposition 5.2.4 will be called a Whittaker function. A more formal definition will be given in Section 5.4.

Proof First of all, the integral on the right-hand side of (5.2.3) is an integral over $U_n(\mathbb{Z})\backslash U_n(\mathbb{R})$. Since both ϕ and ψ_m are invariant under $U_n(\mathbb{Z})$, the integral is independent of the choice of fundamental domain for $U_n(\mathbb{Z})\backslash U_n(\mathbb{R})$.

Every $z \in \mathfrak{h}^n$ can be written in the form $z = x \cdot y$, as in the beginning of Section 5.1. In the integral (5.2.3), we make the change of variables $u \to u \cdot x^{-1}$. It follows that

$$\tilde{\phi}_m(z) = \int_0^1 \cdots \int_0^1 \phi(u \cdot y)\, \overline{\psi_m(u \cdot x^{-1})} \prod_{1 \leq i < j \leq n} du_{i,j}.$$

But (5.2.2) implies that $\overline{\psi_m(u \cdot x^{-1})} = \overline{\psi_m(u)} \cdot \overline{\psi_m(x^{-1})} = \overline{\psi_m(u)} \cdot \psi_m(x)$. This proves that for $u \in U_n(\mathbb{R})$, $\tilde{\phi}(u \cdot z) = \psi_m(u \cdot x)\tilde{\phi}(y) = \psi_m(u)\tilde{\phi}(z)$. The second part is an immediate consequence of the fact that ϕ is an eigenfunction of every $D \in \mathfrak{D}^n$ with eigenvalue λ_D.

To prove (3), we use the Cauchy–Schwartz inequality and the fact that ϕ is in $\mathcal{L}^2$ to deduce that

$$\int\limits_{\Sigma_{\frac{\sqrt{3}}{2}, \frac{1}{2}}} |\tilde{\phi}_m(z)|^2 \, d^*z < \int_0^1 \cdots \int_0^1 \int\limits_{\Sigma_{\frac{\sqrt{3}}{2}, \frac{1}{2}}} |\phi(u \cdot z)|^2 \, d^*z \prod_{1 \leq i < j \leq n} du_{i,j}$$

$$< \int\limits_{\Sigma_{\frac{\sqrt{3}}{2}, \frac{1}{2}}} |\phi(z)|^2 \, d^*z \; < \; \infty.$$

□

5.3 Fourier expansions on $SL(n,\mathbb{Z})\backslash\mathfrak{h}^n$

The classical Fourier expansion theorem states that every smooth periodic function ϕ on $\mathbb{Z}\backslash\mathbb{R}$ has a Fourier expansion

$$\phi(x) = \sum_{m\in\mathbb{Z}} \tilde{\phi}_m(x), \tag{5.3.1}$$

where

$$\tilde{\phi}_m(x) = \int_0^1 \phi(u+x)e^{-2\pi imu}\,du.$$

We seek to generalize (5.3.1) to smooth automorphic functions on $SL(n,\mathbb{Z})\backslash\mathfrak{h}^n$.

Theorem 5.3.2 *For $n \geq 2$, let U_n denote the group of $n \times n$ upper triangular matrices with 1s on the diagonal as in Section 5.2. Let ϕ be a Maass form for $SL(n,\mathbb{Z})$. Then for all $z \in SL(n,\mathbb{Z})\backslash\mathfrak{h}^n$*

$$\phi(z) = \sum_{\gamma\,\in\, U_{n-1}(\mathbb{Z})\backslash SL(n-1,\mathbb{Z})} \sum_{m_1\neq 0} \sum_{m_2=1}^{\infty} \cdots \sum_{m_{n-1}=1}^{\infty} \tilde{\phi}_{(m_1,\ldots,m_{n-1})}\left(\begin{pmatrix}\gamma & \\ & 1\end{pmatrix} z\right),$$

where the sum is independent of the choice of coset representatives γ and

$$\tilde{\phi}_{(m_1,\ldots,m_{n-1})}(z) := \int_0^1 \cdots \int_0^1 \phi(u\cdot z)\, e^{-2\pi i\left(m_1u_{1,2}+m_2u_{2,3}+\cdots+m_{n-1}u_{n-1,n}\right)}\, d^*u,$$

with $u \in U_n(\mathbb{R})$ given by (5.2.1) and $d^*u = \prod_{1\leq i<j\leq n} du_{i,j}$.

If ϕ satisfies conditions (1), (2), but does not satisfy condition (3) of Definition 5.1.3, then the Fourier expansion takes the form

$$\phi(z) = \sum_{\gamma\in U_{n-1}(\mathbb{Z})\backslash SL(n-1,\mathbb{Z})} \sum_{m_1=-\infty}^{\infty} \sum_{m_2=0}^{\infty} \cdots \sum_{m_{n-1}=0}^{\infty} \tilde{\phi}_{(m_1,\ldots,m_{n-1})}\left(\begin{pmatrix}\gamma & \\ & 1\end{pmatrix} z\right).$$

The proof of Theorem 5.3.2 makes use of an elementary lemma in group theory which we shall straightaway state and prove.

Lemma 5.3.3 *Let $C \subseteq B \subseteq A$ be groups. Let $f : C\backslash A \to \mathbb{C}$ be any function such that the sum, $\sum_{\gamma\in C\backslash A} f(\gamma)$, converges absolutely. Then*

$$\sum_{\gamma\in C\backslash A} f(\gamma) = \sum_{\delta'\in C\backslash B}\ \sum_{\delta\in B\backslash A} f(\delta'\delta).$$

Proof It is clear that if $\delta \in B\backslash A$ and $\delta' \in C\backslash B$ then $\delta'\delta \in C\backslash A$. On the other hand, every $\gamma \in C\backslash A$ can be written in the form $\gamma = ca$ for $c \in C, a \in A$. If we now set $\delta' = c \cdot 1 \in C\backslash B$ and $\delta = 1 \cdot a \in B\backslash A$, then we have expressed $\gamma = \delta'\delta$. □

Proof of Theorem 5.3.2 Since ϕ is automorphic for $SL(n, \mathbb{Z})$, we have

$$\phi\left(\begin{pmatrix} 1 & & & & m_1 \\ & 1 & & & m_2 \\ & & \ddots & & \vdots \\ & & & 1 & m_{n-1} \\ & & & & 1 \end{pmatrix} \cdot z\right) = \phi(z)$$

for all $m_1, m_2, \ldots, m_{n-1} \in \mathbb{Z}$. It then follows from classical one–dimensional Fourier theory that

$$\phi(z) = \sum_{m_1,\ldots,m_{n-1} \in \mathbb{Z}} \int_0^1 \cdots \int_0^1 \phi\left(\begin{pmatrix} 1 & & & & v_1 \\ & 1 & & & v_2 \\ & & \ddots & & \vdots \\ & & & 1 & v_{n-1} \\ & & & & 1 \end{pmatrix} \cdot z\right)$$
$$\times e^{-2\pi i(m_1 v_1 + \cdots + m_{n-1} v_{n-1})}\, dv_1 \cdots dv_{n-1}. \tag{5.3.4}$$

Here, we are simply using the fact that the matrices

$$\begin{pmatrix} 1 & & & & 1 \\ & 1 & & & 0 \\ & & \ddots & & \vdots \\ & & & 1 & 0 \\ & & & & 1 \end{pmatrix}, \begin{pmatrix} 1 & & & & 0 \\ & 1 & & & 1 \\ & & \ddots & & \vdots \\ & & & 1 & 0 \\ & & & & 1 \end{pmatrix}, \ldots, \begin{pmatrix} 1 & & & & 0 \\ & 1 & & & 0 \\ & & \ddots & & \vdots \\ & & & 1 & 1 \\ & & & & 1 \end{pmatrix}$$

commute with each other and generate the abelian group of all matrices of the form

$$\left\{\begin{pmatrix} 1 & & & & m_1 \\ & 1 & & & m_2 \\ & & \ddots & & \vdots \\ & & & 1 & m_{n-1} \\ & & & & 1 \end{pmatrix} \middle|\ m_1, \ldots, m_{n-1} \in \mathbb{Z}\right\}.$$

□

We now rewrite (5.3.4) with a more compact notation. For

$$m = (m_1, \ldots, m_{n-1}) \in \mathbb{Z}^{n-1}, \quad v = \begin{pmatrix} 1 & & & & v_1 \\ & 1 & & & v_2 \\ & & \ddots & & \vdots \\ & & & 1 & v_{n-1} \\ & & & & 1 \end{pmatrix}, \quad d^*v = \prod_{i=1}^{n-1} dv_i,$$

define

$$\hat{\phi}_m(z) := \int_0^1 \cdots \int_0^1 \phi(vz) e^{-2\pi i \langle v, m\rangle}\, d^*v,$$

where $\langle v, m\rangle = \sum_{i=1}^{n-1} v_i m_i$. Note the difference between $\hat{\phi}$ and $\tilde{\phi}$, for example, $\tilde{\phi}$ involves integration with repect to all the variables $u_{i,j}$ with $1 \le i < j \le n-1$. With this notation, (5.3.4) becomes

$$\phi(z) = \sum_{m \in \mathbb{Z}^{n-1}} \hat{\phi}_m(z). \tag{5.3.5}$$

The Fourier expansion (5.3.5) does not make use of the fact that ϕ is automorphic for all $SL(n,\mathbb{Z})$. To proceed further, we need the following lemma.

Lemma 5.3.6 *Let $n > 2$. Fix an integer $M \neq 0$, and let $\gamma \in SL(n-1,\mathbb{Z})$. Then*

$$\hat{\phi}_{Me_{n-1}\gamma}(z) = \hat{\phi}_{(0,\ldots,0,M)}\left(\begin{pmatrix} \gamma & 0 \\ 0 & 1 \end{pmatrix} \cdot z\right),$$

where $e_{n-1} = (0, \ldots, 0, 1)$ lies in $\mathbb{Z}^{n-1}$.

Proof Let

$$\gamma = \begin{pmatrix} a_{1,1} & a_{1,2} & \cdots & a_{1,n-1} \\ \vdots & \vdots & & \vdots \\ a_{n-2,1} & a_{n-2,2} & \cdots & a_{n-2,n-1} \\ \gamma_1 & \gamma_2 & \cdots & \gamma_{n-1} \end{pmatrix},$$

and

$$v = \begin{pmatrix} 1 & & & & v_1 \\ & 1 & & & v_2 \\ & & \ddots & & \vdots \\ & & & 1 & v_{n-1} \\ & & & & 1 \end{pmatrix}, \qquad v' = \begin{pmatrix} 1 & & & & v'_1 \\ & 1 & & & v'_2 \\ & & \ddots & & \vdots \\ & & & 1 & v'_{n-1} \\ & & & & 1 \end{pmatrix}.$$

Then we have the identity

$$\begin{pmatrix} \gamma & 0 \\ 0 & 1 \end{pmatrix} \cdot v = v' \cdot \begin{pmatrix} \gamma & 0 \\ 0 & 1 \end{pmatrix} \tag{5.3.7}$$

where

$$\begin{aligned} v_1' &= a_{1,1}v_1 + \cdots + a_{1,n-1}v_{n-1}, \\ v_2' &= a_{2,1}v_1 + \cdots + a_{2,n-1}v_{n-1}, \\ &\vdots \\ v_{n-2}' &= a_{n-2,1}v_1 + \cdots + a_{n-2,n-1}v_{n-1} \\ v_{n-1}' &= \gamma_1 v_1 + \cdots + \gamma_{n-1}v_{n-1}. \end{aligned}$$

Now, since $\gamma \in SL(n-1, \mathbb{Z})$, we have $\phi(vz) = \phi\left(\begin{pmatrix} \gamma & 0 \\ 0 & 1 \end{pmatrix} vz\right)$. It then follows from (5.3.7) and a simple change of variables that

$$\begin{aligned} \hat{\phi}_{(M\gamma_1,\ldots,M\gamma_{n-1})}(z) &= \hat{\phi}_{Me_{n-1}\gamma}(z) \\ &= \int_0^1 \cdots \int_0^1 \phi\left(\begin{pmatrix} \gamma & 0 \\ 0 & 1 \end{pmatrix} vz\right) e^{-2\pi i M(\gamma_1 v_1 + \cdots + \gamma_{n-1}v_{n-1})}\, dv_1 \cdots dv_{n-1} \\ &= \int_0^1 \cdots \int_0^1 \phi\left(v'\begin{pmatrix} \gamma & 0 \\ 0 & 1 \end{pmatrix} z\right) e^{-2\pi i M(\gamma_1 v_1 + \cdots + \gamma_{n-1}v_{n-1})}\, dv_1 \cdots dv_{n-1} \\ &= \int_0^1 \cdots \int_0^1 \phi\left(v'\begin{pmatrix} \gamma & 0 \\ 0 & 1 \end{pmatrix} z\right) e^{-2\pi i M\cdot v_{n-1}'}\, dv_1' \cdots dv_{n-1}' \\ &= \hat{\phi}_{(0,\ldots,0,M)}\left(\begin{pmatrix} \gamma & 0 \\ 0 & 1 \end{pmatrix} \cdot z\right). \end{aligned}$$

□

For $n > 2$, the group $SL(n-1, \mathbb{Z})$ acts on $\mathbb{Z}^{n-1}$ with two orbits:

$$\{0\}, \quad \mathbb{Z}^{n-1} - \{0\} = \mathbb{Z}^+ \cdot e_{n-1} \cdot SL(n-1, \mathbb{Z}),$$

where $\mathbb{Z}^+$ denotes the positive integers. The second orbit above is a consequence of the fact that every non-zero $m \in \mathbb{Z}^{n-1}$ can be expressed in the form

$$\begin{aligned} m = (m_1, \ldots, m_{n-1}) &= M \cdot (\gamma_1, \ldots, \gamma_{n-1}) \\ &= M \cdot e_{n-1} \cdot \begin{pmatrix} & & * & \\ \gamma_1 & \gamma_2 & \cdots & \gamma_{n-1} \end{pmatrix}, \end{aligned}$$

where $M = \gcd(m_1, \ldots, m_{n-1})$, $M > 0$, and $m_j = M\gamma_j$ (for $j = 1, 2, \ldots, n-1$). The stabilizer of e_{n-1} in $SL(n-1, \mathbb{Z})$ (under right multiplication) is $P_{n-1}(\mathbb{Z})$ where

$$P_{n-1}(\mathbb{Z}) = \left\{ \begin{pmatrix} & & * & & \\ 0 & 0 & \cdots & 0 & 1 \end{pmatrix} \right\},$$

and $*$ denotes arbitrary integer entries.

Remark 5.3.8 The above argument breaks down when $n = 2$. In this case, the orbit consists of all integers $M \neq 0$.

It now follows from this discussion and Lemma 5.3.6 that we may rewrite (5.3.5) as follows:

$$\begin{aligned}\phi(z) &= \hat{\phi}_{(0,\dots,0)}(z) + \sum_{M=1}^{\infty} \sum_{\gamma \in P_{n-1}(\mathbb{Z})\backslash SL(n-1,\mathbb{Z})} \hat{\phi}_{M\cdot e_{n-1}\cdot\gamma}(z) \\ &= \hat{\phi}_{(0,\dots,0)}(z) + \sum_{M=1}^{\infty} \sum_{\gamma \in P_{n-1}(\mathbb{Z})\backslash SL(n-1,\mathbb{Z})} \hat{\phi}_{(0,\dots,0,M)}\left(\begin{pmatrix} \gamma & 0 \\ 0 & 1 \end{pmatrix} \cdot z\right).\end{aligned} \tag{5.3.9}$$

The fact that ϕ is a Maass form implies that $\hat{\phi}_{(0,\dots,0)}(z) = 0$. Replacing M by m_{n-1}, and setting $P_{n-1} = P_{n-1}(\mathbb{Z})$, $SL_{n-1} = SL(n-1,\mathbb{Z})$, we may, therefore, rewrite (5.3.9) in the form

$$\phi(z) = \sum_{m_{n-1}=1}^{\infty} \sum_{\gamma \in P_{n-1}\backslash SL_{n-1}} \int_0^1 \cdots \int_0^1 \phi\left(u \cdot \begin{pmatrix} \gamma & \\ & 1 \end{pmatrix} \cdot z\right) \times e^{-2\pi i m_{n-1}\cdot u_{n-1,n}}\, d^*u, \tag{5.3.10}$$

where

$$u = \begin{pmatrix} 1 & & & & u_{1,n} \\ & 1 & & & u_{2,n} \\ & & \ddots & & \vdots \\ & & & 1 & u_{n-1,n} \\ & & & & 1 \end{pmatrix}, \qquad d^*u = \prod_{j=1}^{n-1} du_{j,n}.$$

Lemma 5.3.11 *The function*

$$\hat{\phi}_{(0,\dots,0,M)}(z) = \int_0^1 \cdots \int_0^1 \phi(v \cdot z) e^{-2\pi i M v_{n-1}}\, d^*v$$

where

$$v = \begin{pmatrix} 1 & & & & v_1 \\ & 1 & & & v_2 \\ & & \ddots & & \vdots \\ & & & 1 & v_{n-1} \\ & & & & 1 \end{pmatrix}, \qquad d^*v = \prod_{i=1}^{n-1} dv_i,$$

is invariant under left multiplication by matrices of the form

$$m = \begin{pmatrix} m_{1,1} & m_{1,2} & \cdots & m_{1,n-1} & 0 \\ m_{2,1} & m_{2,2} & \cdots & m_{2,n-1} & 0 \\ \vdots & \vdots & & \vdots & \vdots \\ m_{n-2,1} & m_{n-2,2} & \cdots & m_{n-2,n-1} & 0 \\ 0 & 0 & \cdots & 1 & 0 \\ 0 & 0 & \cdots & 0 & 1 \end{pmatrix} \in SL(n, \mathbb{Z}).$$

Proof We have

$$\begin{aligned} \hat{\phi}_{(0,\ldots,0,M)}(m \cdot z) &= \int_0^1 \cdots \int_0^1 \phi(v \cdot m \cdot z) e^{-2\pi i M v_{n-1}} \, d^*v \\ &= \int_0^1 \cdots \int_0^1 \phi(m \cdot v' \cdot z) e^{-2\pi i M v_{n-1}} \, d^*v \\ &= \int_0^1 \cdots \int_0^1 \phi(v' \cdot z) e^{-2\pi i M v_{n-1}} \, d^*v, \end{aligned}$$

where

$$v' = \begin{pmatrix} 1 & & & & v'_1 \\ & 1 & & & v'_2 \\ & & \ddots & & \vdots \\ & & & 1 & v'_{n-1} \\ & & & & 1 \end{pmatrix}$$

is chosen so that $v \cdot m = m \cdot v'$. A simple computation shows that we may take

$$\begin{aligned} v'_1 &= m_{1,1}v_1 + m_{1,2}v_2 + \cdots + m_{1,n-1}v_{n-1} \\ v'_2 &= m_{2,1}v_1 + m_{2,2}v_2 + \cdots + m_{2,n-1}v_{n-1} \\ &\vdots \\ v'_{n-2} &= m_{n-2,1}v_1 + m_{n-2,2}v_2 + \cdots + m_{n-2,n-1}v_{n-1} \\ v'_{n-1} &= v_{n-1}. \end{aligned}$$

Finally, if we make the change of variables $v' \to v$, it follows from the above discussion that

$$\begin{aligned}
\hat{\phi}_{(0,\ldots,0,M)}(m \cdot z) &= \int_0^1 \cdots \int_0^1 \phi(v' \cdot z) e^{-2\pi i M v'_{n-1}} \, d^*v \\
&= \int_0^1 \cdots \int_0^1 \phi(v' \cdot z) e^{-2\pi i M v'_{n-1}} \, d^*v' \\
&= \hat{\phi}_{(0,\ldots,0,M)}(z),
\end{aligned}$$

because the Jacobian of the transformation is 1. □

We remark here that in the derivation of (5.3.10), we only used the left invariance of $\phi(z)$ with respect to $P_{n-1}(\mathbb{Z})$. In view of Lemma 5.3.11, we may then reiterate all previous arguments and obtain, instead of (5.3.10), the more general form

$$\begin{aligned}
\phi(z) = \sum_{m_{n-2}=1}^{\infty} \sum_{m_{n-1}=1}^{\infty} \sum_{\gamma_{n-2} \in P_{n-2} \backslash SL_{n-2}} \sum_{\gamma_{n-1} \in P_{n-1} \backslash SL_{n-1}} \int_0^1 \cdots \int_0^1 \\
\times \phi\left(u \cdot \begin{pmatrix} \gamma_{n-2} & & \\ & 1 & \\ & & 1 \end{pmatrix} \cdot \begin{pmatrix} \gamma_{n-1} & \\ & 1 \end{pmatrix} \cdot z \right) \\
\times e^{-2\pi i \left[m_{n-2} u_{n-2,n-1} + m_{n-1} u_{n-1,n} \right]} \, d^*u,
\end{aligned} \tag{5.3.12}$$

where

$$u = \begin{pmatrix} 1 & & & & u_{1,n} \\ & 1 & & & u_{2,n} \\ & & \ddots & & \vdots \\ & & & 1 & u_{n-1,n} \\ & & & & 1 \end{pmatrix} \cdot \begin{pmatrix} 1 & & & u_{1,n-1} & 0 \\ & 1 & & u_{2,n-1} & 0 \\ & & \ddots & \vdots & \vdots \\ & & & u_{n-2,n-1} & 0 \\ & & & 1 & 0 \\ & & & & 1 \end{pmatrix}$$

$$= \begin{pmatrix} 1 & & & u_{1,n-1} & u_{1,n} \\ & 1 & & u_{2,n-1} & u_{2,n} \\ & & \ddots & \vdots & \vdots \\ & & & u_{n-2,n-1} & u_{n-2,n} \\ & & & 1 & u_{n-1,n} \\ & & & & 1 \end{pmatrix},$$

and

$$d^*u = \prod_{1 \le i \le n-2} du_{i,n-1} \cdot \prod_{1 \le j \le n-1} du_{j,n}.$$

Note that by Remark 5.3.8, the sum over m_{n-2} in formula (5.3.12) will range over all $m_{n-2} \neq 0$ when $n = 3$.

Lemma 5.3.13 *For $n \geq 2$, define the following subgroups of $SL_n = SL(n, \mathbb{Z})$:*

$$\tilde{P}_{n,1} = \left\{\begin{pmatrix} * & \cdots & * & * \\ \vdots & \cdots & \vdots & \vdots \\ * & \cdots & * & * \\ 0 & \cdots & 0 & 1 \end{pmatrix}\right\} = P_n(\mathbb{Z}), \quad \tilde{P}_{n,2} = \left\{\begin{pmatrix} * & \cdots & * & * & * \\ \vdots & \cdots & \vdots & \vdots & \vdots \\ * & \cdots & * & * & * \\ 0 & \cdots & 0 & 1 & * \\ 0 & \cdots & 0 & 0 & 1 \end{pmatrix}\right\},$$

$$\tilde{P}_{n,3} = \left\{\begin{pmatrix} * & \cdots & * & * & * \\ \vdots & \cdots & \vdots & \vdots & \vdots \\ * & \cdots & * & * & * \\ 0 & \cdots & 1 & * & * \\ 0 & \cdots & 0 & 1 & * \\ 0 & \cdots & 0 & 0 & 1 \end{pmatrix}\right\}, \quad \ldots, \quad \tilde{P}_{n,n} = \left\{\begin{pmatrix} 1 & & & * \\ & 1 & & \\ & & \ddots & \\ & & & 1 \end{pmatrix}\right\}.$$

Then for $r = 1, 2, \ldots, n-1$, we have

$$P_{n-r}\backslash SL_{n-r} \cong \tilde{P}_{n,r+1}\backslash\tilde{P}_{n,r},$$
$$P_{n-r} \cdot \gamma \mapsto \tilde{P}_{n,r+1} \cdot \begin{pmatrix} \gamma & \\ & I_r \end{pmatrix} \qquad (\gamma \in SL_{n-r}),$$

where I_r denotes the $r \times r$ identity matrix.

Proof We may write

$$\tilde{P}_{n,r} = \left\{\begin{pmatrix} SL_{n-r} & \begin{matrix} * & * & \cdots & * \\ \vdots & \vdots & \cdots & \vdots \\ * & * & \cdots & * \end{matrix} \\ & \begin{matrix} 1 & * & \cdots & * \\ & \ddots & & \vdots \\ & & 1 & * \\ & & & 1 \end{matrix} \end{pmatrix}\right\},$$

$$\tilde{P}_{n,r+1} = \left\{\begin{pmatrix} P_{n-r} & \begin{matrix} * & * & \cdots & * \\ \vdots & \vdots & \cdots & \vdots \\ * & * & \cdots & * \end{matrix} \\ & \begin{matrix} 1 & * & \cdots & * \\ & \ddots & & \vdots \\ & & 1 & * \\ & & & 1 \end{matrix} \end{pmatrix}\right\}.$$

The lemma follows after one notes that a set of coset representatives for the quotient $\tilde{P}_{n,r+1}\backslash\tilde{P}_{n,r}$ is given by

$$\bigcup_{\gamma\in P_{n-r}\backslash SL_{n-r}} \begin{pmatrix} \gamma & \\ & I_r \end{pmatrix}.$$

□

We now apply Lemma 5.3.13 (in the form $P_{n-1-r}\backslash SL_{n-1-r}$ with $r=1$) to the inner sum

$$\sum_{\gamma_{n-2}\in P_{n-2}\backslash SL_{n-2}}$$

in (5.3.12). We can replace $\gamma_{n-2}\in P_{n-2}\backslash SL_{n-2}$ by $\gamma'_{n-2}\in \tilde{P}_{n-1,2}\backslash P_{n-1}$, where

$$\begin{pmatrix} \gamma_{n-2} & & \\ & 1 & \\ & & 1 \end{pmatrix} = \begin{pmatrix} \gamma'_{n-2} & \\ & 1 \end{pmatrix},$$

and where $\tilde{P}_{n-1,2}$ is defined as in Lemma 5.3.13. If we then apply Lemma 5.3.3 to the sums

$$\sum_{\gamma'_{n-2}\in \tilde{P}_{n-1,2}\backslash P_{n-1}} \quad \sum_{\gamma_{n-1}\in P_{n-1}\backslash SL_{n-1}},$$

it follows that

$$\phi(z) = \sum_{m_{n-2}=1}^{\infty}\sum_{m_{n-1}=1}^{\infty}\sum_{\gamma\in\tilde{P}_{n-1,2}\backslash SL_{n-1}} \int_0^1\cdots\int_0^1 \phi\left(u\cdot\begin{pmatrix} \gamma & \\ & 1\end{pmatrix}\cdot z\right)$$
$$\times\, e^{-2\pi i\left[m_{n-2}u_{n-2,n-1}+m_{n-1}u_{n-1,n}\right]}\, d^*u, \quad (5.3.14)$$

where

$$\tilde{P}_{n-1,2} = \left\{\begin{pmatrix} * & * & \cdots & * & * \\ \vdots & \vdots & \cdots & \vdots & \vdots \\ * & * & \cdots & * & * \\ & & & 1 & * \\ & & & & 1 \end{pmatrix}\right\} \subset SL(n-1,\mathbb{Z}),$$

and

$$u = \begin{pmatrix} 1 & & & u_{1,n-1} & u_{1,n} \\ & 1 & & u_{2,n-1} & u_{2,n} \\ & & \ddots & \vdots & \vdots \\ & & & u_{n-2,n-1} & u_{n-2,n} \\ & & & 1 & u_{n-1,n} \\ & & & & 1 \end{pmatrix},$$

$$d^*u = \prod_{1\le i\le n-2} du_{i,n-1} \cdot \prod_{1\le j\le n-1} du_{j,n}.$$

All steps previously taken can be iterated. For example, after one more iteration equation (5.3.14) becomes

$$\phi(z) = \sum_{m_{n-3}=1}^{\infty} \sum_{m_{n-2}=1}^{\infty} \sum_{m_{n-1}=1}^{\infty} \sum_{\gamma\in\tilde{P}_{n-1,3}\backslash SL_{n-1}} \int_0^1 \cdots \int_0^1 \phi\left(u\cdot\begin{pmatrix}\gamma & \\ & 1\end{pmatrix}\cdot z\right)$$
$$\times\, e^{-2\pi i\left[m_{n-3}u_{n-3,n-2}+m_{n-2}u_{n-2,n-1}+m_{n-1}u_{n-1,n}\right]}\, d^*u, \quad (5.3.15)$$

where

$$\tilde{P}_{n-1,3} = \left\{\begin{pmatrix} * & * & \cdots & * & * \\ \vdots & \vdots & \cdots & \vdots & \vdots \\ * & * & \cdots & * & * \\ & & 1 & * & * \\ & & & 1 & * \\ & & & & 1 \end{pmatrix}\right\} \subset SL(n-1,\mathbb{Z}),$$

and

$$u = \begin{pmatrix} 1 & & & & u_{1,n-2} & u_{1,n-1} & u_{1,n} \\ & 1 & & & u_{2,n-2} & u_{2,n-1} & u_{2,n} \\ & & \ddots & & \vdots & \vdots & \vdots \\ & & & 1 & u_{n-3,n-2} & u_{n-3,n-1} & u_{n-3,n} \\ & & & & 1 & u_{n-2,n-1} & u_{n-2,n} \\ & & & & & 1 & u_{n-1,n} \\ & & & & & & 1 \end{pmatrix},$$

$$d^*u = \prod_{1\le i\le n-3} du_{i,n-2} \cdot \prod_{1\le j\le n-2} du_{j,n-1} \cdot \prod_{1\le k\le n-1} du_{k,n}.$$

Theorem 5.3.2 follows from (5.3.15) after continuing this process inductively for $n-2$ steps, and taking into account Remark 5.3.8. □

5.4 Whittaker functions for $SL(n, \mathbb{R})$

For $n \geq 2$, let $\nu = (\nu_1, \nu_2, \ldots, \nu_{n-1}) \in \mathbb{C}^{n-1}$. We have repeatedly used the fact that the function $I_\nu : \mathfrak{h}^n \to \mathbb{C}$ (see Section 2.4) given in (5.1.1), i.e.,

$$I_\nu(z) = \prod_{i=1}^{n-1} \prod_{j=1}^{n-1} y_i^{b_{i,j}\nu_j}, \qquad (\text{with } z \in \mathfrak{h}^n)$$

and

$$b_{i,j} = \begin{cases} ij & \text{if } i + j \leq n, \\ (n-i)(n-j) & \text{if } i + j \geq n, \end{cases}$$

is an eigenfunction of every $SL(n, \mathbb{R})$–invariant differential operator in $\mathfrak{D}^n$. These are not the only possible eigenfunctions, however. For example, we have shown in Section 3.4 that the functions

$$y^\nu, \quad y^{1-\nu}, \quad \sqrt{y}\, K_{\nu-\frac{1}{2}}(2\pi|m|y)e^{2\pi i m x}, \quad \sqrt{y}\, I_{\nu-\frac{1}{2}}(2\pi|m|y)e^{2\pi i m x},$$

(with $m \in \mathbb{Z}$, $m \neq 0$) are all eigenfunctions of $\Delta = -y^2\left(\frac{\partial^2}{\partial x^2} + \frac{\partial^2}{\partial y^2}\right)$ with eigenvalue $\nu(1-\nu)$. Of these four functions, only

$$\sqrt{y}\, K_{\nu-\frac{1}{2}}(2\pi|m|y)e^{2\pi i m x},$$

has good growth properties (exponential decay as $y \to \infty$) and appears in the Fourier expansion of Maass forms. This is the multiplicity one theorem of Section 3.4. We seek to generalize these concepts to the group $SL(n, \mathbb{Z})$ with $n > 2$.

For $n \geq 2$, let $U_n(\mathbb{R})$ denote the group of upper triangular matrices with 1s on the diagonal. Fix $\psi : U_n(\mathbb{R}) \to \mathbb{C}$ to be a character of $U_n(\mathbb{R})$ which, by definition, satisfies the identity

$$\psi(u \cdot v) = \psi(u)\psi(v)$$

for all $u, v \in U_n(\mathbb{R})$.

Definition 5.4.1 *Let $n \geq 2$. An $SL(n, \mathbb{Z})$–**Whittaker function** of type*

$$\nu = (\nu_1, \nu_2, \ldots, \nu_{n-1}) \in \mathbb{C}^{n-1},$$

associated to a character ψ of $U_n(\mathbb{R})$, is a smooth function $W : \mathfrak{h}^n \to \mathbb{C}$ which satisfies the following conditions:

(1) $W(uz) = \psi(u)W(z)$ *(for all $u \in U_n(\mathbb{R})$, $z \in \mathfrak{h}^n$),*
(2) $DW(z) = \lambda_D W(z)$ *(for all $D \in \mathfrak{D}^n$, $z \in \mathfrak{h}^n$),*
(3) $\displaystyle\int_{\Sigma_{\frac{\sqrt{3}}{2},\frac{1}{2}}} |W(z)|^2 \, d^*z < \infty,$

where λ_D is defined by $DI_\nu(z) = \lambda_D I_\nu(z)$, the Siegel set $\Sigma_{\frac{\sqrt{3}}{2},\frac{1}{2}}$ is as in Definition 1.3.1, and the left invariant quotient measure d^*z is given by Proposition 1.5.3.

The primordial example of a Whittaker function for $SL(n, \mathbb{Z})$ is the integral

$$\int_0^1 \cdots \int_0^1 \phi(u \cdot z)\,\overline{\psi(u)} \prod_{1 \le i < j \le n} du_{i,j}$$

given in Proposition 5.2.4. Here $\phi(z)$ is a Maass form for $SL(n, \mathbb{Z})$. This example shows that Whittaker functions occur naturally in the Fourier expansion of Maass forms. The importance of Whittaker functions cannot be underestimated. They are the cornerstone for the entire theory of L–functions.

We shall show in the next section that it is always possible to explicitly construct one non–trivial Whittaker function. Remarkably, this special Whittaker function has good growth properties and is the only Whittaker function that appears in the Fourier expansion of Maass forms (multiplicity one theorem).

5.5 Jacquet's Whittaker function

Whittaker functions for higher rank groups were first studied by Jacquet (1967). The theory was subsequently fully worked out for $GL(3, \mathbb{R})$ in (Bump, 1984), and then for arbitrary real reductive groups in (Wallach, 1988). Jacquet introduced the following explicit construction. For $n \ge 2$, fix

$$m = (m_1, \ldots, m_{n-1}) \in \mathbb{Z}^{n-1}, \qquad \nu = (\nu_1, \nu_2, \ldots, \nu_{n-1}) \in \mathbb{C}^{n-1},$$

and let

$$u = \begin{pmatrix} 1 & u_{1,2} & u_{1,3} & \cdots & u_{1,n} \\ & 1 & u_{2,3} & \cdots & u_{2,n} \\ & & \ddots & & \vdots \\ & & & 1 & u_{n-1,n} \\ & & & & 1 \end{pmatrix} \in U_n(\mathbb{R}).$$

In order to simplify later notation, it is very convenient to relabel the superdiagonal elements

$$u_1 = u_{n-1,n}, \qquad u_2 = u_{n-2,n-1}, \qquad \ldots, \qquad u_{n-1} = u_{1,2}.$$

Define ψ_m to be the character of $U_n(\mathbb{R})$ defined by

$$\psi_m(u) := e^{2\pi i \left[m_1 u_1 + m_2 u_2 + \cdots + m_{n-1} u_{n-1}\right]}.$$

Note that all characters of $U_n(\mathbb{R})$ are of this form.

For $z = xy \in \mathfrak{h}^n$ (as in the beginning of Section 5.1) and $m_i \neq 0$, $(1 \le i \le n-1)$, define

$$W_{\text{Jacquet}}(z;\nu,\psi_m) := \int_{U_n(\mathbb{R})} I_\nu(w_n \cdot u \cdot z)\,\overline{\psi_m(u)}\,d^*u \tag{5.5.1}$$

to be **Jacquet's Whittaker function.** Here ($\lfloor x \rfloor$ denotes the largest integer $\le x$)

$$w_n = \begin{pmatrix} & & & (-1)^{\lfloor n/2 \rfloor} \\ & & 1 & \\ & \cdot^{\cdot^{\cdot}} & & \\ 1 & & & \end{pmatrix} \in SL(n,\mathbb{Z}),$$

$$\int_{U_n(\mathbb{R})} d^*u = \int_{-\infty}^{\infty} \cdots \int_{-\infty}^{\infty} \prod_{1 \le i < j \le n} du_{i,j}.$$

Proposition 5.5.2 *Let $n \ge 2$. Assume that $\Re(\nu_i) > 1/n$ for $i = 1, 2, \ldots, n-1$ and $m_i \neq 0$, $(1 \le i \le n-1)$. Then the integral on the right-hand side of (5.5.1) converges absolutely and uniformly on compact subsets of $\mathfrak{h}^n$ and has meromorphic continuation to all $\nu \in \mathbb{C}^{n-1}$. The function $W_{\text{Jacquet}}(z;\nu,\psi_m)$ is an $SL(n,\mathbb{Z})$–Whittaker function of type ν and character ψ_m. Furthermore, we have the identity*

$$\begin{aligned} W_{\text{Jacquet}}(z;\nu,\psi_m) &= c_{\nu,m} \cdot W_{\text{Jacquet}}\left(Mz;\nu,\psi_{\frac{m_1}{|m_1|},\frac{m_2}{|m_2|},\ldots,\frac{m_{n-1}}{|m_{n-1}|}}\right) \\ &= c_{\nu,m} \cdot \psi_m(x) \cdot W_{\text{Jacquet}}(My;\nu,\psi_{1,\ldots,1}), \end{aligned}$$

where $c_{\nu,m} \neq 0$ (depends only on ν, m) and $M =$

$$\begin{pmatrix} |m_1 m_2 \cdots m_{n-1}| & & & \\ & \ddots & & \\ & & |m_1 m_2| & & \\ & & & |m_1| & \\ & & & & 1 \end{pmatrix}.$$

Remark The reader may verify that $c_{\nu,m} = \prod_{i=1}^{n-1} |m_i|^{\sum_{j=1}^{n-1} b_{i,j}\nu_j - i(n-i)}$. For example when the dimension $n = 2$, we have $c_{2,\nu} = |m_1|^{\nu_1 - 1}$, whereas for "$n = 3$" the coefficient is $c_{3,\nu} = |m_1|^{\nu_1 + 2\nu_2 - 2}|m_2|^{2\nu_1 + \nu_2 - 2}$.

Proof We shall defer the proof of the convergence and the meromorphic continuation of the integral until later. At this point, we show that $W_{\text{Jacquet}}(z;\nu,\psi_m)$ satisfies Definition 5.4.1 (1) and (2) of a Whittaker function. First of all, note that if

$$a = \begin{pmatrix} 1 & a_{1,2} & a_{1,3} & \cdots & a_{1,n} \\ & 1 & a_{2,3} & \cdots & a_{2,n} \\ & & \ddots & & \vdots \\ & & & 1 & a_{n-1,n} \\ & & & & 1 \end{pmatrix} \in U_n(\mathbb{R}),$$

then after changing variables,

$$\begin{aligned} W_{\text{Jacquet}}(az; \nu, \psi_m) &= \int_{U_n(\mathbb{R})} I_\nu(w_n \cdot u \cdot a \cdot z)\, \overline{\psi_m(u)}\, d^*u \\ &= \int_{U_n(\mathbb{R})} I_\nu(w_n \cdot u \cdot z)\, \overline{\psi_m(u \cdot a^{-1})}\, d^*u \\ &= \psi_m(a)\, W_{\text{Jacquet}}(z; \nu, \psi_m). \end{aligned}$$

Second, using the fact that every differential operator $D \in \mathfrak{D}^n$ is invariant under left multiplication by $SL(n, \mathbb{R})$, it follows from the definition of λ_D given in Definition 5.4.1, that

$$D I_\nu(w_n \cdot u \cdot z) = \lambda_D I_\nu(w_n \cdot u \cdot z).$$

Consequently,

$$\begin{aligned} D\, W_{\text{Jacquet}}(z; \nu, \psi_m) &= \int_{U_n(\mathbb{R})} D\left(I_\nu(w_n \cdot u \cdot z)\right) \overline{\psi_m(u)}\, d^*u \\ &= \lambda_D \int_{U_n(\mathbb{R})} I_\nu(w_n \cdot u \cdot z)\, \overline{\psi_m(u)}\, d^*u \\ &= \lambda_D \cdot W_{\text{Jacquet}}(z; \nu, \psi_m). \end{aligned}$$

We have thus proved Proposition 5.5.2 under the assumption that the integral (5.5.1) converges absolutely and uniformly on compact subsets of $\mathfrak{h}^n$ to an $\mathcal{L}^2$ function on the Siegel set $\Sigma_{\frac{\sqrt{3}}{2}, \frac{1}{2}}$.

Next, we prove the identity

$$W_{\text{Jacquet}}(z; \nu, \psi_m) = c_{\nu,m} \cdot W_{\text{Jacquet}}\big(Mz; \nu, \psi_{\epsilon_1, \epsilon_2, \ldots, \epsilon_{n-1}}\big),$$

with $\epsilon_i = m_i/|m_i|$, $(i = 1, 2, \ldots, n-1)$. We have, after making the transformations

$$u_1 \to |m_1| u_1, \qquad u_2 \to |m_2| u_2, \qquad \ldots, \qquad u_{n-1} \to |m_{n-1}| u_{n-1},$$

that

$$\begin{aligned}
W_{\text{Jacquet}}\big(Mz; \nu, \psi_{\epsilon_1,\epsilon_2,\ldots,\epsilon_{n-1}}\big) &= \int_{U_n(\mathbb{R})} I_\nu(w_n \cdot u \cdot Mz) e^{-2\pi i[\epsilon_1 u_1 + \cdots + \epsilon_{n-1}u_{n-1}]} d^*u \\
&= \prod_{i=1}^{n-1} |m_i| \int_{U_n(\mathbb{R})} I_\nu(w_n \cdot Mu \cdot z) e^{-2\pi i[|m_1|\epsilon_1 u_1 + \cdots + |m_{n-1}|\epsilon_{n-1}u_{n-1}]} d^*u \\
&= \prod_{i=1}^{n-1} |m_i| \int_{U_n(\mathbb{R})} I_\nu(w_n M w_n \cdot w_n u z) e^{-2\pi i[m_1 u_1 + \cdots + m_{n-1}u_{n-1}]} d^*u \\
&= c_{\nu,m} \cdot W_{\text{Jacquet}}(z; \nu, \psi_m),
\end{aligned}$$

for some constant $c_{\nu,m} \in \mathbb{C}$. Now,

$$W_{\text{Jacquet}}(Mz; \nu, \psi_{\epsilon_1,\epsilon_2,\ldots,\epsilon_{n-1}}) = \psi_m(x) \cdot W_{\text{Jacquet}}(My; \nu, \psi_{\epsilon_1,\epsilon_2,\ldots,\epsilon_{n-1}}).$$

To complete the proof of the identity in Proposition 5.5.2 it remains to show that

$$W_{\text{Jacquet}}(My; \nu, \psi_{\epsilon_1,\epsilon_2,\ldots,\epsilon_{n-1}}) = W_{\text{Jacquet}}(My; \nu, \psi_{1,\ldots,1}).$$

Note that this identity holds because My is a diagonal matrix with positive entries. To prove it, consider (for $j = 1, 2, \ldots, n-1$) the $(n-j+1)$th row:

$$(0, \ldots, 0, 1, u_j, \; u_{n-j+1,n-j+2}, \quad \ldots, \quad u_{n-j+1,n}),$$

of the matrix u. It is easy to see that for each $1 \le j \le n-1$, we can make the transformation

$$u_j \mapsto \epsilon_j u_j,$$

by letting

$$u \mapsto \delta_j \cdot u \cdot \delta_j,$$

where δ_j is a diagonal matrix with 1s along the diagonal except at the $(n-j+1)$th row where there is an ϵ_j. Note that the other $u_{\ell,k}$ with $1 \le \ell \le k-2 \le n-2$ may also be transformed by ϵ_j, but, as we shall soon see, this will not be relevant.

If we make the above transformations in the integral for the Whittaker function, then that integral takes the form:

$$W_{\text{Jacquet}}(My;\nu,\psi_{\epsilon_1,\ldots,\epsilon_{n-1}}) = \int_{U_n(\mathbb{R})} I_\nu(w_n\cdot u\cdot My)e^{-2\pi i[\epsilon_1u_1+\cdots+\epsilon_{n-1}u_{n-1}]}\,d^*u$$
$$= \int_{U_n(\mathbb{R})} I_\nu(w_n\cdot\delta_j u\delta_j\cdot My)e^{-2\pi i[\epsilon_1u_1+\cdots+u_j+\cdots+\epsilon_{n-1}u_{n-1}]}\,d^*u$$
$$= \int_{U_n(\mathbb{R})} I_\nu(\delta_j w_n\cdot u\cdot My\delta_j)e^{-2\pi i[\epsilon_1u_1+\cdots+u_j+\cdots+\epsilon_{n-1}u_{n-1}]}\,d^*u$$
$$= \int_{U_n(\mathbb{R})} I_\nu(w_n\cdot u\cdot My)e^{-2\pi i[\epsilon_1u_1+\cdots+u_j+\cdots+\epsilon_{n-1}u_{n-1}]}\,d^*u$$
$$= W_{\text{Jacquet}}(My;\nu,\psi_{\epsilon_1,\ldots,\underbrace{1}_{j\text{th position}},\ldots,\epsilon_{n-1}}).$$

One may do the above procedure for each $j = 1,\ldots,n-1$. In the end, we prove the required identity.

The proof of the absolute convergence and meromorphic continuation of the integral (5.5.1) is much more difficult. We shall prove it now for $n = 2$. In this case we may take $z = \begin{pmatrix} y & x\\ 0 & 1\end{pmatrix}$ and $\nu\in\mathbb{C}$. It follows that

$$|W_{\text{Jacquet}}(z;\nu,\psi_m)| \le \int_{-\infty}^{\infty} I_{\text{Re}(\nu)}\left(\begin{pmatrix} 0 & -1\\ 1 & 0\end{pmatrix}\cdot\begin{pmatrix} 1 & u\\ 0 & 1\end{pmatrix}\cdot\begin{pmatrix} y & x\\ 0 & 1\end{pmatrix}\right)du$$
$$= \int_{-\infty}^{\infty}\left(\frac{y}{(x+u)^2+y^2}\right)^{\text{Re}(\nu)} du$$
$$= y^{1-\text{Re}(\nu)}\int_{-\infty}^{\infty}\frac{du}{(u^2+1)^{\text{Re}(\nu)}}, \tag{5.5.3}$$

which converges absolutely for $\text{Re}(\nu) > 1/2$. For $n = 2$, the meromorphic continuation of (5.5.1) is obtained by direct computation of the integral

$W_{\text{Jacquet}}(z;\nu,\psi_m)$. With the choice $\psi_m\left(\begin{pmatrix}1 & u\\ 0 & 1\end{pmatrix}\right)=e^{2\pi i m u}$, we have

$$\begin{aligned} W_{\text{Jacquet}}(z;\nu,\psi_m) &= \int_{-\infty}^{\infty}\left(\frac{y}{(x+u)^2+y^2}\right)^{\nu} e^{-2\pi i m u}\,du\\ &= e^{2\pi i m x} y^{1-\nu}\int_{-\infty}^{\infty}\frac{e^{-2\pi i m u y}}{(u^2+1)^{\nu}}\,du\\ &= \frac{2|m|^{\nu-\frac{1}{2}}\pi^{\nu}}{\Gamma(\nu)}\sqrt{y}\,K_{\nu-\frac{1}{2}}(2\pi|m|y)\cdot e^{2\pi i m x}. \end{aligned} \tag{5.5.4}$$

Since the Bessel function, $K_s(y)$, satisfies the functional equation $K_s(y)=K_{-s}(y)$, we see that the Whittaker function satisfies the functional equation

$$\begin{aligned} W^*_{\text{Jacquet}}(z;\nu,\psi_m) &:= |m\pi|^{-\nu}\Gamma(\nu)W_{\text{Jacquet}}(z;\nu,\psi_m)\\ &= W^*_{\text{Jacquet}}(z;1-\nu,\psi_m). \end{aligned} \tag{5.5.5}$$

□

The proof of the absolute convergence and meromorphic continuation of (5.5.1) for the case $n>2$, is presented in Section 5.8. We shall deduce it using properties of norms of exterior products of vectors in $\mathbb{R}^n$. The theory of exterior powers of $\mathbb{R}^n$ is briefly reviewed in the next section.

5.6 The exterior power of a vector space

Basic references for this material are: (Bourbaki, 1998a, 2003), (Edelen, 1985), (Brown, 1988).

For $n=1,2,\ldots,\ \ell=1,2,\ldots,$ let $\otimes^{\ell}(\mathbb{R}^n)$ denote the ℓth tensor product of the vector space $\mathbb{R}^n$ (considered as a vector space over $\mathbb{R}$). The vector space $\otimes^{\ell}(\mathbb{R}^n)$ is generated by all elements of type

$$v_1\otimes v_2\otimes\cdots\otimes v_\ell,$$

with $v_i\in\mathbb{R}^n$ for $i=1,2,\ldots,\ell$.

We define

$$\Lambda^{\ell}(\mathbb{R}^n)=\otimes^{\ell}(\mathbb{R}^n)/\mathfrak{a}_\ell,$$

where $\mathfrak{a}_\ell$ denotes the vector subspace of $\otimes^{\ell}(\mathbb{R}^n)$ generated by all elements of type

$$v_1\otimes v_2\otimes\cdots\otimes v_\ell,$$

where $v_i = v_j$ for some $i \neq j$. It is not hard to show (see (Bourbaki, 1998a, 2003)) that $\mathbf{\Lambda}^\ell(\mathbb{R}^n)$ can be realized as the vector space (over $\mathbb{R}$) generated by all elements of the form

$$v_1 \wedge v_2 \wedge \cdots \wedge v_\ell$$

where the wedge product, $\wedge$, satisfies the rules

$$v \wedge v = 0, \qquad v \wedge w = -w \wedge v,$$
$$(a_1 v_1 + a_2 v_2) \wedge w = a_1 v_1 \wedge w + a_2 v_2 \wedge w,$$

for all $v, v_1, v_2, w \in \mathbb{R}^n$ and $a_1, a_2 \in \mathbb{R}$.

Example 5.6.1 Consider the wedge product in $\mathbb{R}^2$, where the canonical basis for $\mathbb{R}^2$ is taken to be $e_1 = (1, 0)$, $e_2 = (0, 1)$. Then we have

$$(a_1 e_1 + a_2 e_2) \wedge (b_1 e_1 + b_2 e_2) = (a_1 b_2 - a_2 b_1) e_1 \wedge e_2.$$

There is a canonical inner product, $\langle\ ,\ \rangle : \mathbb{R}^n \times \mathbb{R}^n \to \mathbb{R}$, given by

$$\langle v, w \rangle := v \cdot {}^t w, \tag{5.6.2}$$

for all $v, w \in \mathbb{R}^n$. It easily follows that we may extend this inner product to an inner product $\langle\ ,\ \rangle_{\otimes^\ell}$ on $\otimes^\ell(\mathbb{R}^n) \times \otimes^\ell(\mathbb{R}^n)$ by defining

$$\langle v, w \rangle_{\otimes^\ell} := \prod_{i=1}^{\ell} \langle v_i, w_i \rangle, \tag{5.6.3}$$

for all $v = v_1 \otimes v_2 \otimes \cdots \otimes v_\ell$, $w = w_1 \otimes w_2 \otimes \cdots \otimes w_\ell \in \otimes^\ell(\mathbb{R}^n)$. Note that this agrees with the canonical inner product on $\otimes^\ell(\mathbb{R}^n) \times \otimes^\ell(\mathbb{R}^n)$.

Our next goal is to define an inner product $\langle\ ,\ \rangle_{\Lambda^\ell}$ on $\mathbf{\Lambda}^\ell(\mathbb{R}^n) \times \mathbf{\Lambda}^\ell(\mathbb{R}^n)$. Let $e_1 = (1, 0, \ldots, 0)$, $\ldots$, $e_n = (0, \ldots, 0, 1)$ denote the canonical basis for $\mathbb{R}^n$, and let

$$a = \sum_{1 \le i_1, \ldots, i_\ell \le n} a_{i_1, \ldots, i_\ell}\, e_{i_1} \wedge \cdots \wedge e_{i_\ell}$$

denote an arbitrary element of $\mathbf{\Lambda}^\ell(\mathbb{R}^n)$.

Consider the map $\phi_\ell : \mathbf{\Lambda}^\ell(\mathbb{R}^n) \to \otimes^\ell(\mathbb{R}^n)$ given by

$$\phi_\ell(a) := \frac{1}{\ell!} \sum_{1 \le i_1, \ldots, i_\ell \le n} a_{i_1, \ldots, i_\ell} \sum_{\sigma \in S_\ell} \mathrm{Sign}(\sigma) \cdot e_{\sigma(i_1)} \otimes e_{\sigma(i_2)} \otimes \cdots \otimes e_{\sigma(i_\ell)}, \tag{5.6.4}$$

where S_ℓ denotes the symmetric group of all permutations of $\{1, 2, \ldots, \ell\}$ and $\mathrm{Sign}(\sigma)$ is plus or minus 1 according to whether $\sigma \in S_\ell$ is an even or odd permutation, i.e., it is a product of an even or odd number of transpositions.

One easily checks that the map (5.6.4) is well defined and shows that $\mathbf{\Lambda}^\ell(\mathbb{R}^n)$ is isomorphic to a subspace of $\otimes^\ell(\mathbb{R}^n)$ generated by

$$e_{i_1} \otimes e_{i_2} \otimes \cdots \otimes e_{i_\ell},$$

where $1 \le i_1, i_2, \ldots, i_\ell \le n$ are distinct integers. We may then define

$$\langle v, w\rangle_{\Lambda^\ell} := \langle \phi_\ell(v), \phi_\ell(w)\rangle_{\otimes^\ell}, \tag{5.6.5}$$

for all $v, w \in \mathbf{\Lambda}^\ell(\mathbb{R}^n)$.

Now, the group $SL(n, \mathbb{R})$ acts on the vector space $\mathbb{R}^n$ by right multiplication. Thus, if $v = (a_1, \ldots, a_n) \in \mathbb{R}^n$ and $g \in SL(n, \mathbb{R})$, then the action is given by $v \cdot g$, where $\cdot$ denotes the multiplication of a row vector by a matrix. This action may be extended to an action $\circ$ of $SL(n, \mathbb{R})$ on $\otimes^\ell(\mathbb{R}^n)$, by defining

$$v \circ g := (v_1 \cdot g) \otimes (v_2 \cdot g) \otimes \cdots \otimes (v_\ell \cdot g),$$

for all $v = v_1 \otimes v_2 \otimes \cdots \otimes v_\ell \in \otimes^\ell(\mathbb{R}^n)$. In view of the isomorphism (5.6.4), one may also define an action $\circ$ of $SL(n, \mathbb{R})$ on $\mathbf{\Lambda}^\ell(\mathbb{R}^n)$ given by

$$v \circ g := (v_1 \cdot g) \wedge (v_2 \cdot g) \wedge \cdots \wedge (v_\ell \cdot g),$$

for all $v = v_1 \wedge v_2 \wedge \cdots \wedge v_\ell \in \mathbf{\Lambda}^\ell(\mathbb{R}^n)$.

We now prove two lemmas which will allow us to construct, using norms on $\mathbf{\Lambda}^\ell(\mathbb{R}^n)$, a function very similar to the function $I_\nu(z)$ as defined in Section 2.4. In fact, the sole purpose of this brief excursion into the theory of exterior powers of a vector space is to ultimately realize Jacquet's Whittaker function as an integral of certain complex powers of norms on exterior product spaces.

Lemma 5.6.6 *Let $k \in O(n, \mathbb{R})$ with $n \ge 2$. Then for $\ell \ge 1$, we have*

$$\langle v, w\rangle_{\Lambda^\ell} = \langle v \circ k, w \circ k\rangle_{\Lambda^\ell},$$
$$||v|| := \sqrt{\langle v, v\rangle_{\Lambda^\ell}} = ||v \circ k||,$$

for all $v, w \in \mathbf{\Lambda}^\ell(\mathbb{R}^n)$. Here $||\ \ ||$ denotes the canonical norm on $\mathbf{\Lambda}^\ell(\mathbb{R}^n)$.

Proof First note that for the inner product on $\mathbb{R}^n$, we have

$$\langle v \cdot k, w \cdot k\rangle = (v \cdot k) \cdot {}^t(w \cdot k) = v \cdot k \cdot {}^tk \cdot {}^tw = v \cdot {}^tw = \langle v, w\rangle,$$

for all $v, w \in \mathbb{R}^n$. It immediately follows from (5.6.3) that

$$\langle v \circ k, w \circ k\rangle_{\otimes^\ell(\mathbb{R}^n)} = \langle v, w\rangle_{\otimes^\ell(\mathbb{R}^n)},$$

for all $v, w \in \otimes^\ell(\mathbb{R}^n)$. Finally, the invariance of the action by k on the inner product can be extended to $\mathbf{\Lambda}^\ell(\mathbb{R}^n)$ by (5.6.5). □

Lemma 5.6.7 *Fix $n \ge 2$. Let $e_1 = (1, 0, \dots, 0), \quad \dots, \quad e_n = (0, \dots, 0, 1)$ denote the canonical basis for $\mathbb{R}^n$. Then for all*

$$u = \begin{pmatrix} 1 & u_{1,2} & u_{1,3} & \cdots & u_{1,n} \\ & 1 & u_{2,3} & \cdots & u_{2,n} \\ & & \ddots & & \vdots \\ & & & 1 & u_{n-1,n} \\ & & & & 1 \end{pmatrix} \in SL(n, \mathbb{R}),$$

and every $1 \le \ell \le n-1$, we have

$$(e_{n-\ell} \wedge \cdots \wedge e_{n-1} \wedge e_n) \circ u \;=\; e_{n-\ell} \wedge \cdots \wedge e_{n-1} \wedge e_n.$$

Proof First of all, we have

$$(e_{n-\ell} \wedge \cdots \wedge e_{n-1} \wedge e_n) \circ u \;=\; (e_{n-\ell} \cdot u) \wedge \cdots \wedge (e_{n-1} \cdot u) \wedge (e_n \cdot u).$$

Since $e_i \cdot u = e_i +$ linear combination of e_j with $j > i$, we see that the extra linear combination is killed in the wedge product. □

Lemma 5.6.8 (Cauchy–Schwartz type inequalities) *Let $n \ge 2$. Then for all $v, w \in \mathbf{\Lambda}^\ell(\mathbb{R}^n)$, we have*

$$|\langle v,\ w\rangle_{\Lambda^\ell}|^2 \;\le\; \langle v,\ v\rangle_{\Lambda^\ell} \cdot \langle w,\ w\rangle_{\Lambda^\ell},$$
$$||v \wedge w||_{\Lambda^\ell} \;\le\; ||v||_{\Lambda^\ell} \cdot ||w||_{\Lambda^\ell}.$$

Proof The classical Cauchy–Schwartz inequality

$$|\langle v,\ w\rangle_{\otimes^\ell}|^2 \;\le\; \langle v,\ v\rangle_{\otimes^\ell} \cdot \langle w,\ w\rangle_{\otimes^\ell}$$

is well known on the tensor product space $\otimes^\ell(\mathbb{R}^n)$. It extends to $\mathbf{\Lambda}^\ell(\mathbb{R}^n)$ by the identity (5.6.5).

To prove the second Cauchy–Schwartz type inequality consider

$$v = \sum_{i_1, i_2, \dots, i_\ell} a_{i_1, i_2, \dots, i_\ell}\, e_{i_1} \otimes e_{i_2} \otimes \cdots \otimes e_{i_\ell} \;\in\; \Lambda^\ell(\mathbb{R}^n),$$
$$w = \sum_{j_1, j_2, \dots, j_\ell} b_{j_1, j_2, \dots, j_\ell}\, e_{j_1} \otimes e_{j_2} \otimes \cdots \otimes e_{j_\ell} \;\in\; \Lambda^\ell(\mathbb{R}^n).$$

Note that we are thinking of v, w as also lying in $\otimes^\ell(\mathbb{R}^n)$. Then

$$||v||^2_{\Lambda^\ell} = \sum_{i_1, i_2, \dots, i_\ell} |a_{i_1, i_2, \dots, i_\ell}|^2, \qquad ||w||^2_{\Lambda^\ell} = \sum_{j_1, j_2, \dots, j_\ell} |b_{j_1, j_2, \dots, j_\ell}|^2,$$

and

$$||v||^2_{\Lambda^\ell}\,||w||^2_{\Lambda^\ell} = \sum_{\substack{i_1,i_2,\ldots,i_\ell\\ j_1,j_2,\ldots,j_\ell}} |a_{i_1,i_2,\ldots,i_\ell}|^2 |b_{j_1,j_2,\ldots,j_\ell}|^2.$$

Now

$$\begin{aligned} v\wedge w &= \sum_{\substack{i_1,i_2,\ldots,i_\ell\\ j_1,j_2,\ldots,j_\ell}} a_{i_1,i_2,\ldots,i_\ell}\, b_{j_1,j_2,\ldots,j_\ell}\big(e_{i_1}\otimes e_{i_2}\otimes\cdots\otimes e_{i_\ell}\big)\wedge\big(e_{j_1}\otimes e_{j_2}\otimes\cdots\otimes e_{j_\ell}\big)\\ &= \frac{1}{(\ell!)^2}\sum_{\substack{i_1,i_2,\ldots,i_\ell\\ j_1,j_2,\ldots,j_\ell}} a_{i_1,i_2,\ldots,i_\ell}\, b_{j_1,j_2,\ldots,j_\ell}\sum_{\sigma\in S_\ell}\sum_{\sigma'\in S_\ell}\mathrm{Sign}(\sigma)\,\mathrm{Sign}(\sigma')\\ &\qquad\times\big(e_{\sigma(i_1)}\otimes\cdots\otimes e_{i_\sigma(\ell)}\big)\otimes\big(e_{\sigma'(j_1)}\otimes\cdots\otimes e_{\sigma'(j_\ell)}\big). \end{aligned}$$

The result now follows because $||u+u'||^2_{\otimes^\ell}\le ||u||^2_{\otimes^\ell} + ||u'||^2_{\otimes^\ell}$ for all $u,u'\in\otimes^\ell(\mathbb{R}^n)$. □

5.7 Construction of the I_ν function using wedge products

We now construct, using wedge products and norms on $\mathbf{\Lambda}^\ell(\mathbb{R}^n)$, the function $I_\nu(z)$ as defined in Section 5.4. Recall from Section 1.2 that every $z\in\mathfrak{h}^n$ can be uniquely written in the form

$$z = \begin{pmatrix} 1 & x_{1,2} & x_{1,3} & \cdots & x_{1,n}\\ & 1 & x_{2,3} & \cdots & x_{2,n}\\ & & \ddots & & \vdots\\ & & & 1 & x_{n-1,n}\\ & & & & 1\end{pmatrix}\cdot\begin{pmatrix} y_1y_2\cdots y_{n-1} & & & & \\ & y_1y_2\cdots y_{n-2} & & & \\ & & \ddots & & \\ & & & y_1 & \\ & & & & 1\end{pmatrix}. \tag{5.7.1}$$

Lemma 5.7.2 *For $n\ge 2$, let z be given by (5.7.1) and let $||\ ||$ denote the norm on $\Lambda^\ell(\mathbb{R}^n)$ as in Lemma 5.6.6. Then for $\nu=(\nu_1,\ldots,\nu_{n-1})\in\mathbb{C}^{n-1}$, we have the identity*

$$I_\nu(z) := \left(\prod_{i=0}^{n-2}\Big|\Big|(e_{n-i}\wedge\cdots\wedge e_{n-1}\wedge e_n)\circ z\Big|\Big|^{-n\nu_{n-i-1}}\right)\cdot|\mathrm{Det}(z)|^{\sum\limits_{i=1}^{n-1} i\,\nu_{n-i}}.$$

Proof It follows from Lemma 5.6.6 that the function

$$I_\nu^*(z) := \left(\prod_{i=0}^{n-2} \left\|(e_{n-i} \wedge \cdots \wedge e_{n-1} \wedge e_n) \circ z\right\|^{-n\nu_{n-i-1}}\right) \cdot |\mathrm{Det}(z)|^{\sum_{i=1}^{n-1} i\,\nu_{n-i}},$$

is invariant under the transformation

$$z \to z \cdot k$$

for all $k \in SO(n, \mathbb{R})$. Here, we have used the fact that $\mathrm{Det}(z \cdot k) = \mathrm{Det}(z)$ since $\mathrm{Det}(k) = 1$. It also follows from Lemma 5.6.7 that $I_\nu^*(z) = I_\nu^*(y)$ with

$$y = \begin{pmatrix} y_1 y_2 \cdots y_{n-1} & & & & \\ & y_1 y_2 \cdots y_{n-2} & & & \\ & & \ddots & & \\ & & & y_1 & \\ & & & & 1 \end{pmatrix}.$$

One also easily checks that

$$I_\nu^*(az) = I_\nu^*(z)$$

for $a \in \mathbb{R}^\times$. Thus $I_\nu^*(z)$ is well defined on $\mathfrak{h}^n$. To show that $I_\nu^*(y) = I_\nu(y)$, note that

$$e_{n-i} \cdot y = y_1 \cdots y_i, \qquad \cdots, \qquad e_{n-1} \cdot y = y_1, \qquad e_n \cdot y = 1.$$

Consequently,

$$\begin{aligned} \left\|(e_{n-i} \wedge \cdots \wedge e_{n-1} \wedge e_n) \circ y\right\|^{-n\nu_{n-i-1}} &= \big(y_1 \cdot (y_1 y_2) \cdots (y_1 \cdots y_i)\big)^{-n\nu_{n-i-1}} \\ &= \left(\prod_{\ell=1}^{i} y_\ell^{i+1-\ell}\right)^{-n\nu_{n-i-1}}. \end{aligned}$$

Furthermore,

$$|\mathrm{Det}(z)|^{\sum_{i=1}^{n-1} i\,\nu_{n-i}} = \left(\prod_{\ell=1}^{n-1} y_\ell^{n-\ell}\right)^{\sum_{i=1}^{n-1} i\,\nu_{n-i}},$$

from which the result follows by a brute force computation. □

In order to demonstrate the power of the exterior algebra approach, we explicitly compute the integral for the Jacquet Whittaker function on $GL(3, \mathbb{R})$.

Example 5.7.3 ($GL(3, \mathbb{R})$ Whittaker function) Let $n = 3$, $\nu = (\nu_1, \nu_2) \in \mathbb{C}^2$, $u = \begin{pmatrix} 1 & u_2 & u_3 \\ & 1 & u_1 \\ & & 1 \end{pmatrix}$, $y = \begin{pmatrix} y_1 y_2 & & \\ & y_1 & \\ & & 1 \end{pmatrix}$, and for fixed $m = (m_1, m_2) \in \mathbb{Z}^2$

define $\psi_m(u) = e^{2\pi i(m_1u_1+m_2u_2)}$. Then we have the explicit integral representation:

$$W_{\text{Jacquet}}(y; \nu, \psi_m) = y_1^{\nu_1+2\nu_2} y_2^{2\nu_1+\nu_2} \int_{-\infty}^{\infty}\int_{-\infty}^{\infty}\int_{-\infty}^{\infty} \left[y_1^2y_2^2 + u_1^2y_2^2 + (u_1u_2 - u_3)^2\right]^{-3\nu_1/2} \times \left[y_1^2y_2^2 + u_2^2y_1^2 + u_3^2\right]^{-3\nu_2/2} e^{-2\pi i(m_1u_1+m_2u_2)}\, du_1du_2du_3.$$

Proof In view of (5.5.1), it is enough to compute $I_\nu(w_3uy)$. It follows from Lemma 5.7.2 that

$$I_\nu(w_3uy) = ||e_3w_3uy||^{-3\nu_2} \cdot ||(e_2w_3uy) \wedge (e_3w_3uy)||^{-3\nu_1} \cdot \left(y_1^2y_2\right)^{\nu_2+2\nu_1}. \tag{5.7.4}$$

We compute

$$\begin{aligned} e_3w_3uy &= (0, 0, 1)\begin{pmatrix} & & -1 \\ & 1 & \\ 1 & & \end{pmatrix}\begin{pmatrix} 1 & u_2 & u_3 \\ & 1 & u_1 \\ & & 1 \end{pmatrix}\begin{pmatrix} y_1y_2 & & \\ & y_1 & \\ & & 1 \end{pmatrix} \\ &= (1, 0, 0)\begin{pmatrix} y_1y_2 & u_2y_1 & u_3 \\ & y_1 & u_1 \\ & & 1 \end{pmatrix} \\ &= y_1y_2e_1 + u_2y_1e_2 + u_3e_3. \end{aligned}$$

Hence

$$||e_3w_3uy|| = \left[y_1^2y_2^2 + u_2^2y_1^2 + u_3^2\right]^{\frac{1}{2}}.$$

Similarly,

$$\begin{aligned} e_2w_3uy &= (0, 1, 0)\begin{pmatrix} & & -1 \\ & 1 & \\ 1 & & \end{pmatrix}\begin{pmatrix} 1 & u_2 & u_3 \\ & 1 & u_1 \\ & & 1 \end{pmatrix}\begin{pmatrix} y_1y_2 & & \\ & y_1 & \\ & & 1 \end{pmatrix} \\ &= (0, 1, 0)\begin{pmatrix} y_1y_2 & u_2y_1 & u_3 \\ & y_1 & u_1 \\ & & 1 \end{pmatrix} \\ &= y_1e_2 + u_1e_3, \end{aligned}$$

so that

$$(e_2w_3uy)\wedge(e_3w_3uy) = y_1^2y_2e_1\wedge e_2 + (u_1u_2y_1 - u_3y_1)e_2\wedge e_3 + u_1y_1y_2e_1\wedge e_3.$$

Consequently

$$||(e_2 w_3 u y) \wedge (e_3 w_3 u y)|| = y_1 \left[y_1^2 y_2^2 + (u_1 u_2 - u_3)^2 + u_1^2 y_2^2 \right]^{\frac{1}{2}}.$$

The result follows upon substituting the results of the above computations into (5.7.4). □

5.8 Convergence of Jacquet's Whittaker function

Assume $\mathrm{Re}(\nu_i) > 1/n$ (for $i = 1, 2, \ldots, n-1$.) The absolute convergence of the integral (5.5.1) has already been proved for the case $n = 2$ in (5.5.4). The proof for all $n \geq 2$ is based on induction on n and will now be given.

Recall the notation:

$$w_n = \begin{pmatrix} & & & (-1)^{\lfloor n/2 \rfloor} \\ & & 1 & \\ & \cdots & & \\ 1 & & & \end{pmatrix} \in SL(n, \mathbb{Z}),$$

$$u = \begin{pmatrix} 1 & u_{1,2} & u_{1,3} & \cdots & u_{1,n} \\ & 1 & u_{2,3} & \cdots & u_{2,n} \\ & & \ddots & & \vdots \\ & & & 1 & u_{n-1,n} \\ & & & & 1 \end{pmatrix}, \quad y = \begin{pmatrix} y_1 y_2 \cdots y_{n-1} & & & & \\ & y_1 y_2 \cdots y_{n-2} & & & \\ & & \ddots & & \\ & & & y_1 & \\ & & & & 1 \end{pmatrix}.$$

It is enough to prove the absolute convergence of

$$W_{\text{Jacquet}}(y; \nu, \psi) := \int_{U_n(\mathbb{R})} I_\nu(w_n \cdot u \cdot y)\, \overline{\psi(u)}\, d^*u. \tag{5.8.1}$$

Now, it follows from the Cauchy–Schwartz inequality, Lemma 5.6.8, and Lemma 5.7.2, that

$$|I_\nu(w_n u y)| \ll_y \|e_n \cdot w_n u y\|^{-nV} \left(\prod_{i=1}^{n-2} \left\| (e_{n-i} \wedge \cdots \wedge e_{n-1}) \circ w_n u y \right\|^{-n\Re(\nu_{n-i-1})} \right), \tag{5.8.2}$$

where

$$V = \sum_{i=0}^{n-2} \Re(\nu_{n-i-1}).$$

The $\ll$ constant in (5.8.2) is independent of u and depends only on y. It appears because $\mathrm{Det}(w_n u y) = \mathrm{Det}(y)$ does not depend on u, but only on y. In view of the identity

$$\begin{aligned}||e_n\cdot w_n u y|| &= ||e_1\cdot uy||\\ &= \left[(y_1\cdots y_{n-1})^2+(u_{1,2}y_1\cdots y_{n-2})^2+\cdots+(u_{1,n-1}y_1)^2+(u_{1,n})^2\right]^{\frac{1}{2}},\end{aligned}$$

it immediately follows from (5.8.2) that (5.8.1) converges absolutely for $\mathrm{Re}(\nu_i)>1/n$ $(1\le i\le n-1)$ if the following two integrals converge absolutely:

$$\int_{-\infty}^{\infty}\cdots\int_{-\infty}^{\infty}\left[(y_1\cdots y_{n-1})^2+(u_{1,2}y_1\cdots y_{n-2})^2+\cdots+(u_{1,n})^2\right]^{-nV/2}\prod_{k=1}^{n}du_{1,k},\tag{5.8.3}$$

$$\int_{-\infty}^{\infty}\cdots\int_{-\infty}^{\infty}\left(\prod_{i=1}^{n-2}\left\|(e_{n-i}\wedge\cdots\wedge e_{n-1})\circ w_n u y\right\|^{-n\nu_{n-i-1}}\right)\prod_{1<i<j\le n}du_{i,j}.\tag{5.8.4}$$

Clearly, the first integral converges absolutely if $\mathrm{Re}(\nu_i)>1/n$, for $1\le i\le n-1$. Furthermore, if $1<i<n$, then $e_i\cdot w_n=e_\ell$ for some $\ell\ne 1$. This immediately implies that

$$e_i\cdot w_n u y = e_i\cdot w_n u' y'$$

where

$$u'=\begin{pmatrix}1&0&0&\cdots&0\\ &1&u_{2,3}&\cdots&u_{2,n}\\ &&\ddots&&\vdots\\ &&&1&u_{n-1,n}\\ &&&&1\end{pmatrix},\quad y'=\begin{pmatrix}1&&&\\ &y_1y_2\cdots y_{n-2}&&\\ &&\ddots&\\ &&&y_1&\\ &&&&1\end{pmatrix}.$$

It follows that the second integral (5.8.4) may be rewritten as

$$\int_{-\infty}^{\infty}\cdots\int_{-\infty}^{\infty}\left(\prod_{i=1}^{n-2}\left\|(e_{n-i}\wedge\cdots\wedge e_{n-1})\circ w_n u' y'\right\|^{-n\nu_{n-i-1}}\right)\prod_{1<i<j\le n}du_{i,j}.\tag{5.8.5}$$

The remarkable thing is that the integral in (5.8.5) can be interpreted as a Jacquet Whittaker function for $SL(n-1,\mathbb{Z})$. This allows us to apply induction from which the absolute convergence of (5.8.1) follows.

Recall the definition of w_n given just after (5.5.1). To see that the integral (5.8.5) is a Jacquet Whittaker function for $SL(n-1,\mathbb{Z})$, we use the matrix identity:

$$w_n = \begin{pmatrix} 0 & 1 & & & \\ & 0 & 1 & & \\ & & \ddots & \ddots & \\ & & & 0 & 1 \\ 1 & & & & 0 \end{pmatrix} \cdot \begin{pmatrix} 1 & & & & 0 \\ & & & 0 & (-1)^{\lfloor n/2 \rfloor} \\ & & \cdot^{\cdot^{\cdot}} & \cdot^{\cdot^{\cdot}} & \\ & 0 & 1 & & \\ 0 & 1 & & & \end{pmatrix}.$$

Further,

$$e_\ell \cdot \begin{pmatrix} 0 & 1 & & & \\ & 0 & 1 & & \\ & & \ddots & \ddots & \\ & & & 0 & 1 \\ 1 & & & & 0 \end{pmatrix} = e_{\ell+1}$$

for $1 \le \ell \le n-1$. It immediately follows from these remarks that (5.8.5) may be rewritten as

$$\int_{-\infty}^{\infty} \cdots \int_{-\infty}^{\infty} \left(\prod_{i=1}^{n-2} \left\| (e_{n+1-i} \wedge \cdots \wedge e_n) \circ w_n' u' y' \right\|^{-n\nu_{n-i-1}} \right) \prod_{1<i<j\le n} du_{i,j}, \tag{5.8.6}$$

where

$$w_n' = \begin{pmatrix} 1 & & & & 0 \\ & & & 0 & (-1)^{\lfloor n/2 \rfloor} \\ & & \cdot^{\cdot^{\cdot}} & \cdot^{\cdot^{\cdot}} & \\ & 0 & 1 & & \\ 0 & 1 & & & \end{pmatrix} = \begin{pmatrix} 1 & \\ & w_{n-1} \end{pmatrix}.$$

Furthermore, we may write

$$u' = \begin{pmatrix} 1 & \\ & \mu \end{pmatrix}, \qquad y' = \begin{pmatrix} 1 & \\ & \eta \end{pmatrix},$$

with $\mu \in U_{n-1}(\mathbb{R})$ and η a diagonal matrix in $GL(n-1, \mathbb{R})$. With these observations, one may deduce that

$$\prod_{i=1}^{n-2} \left\| (e_{n+1-i} \wedge \cdots \wedge e_n) \circ w_n' u' y' \right\|^{-n\nu_{n-i-1}}$$
$$= \left(\prod_{i=0}^{(n-1)-2} \left\| (e_{n-1-i} \wedge \cdots \wedge e_{n-1}) \circ w_{n-1} \mu\eta \right\|^{-(n-1)\nu_{n-2-i}} \right)^{n/(n-1)}.$$

It follows that

$$\int_{-\infty}^{\infty}\cdots\int_{-\infty}^{\infty}\left(\prod_{i=1}^{n-2}\left\|(e_{n+1-i}\wedge\cdots\wedge e_n)\circ w_n'u'y'\right\|^{-n\nu_{n-i-1}}\right)\prod_{1<i<j\le n}du_{i,j}$$
$$\ll_y \int_{-\infty}^{\infty}\cdots\int_{-\infty}^{\infty}\left|I_{(n\nu/(n-1)),n-1}(w_{n-1}\mu\eta)\right|d\mu,$$

where $I_{\nu,n-1}$ denotes the I_ν function (as defined in Section 5.4) for $GL(n-1,\mathbb{R})$. By induction, we obtain the absolute convergence in the region $\mathrm{Re}(\nu)>1/n$.

5.9 Functional equations of Jacquet's Whittaker function

In order to explicitly state the group of functional equations satisfied by Jacquet's Whittaker function, we need to introduce some preliminary notation. Fix an integer $n\ge 2$, and let W_n denote the Weyl group of $SL(n,\mathbb{Z})$ consisting of all $n\times n$ matrices in $SL(n,\mathbb{Z})$ which have exactly one ± 1 in each row and column. For each fixed $w\in W_n$, and every $\nu-\frac{1}{n}=\left(\nu_1-\frac{1}{n},\ \ldots,\ \nu_{n-1}-\frac{1}{n}\right)\in\mathbb{C}^{n-1}$, let us define ν' so that $\nu'-\frac{1}{n}=\left(\nu_1'-\frac{1}{n},\ldots,\nu_{n-1}'-\frac{1}{n}\right)\in\mathbb{C}^{n-1}$ satisfies

$$I_{\nu-\frac{1}{n}}(y)=I_{\nu'-\frac{1}{n}}(wy), \tag{5.9.1}$$

for all

$$y=\begin{pmatrix} y_1y_2\cdots y_{n-1} & & & & \\ & y_1y_2\cdots y_{n-2} & & & \\ & & \ddots & & \\ & & & y_1 & \\ & & & & 1\end{pmatrix}.$$

Definition 5.9.2 *Fix an integer $n\ge 2$, and let ψ be a character of $U_n(\mathbb{R})$. We define*

$$W^*_{\text{Jacquet}}(z;\nu,\psi)=W_{\text{Jacquet}}(z;\nu,\psi)\cdot\prod_{j=1}^{n-1}\prod_{j\le k\le n-1}\pi^{-\frac{1}{2}-v_{j,k}}\Gamma\left(\frac{1}{2}+v_{j,k}\right),$$

where

$$v_{j,k}=\sum_{i=0}^{j-1}\frac{n\nu_{n-k+i}-1}{2},$$

and $W_{\text{Jacquet}}(z;\nu,\psi)$ denotes Jacquet's Whittaker function (5.5.1).

We may now state the functional equations of the Whittaker functions.

Theorem 5.9.3 *Fix an integer $n \geq 2$, and fix $\psi = \psi_{1,1,\ldots,1}$ so that*

$$\psi(u) = e^{2\pi i \sum\limits_{\ell=1}^{n-1} u_{\ell,\ell+1}},$$

for $u \in U_n(\mathbb{R})$. Then the Whittaker function $W^*_{\text{Jacquet}}(z; \nu, \psi)$ has a holomorphic continuation to all $\nu \in \mathbb{C}^{n-1}$. For each $w \in W_n$, let ν, ν' satisfy (5.9.1). Then we have the functional equation

$$W^*_{\text{Jacquet}}(z; \nu, \psi) = W^*_{\text{Jacquet}}(z; \nu', \psi).$$

Remarks It is clear that $W^*_{\text{Jacquet}}(z; \nu, \psi)$ and $W^*_{\text{Jacquet}}(z; \nu', \psi)$ are both Whittaker functions of type ν and character ψ. It follows from Shalika's multiplicity one theorem (Shalika, 1974) that the functional equation must hold up to a constant depending on ν. The assumption that $\psi(u) = \psi_{1,1,\ldots,1}$ is not restrictive because Proposition 5.5.2 tells us that there is a simple identity relating $W^*_{\text{Jacquet}}(z; \nu, \psi_m)$ and $W^*_{\text{Jacquet}}(Mz; \nu, \psi_{1,1,\ldots,1})$.

Before giving the proof of the functional equation, we will obtain explicit versions of the functional equation $\nu \mapsto \nu'$ given by (5.9.1). The Weyl group W_n is generated by the simple reflections

$$\sigma_i = \begin{pmatrix} I_{n-i-1} & & & \\ & 0 & -1 & \\ & 1 & 0 & \\ & & & I_{i-1} \end{pmatrix}, \qquad (i = 1, 2, \ldots, n-1),$$

where I_a denotes the $a \times a$ identity matrix. We adopt the convention that I_0 is the empty set so that $\sigma_1 = \begin{pmatrix} I_{n-2} & & \\ & 0 & -1 \\ & 1 & 0 \end{pmatrix}$ and $\sigma_{n-1} = \begin{pmatrix} 0 & -1 & \\ 1 & 0 & \\ & & I_{n-2} \end{pmatrix}$.

Since σ_i $(i = 1, \ldots, n-1)$ generate W_n, it is enough to give the functional equation $\nu \mapsto \nu'$ for the simple reflections $w = \sigma_i$. Fix an integer i with $1 \leq i \leq n-1$. In this case, ν' is defined by the equation

$$I_{\nu-\frac{1}{n}}(y) = I_{\nu'-\frac{1}{n}}(\sigma_i y) = I_{\nu'-\frac{1}{n}}\left(\sigma_i y \sigma_i^{-1}\right) = I_{\nu'-\frac{1}{n}}(y') \tag{5.9.4}$$

where

$$y' = \begin{pmatrix} y_1\cdots y_{n-1} & & & & & \\ & \ddots & & & & \\ & & y_1\cdots y_i & & & \\ & & & y_1\cdots y_i y_{i+1} & & \\ & & & & \ddots & \\ & & & & & y_1 & \\ & & & & & & 1 \end{pmatrix}$$

is the diagonal matrix y with the $(n-i)$th and $(n-i+1)$th rows interchanged. It follows that if we put y' in Iwasawa form

$$y' = \begin{pmatrix} y'_1\cdots y'_{n-1} & & & \\ & \ddots & & \\ & & y'_1 & \\ & & & 1 \end{pmatrix},$$

then we must have

$$y'_\ell = \begin{cases} y_\ell & \text{if } \ell \neq i-1, i, i+1, \\ y_\ell y_{\ell+1} & \text{if } \ell = i-1, \\ y_\ell^{-1} & \text{if } \ell = i, \\ y_{\ell-1} y_\ell & \text{if } \ell = i+1, \end{cases}$$

for $1 \le i \le n-1$ and $1 \le \ell \le n-1$. Equation (5.9.4), together with the definition of the I_ν function given in Definition 5.4.1 imply (for $1 < i \le n-1$) the following system of linear equations:

$$\begin{aligned} \sum_{j=1}^{n-1} b_{\ell,j}\left(\nu_j - \frac{1}{n}\right) &= \sum_{j=1}^{n-1} b_{\ell,j}\left(\nu'_j - \frac{1}{n}\right), \qquad (\text{for } \ell \neq i), \\ \sum_{j=1}^{n-1} b_{i,j}\left(\nu_j - \frac{1}{n}\right) &= \sum_{j=1}^{n-1} (b_{i-1,j} - b_{i,j} + b_{i+1,j}) \cdot \left(\nu'_j - \frac{1}{n}\right). \end{aligned} \tag{5.9.5}$$

For each $1 \le i \le n-1$, the system of linear equations (5.9.5) has the solution

$$\begin{aligned} \nu'_{n-i-1} &= -\frac{1}{n} + \nu_{n-i-1} + \nu_{n-i}, \qquad (\text{if } i \neq n-1),\\ \nu'_{n-i} &= \frac{2}{n} - \nu_{n-i},\\ \nu'_{n-i+1} &= -\frac{1}{n} + \nu_{n-i} + \nu_{n-i+1}, \qquad (\text{if } i \neq 1),\\ \nu'_\ell &= \nu_\ell \qquad (\ell \neq n-i-1, n-i, n-i+1). \end{aligned} \tag{5.9.6}$$

Example 5.9.7 For $n = 2, 3$ the functional equations (5.9.6) take the explicit form:

$$\begin{aligned} &\nu'_1 = 1 - \nu_1, && (n = 2),\\ &\nu'_2 = \frac{2}{3} - \nu_2, \quad \nu'_1 = -\frac{1}{3} + \nu_1 + \nu_2, && (n = 3,\ i = 1),\\ &\nu'_1 = \frac{2}{3} - \nu_1, \quad \nu'_2 = -\frac{1}{3} + \nu_1 + \nu_2, && (n = 3,\ i = 2). \end{aligned}$$

It immediately follows from these computations that Theorem 5.9.3 can be put in the following more explicit form.

Theorem 5.9.8 *Fix an integer $n \ge 2$ and fix an integer i with $1 \le i \le n-1$. Let ν, ν' satisfy (5.9.6). Then we have the functional equations:*

$$W^*_{\text{Jacquet}}(z; \nu, \psi) = W^*_{\text{Jacquet}}(z; \nu', \psi),$$

$$\begin{aligned} W_{\text{Jacquet}}(z; \nu, \psi) &= \frac{\pi^{-\frac{n}{2}\nu'_{n-i}}}{\pi^{-\frac{n}{2}\nu_{n-i}}} \cdot \frac{\Gamma\left(\frac{n\nu'_{n-i}}{2}\right)}{\Gamma\left(\frac{n\nu_{n-i}}{2}\right)} \cdot W_{\text{Jacquet}}(z; \nu', \psi)\\ &= \frac{\pi^{-\left(1-\frac{n}{2}\nu_{n-i}\right)}}{\pi^{-\frac{n}{2}\nu_{n-i}}} \cdot \frac{\Gamma\left(1 - \frac{n\nu_{n-i}}{2}\right)}{\Gamma\left(\frac{n\nu_{n-i}}{2}\right)} \cdot W_{\text{Jacquet}}(z; \nu', \psi). \end{aligned}$$

Proof Fix a simple reflection σ_i with $1 \le i \le n-1$. The second functional equation in Theorem 5.9.8 is equivalent to the first and can be obtained by computing the affect of the transformations (5.9.6) on the $v_{j,k}$ given in Definition 5.9.2. It remains to prove the second functional equation in Theorem 5.9.8. Recall that

$$w_n = \begin{pmatrix} & & & (-1)^{\lfloor n/2 \rfloor} \\ & & 1 & \\ & \cdot^{\cdot^{\cdot}} & & \\ 1 & & & \end{pmatrix} \in SL(n, \mathbb{Z}),$$

where $\lfloor x \rfloor$ denotes the smallest integer $\le x$. Define

$$w_i := \sigma_i^{-1} w_n.$$

The group $N = U_n(\mathbb{R})$ of upper triangular matrices with coefficients in $\mathbb{R}$ and 1s on the diagonal decomposes as

$$N = \left(w_i^{-1} N_i w_i^{-1}\right) \cdot N_i' \tag{5.9.9}$$

where N_i is a one-dimensional subgroup with 1s on the diagonal, an arbitrary real number at position $\{n-i, n-i+1\}$, and zeros elsewhere, and N_i' is the subgroup of N with a zero at position $\{n-i, n-i+1\}$.

Lemma 5.9.10 *For $1 \le i \le n-1$, we have $w_i^{-1} N_i w_i^{-1} = N_{n-i}$.*

Proof Note that if

$$n_i = \begin{pmatrix} 1 & & & & \\ & \ddots & & & \\ & & 1 & a & \\ & & & \ddots & \\ & & & & 1 \end{pmatrix} \in N_i$$

then

$$\sigma_i n_i \sigma_i^{-1} = \begin{pmatrix} 1 & & & & \\ & \ddots & & & \\ & -a & 1 & & \\ & & & \ddots & \\ & & & & 1 \end{pmatrix}.$$

Further, conjugating this by w_n moves the $-a$ to the position $\{i, i+1\}$. □

It follows from (5.9.9) that we may factor every $u \in U_n(\mathbb{R}) = N$ as $u = n_i \cdot n_i'$ with $n_i \in w_i^{-1} N_i w_i$ and $n_i' \in N_i'$. For example, we may take

$$\sigma_i = \begin{pmatrix} 1 & 0 & 0 & 0 & 0 \\ 0 & 0 & -1 & 0 & 0 \\ 0 & 1 & 0 & 0 & 0 \\ 0 & 0 & 0 & 1 & 0 \\ 0 & 0 & 0 & 0 & 1 \end{pmatrix},$$

$$N_i = \left\{ \begin{pmatrix} 1 & 0 & 0 & 0 & 0 \\ 0 & 1 & * & 0 & 0 \\ 0 & 0 & 1 & 0 & 0 \\ 0 & 0 & 0 & 1 & 0 \\ 0 & 0 & 0 & 0 & 1 \end{pmatrix} \right\}, \qquad w_i^{-1} N_i w_i = \left\{ \begin{pmatrix} 1 & 0 & 0 & 0 & 0 \\ 0 & 1 & 0 & 0 & 0 \\ 0 & 0 & 1 & * & 0 \\ 0 & 0 & 0 & 1 & 0 \\ 0 & 0 & 0 & 0 & 1 \end{pmatrix} \right\}$$

$$N_i' = \left\{ \begin{pmatrix} 1 & * & * & * & * \\ 0 & 1 & * & * & * \\ 0 & 0 & 1 & 0 & * \\ 0 & 0 & 0 & 1 & * \\ 0 & 0 & 0 & 0 & 1 \end{pmatrix} \right\}.$$

Furthermore, by Lemma 5.9.10, we may express Jacquet's Whittaker function (5.5.1) in the form

$$\begin{aligned} W_{\text{Jacquet}}(z; \nu, \psi) &= \int_{U_n(\mathbb{R})} I_\nu(w_n \cdot u \cdot z) \, \overline{\psi(u)} \, d^*u \\ &= \int_{N_i'} \left[\int_{N_i} I_\nu(\sigma_i n_i w_i n_i' z) \bar{\psi}(n_i) \, dn_i \right] \bar{\psi}(n_i') \, dn_i'. \end{aligned}$$

The inner integral above (over the region N_i) is a Whittaker function for the group $SL(2, \mathbb{R})$, and has a functional equation of type (5.5.5). This functional equation is independent of the choice of $w_i n_i' z$, and is precisely what is needed to complete the proof of Theorem 5.9.8. We shall find the functional equation by examining the case when $w_i n_i' z$ is the identity matrix.

We first compute $I_\nu(\sigma_i n_i)$ for

$$\sigma_i = \begin{pmatrix} I_{n-i-1} & & & \\ & 0 & -1 & \\ & 1 & 0 & \\ & & & I_{i-1}, \end{pmatrix}, \qquad n_i = \begin{pmatrix} I_{n-i-1} & & & \\ & 1 & u & \\ & 0 & 1 & \\ & & & I_{i-1}, \end{pmatrix}.$$

It follows that

$$\sigma_i n_i = \begin{pmatrix} I_{n-i-1} & & & \\ & (u^2+1)^{-\frac{1}{2}} & -u/(u^2+1)^{\frac{1}{2}} & \\ & 0 & (u^2+1)^{\frac{1}{2}} & \\ & & & I_{i-1}, \end{pmatrix} \cdot k$$

for some orthogonal matrix k. If we then put $\sigma_i n_i$ in standard Iwasawa form, we have

$$\sigma_i n_i = \begin{pmatrix} I_{n-i-1} & & & \\ & 1 & -u/(u^2+1) & \\ & 0 & 1 & \\ & & & I_{i-1}, \end{pmatrix} \cdot \begin{pmatrix} a_1 \cdots a_{n-1} & & & \\ & \ddots & & \\ & & a_1 & \\ & & & 1 \end{pmatrix}$$

with

$$a_\ell = \begin{cases} 1 & \text{if } \ell \neq i-1, i, i+1 \\ (u^2+1)^{\frac{1}{2}} & \text{if } \ell = i-1 \\ (u^2+1)^{-1} & \text{if } \ell = i \\ (u^2+1)^{\frac{1}{2}} & \text{if } \ell = i+1. \end{cases}$$

Consequently

$$I_\nu(\sigma_i n_i) = (u^2+1)^{\sum\limits_{j=1}^{n-1}\left(b_{i-1,j}-2b_{i,j}+b_{i+1,j}\right)\frac{\nu_j}{2}}.$$

The functional equation now follows from (5.5.5). □

5.10 Degenerate Whittaker functions

Jacquet's Whittaker function was constructed by integrating $I_\nu(w_n z)$ where w_n is the so-called long element of the Weyl group as in (5.5.1). Since the I_ν function is an eigenfunction of the invariant differential operators, its integral inherits all those properties and gives us a Whittaker function.

It is natural to try this type of construction with other Weyl group elements besides the long element w_n. To get a feel for what is going on, let us try to do this on $GL(4)$. As an example, we shall consider the Weyl group element

$$w = \begin{pmatrix} 1 & 0 & 0 & 0 \\ 0 & 0 & 0 & 1 \\ 0 & 1 & 0 & 0 \\ 0 & 0 & 1 & 0 \end{pmatrix}.$$

Let

$$u = \begin{pmatrix} 1 & u_{1,2} & u_{1,3} & u_{1,4} \\ 0 & 1 & u_{2,3} & u_{2,4} \\ 0 & 0 & 1 & u_{3,4} \\ 0 & 0 & 0 & 1 \end{pmatrix}, \qquad y = \begin{pmatrix} y_1y_2y_3 & & & \\ & y_1y_2 & & \\ & & y_1 & \\ & & & 1 \end{pmatrix}.$$

With a brute force computation, one sees that for $\nu = (\nu_1, \nu_2, \nu_3) \in \mathbb{C}^3$ we have

$$\begin{aligned} I_\nu(wuy) = &\left(\frac{y_2^2\left(u_{3,4}^2 + y_1^2\right)}{\left((u_{2,4} - u_{2,3}u_{3,4})^2 + \left(u_{3,4}^2 + y_1^2\right)y_2^2\right)^2} \right)^{\nu_1+2\nu_2+\nu_3} \\ &\times \left(\frac{y_1^2\left((u_{2,4} - u_{2,3}u_{3,4})^2 + \left(u_{3,4}^2 + y_1^2\right)y_2^2\right)}{\left(u_{3,4}^2 + y_1^2\right)^2} \right)^{(\nu_1+2\nu_2+3\nu_3)/2} \\ &\times \left(\left((u_{2,4} - u_{2,3}u_{3,4})^2 + \left(u_{3,4}^2 + y_1^2\right)y_2^2\right)y_3\right)^{(3\nu_1+2\nu_2+\nu_3)/2}. \end{aligned}$$

Clearly, the function does not involve the variables $u_{1,2}$, $u_{1,3}$, $u_{1,4}$, so it is not possible to integrate the function $I_\nu(wuy)$ over the entire u space. We may only consider some type of partial integral which does not involve all the u-variables.

We leave it to the reader to work out a general theory of degenerate Whittaker functions and only briefly indicate how to define these objects. Let $U = U_n(\mathbb{R})$ denote the group of upper triangular $n \times n$ matrices with real coefficients and 1s on the diagonal (upper triangular unipotent matrices). For each element w in the Weyl group of $SL(n, \mathbb{R})$ define

$$U_w := (w^{-1} \cdot U \cdot w) \cap U, \qquad \bar{U}_w = (w^{-1} \cdot {}^t U \cdot w) \cap U.$$

For example, if

$$w = \begin{pmatrix} 1 & 0 & 0 & 0 \\ 0 & 0 & 0 & 1 \\ 0 & 1 & 0 & 0 \\ 0 & 0 & 1 & 0 \end{pmatrix}$$

then

$$U_w = \begin{pmatrix} 1 & * & * & * \\ & 1 & * & \\ & & 1 & \\ & & & 1 \end{pmatrix}, \qquad \bar{U}_w = \begin{pmatrix} 1 & & & \\ & 1 & & * \\ & & 1 & * \\ & & & 1 \end{pmatrix}.$$

We may think of $\bar{U}_w$ as the group opposite or complementary to U_w in the upper triangular unipotent matrices. These spaces have natural Lebesgue measures. For example in the above situation we may write every element $u \in \bar{U}_w$ in the form $u = \begin{pmatrix} 1 & & & \\ & 1 & & u_{2,4} \\ & & 1 & u_{3,4} \\ & & & 1 \end{pmatrix}$, with $u_{2,4}, u_{3,4} \in \mathbb{R}$. The natural measure, du^*, on the space $\bar{U}_w$ is $du^* = du_{2,4}\, du_{3,4}$.

Definition 5.10.1 *For $n \geq 2$, let $z \in \mathfrak{h}^n$, $\nu = (\nu_1, \nu_2, \ldots, \nu_{n-1}) \in \mathbb{C}^{n-1}$, and let ψ_m be a character as in (5.5.1). Then the degenerate Whittaker function associated to w is defined to be*

$$\int_{\bar{U}_w} I_\nu(wuz)\overline{\psi_m(u)}\, d^*u,$$

where d^*u is the natural measure on $\bar{U}_w$.

Remark It may be shown that the degenerate Whittaker function can be meromorphically continued and satisfies the same group of functional equations as Jacquet's Whittaker function as given in Theorem 5.9.8.

GL(n)pack functions The following **GL(n)pack** functions, described in the appendix, relate to the material in this chapter:

FunctionalEquation	IFun	ModularGenerators
Wedge	d	Whittaker
WhittakerGamma	WMatrix.	

6

Automorphic forms and L-functions for $SL(3, \mathbb{Z})$

6.1 Whittaker functions and multiplicity one for $SL(3, \mathbb{Z})$

The generalized upper half-plane $\mathfrak{h}^3$ was introduced in Example 1.2.4 and consists of all matrices $z = x \cdot y$ with

$$x = \begin{pmatrix} 1 & x_{1,2} & x_{1,3} \\ 0 & 1 & x_{2,3} \\ 0 & 0 & 1 \end{pmatrix}, \qquad y = \begin{pmatrix} y_1 y_2 & 0 & 0 \\ 0 & y_1 & 0 \\ 0 & 0 & 1 \end{pmatrix},$$

where $x_{1,2}, x_{1,3}, x_{2,3} \in \mathbb{R}$, $y_1, y_2 > 0$. A basis for the ring $\mathcal{D}^3$, of differential operators in $\partial/\partial x_{1,2}, \partial/\partial x_{1,3}, \partial/\partial x_{2,3}, \partial/\partial y_1, \partial/\partial y_2$ which commute with $GL(3, \mathbb{R})$, can be computed by Proposition 2.3.3, and is given by (see also (Bump, 1984)):

$$\begin{aligned} \Delta_1 = {} & y_1^2 \frac{\partial^2}{\partial y_1^2} + y_2^2 \frac{\partial^2}{\partial y_2^2} - y_1 y_2 \frac{\partial^2}{\partial y_1 \partial y_2} + y_1^2 \left(x_{1,2}^2 + y_2^2\right) \frac{\partial^2}{\partial x_{1,3}^2} \\ & + y_1^2 \frac{\partial^2}{\partial x_{2,3}^2} + y_2^2 \frac{\partial^2}{\partial x_{1,2}^2} + 2 y_1^2 x_{1,2} \frac{\partial^2}{\partial x_{2,3} \partial x_{1,3}}, \end{aligned} \tag{6.1.1}$$

$$\begin{aligned} \Delta_2 = {} & -y_1^2 y_2 \frac{\partial^3}{\partial y_1^2 \partial y_2} + y_1 y_2^2 \frac{\partial^3}{\partial y_1 \partial y_2^2} - y_1^3 y_2^2 \frac{\partial^3}{\partial x_{1,3}^2 \partial y_1} + y_1 y_2^2 \frac{\partial^3}{\partial x_{1,2}^2 \partial y_1} \\ & - 2 y_1^2 y_2 x_{1,2} \frac{\partial^3}{\partial x_{2,3} \partial x_{1,3} \partial y_2} + \left(-x_{1,2}^2 + y_2^2\right) y_1^2 y_2 \frac{\partial^3}{\partial x_{1,3}^2 \partial y_2} \\ & - y_1^2 y_2 \frac{\partial^3}{\partial x_{2,3}^2 \partial y_2} + 2 y_1^2 y_2^2 \frac{\partial^3}{\partial x_{2,3} \partial x_{1,2} \partial x_{1,3}} + 2 y_1^2 y_2^2 x_{1,2} \frac{\partial^3}{\partial x_{1,2} \partial x_{1,3}^2} \\ & + y_1^2 \frac{\partial^2}{\partial y_1^2} - y_2^2 \frac{\partial^2}{\partial y_2^2} + 2 y_1^2 x_{1,2} \frac{\partial^2}{\partial x_{2,3} \partial x_{1,3}} + \left(x_{1,2}^2 + y_2^2\right) y_1^2 \frac{\partial^2}{\partial x_{1,3}^2} \\ & + y_1^2 \frac{\partial^2}{\partial x_{2,3}^2} - y_2^2 \frac{\partial^2}{\partial x_{1,2}^2}. \end{aligned}$$

Let $m = (m_1, m_2)$ with $m_1, m_2 \in \mathbb{Z}$ and let $\nu = (\nu_1, \nu_2)$, with $\nu_1, \nu_2 \in \mathbb{C}$. The Jacquet Whittaker function for $SL(3,\mathbb{Z})$ was introduced in (5.5.1), and takes the form

$$W_{\text{Jacquet}}(z, \nu, \psi_m) = \int_{-\infty}^{\infty}\int_{-\infty}^{\infty}\int_{-\infty}^{\infty} I_\nu(w_3 \cdot u \cdot z)\overline{\psi_m(u)}\, du_{1,2}du_{1,3}du_{2,3}, \qquad (6.1.2)$$

where

$$w_3 = \begin{pmatrix} & & 1 \\ & -1 & \\ 1 & & \end{pmatrix}, \qquad u = \begin{pmatrix} 1 & u_{1,2} & u_{1,3} \\ & 1 & u_{2,3} \\ & & 1 \end{pmatrix},$$

and

$$\psi_m(u) = e^{2\pi i(m_1u_{2,3}+m_2u_{1,2})}, \qquad I_\nu(z) = y_1^{\nu_1+2\nu_2}y_2^{2\nu_1+\nu_2}.$$

It was shown in Section 5.8, 5.9 that $W_{\text{Jacquet}}(z, \nu, \psi_{1,1})$ has meromorphic continuation to all $\nu_1, \nu_2 \in \mathbb{C}$ and satisfies the functional equations:

$$\begin{aligned} W^*_{\text{Jacquet}}(z, (\nu_1, \nu_2), \psi_{1,1}) &= W^*_{\text{Jacquet}}\left(z, \left(\nu_1 + \nu_2 - \tfrac{1}{3}, \tfrac{2}{3} - \nu_2\right), \psi_{1,1}\right) \\ &= W^*_{\text{Jacquet}}\left(z, \left(\tfrac{2}{3} - \nu_1, \nu_1 + \nu_2 - \tfrac{1}{3}, \right), \psi_{1,1}\right), \end{aligned}$$

where

$$\begin{aligned} W^*_{\text{Jacquet}}(z, (\nu_1, \nu_2), \psi_{1,1}) &= \pi^{\frac{1}{2}-3\nu_1-3\nu_2}\Gamma\left(\frac{3\nu_1}{2}\right)\Gamma\left(\frac{3\nu_2}{2}\right) \\ &\quad\times \Gamma\left(\frac{3\nu_1+3\nu_2-1}{2}\right) W_{\text{Jacquet}}(z, \nu, \psi_{1,1}). \end{aligned}$$

Vinogradov and Takhtadzhyan (1982) and Stade (1990) have obtained the following very explicit integral representation

$$\begin{aligned} &W^*_{\text{Jacquet}}(y, (\nu_1, \nu_2), \psi_{1,1}) \\ &= 4y_1^{1+(\nu_1-\nu_2)/2}y_2^{1-(\nu_1-\nu_2)/2}\int_0^\infty K_{\frac{3\nu_1+3\nu_2-2}{2}}\left(2\pi y_2\sqrt{1+u^{-2}}\right) \\ &\quad\times K_{\frac{3\nu_1+3\nu_2-2}{2}}\left(2\pi y_1\sqrt{1+u^2}\right) u^{\frac{3\nu_1-3\nu_2}{2}}\,\frac{du}{u}. \end{aligned} \qquad (6.1.3)$$

Using the above representation, or alternatively following Bump (1984), one may obtain the double Mellin transform pair

$$\tilde{W}_{\text{Jacquet}}(s, \nu) := \int_0^\infty \int_0^\infty W^*_{\text{Jacquet}}(y, (\nu_1, \nu_2), \psi_{1,1})\, y_1^{s_1-1} y_2^{s_2-1} \frac{dy_1}{y_1} \frac{dy_2}{y_2}$$

$$= \frac{\pi^{-s_1-s_2}}{4} \cdot G(s_1, s_2), \tag{6.1.4}$$

$$W^*_{\text{Jacquet}}(y, (\nu_1, \nu_2), \psi_{1,1})$$

$$= \frac{1}{(4\pi^2 i)^2} \int_{2-i\infty}^{2+i\infty} \int_{2-i\infty}^{2+i\infty} G(s_1, s_2)(\pi y_1)^{1-s_1}(\pi y_2)^{1-s_2} ds_1 ds_2,$$

where

$$G(s_1, s_2) = \frac{\Gamma\left(\frac{s_1+\alpha}{2}\right)\Gamma\left(\frac{s_1+\beta}{2}\right)\Gamma\left(\frac{s_1+\gamma}{2}\right)\Gamma\left(\frac{s_2-\alpha}{2}\right)\Gamma\left(\frac{s_2-\beta}{2}\right)\Gamma\left(\frac{s_2-\gamma}{2}\right)}{\Gamma\left(\frac{s_1+s_2}{2}\right)}, \tag{6.1.5}$$

and

$$\alpha = -\nu_1 - 2\nu_2 + 1, \qquad \beta = -\nu_1 + \nu_2, \qquad \gamma = 2\nu_1 + \nu_2 - 1.$$

The Gamma function,

$$\Gamma(s) = \int_0^\infty e^{-u} u^s \frac{du}{u},$$

is uniquely characterized by its functional equation $\Gamma(s+1) = s\Gamma(s)$, growth conditions, and the initial condition $\Gamma(1) = 1$. This is the well-known Bohr–Mollerup theorem (Conway, 1973). A simple proof of the Bohr–Mollerup theorem can be obtained by assuming that if there exists another such function $F(s)$ then $F(s)/\Gamma(s)$ would have to be a periodic function. From the periodicity and the growth conditions, one can conclude (Ahlfors, 1966) that $F(s)/\Gamma(s)$ must be the constant function. Remarkably, this method of proof generalizes to $SL(3, \mathbb{Z})$ with periodic functions replaced by doubly periodic functions. The following proof of multiplicity one was found by Diaconu and Goldfeld. It is not clear if it can be generalized to $SL(n, \mathbb{Z})$ with $n > 3$.

Theorem 6.1.6 (Multiplicity one) *Fix $\nu = (\nu_1, \nu_2) \in \mathbb{C}^2$. Let $\Psi_\nu(z)$ be an $SL(3, \mathbb{Z})$ Whittaker function of type ν associated to a character ψ as in Definition 5.4.1. Assume that $\Psi_\nu(z)$ has sufficient decay in y_1, y_2 so that*

$$\int_0^\infty \int_0^\infty y_1^{\sigma_1} y_2^{\sigma_2} \, |\Psi_\nu(y)| \, \frac{dy_1 dy_2}{y_1 y_2}$$

converges for sufficiently large σ_1, σ_2. Then

$$\Psi_\nu(z) = c \cdot W_{\text{Jacquet}}(z, \nu, \psi),$$

for some constant $c \in \mathbb{C}$.

Proof It is enough to prove the theorem for the case when $\psi(x) = e^{\pi i(x_{2,3}+x_{1,2})}$ since, in general, $\Psi_\nu(z) = \Psi_\nu(y)\psi(x)$ for $z = xy \in \mathfrak{h}^3$. For $s = (s_1, s_2) \in \mathbb{C}^2$, consider the double Mellin transform

$$\tilde{\Psi}_\nu(s) = \int_0^\infty\int_0^\infty y_1^{s_1} y_2^{s_2}\Psi_\nu(y)\,\frac{dy_1 dy_2}{y_1 y_2}$$

which is well defined for $\Re(s_1), \Re(s_2)$ sufficiently large by the assumption in our theorem. Define

$$J_s(z) := \psi(x) y_1^{s_1} y_2^{s_2}$$

for $z = xy \in \mathfrak{h}^3$.

Define the inner product, $\langle , \rangle$, on $\mathcal{L}^2(U(\mathbb{Z})\backslash\mathfrak{h}^3)$ by

$$\langle f, g\rangle = \int_0^\infty\int_0^\infty\int_0^1\int_0^1\int_0^1 f(z)\overline{g(z)}\, dx_{1,2}\, dx_{1,3}\, dx_{2,3}\,\frac{dy_1 dy_2}{(y_1 y_2)^3},$$

for all $f, g \in \mathcal{L}^2\left(U(\mathbb{Z})\backslash\mathfrak{h}^3\right)$. Taking $f(z) = \Psi_\nu(z)$ and $g(z) = J_{\bar{s}}(z)$, it follows that

$$\langle \Psi_\nu, J_{\bar{s}}\rangle = \int_0^\infty\int_0^\infty \Psi_\nu(y)\cdot y_1^{s_1-2} y_2^{s_2-2}\,\frac{dy_1 dy_2}{y_1 y_2} = \tilde{\Psi}_\nu(s^*)$$

where $s^* = (s_1^*, s_2^*)$ with $s_1^* = s_1 - 2$, and $s_2^* = s_2 - 2$.

Let $\mathcal{D}$ denote the polynomial ring over $\mathbb{C}$ of $GL(3,\mathbb{R})$ invariant differential operators generated by Δ_1, Δ_2 given in (6.1.1). Then since $\Psi_\nu(z)$ is an eigenfunction of every $D \in \mathcal{D}$, we may write

$$D\Psi_\nu = \lambda_\nu(D)\Psi_\nu \tag{6.1.7}$$

for some $\lambda_\nu(D) \in \mathbb{R}$. Since D is a self-adjoint operator with respect to the above inner product, it follows that

$$\lambda_\nu(D)\tilde{\Psi}_\nu(s^*) = \langle D\Psi_\nu, J_{\bar{s}}\rangle = \langle \Psi_\nu, DJ_{\bar{s}}\rangle. \tag{6.1.8}$$

Lemma 6.1.9 *Let $\Delta_1, \Delta_2 \in \mathcal{D}$ be given by (6.1.1). Let $s = (s_1, s_2) \in \mathbb{C}^2$. We have*

$$\Delta_i J_s(z) = \sum_{\sigma=(\sigma_1,\sigma_2)\in S} c(s,\sigma,\Delta_i) J_{s+\sigma}(z), \qquad (i = 1, 2),$$

where the sum ranges over the finite set $S = \{(0,0), (0,2), (2,0)\}$, and

$$\begin{aligned} c(s,(0,0),\Delta_i) &= \lambda_{s'}(\Delta_i), \qquad (i = 1, 2),\\ c(s,(2,0),\Delta_1) &= c(s,(0,2),\Delta_1) = -(2\pi i)^2,\\ c(s,(2,0),\Delta_2) &= (2\pi i)^2(1 - s_2),\\ c(s,(0,2),\Delta_2) &= (2\pi i)^2(s_1 - 1), \end{aligned}$$

where $s' = \left(\frac{2s_2 - s_1}{3}, \frac{2s_1 - s_2}{3}\right)$.

The proof of Lemma 6.1.9, first obtained by Friedberg and Goldfield (1993), is given by a simple brute force computation which we omit. Note that the map $s \mapsto s'$ denotes the linear transformation of $\mathbb{C}^2$ such that $y_1^{s_1} y_2^{s_2} = I_{s'}(z)$ with $I_w(z) = y_1^{w_1+2w_2} y_2^{2w_1+w_2}$ for all $w = (w_1, w_2) \in \mathbb{C}^2$. □

It immediately follows from (6.1.8) and Lemma 6.1.9 that

$$\lambda_\nu(\Delta_i)\tilde{\Psi}_\nu(s^*) = \sum_{\sigma=(\sigma_1,\sigma_2)\in S} c(s,\sigma,\Delta_i)\tilde{\Psi}_\nu(s^* + \sigma), \tag{6.1.10}$$

for $i = 1, 2$, and where $s^* = (s_1 - 2, s_2 - 2)$ as before. From (6.1.10), we obtain

$$\begin{aligned} (\lambda_\nu(\Delta_1) - \lambda_{s'}(\Delta_1))\tilde{\Psi}_\nu(s^*) &= 4\pi^2(\tilde{\Psi}_\nu(s^* + (2,0)) \;+\; \tilde{\Psi}_\nu(s^* + (0,2)))\\ (\lambda_\nu(\Delta_2) - \lambda_{s'}(\Delta_2))\tilde{\Psi}_\nu(s^*) &= -4\pi^2((1 - s_2)\tilde{\Psi}_\nu(s^* + (2,0))\\ &\qquad + \;(s_1 - 1)\tilde{\Psi}_\nu(s^* + (0,2))). \end{aligned}$$

Consequently, $\tilde{\Psi}_\nu(s)$ must satisfy the shift equations

$$\tilde{\Psi}_\nu(s) = A(s)\tilde{\Psi}_\nu(s + (2,0)) = B(s)\tilde{\Psi}_\nu(s + (0,2)),$$

for certain meromorphic functions $A(s)$, $B(s)$.

Stirling's formula for the Gamma function (Whittaker and Watson, 1935) tells us that

$$|t|^{\sigma-\frac{1}{2}} e^{-\frac{\pi}{2}|t|} \ll |\Gamma(\sigma + it)| \ll |t|^{\sigma-\frac{1}{2}} e^{-\frac{\pi}{2}|t|} \tag{6.1.11}$$

for $\sigma, t \in \mathbb{R}$ and $|t|$ sufficiently large. If we combine (6.1.11) with the Mellin transform (6.1.5), we obtain

$$\tilde{W}_{\text{Jacquet}}((s_1, s_2), \nu) \underset{\gg}{\ll} |t_1|^{\sigma_1 + \frac{1}{2}\Re(\alpha+\beta+\gamma-s_2-2)} e^{-\frac{\pi}{2}|t_1|} \tag{6.1.12}$$

for $s_1 = \sigma_1 + it_1, s_2 = \sigma_2 + it_2$ and with $|t_1| \to \infty$ and s_2 fixed. We also have a similar estimate for s_1 fixed and $|t_2| \to \infty$.

Let us define the quotient function

$$F_\nu(s) := \frac{\tilde{\Psi}_\nu(s)}{\tilde{W}_{\text{Jacquet}}(s, \nu)}.$$

If we fix s_2 and let $\Re(s_1), \Re(s_2)$ be sufficiently large, then (6.1.12) implies that

$$F_\nu((s_1, s_2)) \ll e^{\frac{\pi}{2}|t_1|} \tag{6.1.13}$$

as $t_1 \to \infty$. Similarly, for s_1 fixed and $\Re(s_1), \Re(s_2)$ sufficiently large, we have

$$F_\nu((s_1, s_2)) \ll e^{\frac{\pi}{2}|t_2|} \tag{6.1.14}$$

as $t_2 \to \infty$.

Note that the shift equations imply that

$$F_\nu((s_1, s_2)) = F_\nu((s_1 + 2, s_2)) = F_\nu((s_1, s_2 + 2)). \tag{6.1.15}$$

Since $F_\nu(s)$ is holomorphic for $\Re(s_1), \Re(s_2)$ sufficiently large, it follows that $F_\nu(s)$ is entire. If we fix s_2 and consider $F_\nu((s_1, s_2))$ as a function of s_1, it is an immediate consequence of (6.1.15) that $F_\nu((s_1, s_2))$ will be periodic (of period 2) in s_1. Thus, for fixed s_2, $F_\nu((s_1, s_2))$ will be a function of $z_1 = e^{\pi i s_1}$ and will have a Laurent expansion in the variable z_1 of the form

$$F_\nu((s_1, s_2)) = \sum_{n=-\infty}^{\infty} c_n(s_2) z_1^n,$$

where the coefficients $c_n(s_2)$ are entire and periodic of period 2. If we fix $s_2 \in \mathbb{C}$, then

$$\int_{\Re(s_1)=0}^{1} |F_\nu((s_1, s_2))|^2 \, ds_1 = \sum_{n=-\infty}^{\infty} |c_n(s_2)|^2 \, e^{-2\pi n t_1} \geq |c_k(s_2)|^2 \, e^{-2\pi k t_1}$$

for every $k = 0, \pm 1, \pm 2, \ldots$

Therefore, we have the Fourier expansion

$$F_\nu((s_1, s_2)) = \sum_{m=-\infty}^{\infty} \sum_{n=-\infty}^{\infty} a_{m,n} e^{\pi i m s_1} e^{\pi i n s_2}.$$

The bound (6.1.13) implies that

$$\int_{\Re(s_1)=0}^{1} |F_\nu((s_1, s_2))|^2 \, ds_1 \ll e^{\pi |t_1|},$$

for $|t_1| \to \infty$ and s_2 fixed. This implies that $c_k(s_2) = 0$ for $k = \pm 1, \pm 2, \ldots$ Similarly,

$$\int_{\Re(s_2)=0}^{1} |F_\nu((s_1, s_2))|^2 \, ds_2 \;\geq\; |a_{0,j}|^2 e^{-2\pi j t_2}$$

for $j = 0, \pm 1, \pm 2, \ldots$ We also have by (6.1.14) that

$$\int_{\Re(s_2)=0}^{1} |F_\nu((s_1, s_2))|^2 \, ds_2 \;\ll\; e^{\pi|t_2|},$$

for $|t_2| \to \infty$ and s_1 fixed. It follows that $a_{0,j} = 0$ for $j = \pm 1, \pm 2, \ldots$ Thus $F_\nu((s_1, s_2))$ must be a constant. This completes the proof of Theorem 6.1.6.

6.2 Maass forms for $SL(3, \mathbb{Z})$

We want to study Maass forms for $SL(3, \mathbb{Z})$. To this end, let us recall Theorem 5.3.2. If U_n denotes the group of $n \times n$ upper triangular matrices with 1s on the diagonal as in Section 5.2, and ϕ is a Maass form for $SL(3, \mathbb{Z})$, then for all $z \in SL(3, \mathbb{Z})\backslash\mathfrak{h}^3$

$$\phi(z) = \sum_{\gamma \in U_2(\mathbb{Z})\backslash SL(2,\mathbb{Z})} \; \sum_{m_1=1}^{\infty} \; \sum_{\substack{m_2=-\infty \\ m_2 \neq 0}}^{\infty} \tilde{\phi}_{(m_1,m_2)}\left(\begin{pmatrix} \gamma & \\ & 1 \end{pmatrix} z\right),$$

where the sum is independent of the choice of coset representatives γ and

$$\tilde{\phi}_{(m_1,m_2)}(z) := \int_0^1 \int_0^1 \int_0^1 \phi(u \cdot z)\, e^{-2\pi i(m_1u_1+m_2u_2)} \, d^*u,$$

with

$$u = \begin{pmatrix} 1 & u_{1,2} & u_{1,3} \\ & 1 & u_{2,3} \\ & & 1 \end{pmatrix} = \begin{pmatrix} 1 & u_2 & u_{1,3} \\ & 1 & u_1 \\ & & 1 \end{pmatrix} \in U_3(\mathbb{R})$$

and $d^*u = du_1 du_2 du_{1,3}$. Note that we have relabeled the super diagonal elements $u_1 = u_{2,3}$, $u_2 = u_{1,2}$ as in Proposition 5.5.2.

Now, we have shown that $\tilde{\phi}_{(m_1,m_2)}(z)$ is a Whittaker function. Further, $\tilde{\phi}_{(m_1,m_2)}$ will inherit the growth properties of the Maass form ϕ and will satisfy the conditions of Theorem 6.1.6. The multiplicity one Theorem 6.1.6 tells us that only the Jacquet Whittaker function (6.1.2) can occur in the Fourier expansion of a Maass form for $SL(3, \mathbb{Z})$, and that $\tilde{\phi}_{(m_1,m_2)}$ must be a constant multiple of the

Jacquet Whittaker function. It follows from Theorem 5.3.2 and Proposition 5.5.2 that if ϕ is a Maass form of type $\nu = (\nu_1, \nu_2) \in \mathbb{C}^2$ for $SL(3,\mathbb{Z})$ then

$$\phi(z) = \sum_{\gamma \in U_2(\mathbb{Z})\backslash SL(2,\mathbb{Z})} \sum_{m_1=1}^{\infty} \sum_{m_2 \neq 0} \frac{A(m_1, m_2)}{|m_1 m_2|} \times W_{\text{Jacquet}}\left(\begin{pmatrix} |m_1 m_2| & & \\ & m_1 & \\ & & 1 \end{pmatrix} \begin{pmatrix} \gamma & \\ & 1 \end{pmatrix} z,\ \nu,\ \psi_{1, \frac{m_2}{|m_2|}} \right), \tag{6.2.1}$$

where $A(m_1, m_2) \in \mathbb{C}$ and $\psi_{\epsilon_1,\epsilon_2}\left(\begin{pmatrix} 1 & u_2 & u_{1,3} \\ & 1 & u_1 \\ & & 1 \end{pmatrix} \right) = e^{2\pi i(\epsilon_1 u_1 + \epsilon_2 u_2)}$.

The particular normalization $A(m_1, m_2)/|m_1 m_2|$ is chosen so that later formulae are as simple as possible.

Lemma 6.2.2 (Fourier coefficients are bounded) *Let ϕ be a Maass form for $SL(3,\mathbb{Z})$ as in (6.2.1). Then for all integers $m_1 \geq 1$, $m_2 \neq 0$,*

$$\frac{A(m_1, m_2)}{|m_1 m_2|} = \mathcal{O}(1).$$

Proof Let $z = \begin{pmatrix} 1 & x_2 & x_{1,3} \\ & 1 & x_1 \\ & & 1 \end{pmatrix} \begin{pmatrix} y_1 y_2 & & \\ & y_1 & \\ & & 1 \end{pmatrix}$. A simple computation shows that

$$\frac{A(m_1, m_2)}{|m_1 m_2|} \cdot W_{\text{Jacquet}}\left(\begin{pmatrix} |m_1 m_2| y_1 y_2 & & \\ & m_1 y_1 & \\ & & 1 \end{pmatrix},\ \nu,\ \psi_{1, \frac{m_2}{|m_2|}} \right)$$
$$= \int_0^1 \int_0^1 \int_0^1 \phi(z)\, e^{-2\pi i [m_1 x_1 + m_2 x_2]}\, dx_1 dx_2 dx_{1,3}.$$

We choose $y_1 = |m_1|^{-1} c_1$, $y_2 = |m_2|^{-2} c_2$, where c_1, c_2 are so chosen that

$$W_{\text{Jacquet}}\left(\begin{pmatrix} c_1 c_2 & & \\ & c_1 & \\ & & 1 \end{pmatrix},\ \nu,\ \psi_{1, \frac{m_2}{|m_2|}} \right) \neq 0.$$

Since ϕ is bounded everywhere, this implies that

$$A(m_1, m_2) = \mathcal{O}\, |m_1 m_2|) .$$

□

6.3 The dual and symmetric Maass forms

Let $\phi(z)$ be a Maass form for $SL(3,\mathbb{Z})$ as in (6.2.1). We shall now define, $\tilde{\phi}(z)$, the dual Maass form associated to ϕ which plays an important role in automorphic form theory.

Proposition 6.3.1 *Let $\phi(z)$ be a Maass form of type $(\nu_1,\nu_2)\in\mathbb{C}^2$ as in (6.2.1). Then*

$$\tilde{\phi}(z) := \phi\big(w\cdot {}^t(z^{-1})\cdot w\big), \qquad w = \begin{pmatrix} & & 1\\ & -1 & \\ 1 & & \end{pmatrix},$$

is a Maass form of type (ν_2,ν_1) for $SL(3,\mathbb{Z})$. The Maass form $\tilde{\phi}$ is called the dual Maass form. If $A(m_1,m_2)$ is the (m_1,m_2)th Fourier coefficient of ϕ then $A(m_2,m_1)$ is the corresponding Fourier coefficient of $\tilde{\phi}$.

Proof First, for every $\gamma\in SL(3,\mathbb{Z})$,

$$\tilde{\phi}(\gamma z) = \phi\big(w\cdot {}^t((\gamma z)^{-1})\cdot w\big) = \phi\big(\gamma' w\cdot {}^t(z^{-1})\cdot w\big) = \tilde{\phi}(z)$$

since $\gamma' = w\cdot {}^t(\gamma^{-1})\cdot w\in SL(3,\mathbb{Z})$. Thus $\tilde{\phi}$ satisfies the automorphic condition (1) of Definition 5.1.3 of a Maass form.

Next, note that if

$$z = \begin{pmatrix} 1 & x_2 & x_{1,3}\\ & 1 & x_1\\ & & 1\end{pmatrix}\cdot\begin{pmatrix} y_1y_2 & & \\ & y_1 & \\ & & 1\end{pmatrix},$$

then

$$w\cdot {}^t(z^{-1})\cdot w = \begin{pmatrix} 1 & x_1 & x_1x_2 - x_{1,3}\\ & 1 & x_2\\ & & 1\end{pmatrix}\cdot\begin{pmatrix} y_1y_2 & & \\ & y_2 & \\ & & 1\end{pmatrix}. \tag{6.3.2}$$

It easily follows that

$$\int_0^1\int_0^1 \tilde{\phi}(z)\, dx_2 dx_{1,3} = \int_0^1\int_0^1 \tilde{\phi}(z)\, dx_1 dx_{1,3} = 0.$$

Thus $\tilde{\phi}$ satisfies the cuspidality condition (3) of Definition 5.1.3.

Now

$$I_{\nu_1,\nu_2}(z) = y_1^{\nu_1+2\nu_2}y_2^{2\nu_1+\nu_2} = I_{\nu_2,\nu_1}\big(w\cdot {}^t(z^{-1})\cdot w\big)$$

since the involution $z\to w\cdot {}^t(z^{-1})\cdot w$ interchanges y_1 and y_2. It then follows from (6.1.1), using the chain rule, that $\tilde{\phi}$ is a Maass form of type (ν_2,ν_1).

Finally, it follows from the identity (6.3.2) that if we integrate

$$\int_0^1 \int_0^1 \int_0^1 \tilde{\phi}(z)\, e^{-2\pi i[m_1x_1+m_2x_2]}\, dx_1 dx_2\, dx_{1,3}$$

to pick off the (m_1, m_2)th Fourier coefficient, then because x_1 and x_2 are interchanged we will actually get $A(m_2, m_1)$. □

In the $SL(2, \mathbb{Z})$ theory, the notions of even and odd Maass forms (see Section 3.9) played an important role. If $a(n)$ is the nth Fourier coefficient of an $SL(2, \mathbb{Z})$ Maass form then $a(n) = \pm a(-n)$ depending on whether the Maass form is even or odd. We shall see that there is a quite different situation in the case of $SL(3, \mathbb{Z})$ and that there are no odd Maass forms in this case. The cognoscenti will recognize that there are no odd Maass forms on $SL(3, \mathbb{Z})$ because our definition of Maass form requires a trivial central character.

Consider a diagonal matrix δ of the form

$$\delta := \begin{pmatrix} \delta_1\delta_2 & & \\ & \delta_1 & \\ & & 1 \end{pmatrix}$$

where $\delta_1, \delta_2 \in \{+1, -1\}$. We define an operator T_δ which maps Maass forms to Maass forms, and is given by

$$T_\delta\phi(z) := \phi(\delta z\delta).$$

Note that

$$T_\delta\,\phi\left(\begin{pmatrix} 1 & x_2 & x_{1,3} \\ & 1 & x_1 \\ & & 1 \end{pmatrix} \cdot \begin{pmatrix} y_1y_2 & & \\ & y_1 & \\ & & 1 \end{pmatrix}\right)$$
$$= \phi\left(\begin{pmatrix} 1 & x_2\delta_2 & x_{1,3}\delta_1\delta_2 \\ & 1 & x_1\delta_1 \\ & & 1 \end{pmatrix} \cdot \begin{pmatrix} y_1y_2 & & \\ & y_1 & \\ & & 1 \end{pmatrix}\right). \tag{6.3.3}$$

Clearly $(T_\delta)^2$ is the identity transformation, so the eigenvalues of T_δ can only be ± 1.

Definition 6.3.4 *A Maass form ϕ of type $\nu = (\nu_1, \nu_2) \in \mathbb{C}^2$ for $SL(3, \mathbb{Z})$ is said to be* **symmetric** *if $T_\delta\phi = \pm\phi$ for all T_δ as in (6.3.3).*

We shall now show that every Maass form ϕ for $SL(3, \mathbb{Z})$ is even, i.e., $T_\delta\phi = \phi$, for all $T_\delta = \begin{pmatrix} \delta_1\delta_2 & & \\ & \delta_1 & \\ & & 1 \end{pmatrix}$ with $\delta_1, \delta_2 \in \{+1, -1\}$. The reason is that a Maass form $\phi(z)$ is invariant under left multiplication by elements in $SL(3,\mathbb{Z})$, in particular by the elements:

$$\begin{pmatrix} -1 & & \\ & -1 & \\ & & 1 \end{pmatrix}, \quad \begin{pmatrix} -1 & & \\ & 1 & \\ & & -1 \end{pmatrix}, \quad \begin{pmatrix} 1 & & \\ & -1 & \\ & & -1 \end{pmatrix}.$$

It is also invariant by the central element

$$\begin{pmatrix} -1 & & \\ & -1 & \\ & & -1 \end{pmatrix}.$$

Since these elements generate all the possible T_δ, this proves our assertion.

Proposition 6.3.5 *Let ϕ be a Maass form of type $\nu = (\nu_1, \nu_2) \in \mathbb{C}^2$ for $SL(3,\mathbb{Z})$ with Fourier–Whittaker expansion*

$$\phi(z) = \sum_{\gamma \in U_2(\mathbb{Z})\backslash SL(2,\mathbb{Z})} \sum_{m_1=1}^{\infty} \sum_{m_2 \neq 0} \frac{A(m_1,m_2)}{|m_1 m_2|}$$
$$\times W_{\text{Jacquet}}\left(\begin{pmatrix} |m_1 m_2| & & \\ & m_1 & \\ & & 1 \end{pmatrix} \begin{pmatrix} \gamma & \\ & 1 \end{pmatrix} z,\ \nu,\ \psi_{1,\frac{m_2}{|m_2|}} \right),$$

as in (6.2.1). Then for all $m_1 \geq 1$ and $m_2 \neq 0$,

$$A(m_1, m_2) = A(m_1, -m_2).$$

Proof Let $T_\delta = \begin{pmatrix} -1 & & \\ & 1 & \\ & & 1 \end{pmatrix}$. Then since $\delta z \delta$ transforms $x_2 \to -x_2$ and $x_{1,3} \to -x_{1,3}$ it easily follows that

$$\int_0^1 \int_0^1 \int_0^1 T_\delta \phi(z) e^{-2\pi i m_1 x_1} e^{-2\pi i m_2 x_2}\, dx_1 dx_2 dx_{1,3}$$

picks off the $A(m_1, -m_2)$ coefficient of $\phi(z)$; and this equals $A(m_1, m_2)$ because $T_\delta \phi(z) = \phi(z)$. □

6.4 Hecke operators for $SL(3,\mathbb{Z})$

We recall the general definition of Hecke operators given in Definition 3.10.5. Consider a group G that acts continuously on a topological space X. Let Γ

be a discrete subgroup of G. For every g in $C_G(\Gamma)$, the commensurator of Γ in G, (i.e., $(g^{-1}\Gamma g) \cap \Gamma$ has finite index in both Γ and $g^{-1}\Gamma g$) we have a decomposition of a double coset into disjoint right cosets of the form

$$\Gamma g \Gamma = \bigcup_i \Gamma \alpha_i. \tag{6.4.1}$$

For each such g, the Hecke operator $T_g : \mathcal{L}^2(\Gamma\backslash X) \to \mathcal{L}^2(\Gamma\backslash X)$ is defined by

$$T_g f(x) = \sum_i f(\alpha_i x),$$

where $f \in \mathcal{L}^2(\Gamma\backslash X)$, $x \in X$, and α_i are given by (6.4.1). The Hecke ring consists of all formal sums

$$\sum_k c_k T_{g_k}$$

with integer coefficients c_k and g_k in a semigroup Δ as in Definition 3.10.8. Since two double cosets are either identical or totally disjoint, it follows that unions of double cosets are associated to elements in the Hecke ring. Finally, we recall Theorem 3.10.10 which states that the Hecke ring is commutative if there exists an antiautomorphism $g \mapsto g^*$ (i.e., $(gh)^* = h^*g^*$) for which $\Gamma^* = \Gamma$ and $(\Gamma g \Gamma)^* = \Gamma g \Gamma$ for every $g \in \Delta$.

We now specialize to the case where

$$G = GL(3, \mathbb{R}), \quad \Gamma = SL(3, \mathbb{Z}), \quad X = GL(3, \mathbb{R})/(O(3, \mathbb{R}) \cdot \mathbb{R}^\times) = \mathfrak{h}^3.$$

For every triple of positive integers m_0, m_1, m_2, the matrix

$$\begin{pmatrix} m_0 m_1 m_2 & & \\ & m_0 m_1 & \\ & & m_0 \end{pmatrix} \in C_G(\Gamma),$$

the commensurator of Γ in G (defined in (3.10.2)). We define Δ to be the semigroup generated by all such matrices. As in the case of $SL(2, \mathbb{Z})$, we have the antiautomorphism

$$g \mapsto {}^t g, \qquad g \in \Delta,$$

where ${}^t g$ denotes the transpose of the matrix g. It is again clear that the conditions of Theorem 3.10.10 are satisfied so that the Hecke ring is commutative.

The following lemma is analogous to Lemma 3.12.1, which came up in the $SL(2, \mathbb{Z})$ situation.

Lemma 6.4.2 *Fix a positive integer $n \ge 1$. Define the set*

$$S_n := \left\{ \begin{pmatrix} a & b_1 & c_1 \\ 0 & b & c_2 \\ 0 & 0 & c \end{pmatrix} \;\middle|\; \begin{matrix} a,b,c \ge 1 \\ abc = n \\ 0 \le b_1 < b,\; 0 \le c_1, c_2 < c \end{matrix} \right\}.$$

Then one has the disjoint partition

$$\bigcup_{m_0^3 m_1^2 m_2 = n} \Gamma \begin{pmatrix} m_0m_1m_2 & & \\ & m_0m_1 & \\ & & m_0 \end{pmatrix} \Gamma = \bigcup_{\alpha \in S_n} \Gamma\alpha. \tag{6.4.3}$$

Proof First of all we claim the decomposition is disjoint. If not, there exists

$$\begin{pmatrix} \gamma_{1,1} & \gamma_{1,2} & \gamma_{1,3} \\ \gamma_{2,1} & \gamma_{2,2} & \gamma_{2,3} \\ \gamma_{3,1} & \gamma_{3,2} & \gamma_{3,3} \end{pmatrix} \in \Gamma$$

such that

$$\begin{pmatrix} \gamma_{1,1} & \gamma_{1,2} & \gamma_{1,3} \\ \gamma_{2,1} & \gamma_{2,2} & \gamma_{2,3} \\ \gamma_{3,1} & \gamma_{3,2} & \gamma_{3,3} \end{pmatrix} \cdot \begin{pmatrix} a & b_1 & c_1 \\ 0 & b & c_2 \\ 0 & 0 & c \end{pmatrix} = \begin{pmatrix} a' & b_1' & c_1' \\ 0 & b' & c_2' \\ 0 & 0 & c' \end{pmatrix}. \tag{6.4.4}$$

This implies that $\gamma_{2,1} = \gamma_{3,1} = \gamma_{3,2} = 0$. Consequently, $\gamma_{1,1}a = a'$, $\gamma_{2,2}b = b'$, and $\gamma_{3,3}c = c'$. But $\gamma_{1,1}\gamma_{2,2}\gamma_{3,3} = 1$ and $a, b, c, a', b', c' \ge 1$. It easily follows that $\gamma_{1,1} = \gamma_{2,2} = \gamma_{3,3} = 1$. Note that the above shows that $a' = a$, $b' = b$, $c' = c$. Therefore, (6.4.4) takes the form

$$\begin{pmatrix} 1 & \gamma_{1,2} & \gamma_{1,3} \\ & 1 & \gamma_{2,3} \\ & & 1 \end{pmatrix} \cdot \begin{pmatrix} a & b_1 & c_1 \\ 0 & b & c_2 \\ 0 & 0 & c \end{pmatrix} = \begin{pmatrix} a & b_1' & c_1' \\ 0 & b & c_2' \\ 0 & 0 & c \end{pmatrix}.$$

Since $0 \le b_1, b_1' < b$ and $0 \le c_1, c_2, c_1', c_2' < c$, one concludes that $\gamma_{1,2} = \gamma_{1,3} = \gamma_{2,3} = 0$, and the decomposition is disjoint as claimed.

Now, by Theorem 3.11.2, every element on the right-hand side of (6.4.3) can be put into Smith normal form, so must occur as an element on the left-hand side of (6.4.3). Similarly, by Theorem 3.11.1, every element on the left-hand side of (6.4.3) can be put into Hermite normal form, so must occur as an element on the right-hand side of (6.4.3). This proves the equality of the two sides of (6.4.3). □

By analogy with the $SL(2,\mathbb{Z})$ situation (see (3.12.3)), it follows that for every integer $n \ge 1$, we have a Hecke operator T_n acting on the space of square integrable automorphic forms $f(z)$ with $z \in \mathfrak{h}^3$. The action is given by the formula

$$T_n f(z) = \frac{1}{n} \sum_{\substack{abc=n \\ 0\le c_1,c_2<c \\ 0\le b_1<b}} f\left(\begin{pmatrix} a & b_1 & c_1 \\ 0 & b & c_2 \\ 0 & 0 & c \end{pmatrix} \cdot z\right). \tag{6.4.5}$$

Note the normalizing factor of $1/n$ which was chosen to simplify later formulae. Clearly, T_1 is just the identity operator.

The $\mathbb{C}$-vector space $\mathcal{L}^2(\Gamma\backslash\mathfrak{h}^3)$ has a natural inner product, denoted $\langle , \rangle$, and defined by

$$\langle f, g\rangle = \int_{\Gamma\backslash\mathfrak{h}^3} f(z)\overline{g(z)}\, d^*z,$$

for all $f, g \in \mathcal{L}^2(\Gamma\backslash\mathfrak{h}^3)$, $z = \begin{pmatrix} 1 & x_{1,2} & x_{1,3} \\ 0 & 1 & x_{2,3} \\ 0 & 0 & 1 \end{pmatrix} \cdot \begin{pmatrix} y_1y_2 & 0 & 0 \\ 0 & y_1 & 0 \\ 0 & 0 & 1 \end{pmatrix} \in \mathfrak{h}^3$,

and where

$$d^*z = dx_{1,2}\, dx_{1,3}\, dx_{2,3}\, \frac{dy_1 dy_2}{(y_1y_2)^3}$$

denotes the left invariant measure given in Proposition 1.5.3.

In the case of $SL(2,\mathbb{Z})$, we showed in Theorem 3.12.4 that the Hecke operators are self-adjoint with respect to the Petersson inner product. For $SL(n,\mathbb{Z})$ with $n \ge 3$, it is no longer true that the Hecke operators are self-adjoint. What happens is that the adjoint operator is again a Hecke operator and, therefore, the Hecke operator commutes with its adjoint, which means that it is a normal operator.

Theorem 6.4.6 (Hecke operators are normal operators) *Consider the Hecke operators $T_n, (n = 1, 2, \ldots)$ defined in (6.4.5). Let T_n^* be the adjoint operator which satisfies*

$$\langle T_n f,\ g\rangle = \langle f,\ T_n^* g\rangle$$

for all $f, g \in \mathcal{L}^2(\Gamma\backslash\mathfrak{h}^3)$. Then T_n^* is another Hecke operator which commutes with T_n so that T_n is a normal operator. Explicitly, T_n^* is associated to the following union of double cosets:

$$\bigcup_{m_0^3 m_1^2 m_2 = n} \Gamma \begin{pmatrix} m_0^2 m_1^2 m_2 & & \\ & m_0^2 m_1 m_2 & \\ & & m_0^2 m_1 \end{pmatrix} \Gamma. \tag{6.4.7}$$

Proof It follows from (6.4.3), and also from the fact that transposition is an antiautomorphism (as in the proof of Theorem 3.10.10), that

$$\bigcup_{m_0^3 m_1^2 m_2 = n} \Gamma \begin{pmatrix} m_0 m_1 m_2 & & \\ & m_0 m_1 & \\ & & m_0 \end{pmatrix} \Gamma = \bigcup_{\alpha \in S_n} \Gamma\alpha = \bigcup_{\alpha \in S_n} \alpha\Gamma. \tag{6.4.8}$$

Since the action of the Hecke operator is independent of the choice of right coset decomposition, we obtain

$$\begin{aligned} \langle T_n f, g\rangle &= \frac{1}{n} \iint_{\Gamma\backslash\mathfrak{h}^3} \sum_{\alpha\in S_n} f(\alpha z)\,\overline{g(z)}\, d^*z \\ &= \frac{1}{n} \iint_{\Gamma\backslash\mathfrak{h}^3} f(z) \sum_{\alpha\in S_n} \overline{g(\alpha^{-1}z)}\, d^*z \\ &= \frac{1}{n} \iint_{\Gamma\backslash\mathfrak{h}^3} f(z) \sum_{\alpha\in S_n} \overline{g\left(\begin{pmatrix} n & & \\ & n & \\ & & n \end{pmatrix}\alpha^{-1}z\right)}\, d^*z, \end{aligned} \tag{6.4.9}$$

after making the change of variables $z \to \alpha^{-1}z$. Multiplying by the diagonal matrix $\begin{pmatrix} n & & \\ & n & \\ & & n \end{pmatrix}$ above does not change anything because g is well defined on $\mathfrak{h}^3$.

Now, it follows from (6.4.8) that for $\omega = \begin{pmatrix} & & 1 \\ & -1 & \\ 1 & & \end{pmatrix}$, we have

$$\begin{aligned} \bigcup_{\alpha\in S_n} \Gamma\alpha^{-1} &= \bigcup_{m_0^3 m_1^2 m_2 = n} \Gamma\cdot\omega \begin{pmatrix} m_0 m_1 m_2 & & \\ & m_0 m_1 & \\ & & m_0 \end{pmatrix}^{-1} \omega^{-1}\cdot\Gamma \\ &= \bigcup_{m_0^3 m_1^2 m_2 = n} \Gamma\cdot \begin{pmatrix} m_0^{-1} & & \\ & (m_0 m_1)^{-1} & \\ & & (m_0 m_1 m_2)^{-1} \end{pmatrix}\cdot\Gamma. \end{aligned} \tag{6.4.10}$$

Finally, if we multiply both sides of (6.4.10) by the diagonal matrix $\begin{pmatrix} n & & \\ & n & \\ & & n \end{pmatrix}$, with $n = m_0^3 m_1^2 m_2$, it follows that the adjoint Hecke operator defined by (6.4.9) is, in fact, associated to the union of double cosets given in (6.4.7). This completes the proof. □

The Hecke operators commute with the differential operators Δ_1, Δ_2 given in (6.1.1) and they also commute with the operators T_δ given in (6.3.3). It follows by standard methods in functional analysis, that we may simultaneously diagonalize the space $\mathcal{L}^2(SL(3,\mathbb{Z})\backslash\mathfrak{h}^3)$ by all these operators. We shall be interested in studying Maass forms which are eigenfunctions of the full Hecke ring of all such operators. The following theorem is analogous to Theorem 3.12.8 which came up in the $SL(2,\mathbb{Z})$ situation.

Theorem 6.4.11 (Multiplicativity of the Fourier coefficients) *Consider*

$$f(z) = \sum_{\gamma\in U_2(\mathbb{Z})\backslash SL(2,\mathbb{Z})} \sum_{m_1=1}^{\infty} \sum_{m_2\neq 0} \frac{A(m_1,m_2)}{|m_1 m_2|}$$
$$\times W_{\text{Jacquet}}\left(\begin{pmatrix} |m_1 m_2| & & \\ & m_1 & \\ & & 1 \end{pmatrix} \begin{pmatrix} \gamma & \\ & 1 \end{pmatrix} z,\ \nu,\ \psi_{1,\frac{m_2}{|m_2|}} \right),$$

a Maass form for $SL(3,\mathbb{Z})$, as in (6.2.1). Assume that f is an eigenfunction of the full Hecke ring. If $A(1,1) = 0$, then f vanishes identically. Assume $f \neq 0$ and it is normalized so that $A(1,1) = 1$. Then

$$T_n f = A(n,1)\cdot f, \qquad \forall\, n = 1,2,\ldots$$

Furthermore, we have the following multiplicativity relations

$$A(m_1 m_1', m_2 m_2') = A(m_1,m_2)\cdot A(m_1',m_2'), \qquad \textit{if } (m_1 m_2, m_1' m_2') = 1,$$
$$A(n,1)A(m_1,m_2) = \sum_{\substack{d_0 d_1 d_2 = n\\ d_1|m_1\\ d_2|m_2}} A\left(\frac{m_1 d_0}{d_1}, \frac{m_2 d_1}{d_2}\right),$$
$$A(1,n)A(m_1,m_2) = \sum_{\substack{d_0 d_1 d_2 = n\\ d_1|m_1\\ d_2|m_2}} A\left(\frac{m_1 d_2}{d_1}, \frac{m_2 d_0}{d_2}\right),$$
$$A(m_1,1)A(1,m_2) = \sum_{d|(m_1,m_2)} A\left(\frac{m_1}{d}, \frac{m_2}{d}\right).$$

Proof Let $z = x \cdot y$ with

$$x = \begin{pmatrix} 1 & x_2 & x_{1,3} \\ & 1 & x_1 \\ & & 1 \end{pmatrix}, \qquad y = \begin{pmatrix} y_1 y_2 & & \\ & y_1 & \\ & & 1 \end{pmatrix}.$$

In view of Theorem 5.3.2 and (6.2.1), we may write (for $m_1 \geq 1, m_2 \neq 0$)

$$\int_0^1 \int_0^1 \int_0^1 f(z) e^{-2\pi i (m_1 x_1 + m_2 x_2)} \, dx_{1,3} \, dx_1 \, dx_2$$
$$= \frac{A(m_1, m_2)}{|m_1 m_2|} \cdot W_{\text{Jacquet}} \left(\begin{pmatrix} |m_1 m_2| y_1 y_2 & & \\ & m_1 y_1 & \\ & & 1 \end{pmatrix}, \, \nu, \, \psi_{1, \frac{m_2}{|m_2|}} \right). \tag{6.4.12}$$

If f is an eigenfunction of the Hecke operator T_n defined by (6.4.5), then we have $T_n f(z) = \lambda_n f(z)$ for some eigenvalue λ_n. We can compute λ_n directly using a variation of (6.4.12). We begin by considering

$$\frac{1}{n^3} \int_0^n \int_0^n \int_0^n T_n f(z) e^{-2\pi i (m_1 x_1 + m_2 x_2)} \, dx_{1,3} \, dx_1 \, dx_2$$
$$= \lambda_n \frac{A(m_1, m_2)}{|m_1 m_2|} \cdot W_{\text{Jacquet}} \left(\begin{pmatrix} |m_1 m_2| y_1 y_2 & & \\ & m_1 y_1 & \\ & & 1 \end{pmatrix}, \, \nu, \, \psi_{1, \frac{m_2}{|m_2|}} \right)$$
$$= \frac{1}{n^4} \sum_{abc = n} \sum_{\substack{0 \leq c_1, c_2 < c \\ 0 \leq b_1 < b}} \int_0^n \int_0^n \int_0^n f\left(\begin{pmatrix} a & b_1 & c_1 \\ 0 & b & c_2 \\ 0 & 0 & c \end{pmatrix} \begin{pmatrix} 1 & x_2 & x_{1,3} \\ 0 & 1 & x_1 \\ 0 & 0 & 1 \end{pmatrix} \cdot y \right)$$
$$\times e^{-2\pi i (m_1 x_1 + m_2 x_2)} \, dx_{1,3} \, dx_1 \, dx_2. \tag{6.4.13}$$

Next, if we let

$$\begin{pmatrix} a & b_1 & c_1 \\ 0 & b & c_2 \\ 0 & 0 & c \end{pmatrix} \begin{pmatrix} 1 & x_2 & x_{1,3} \\ 0 & 1 & x_1 \\ 0 & 0 & 1 \end{pmatrix} = \begin{pmatrix} 1 & \alpha_2 & \alpha_{1,3} \\ & 1 & \alpha_1 \\ & & 1 \end{pmatrix} \begin{pmatrix} a & & \\ & b & \\ & & c \end{pmatrix},$$

then we may solve for $\alpha_1, \alpha_2, \alpha_{1,3}$, by considering

$$\begin{pmatrix} a & a x_2 + b_1 & a x_{1,3} + b_1 x_1 + c_1 \\ & b & b x_1 + c_2 \\ & & c \end{pmatrix} = \begin{pmatrix} a & b\alpha_2 & c\alpha_{1,3} \\ & b & c\alpha_1 \\ & & c \end{pmatrix}.$$

It follows that the right-hand side of (6.4.13) can be expressed in the form

$$\frac{1}{n^4}\sum_{abc=n}\sum_{\substack{0\le c_1,c_2<c\\ 0\le b_1<b}}\int_0^n\int_0^n\int_0^n$$
$$\times f\left(\begin{pmatrix}1 & \frac{ax_2+b_1}{b} & \frac{ax_{1,3}+b_1x_1+c_1}{c}\\ 0 & 1 & \frac{bx_1+c_2}{c}\\ 0 & 0 & 1\end{pmatrix}\cdot\begin{pmatrix}ay_1y_2 & & \\ & by_1 & \\ & & c\end{pmatrix}\right)$$
$$\times e^{-2\pi i(m_1x_1+m_2x_2)}\,dx_{1,3}\,dx_2\,dx_1,$$

which after the elementary transformations

$$x_2'=\frac{ax_2+b_1}{b},\qquad x_1'=\frac{bx_1+c_2}{c},\qquad x_{1,3}'=\frac{ax_{1,3}+b_1x_1+c_1}{c},$$

becomes

$$\frac{1}{n^4}\sum_{abc=n}\frac{c}{a}\frac{b}{a}\frac{c}{b}\sum_{\substack{0\le c_1,c_2<c\\ 0\le b_1<b}}\int_{\frac{c_2}{c}}^{ab^2+\frac{c_2}{c}}\int_{\frac{b_1}{b}}^{a^2c+\frac{b_1}{b}}\int_{\frac{b_1x_1+c_1}{c}}^{a^2b+\frac{b_1x_1+c_1}{c}}$$
$$\times f\left(\begin{pmatrix}1 & x_2' & x_{1,3}'\\ 0 & 1 & x_1'\\ 0 & 0 & 1\end{pmatrix}\cdot\begin{pmatrix}ay_1y_2 & & \\ & by_1 & \\ & & c\end{pmatrix}\right)$$
$$\times e^{2\pi i\left(\frac{m_1c_2}{b}+\frac{m_2b_1}{a}\right)}e^{-2\pi i\left(m_1\frac{cx_1'}{b}+m_2\frac{bx_2'}{a}\right)}\,dx_{1,3}'\,dx_2'\,dx_1'.$$

In view of the fact that the integrand above is periodic and does not change under transformations of the form

$$x_1'\to x_1'+ab^2,\qquad x_2'\to x_2'+a^2c,\qquad x_{1,3}'\to x_{1,3}'+1,$$

we immediately deduce that the above integral is the same as

$$\frac{1}{n^4}\sum_{abc=n}\frac{c^2}{a^2}\int_0^{ab^2}\int_0^{a^2c}\int_0^{a^2b} f\left(\begin{pmatrix}1 & x_2 & x_{1,3}\\ 0 & 1 & x_1\\ 0 & 0 & 1\end{pmatrix}\cdot\begin{pmatrix}ay_1y_2 & & \\ & by_1 & \\ & & c\end{pmatrix}\right)$$
$$\times\left(\sum_{\substack{0\le c_1,c_2<c\\ 0\le b_1<b}} e^{2\pi i\left(\frac{m_1c_2}{b}+\frac{m_2b_1}{a}\right)}\right)e^{-2\pi i\left(m_1\frac{cx_1}{b}+m_2\frac{bx_2}{a}\right)}\,dx_{1,3}\,dx_2\,dx_1. \quad (6.4.14)$$

But the above integral vanishes unless $b|m_1c$ and $a|m_2b$. Furthermore, in the case that $b|m_1c$ and $a|m_2b$, we have

$$\sum_{\substack{0\le c_1,c_2<c\\0\le b_1<b}} e^{2\pi i\left(\frac{m_1c_2}{b}+\frac{m_2b_1}{a}\right)} = \begin{cases} c^2b & \text{if } b|m_1, a|m_2,\\ 0 & \text{otherwise.}\end{cases}$$

Consequently, it follows that our triple integral (6.4.14) may be written in the form:

$$\frac{1}{n^4}\sum_{\substack{abc=n\\ b|m_1,\, a|m_2}} \frac{c^2}{a^2}\cdot ab^2\cdot a^2c\cdot a^2b\cdot c^2b$$
$$\times\int_0^1\int_0^1\int_0^1 f\left(\begin{pmatrix}1 & x_2 & x_{1,3}\\ 0 & 1 & x_1\\ 0 & 0 & 1\end{pmatrix}\cdot\begin{pmatrix}ay_1y_2 & & \\ & by_1 & \\ & & c\end{pmatrix}\right)$$
$$\times e^{-2\pi i\left(m_1\frac{cx_1}{b}+m_2\frac{bx_2}{a}\right)}\,dx_{1,3}\,dx_2\,dx_1.$$

Here we have used the fact that if $F:\mathbb{R}\to\mathbb{C}$ is a periodic integrable function satisfying $F(x+1)=F(x)$, then for any integer $M\ge 1$, we have

$$\int_0^M F(x)\,dx = M\cdot\int_0^1 F(x)\,dx.$$

Finally, the triple integral above can be evaluated with (6.4.12) and has the value

$$\frac{A\left(\frac{m_1c}{b},\,\frac{m_2b}{a}\right)}{\left|\frac{m_1c}{b}\cdot\frac{m_2b}{a}\right|}\cdot W_{\text{Jacquet}}\left(\begin{pmatrix}|m_1m_2|y_1y_2c & & \\ & m_1y_1c & \\ & & c\end{pmatrix},\ \nu,\ \psi_{1,\frac{m_2}{|m_2|}}\right)$$
$$=\frac{A\left(\frac{m_1c}{b},\,\frac{m_2b}{a}\right)}{\left|\frac{m_1c}{b}\cdot\frac{m_2b}{a}\right|}\cdot W_{\text{Jacquet}}\left(\begin{pmatrix}|m_1m_2|y_1y_2 & & \\ & m_1y_1 & \\ & & 1\end{pmatrix},\ \nu,\ \psi_{1,\frac{m_2}{|m_2|}}\right),$$

from which it follows from (6.4.13) and (6.4.14) that

$$\lambda_n\frac{A(m_1,m_2)}{|m_1m_2|}\cdot W_{\text{Jacquet}}\left(\begin{pmatrix}|m_1m_2|y_1y_2 & & \\ & m_1y_1 & \\ & & 1\end{pmatrix},\ \nu,\ \psi_{1,\frac{m_2}{|m_2|}}\right)$$
$$=\frac{1}{n^4}\sum_{\substack{abc=n\\ b|m_1,\, a|m_2}} \frac{c^2}{a^2}\cdot ab^2\cdot a^2c\cdot a^2b\cdot c^2b\cdot\frac{A\left(\frac{m_1c}{b},\,\frac{m_2b}{a}\right)}{\left|\frac{m_1c}{b}\cdot\frac{m_2b}{a}\right|}$$
$$\times W_{\text{Jacquet}}\left(\begin{pmatrix}|m_1m_2|y_1y_2 & & \\ & m_1y_1 & \\ & & 1\end{pmatrix},\ \nu,\ \psi_{1,\frac{m_2}{|m_2|}}\right).$$

If we cancel the Whittaker functions on both sides of the above identity and simplify the expressions, we obtain

$$\lambda_n A(m_1,m_2) = \sum_{\substack{abc=n \\ b|m_1,\ a|m_2}} A\left(\frac{m_1 c}{b}, \frac{m_2 b}{a}\right). \tag{6.4.15}$$

We now explore the consequences of the assumption that $A(1,1)=0$. It follows easily from (6.4.15) that $A(n,1)=0$ for all integers n, and then the left-hand side of (6.4.15) vanishes for all n, m_1 as long as $m_2=1$. By choosing $m_2=1, m_1=p, n=p$ one obtains $A(1,p)=0$. Arguing inductively, we may choose $m_2=1, m_1=p, n=p^\ell$ for $\ell=1,2,\ldots$ from which one can conclude that $A(p^\ell,p)=0$ for all $\ell=0,1,2,\ldots$ One then obtains that the left-hand side of (6.4.15) vanishes as long as $m_1=p$. One can continue in the same manner to show that $A(p^i,p^j)=0$ for all non-negative integers i,j. One may then proceed to products of two primes, products of three primes, etc. to eventually obtain that if $A(1,1)=0$ then all coefficients $A(m,n)$ must vanish.

If $f\neq 0$ then we may assume it is normalized so that $A(1,1)=1$. If we now choose $m_1=m_2=1$, it immediately follows from (6.4.15) that $\lambda_n=A(n,1)$. Substituting this into (6.4.15) proves the identity

$$A(n,1)A(m_1,m_2) = \sum_{\substack{abc=n \\ b|m_1,\ a|m_2}} A\left(\frac{m_1 c}{b}, \frac{m_2 b}{a}\right).$$

The rest of the proof of Theorem 6.4.11 follows easily. □

6.5 The Godement–Jacquet L-function

Let

$$f(z) = \sum_{\gamma\in U_2(\mathbb{Z})\backslash SL(2,\mathbb{Z})} \sum_{m_1=1}^{\infty}\sum_{m_2\neq 0} \frac{A(m_1,m_2)}{|m_1 m_2|} \times W_{\text{Jacquet}}\left(\begin{pmatrix} |m_1 m_2| & & \\ & m_1 & \\ & & 1\end{pmatrix}\begin{pmatrix}\gamma & \\ & 1\end{pmatrix} z,\ \nu,\ \psi_{1,\frac{m_2}{|m_2|}}\right),$$

be a non-zero Maass form for $SL(3,\mathbb{Z})$, normalized so that $A(1,1)=1$, which is a simultaneous eigenfunction of all the Hecke operators as in Theorem 6.4.11. We want to build an L-function out of the Fourier coefficients of f. Lemma 6.2.2 tells us that we may form absolutely convergent Dirichlet series in a suitable half-plane.

The sum over $\gamma \in U_2(\mathbb{Z})\backslash SL(2,\mathbb{Z})$ in the Fourier expansion of f creates seemingly insurmountable complications, and it is not possible to simply set

$$x_{1,2} = x_{1,3} = x_{2,3} = 0,$$

and then take the double Mellin transform in y_1, y_2 which would be the analogue of what we did to create L-functions in the $SL(2,\mathbb{Z})$ situation. The ingenious construction of the L-functions and the proof of their functional equations was first obtained by Godement and Jacquet (1972) and is based on Tate's thesis (Tate, 1950). The original construction of Godement and Jacquet did not use Whittaker models. We follow here a different method of construction as in (Jacquet and Piatetski-Shapiro and Shalika, 1979). The Godement–Jacquet L-function is also commonly referred to as the **standard L-function.**

By Theorem 6.4.11, the Fourier coefficients, $A(m_1, m_2)$, of f must satisfy the multiplicativity relations

$$A(m_1m_1', m_2m_2') = A(m_1, m_2) \cdot A(m_1', m_2'), \qquad \text{if } (m_1m_2, m_1'm_2') = 1,$$

$$A(1,n)A(m_1,m_2) = \sum_{\substack{d_0d_1d_2=n \\ d_1|m_1 \\ d_2|m_2}} A\left(\frac{m_1d_2}{d_1}, \frac{m_2d_0}{d_2}\right),$$

$$A(m_1,1)A(1,m_2) = \sum_{d|(m_1,m_2)} A\left(\frac{m_1}{d}, \frac{m_2}{d}\right).$$

It follows that

$$A(p,1)A(1,p^k) = \sum_{d|p} A\left(\frac{p}{d}, \frac{p^k}{d}\right) = A(1,p^{k-1}) + A(p,p^k)$$

$$A(1,p)A(1,p^{k+1}) = \sum_{\substack{d_0d_2=p \\ d_2|p^{k+1}}} A\left(d_2, \frac{p^{k+1}d_0}{d_2}\right) = A(1,p^{k+2}) + A(p,p^k),$$

with the understanding that $A(1, p^{-1}) = 0$.

Therefore,

$$\frac{A(p,1)A(1,p^k) - A(1,p)A(1,p^{k+1})}{p^{ks}} = \frac{A(1,p^{k-1}) - A(1,p^{k+2})}{p^{ks}}. \tag{6.5.1}$$

If we define

$$\phi_p(s) := \sum_{k=0}^{\infty} \frac{A(1,p^k)}{p^{ks}},$$

then, after summing over k, equation (6.5.1) implies that

$$A(p,1)\cdot\phi_p(s) - A(1,p)\cdot[\phi_p(s)p^s - p^s]$$
$$= \phi_p(s)p^{-s} - [\phi_p(s)p^{2s} - p^{2s} - A(1,p)p^s].$$

Multiplying through by p^{-2s} in the above and solving for $\phi_p(s)$ yields

$$\phi_p(s) = \left(1 - A(1,p)p^{-s} + A(p,1)p^{-2s} - p^{-3s}\right)^{-1}.$$

In a manner completely analogous to the situation of $SL(2,\mathbb{Z})$, as in Definition 3.13.3, it is natural to make the following definition.

Definition 6.5.2 *Let $s\in\mathbb{C}$ with $\Re(s)>2$, and let $f(z)$ be a Maass form for $SL(3,\mathbb{Z})$ as in Theorem 6.4.11. We define the* **Godement–Jacquet L-function** *$L_f(s)$ (termed the L-function associated to f) by the absolutely convergent series*

$$L_f(s) = \sum_{n=1}^{\infty} A(1,n)n^{-s} = \prod_p \left(1 - A(1,p)p^{-s} + A(p,1)p^{-2s} - p^{-3s}\right)^{-1}.$$

Remark It is clear that the L-function associated to the dual Maass form $\tilde{f}$ takes the form

$$L_{\tilde{f}}(s) = \sum_{n=1}^{\infty} A(n,1)n^{-s} = \prod_p \left(1 - A(p,1)p^{-s} + A(1,p)p^{-2s} - p^{-3s}\right)^{-1}.$$

By analogy with the $GL(2)$ situation, we would like to construct the L-function $L_f(s)$ as a Mellin transform of the Maass form f. Before taking the Mellin transform, it is necessary to kill the sum over $GL(2)$ in the Fourier–Whittaker expansion (6.2.1). The procedure to do this uses an auxilliary integral which requires some preliminary preparation.

Set

$$M = \begin{pmatrix} |m_2m_1| & & \\ & |m_1| & \\ & & 1\end{pmatrix},\quad x = \begin{pmatrix} 1 & x_{1,2} & x_{1,3} \\ & 1 & x_{2,3} \\ & & 1\end{pmatrix},\quad y = \begin{pmatrix} y_1y_2 & & \\ & y_1 & \\ & & 1\end{pmatrix}.$$

A simple computation gives

$$M\cdot x = \begin{pmatrix} 1 & |m_2|x_{1,2} & |m_1m_2|x_{1,3} \\ & 1 & |m_1|x_{2,3} \\ & & 1\end{pmatrix}\cdot M.$$

To simplify the subsequent notation, it is very convenient to set

$$x_1 := x_{2,3},\qquad x_2 := x_{1,2},$$

so that the super diagonal elements of the matrix x are x_1, x_2. We also define

$$z_1 := x_1 + iy_1, \qquad z_2 := x_2 + iy_2.$$

It follows from Definition 5.4.1 (1), that that for any integers ϵ_1, ϵ_2, the Jacquet Whittaker function satisfies

$$\begin{aligned} &W_{\text{Jacquet}}(Mz,\ \nu,\ \psi_{\epsilon_1,\epsilon_2}) \\ &\quad = e^{2\pi i\left[|m_1|\epsilon_1 x_1 + |m_2|\epsilon_2 x_2\right]} \cdot W_{\text{Jacquet}}\left(M\begin{pmatrix} y_1y_2 & & \\ & y_1 & \\ & & 1\end{pmatrix},\ \nu,\ \psi_{\epsilon_1,\epsilon_2}\right). \end{aligned} \tag{6.5.3}$$

Further, for any $SL(2,\mathbb{Z})$ matrix $\begin{pmatrix} a & b\\ c & d\end{pmatrix}$, we may put the $GL(3)$ matrix, $\begin{pmatrix} a & b & \\ c & d & \\ & & 1\end{pmatrix}\begin{pmatrix} y_1y_2 & y_1x_2 & x_{1,3}\\ & y_1 & x_1\\ & & 1\end{pmatrix}$ into Iwasawa form:

$$\begin{pmatrix} a & b & \\ c & d & \\ & & 1\end{pmatrix}\begin{pmatrix} y_1y_2 & y_1x_2 & x_{1,3}\\ & y_1 & x_1\\ & & 1\end{pmatrix} \equiv \begin{pmatrix} y_1'y_2' & y_1'x_2' & x_{1,3}'\\ & y_1' & x_1'\\ & & 1\end{pmatrix} \ (\text{mod } Z_3 O(3,\mathbb{R})), \tag{6.5.4}$$

where

$$\begin{aligned} &z_2' = x_2' + iy_2' = \frac{az_2 + b}{cz_2 + d}, \qquad y_1' = |cz_2 + d|y_1,\\ &x_1' = cx_{1,3} + dx_1, \qquad x_{1,3}' = ax_{1,3} + bx_1. \end{aligned}$$

It immediately follows from (6.5.3) and (6.5.4) that

$$\begin{aligned} &W_{\text{Jacquet}}\left(M\begin{pmatrix} a & b & \\ c & d & \\ & & 1\end{pmatrix} z,\ \nu,\ \psi_{\epsilon_1,\epsilon_2}\right)\\ &= e^{2\pi i\left[|m_1|\epsilon_1(cx_{1,3}+dx_1) + |m_2|\epsilon_2\Re\left(\frac{az_2+b}{cz_2+d}\right)\right]}\\ &\quad \times W_{\text{Jacquet}}\left(M\begin{pmatrix} \frac{y_1y_2}{|cz_2+d|} & & \\ & y_1\cdot|cz_2+d| & \\ & & 1\end{pmatrix},\ \nu,\ \psi_{\epsilon_1,\epsilon_2}\right). \end{aligned} \tag{6.5.5}$$

Lemma 6.5.6 *For all $\nu \in \mathbb{C}^2$, $\epsilon_1, \epsilon_2 \in \{\pm 1\}$, and any matrix $y = \begin{pmatrix} y_1y_2 & & \\ & y_1 & \\ & & 1 \end{pmatrix}$ with $y_1, y_2 > 0$, we have*

$$W_{\text{Jacquet}}(y,\ \nu,\ \psi_{\epsilon_1,\epsilon_2}) = W_{\text{Jacquet}}(y,\ \nu,\ \psi_{1,1}).$$

Proof Recall the definition

$$W_{\text{Jacquet}}(y,\ \nu,\ \psi_{\epsilon_1,\epsilon_2}) = \int_{-\infty}^{\infty}\int_{-\infty}^{\infty}\int_{-\infty}^{\infty} I_\nu\left(\begin{pmatrix} & & 1 \\ & -1 & \\ 1 & & \end{pmatrix}\begin{pmatrix} 1 & u_2 & u_{1,3} \\ & 1 & u_1 \\ & & 1 \end{pmatrix} y\right)$$
$$\times e^{-2\pi i(\epsilon_1 u_1 + \epsilon_2 u_2)}\, du_1 du_2 du_{1,3}$$
$$= \int_{-\infty}^{\infty}\int_{-\infty}^{\infty}\int_{-\infty}^{\infty} \left(\frac{y_1\sqrt{(u_{1,3}-u_1u_2)^2 + \left(u_1^2+y_1^2\right)y_2^2}}{u_{1,3}^2 + y_1^2\left(u_2^2+y_2^2\right)}\right)^{\nu_1+2\nu_2}$$
$$\times \left(\frac{y_2\sqrt{u_{1,3}^2 + y_1^2\left(u_2^2+y_2^2\right)}}{(u_{1,3}-u_1u_2)^2 + \left(u_1^2+y_1^2\right)y_2^2}\right)^{2\nu_1+\nu_2}$$
$$\times e^{-2\pi i(\epsilon_1 u_1 + \epsilon_2 u_2)}\, du_1 du_2 du_{1,3}.$$

To complete the proof of the lemma, we simply make the transformation

$$u_1 \to \epsilon_1 u_1, \qquad u_2 \to \epsilon_2 u_2, \qquad u_{1,3} \to \epsilon_1\epsilon_2 u_{1,3}.$$

□

Finally, we obtain the following theorem, which is the basis for the construction of the L-function $L_f(s)$ (given in Definition 6.5.2) as a Mellin transform.

Theorem 6.5.7 *Let $f(z)$ be a Maass form of type ν for $SL(3,\mathbb{Z})$ as in (6.2.1). Then we have the representation*

$$f(z) = \sum_{\begin{pmatrix} a & b \\ c & d \end{pmatrix} \in U_2(\mathbb{Z})\backslash SL(2,\mathbb{Z})} \sum_{m_1=1}^{\infty}\sum_{m_2\neq 0} \frac{A(m_1,m_2)}{|m_1m_2|} \cdot e^{2\pi i\left[m_1(cx_{1,3}+dx_1)+m_2\Re\left(\frac{az_2+b}{cz_2+d}\right)\right]}$$
$$\times W_{\text{Jacquet}}\left(\begin{pmatrix} |m_1m_2| & & \\ & m_1 & \\ & & 1 \end{pmatrix}\begin{pmatrix} \frac{y_1y_2}{|cz_2+d|} & & \\ & y_1\cdot|cz_2+d| & \\ & & 1 \end{pmatrix},\ \nu,\ \psi_{1,1}\right).$$

Proof The proof follows from Theorem 6.4.11, (6.5.5) and Lemma 6.5.6. □

Corollary 6.5.8 *Let $f(z)$ be a Maass form of type ν for $SL(3,\mathbb{Z})$ as in (6.2.1). Then*

$$\int_0^1\int_0^1 f\left(\begin{pmatrix}1 & & u_3\\ & 1 & u_1\\ & & 1\end{pmatrix}\cdot z\right) e^{-2\pi i u_1}\,du_1 du_3$$

$$= \sum_{m_2\neq 0} \frac{A(1,m_2)}{|m_2|} e^{2\pi i(x_1+m_2x_2)} \cdot W_{\rm Jacquet}\left(\begin{pmatrix}|m_2|y_1y_2 & & \\ & y_1 & \\ & & 1\end{pmatrix}, \nu, \psi_{1,1}\right).$$

Proof By Theorem 6.5.7, we see that

$$\int_0^1\int_0^1 f\left(\begin{pmatrix}1 & & u_3\\ & 1 & u_1\\ & & 1\end{pmatrix}\cdot z\right) e^{-2\pi i u_1}\,du_1 du_3$$

$$= \sum_{\begin{pmatrix}a & b\\ c & d\end{pmatrix}\in U_2(\mathbb{Z})\backslash SL(2,\mathbb{Z})} \sum_{m_1=1}^{\infty}\sum_{m_2\neq 0} \frac{A(m_1,m_2)}{|m_1m_2|}$$

$$\times \int_0^1\int_0^1 e^{2\pi i\left[m_1(c(u_3+x_{1,3})+d(u_1+x_1))+m_2\Re\frac{az_2+b}{cz_2+d}\right]} e^{-2\pi i u_1}\,du_1du_3$$

$$\times W_{\rm Jacquet}\left(\begin{pmatrix}|m_1m_2| & & \\ & m_1 & \\ & & 1\end{pmatrix}\begin{pmatrix}\frac{y_1y_2}{|cz_2+d|} & & \\ & y_1\cdot|cz_2+d| & \\ & & 1\end{pmatrix}, \nu, \psi_{1,1}\right).$$

The integrals

$$\int_0^1 e^{2\pi i m_1 c u_3}\,du_3, \qquad \int_0^1 e^{2\pi i(m_1d-1)u_1}\,du_1,$$

vanish unless $c=0$ and $m_1d=1$, in which case they take the value 1. The proof of Corollary 6.5.8 follows immediately from this. □

Our next objective is to construct L-functions associated to a Maass form and show that they satisfy functional equations. In view of the $SL(3,\mathbb{Z})$ Hecke theory we have shown that the natural definition of the L-function associated to a Maass form f is given by Definition 6.5.2:

$$L_f(s) = \sum_{n=1}^{\infty} A(1,n)n^{-s} = \prod_p \left(1 - A(1,p)p^{-s} + A(p,1)p^{-2s} - p^{-3s}\right)^{-1}.$$

This L-function appears naturally in a Mellin transform applied to the integral in Corollary 6.5.8. It, therefore, seems prudent to consider

$$\int_0^\infty\int_0^\infty\int_0^1\int_0^1 f\left(\begin{pmatrix}1 & & u_3\\ & 1 & u_1\\ & & 1\end{pmatrix}\cdot\begin{pmatrix}y_1y_2 & & \\ & y_1 & \\ & & 1\end{pmatrix}\right) e^{-2\pi i u_1}\, du_1 du_3\, y_1^{s_1} y_2^{s_2}\, \frac{dy_1 dy_2}{y_1 y_2},$$

where the inner double integral $\int_0^1\int_0^1$ has the sole function of picking off the Fourier coefficients $A(m_1, m_2)$ of f with $m_1 = 1$. This is the analogue of the Mellin transform

$$\int_0^\infty f\left(\begin{pmatrix}y & \\ & 1\end{pmatrix}\right) y^s \frac{dy}{y}$$

which occurs in the $SL(2, \mathbb{Z})$ theory in Section 3.13. The functional equation in the $SL(2, \mathbb{Z})$ case arises from the symmetry $f(z) = f(\omega\cdot z)$, where $\omega = \begin{pmatrix} & -1\\ 1 & \end{pmatrix}$.

One is, thus, highly motivated to try to generalize this idea to $SL(3, \mathbb{Z})$ by considering symmetries $f(z) = f(\omega\cdot z)$ where ω is in the Weyl group. Curiously, the choice $\omega = \begin{pmatrix} & & -1\\ & 1 & \\ 1 & & \end{pmatrix}$ does not work in the $SL(3, \mathbb{Z})$ situation, but fortunately the choice $\omega = \begin{pmatrix} & 1 & \\ & & 1\\ 1 & & \end{pmatrix}$ does.

We shall now prove Lemma 6.5.9 which contains the symmetry required to obtain the functional equation of the Godement–Jacquet L-function. A new feature which does not appear in the $GL(2)$ theory is the unbalanced nature of this symmetry. One side has a double integral, while the other side has a triple integral!

For the following Lemma 6.5.9, we define for any function $f : \mathfrak{h}^3 \to \mathbb{C}$, its dual function

$$\tilde{f}(z) = f\left(w\,{}^t(z^{-1})w\right), \qquad w = \begin{pmatrix} & & 1\\ & -1 & \\ 1 & & \end{pmatrix},$$

as in Section 6.3. We also introduce

$$w_1 = \begin{pmatrix} & 1 & \\ & & 1 \\ 1 & & \end{pmatrix}, \qquad w_1' = \begin{pmatrix} & & 1 \\ -1 & & \\ & -1 & \end{pmatrix},$$

and

$$z^* = \begin{pmatrix} 1 & & \\ & & -1 \\ & 1 & \end{pmatrix} \cdot {}^t(z^{-1}) \cdot \begin{pmatrix} 1 & & \\ & & 1 \\ & -1 & \end{pmatrix}.$$

If $z = \begin{pmatrix} y_1 y_2 & & x_3 \\ & y_1 & x_1 \\ & & 1 \end{pmatrix}$, then $z^* = y_1^{-1} \begin{pmatrix} y_2^{-1} & & \\ x_3 y_2^{-1} & y_1 & x_1 \\ & & 1 \end{pmatrix}$.

Lemma 6.5.9 *Let $f : \mathfrak{h}^3 \to \mathbb{C}$ be such that $\tilde{f}(z)$ has a Fourier–Whittaker expansion of type (6.2.1). Then for any $z \in \mathfrak{h}^3$, we have the identity*

$$\int_0^1 \int_0^1 f\left(w_1 \cdot \begin{pmatrix} 1 & & u_3 \\ & 1 & u_1 \\ & & 1 \end{pmatrix} \cdot z \right) e^{-2\pi i u_1} \, du_1 du_3$$

$$= \int_0^1 \int_0^1 \tilde{f}\left(w_1' \cdot \begin{pmatrix} 1 & u_1 & -u_3 \\ & 1 & \\ & & 1 \end{pmatrix} \cdot w\, {}^t(z^{-1}) w \right) e^{-2\pi i u_1} \, du_1 du_3$$

$$= \int_{-\infty}^{\infty} \int_0^1 \int_0^1 \tilde{f}\left(\begin{pmatrix} 1 & & u_3 \\ & 1 & u_1 \\ & & 1 \end{pmatrix} \begin{pmatrix} 1 & & \\ u & 1 & \\ & & 1 \end{pmatrix} \cdot z^* \right) e^{-2\pi i u_1} \, du_1 du_3 du.$$

Proof By assumption

$$f(z) = \tilde{f}\big(w \cdot {}^t(z^{-1}) \cdot w\big).$$

Consequently

$$f\left(w_1 \begin{pmatrix} 1 & & u_3 \\ & 1 & u_1 \\ & & 1 \end{pmatrix} \cdot z \right) = \tilde{f}\left(w_1' \begin{pmatrix} 1 & u_1 & -u_3 \\ & 1 & \\ & & 1 \end{pmatrix} \cdot w \cdot {}^t(z^{-1}) \cdot w \right). \tag{6.5.10}$$

Integrating both sides of (6.5.10) with respect to u_1, u_3 gives the first identity in Lemma 6.5.9.

It follows from (6.5.10) that

$$\begin{aligned}
& f\left(w_1\begin{pmatrix}1 & & u_3\\ & 1 & u_1\\ & & 1\end{pmatrix}\cdot z\right)\\
&= \tilde{f}\left(w_1'\begin{pmatrix}1 & u_1 & -u_3\\ & 1 & \\ & & 1\end{pmatrix} w_1'^{-1}\cdot w_1'\cdot w\cdot {}^t(z^{-1})\cdot w w_1'^{-1}\right)\\
&= \tilde{f}\left(\begin{pmatrix}1 & & \\ u_3 & 1 & u_1\\ & & 1\end{pmatrix}\cdot w_1' w\cdot {}^t(z^{-1})\cdot w w_1'^{-1}\right). \qquad (6.5.11)
\end{aligned}$$

Note that $w_1' w = \begin{pmatrix}1 & & \\ & & -1\\ & 1 & \end{pmatrix}$. Recall that $z^* = w_1' w\cdot {}^t(z^{-1})\cdot w w_1'^{-1}$. If we integrate both sides of (6.5.11) then

$$\begin{aligned}
&\int_0^1\int_0^1 f\left(\begin{pmatrix}1 & & u_3\\ & 1 & u_1\\ & & 1\end{pmatrix}\cdot z\right) e^{-2\pi i u_1}\, du_1 du_3\\
&= \int_0^1\int_0^1 \tilde{f}\left(\begin{pmatrix}1 & & \\ & 1 & u_1\\ & & 1\end{pmatrix}\begin{pmatrix}1 & & \\ u_3 & 1 & \\ & & 1\end{pmatrix}\cdot z^*\right) e^{-2\pi i u_1}\, du_1 du_3.
\end{aligned}$$

Finally, the proof may be completed by applying the following Lemma 6.5.12 to the integral above. □

Lemma 6.5.12 *Assume $f : \mathfrak{h}^3 \to \mathbb{C}$ has a Fourier–Whittaker expansion of type (6.2.1). Then for any $z \in \mathfrak{h}^3$, we have the identity*

$$\begin{aligned}
&\int_0^1\int_0^1 f\left(\begin{pmatrix}1 & & \\ & 1 & u_1\\ & & 1\end{pmatrix}\begin{pmatrix}1 & & \\ u & 1 & \\ & & 1\end{pmatrix}\cdot z\right) e^{-2\pi i u_1}\, du_1 du\\
&= \int_{-\infty}^{\infty}\int_0^1\int_0^1 f\left(\begin{pmatrix}1 & & u_3\\ & 1 & u_1\\ & & 1\end{pmatrix}\begin{pmatrix}1 & & \\ u & 1 & \\ & & 1\end{pmatrix}\cdot z\right) e^{-2\pi i u_1}\, du_1 du_3 du.
\end{aligned}$$

Proof For any integer m note that since f is automorphic for $SL(3,\mathbb{Z})$, we have

$$\int_0^1\int_0^1 f\left(\begin{pmatrix}1 & & u_3\\ & 1 & u_1\\ & & 1\end{pmatrix}\begin{pmatrix}1 & & \\ m & 1 & \\ & & 1\end{pmatrix}\cdot z\right) e^{-2\pi i u_1}\, du_1 du_3$$

$$= \int_0^1\int_0^1 f\left(\begin{pmatrix}1 & & \\ m & 1 & \\ & & 1\end{pmatrix}\begin{pmatrix}1 & & u_3\\ & 1 & u_1 - mu_3\\ & & 1\end{pmatrix}\cdot z\right) e^{-2\pi i u_1}\, du_1 du_3$$

$$= \int_0^1\int_0^1 f\left(\begin{pmatrix}1 & & u_3\\ & 1 & u_1\\ & & 1\end{pmatrix}\cdot z\right) e^{-2\pi i u_1} e^{-2\pi i m u_3}\, du_1 du_3. \tag{6.5.13}$$

Furthermore, by Fourier theory

$$f(z) = \sum_{m_1,m_3\in\mathbb{Z}} \int_0^1\int_0^1 f\left(\begin{pmatrix}1 & & u_3\\ & 1 & u_1\\ & & 1\end{pmatrix}\cdot z\right) e^{-2\pi i m_1 u_1} e^{-2\pi i m_3 u_3}\, du_1 du_3,$$

which implies that

$$\int_0^1 f\left(\begin{pmatrix}1 & & \\ & 1 & \xi_1\\ & & 1\end{pmatrix}\cdot z\right) e^{-2\pi i \xi_1}\, d\xi_1$$

$$= \sum_{m_1,m_3\in\mathbb{Z}} \int_0^1\int_0^1\int_0^1 f\left(\begin{pmatrix}1 & & u_3\\ & 1 & u_1+\xi_1\\ & & 1\end{pmatrix}\cdot z\right)$$
$$\times e^{-2\pi i m_1 u_1} e^{-2\pi i m_3 u_3} e^{-2\pi i \xi_1}\, du_1 du_3 d\xi_1$$

$$= \sum_{m_3\in\mathbb{Z}} \int_0^1\int_0^1 f\left(\begin{pmatrix}1 & & u_3\\ & 1 & u_1\\ & & 1\end{pmatrix}\cdot z\right) e^{-2\pi i u_1} e^{-2\pi i m_3 u_3}\, du_1 du_3. \tag{6.5.14}$$

Changing ξ_1 to u_1 in (6.5.14) and combining with (6.5.13) we obtain

$$\int_0^1 f\left(\begin{pmatrix}1 & & \\ & 1 & u_1\\ & & 1\end{pmatrix}\cdot z\right) e^{-2\pi i u_1}\, du_1$$

$$= \sum_{m\in\mathbb{Z}} \int_0^1\int_0^1 f\left(\begin{pmatrix}1 & & u_3\\ & 1 & u_1\\ & & 1\end{pmatrix}\begin{pmatrix}1 & & \\ m & 1 & \\ & & 1\end{pmatrix}\cdot z\right) e^{-2\pi i u_1}\, du_1 du_3.$$

Replacing z by $\begin{pmatrix} 1 & & \\ u & 1 & \\ & & 1 \end{pmatrix} \cdot z$ and then integrating over u in the above identity yields

$$\int_0^1\int_0^1 f\left(\begin{pmatrix} 1 & & \\ & 1 & u_1 \\ & & 1 \end{pmatrix}\cdot\begin{pmatrix} 1 & & \\ u & 1 & \\ & & 1 \end{pmatrix} z\right) e^{-2\pi i u_1}\, du_1 du$$

$$= \sum_{m\in\mathbb{Z}}\int_0^1\int_0^1\int_0^1 f\left(\begin{pmatrix} 1 & & u_3 \\ & 1 & u_1 \\ & & 1 \end{pmatrix}\cdot\begin{pmatrix} 1 & & \\ u+m & 1 & \\ & & 1 \end{pmatrix} z\right) e^{-2\pi i u_1}\, du_1 du_3 du$$

$$= \int_{-\infty}^{\infty}\int_0^1\int_0^1 f\left(\begin{pmatrix} 1 & & u_3 \\ & 1 & u_1 \\ & & 1 \end{pmatrix}\cdot\begin{pmatrix} 1 & & \\ u & 1 & \\ & & 1 \end{pmatrix} z\right) e^{-2\pi i u_1}\, du_1 du_3 du.$$

□

Theorem 6.5.15 *Let f be a Maass form of type $\nu = (\nu_1, \nu_2)$ for $SL(3,\mathbb{Z})$ with dual $\tilde{f}$ as in Proposition 6.3.1. Then $L_f(s)$(respectively $L_{\tilde{f}}(s)$) (given in Definition 6.5.2) have a holomorphic continuation to all $s \in \mathbb{C}$ and satisfy the functional equation*

$$G_\nu(s)L_f(s) = \tilde{G}_\nu(1-s)L_{\tilde{f}}(1-s),$$

where

$$G_\nu(s) = \pi^{-3s/2}\,\Gamma\left(\frac{s+1-2\nu_1-\nu_2}{2}\right)\Gamma\left(\frac{s+\nu_1-\nu_2}{2}\right)\Gamma\left(\frac{s-1+\nu_1+2\nu_2}{2}\right),$$

$$\tilde{G}_{\tilde{\nu}}(s) = \pi^{-3s/2}\,\Gamma\left(\frac{s+1-\nu_1-2\nu_2}{2}\right)\Gamma\left(\frac{s-\nu_1+\nu_2}{2}\right)\Gamma\left(\frac{s-1+2\nu_1+\nu_2}{2}\right).$$

Proof An indirect proof of Theorem 6.5.15 was obtained by Bump (1984) who showed that the functional equation of a Maass form must be the same as that of an Eisenstein series which can be easily derived. We shall follow this method later in this book for the case of $SL(n,\mathbb{R})$ with $n \geq 3$.

A direct proof is much more difficult. We present a proof of Hoffstein and Murty (1989) which makes use of a double Mellin transform. This is different than the proof of Jacquet and Piatetski-Shapiro and Shalika (1979) which utilizes a single Mellin transform. Accordingly, let us set $z = \begin{pmatrix} y_1y_2 & & \\ & y_1 & \\ & & 1 \end{pmatrix}$.

Since we are assuming that f is a Maass form, then we know that $f(w_1 z) = f(z)$, and Lemma 6.5.9 takes the form

$$\int_0^1\int_0^1 f\left(\begin{pmatrix} y_1y_2 & & u_3 \\ & y_1 & u_1 \\ & & 1\end{pmatrix}\right) e^{-2\pi i u_1}\, du_1 du_3$$
$$= \int_{-\infty}^{\infty}\int_0^1\int_0^1 \tilde{f}\left(\begin{pmatrix} 1 & & u_3 \\ & 1 & u_1 \\ & & 1\end{pmatrix}\begin{pmatrix} y_2^{-1} & & \\ uy_2^{-1} & y_1 & \\ & & 1\end{pmatrix}\right) e^{-2\pi i u_1}\, du_1 du_3 du, \tag{6.5.16}$$

The left-hand side of (6.5.16) can be evaluated by Corollary 6.5.8. We have

$$\int_0^1\int_0^1 f\left(\begin{pmatrix} y_1y_2 & & u_3 \\ & y_1 & u_1 \\ & & 1\end{pmatrix}\right) e^{-2\pi i u_1}\, du_1 du_3$$
$$= \sum_{m_2\neq 0} \frac{A(1,m_2)}{|m_2|}\cdot W_{\text{Jacquet}}\left(\begin{pmatrix} |m_2|y_1y_2 & & \\ & y_1 & \\ & & 1\end{pmatrix}, (\nu_1,\nu_2), \psi_{1,1}\right). \tag{6.5.17}$$

In a similar manner, the right-hand side of (6.5.16) can be evaluated to give

$$\int_0^1\int_0^1 f\left(\begin{pmatrix} y_1y_2 & & u_3 \\ & y_1 & u_1 \\ & & 1\end{pmatrix}\right) e^{-2\pi i u_1}\, du_1 du_3$$
$$= \sum_{m_1\neq 0} \frac{A(m_1,1)}{|m_1|}\int_{-\infty}^{\infty} W_{\text{Jacquet}}\left(\begin{pmatrix} |m_1|y_2^{-1} & & \\ uy_2^{-1} & y_1 & \\ & & 1\end{pmatrix}, (\nu_2,\nu_1), \psi_{1,1}\right) du$$
$$= \sum_{m_1\neq 0} \frac{A(m_1,1)}{|m_1|}\int_{-\infty}^{\infty} y_2\cdot W_{\text{Jacquet}}\left(\begin{pmatrix} |m_1|y_2^{-1} & & \\ u & y_1 & \\ & & 1\end{pmatrix}, (\nu_2,\nu_1), \psi_{1,1}\right) du. \tag{6.5.18}$$

□

Note that the right-hand sides of (6.5.17) and (6.5.18) are identical. Then the double Mellin transforms in y_1, y_2 of the right-hand sides of (6.5.17) and (6.5.18) must be the same. For $\Re(s_1)$, $\Re(s_2)$ sufficiently large, the double Mellin transform of the right-hand side of (6.5.17) converges absolutely and is:

$$\int_0^\infty\int_0^\infty\int_0^1\int_0^1 f\left(\begin{pmatrix} y_1y_2 & & u_3\\ & y_1 & u_1\\ & & 1\end{pmatrix}\right) e^{-2\pi i u_1}\, y_1^{s_1-1}y_2^{s_2-1}\, du_1du_3\,\frac{dy_1}{y_1}\frac{dy_2}{y_2}$$

$$= 2L_f(s_2)\cdot\int_0^\infty\int_0^\infty W_{\text{Jacquet}}\left(\begin{pmatrix} y_1y_2 & & \\ & y_1 & \\ & & 1\end{pmatrix},\ (\nu_1,\nu_2),\ \psi_{1,1}\right)$$

$$\times\, y_1^{s_1-1}y_2^{s_2-1}\,\frac{dy_1}{y_1}\frac{dy_2}{y_2}. \tag{6.5.19}$$

Similarly, for $-\Re(s_1), -\Re(s_2)$ sufficiently large, the double Mellin transform of the right-hand side of (6.5.18) converges absolutely and equals:

$$\int_0^\infty\int_0^\infty\int_0^1\int_0^1 f\left(\begin{pmatrix} y_1y_2 & & u_3\\ & y_1 & u_1\\ & & 1\end{pmatrix}\right) e^{-2\pi i u_1}\, y_1^{s_1-1}y_2^{s_2-1}\, du_1du_3\,\frac{dy_1}{y_1}\frac{dy_2}{y_2}$$

$$= 2L_{\tilde f}(1-s_2)\cdot\int_0^\infty\int_0^\infty\left(\int_{-\infty}^\infty W_{\text{Jacquet}}\left(\begin{pmatrix} y_2^{-1} & & \\ u & y_1 & \\ & & 1\end{pmatrix},\ (\nu_2,\nu_1),\ \psi_{1,1}\right)du\right)$$

$$\times\, y_1^{s_1-1}y_2^{s_2}\,\frac{dy_1}{y_1}\frac{dy_2}{y_2}$$

$$= 2L_{\tilde f}(1-s_2)\cdot\int_0^\infty\int_0^\infty\left(\int_{-\infty}^\infty W_{\text{Jacquet}}\left(\begin{pmatrix} y_2 & & \\ u & y_1 & \\ & & 1\end{pmatrix},\ (\nu_2,\nu_1),\ \psi_{1,1}\right)du\right)$$

$$\times\, y_1^{s_1-1}y_2^{-s_2}\,\frac{dy_1}{y_1}\frac{dy_2}{y_2}. \tag{6.5.20}$$

First, the holomorphic continuation of $L_f(s)$ follows by Riemann's trick of breaking the line of integration in the y_2 variable into two pieces [0,1], [1,∞], and then making the transformation $y_2 \mapsto 1/y_2$ in the first piece [0,1]. The proof of the functional equation in Theorem 6.5.15 follows immediately from the following lemma.

Lemma 6.5.21 *The ratio of double Mellin transforms*

$$\frac{\displaystyle\int_0^\infty\int_0^\infty W_{\text{Jacquet}}\left(\begin{pmatrix} y_1y_2 & & \\ & y_1 & \\ & & 1\end{pmatrix},\ (\nu_1,\nu_2),\ \psi_{1,1}\right) y_1^{s_1-1}y_2^{s_2-1}\,\frac{dy_1}{y_1}\frac{dy_2}{y_2}}{\displaystyle\int_0^\infty\int_0^\infty\left(\int_{-\infty}^\infty W_{\text{Jacquet}}\left(\begin{pmatrix} y_2 & & \\ u & y_1 & \\ & & 1\end{pmatrix},\ (\nu_2,\nu_1),\ \psi_{1,1}\right)du\right) y_1^{s_1-1}y_2^{-s_2}\,\frac{dy_1}{y_1}\frac{dy_2}{y_2}}$$

is precisely equal to

$$\frac{G_\nu(s_2)}{\tilde{G}_\nu(1-s_2)}.$$

Proof The triple integral in the denominator of (6.5.21) can be evaluated as follows. We first put the matrix $\begin{pmatrix} y_2 & & \\ u & y_1 & \\ & & 1 \end{pmatrix}$ into Iwasawa form to obtain

$$\begin{pmatrix} y_2 & & \\ u & y_1 & \\ & & 1 \end{pmatrix} = \begin{pmatrix} y_1y_2/\sqrt{u^2+y_1^2} & uy_2/\sqrt{u^2+y_1^2} & \\ & \sqrt{u^2+y_1^2} & \\ & & 1 \end{pmatrix} \quad (\text{mod } O(3,\mathbb{R})\cdot\mathbb{R}^\times).$$

To simplify notation, we also write $W(z)$ instead of $W_{\text{Jacquet}}(z,(\nu_2,\nu_1),\psi_{1,1})$. It follows, after successively making the transformations, $u \mapsto u\cdot y_1$, $y_2 \mapsto y_2\cdot\sqrt{u^2+1}$, $y_1 \mapsto y_1\cdot 1/\sqrt{u^2+1}$, that

$$\begin{aligned}
&\int_0^\infty\int_0^\infty\left(\int_{-\infty}^\infty W\left(\begin{pmatrix} y_2 & & \\ u & y_1 & \\ & & 1 \end{pmatrix}\right)du\right) y_1^{s_1-1}y_2^{-s_2}\,\frac{dy_1}{y_1}\frac{dy_2}{y_2}\\
&= \int_0^\infty\int_0^\infty\left(\int_{-\infty}^\infty W\left(\begin{pmatrix} y_1y_2/\sqrt{u^2+y_1^2} & uy_2/\sqrt{u^2+y_1^2} & \\ & \sqrt{u^2+y_1^2} & \\ & & 1 \end{pmatrix}\right)du\right)\\
&\qquad\qquad \times y_1^{s_1-1}y_2^{-s_2}\,\frac{dy_1}{y_1}\frac{dy_2}{y_2}\\
&= \int_0^\infty\int_0^\infty\left(\int_{-\infty}^\infty W\left(\begin{pmatrix} y_2/\sqrt{u^2+1} & uy_2/\sqrt{u^2+1} & \\ & y_1\sqrt{u^2+1} & \\ & & 1 \end{pmatrix}\right)du\right)\\
&\qquad\qquad \times y_1^{s_1}y_2^{-s_2}\,\frac{dy_1}{y_1}\frac{dy_2}{y_2}\\
&= \int_0^\infty\int_0^\infty\left(\int_{-\infty}^\infty W\left(\begin{pmatrix} y_2 & uy_2 & \\ & y_1 & \\ & & 1 \end{pmatrix}\right)(u^2+1)^{-\frac{s_1+s_2}{2}}\,du\right) y_1^{s_1}y_2^{-s_2}\,\frac{dy_1}{y_1}\frac{dy_2}{y_2}\\
&= \int_0^\infty\int_0^\infty W\left(\begin{pmatrix} y_1y_2 & & \\ & y_1 & \\ & & 1 \end{pmatrix}\right)\left(\int_{-\infty}^\infty e^{2\pi i u y_2}(u^2+1)^{-\frac{s_1+s_2}{2}}\,du\right)\\
&\qquad\qquad \times y_1^{s_1-s_2}y_2^{-s_2}\,\frac{dy_1}{y_1}\frac{dy_2}{y_2}.
\end{aligned}$$

It is well known that

$$\int_{-\infty}^{\infty} e^{2\pi i u y_2}(u^2+1)^{-s}\,du = 2\frac{\pi^s}{\Gamma(s)} y_2^{s-\frac{1}{2}} K_{s-\frac{1}{2}}(2\pi y_2).$$

Consequently, the denominator in Lemma 6.5.21 takes the form:

$$\int_0^\infty\int_0^\infty \left(\int_{-\infty}^{\infty} W\left(\begin{pmatrix} y_2 & & \\ u & y_1 & \\ & & 1\end{pmatrix}\right) du\right) y_1^{s_1-1} y_2^{-s_2}\frac{dy_1}{y_1}\frac{dy_2}{y_2}$$

$$= \frac{2\pi^{(s_1+s_2)/2}}{\Gamma((s_1+s_2)/2)} \int_0^\infty\int_0^\infty W\left(\begin{pmatrix} y_1y_2 & & \\ & y_1 & \\ & & 1\end{pmatrix}\right)$$

$$\times K_{\frac{s_1+s_2-1}{2}}(2\pi y_2)\, y_1^{s_1-s_2} y_2^{\frac{s_1-s_2-1}{2}}\frac{dy_1}{y_1}\frac{dy_2}{y_2}. \tag{6.5.22}$$

The double integral in (6.5.22) was first evaluated by Bump (1984) and is equal to

$$\frac{2^4\pi^{2s_1-s_2-1}\Gamma\left(\frac{1-s_2+\alpha}{2}\right)\Gamma\left(\frac{1-s_2+\beta}{2}\right)\Gamma\left(\frac{1-s_2+\gamma}{2}\right)\Gamma\left(\frac{s_1+\alpha}{2}\right)\Gamma\left(\frac{s_1+\beta}{2}\right)\Gamma\left(\frac{s_1+\gamma}{2}\right)}{\Gamma\left(\frac{s_1+s_2}{2}\right)}, \tag{6.5.23}$$

where

$$\alpha = -\nu_1 - 2\nu_2 + 1,$$
$$\beta = -\nu_1 + \nu_2,$$
$$\gamma = 2\nu_1 + \nu_2 - 1.$$

A number of years later Stade (1990) found another method to obtain (6.5.23).

Finally, we may complete the proof of Lemma 6.5.21 by evaluating the numerator of the expression in Lemma 6.5.21 using (6.1.4) and then explicitly computing the ratio of double Mellin transforms given in Lemma 6.5.21. □

6.6 Bump's double Dirichlet series

In (Bump, 1984) it was shown that if

$$f(z) = \sum_{\gamma\in U_2(\mathbb{Z})\backslash SL(2,\mathbb{Z})}\sum_{m_1=1}^{\infty}\sum_{m_2\neq 0}\frac{A(m_1,m_2)}{|m_1m_2|}$$

$$\times W_{\text{Jacquet}}\left(\begin{pmatrix} |m_1m_2| & & \\ & m_1 & \\ & & 1\end{pmatrix}\begin{pmatrix}\gamma & \\ & 1\end{pmatrix} z,\ \nu,\ \psi_{1,\frac{m_2}{|m_2|}}\right) \tag{6.6.1}$$

is a non-zero Maass form for $SL(3, \mathbb{Z})$, normalized so that $A(1, 1) = 1$, which is a simultaneous eigenfunction of all the Hecke operators as in Theorem 6.4.11, then the double Dirichlet series

$$\sum_{m_1=1}^{\infty}\sum_{m_2=1}^{\infty} \frac{A(m_1, m_2)}{m_1^{s_1} m_2^{s_2}} \tag{6.6.2}$$

has a meromorphic continuation to all $s_1, s_2 \in \mathbb{C}^2$ and satisfies certain functional equations. This result is a consequence of the following proposition.

Proposition 6.6.3 *Let f be a Maass form for $SL(3, \mathbb{Z})$ as in (6.6.1). Then we have the factorization*

$$\sum_{m_1=1}^{\infty}\sum_{m_2=1}^{\infty} \frac{A(m_1, m_2)}{m_1^{s_1} m_2^{s_2}} = \frac{L_{\tilde{f}}(s_1) L_f(s_2)}{\zeta(s_1 + s_2)}.$$

Proof By Theorem 6.4.11, the Fourier coefficients of f satisfy

$$A(m_1, 1)A(1, m_2) = \sum_{d|m_1}\sum_{d|m_2} A\left(\frac{m_1}{d}, \frac{m_2}{d}\right).$$

It follows that

$$\begin{aligned}
\sum_{m_1=1}^{\infty}\sum_{m_2=1}^{\infty} \frac{A(m_1, 1)}{m_1^{s_1}} \frac{A(1, m_2)}{m_2^{s_2}} &= \sum_{m_1=1}^{\infty}\sum_{m_2=1}^{\infty}\sum_{d|m_1}\sum_{d|m_2} A\left(\frac{m_1}{d}, \frac{m_2}{d}\right) m_1^{-s_1} m_2^{-s_2} \\
&= \sum_{d=1}^{\infty} \sum_{\substack{m_1=1 \\ m_1 \equiv 0 \pmod d}}^{\infty} \sum_{\substack{m_2=1 \\ m_2 \equiv 0 \pmod d}}^{\infty} \frac{A(m_1/d, m_2/d)}{m_1^{s_1} m_2^{s_2}} \\
&= \sum_{d=1}^{\infty}\sum_{m_1'=1}^{\infty}\sum_{m_2'=1}^{\infty} \frac{A(m_1', m_2')}{(m_1' d)^{s_1} (m_2' d)^{s_2}} \\
&= \zeta(s_1 + s_2) \cdot \sum_{m_1=1}^{\infty}\sum_{m_2=1}^{\infty} \frac{A(m_1, m_2)}{m_1^{s_1} m_2^{s_2}}.
\end{aligned}$$

□

A direct proof of the meromorphic continuation and functional equation of Bump's double Dirichlet series (6.6.2) has been found by M. Thillainatesan. By Proposition 6.6.3, this gives a new proof of the functional equation of the Godement–Jacquet L-function $L_f(s)$. The new proof of the functional equation of (6.6.2) is based on the following two propositions.

Proposition 6.6.4 *Let f be a Maass form of type ν for $SL(3,\mathbb{Z})$ as in (6.6.1). For $\Re(s_1), \Re(s_2)$ sufficiently large, define*

$$\Lambda(s_1,s_2) = \int_0^\infty\int_0^\infty\int_0^1 f\left(\begin{pmatrix}1&0&u_3\\0&1&0\\0&0&1\end{pmatrix}\begin{pmatrix}y_1y_2&&\\&y_1&\\&&1\end{pmatrix}\right) \times y_1^{s_1-1}y_2^{s_2-1}du_3\frac{dy_1}{y_1}\frac{dy_2}{y_2}.$$

Then

$$\Lambda(s_1,s_2) = \frac{4L_{\tilde f}(s_1)L_f(s_2)}{\zeta(s_1+s_2)}\int_0^\infty\int_0^\infty W_{\text{Jacquet}}\left(\begin{pmatrix}y_1y_2&&\\&y_1&\\&&1\end{pmatrix}, \nu, \psi_{1,1}\right) \times y_1^{s_1-1}y_2^{s_2-1}\frac{dy_1}{y_1}\frac{dy_2}{y_2}.$$

Proposition 6.6.5 *Let f be a Maass form of type ν for $SL(3,\mathbb{Z})$ as in (6.6.1). For $\Re(s_1), -\Re(s_2)$ sufficiently large and $\Re(s_1+s_2) > \frac{1}{2}$, define*

$$\Lambda_1(s_1,s_2) = \int_0^\infty\int_0^\infty\int_0^1 f_1\left(\begin{pmatrix}1&0&u_3\\0&1&0\\0&0&1\end{pmatrix}\begin{pmatrix}y_1y_2&&\\&y_1&\\&&1\end{pmatrix}\right) \times y_1^{s_1-1}y_2^{s_2-1}du_3\frac{dy_1}{y_1}\frac{dy_2}{y_2},$$

where $f_1(z) = \tilde f(w_2{}^t(z^{-1}))$ with $w_2 = \begin{pmatrix}0&0&1\\1&0&0\\0&1&0\end{pmatrix}$. Then

$$\Lambda_1(s_1,s_2) = \frac{4L_{\tilde f}(s_1)L_{\tilde f}(1-s_2)}{\Gamma((s_1+s_2)/2)\zeta(s_1+s_2)}\int_0^\infty\int_0^\infty W_{\text{Jacquet}}\left(\begin{pmatrix}y_1y_2&&\\&y_1&\\&&1\end{pmatrix}, \nu, \psi_{1,1}\right) \times y_1^{s_1-s_2}y_2^{(s_1-s_2-1)/2}K_{\frac{s_1+s_2-1}{2}}(y_2)\frac{dy_1}{y_1}\frac{dy_2}{y_2}.$$

Since f is automorphic, it is clear that $f = f_1$ and $\Lambda(s_1,s_2) = \Lambda_1(s_1,s_2)$. The proof of the functional equation (in Theorem 6.5.15) of the Godement–Jacquet L-function now follows from Lemma 6.5.21 and (6.5.22).

Proof of Proposition 6.6.4 We may substitute the Whittaker expansion in Theorem 6.5.7 into the integral for $\Lambda(s_1,s_2)$. The integral over u_3 forces $c = 0$

and $d = \pm 1$ (see below), and kills the sum over $SL(2,\mathbb{Z})$. We have

$$\Lambda(s_1, s_2)$$

$$= \sum_{\left(\begin{smallmatrix} a & b \\ c & d \end{smallmatrix}\right) \in U_2(\mathbb{Z})\backslash SL(2,\mathbb{Z})} \sum_{m_1=1}^{\infty} \sum_{m_2 \neq 0} \frac{A(m_1, m_2)}{|m_1 m_2|}$$

$$\times \int_0^\infty \int_0^\infty \left(\int_0^1 e^{2\pi i \left[m_1 c u_3 + m_2 \Re\left(\frac{a i y_2 + b}{c i y_2 + d} \right) \right]} du_3 \right)$$

$$\times W_{\text{Jacquet}} \left(\begin{pmatrix} \frac{y_1 y_2 m_1 |m_2|}{|c i y_2 + d|} & & \\ & y_1 m_1 \cdot |c i y_2 + d| & \\ & & 1 \end{pmatrix}, \nu, \psi_{1,1} \right)$$

$$\times y_1^{s_1 - 1} y_2^{s_2 - 1} \frac{dy_1}{y_1} \frac{dy_2}{y_2}$$

$$= 2 \sum_{m_1=1}^{\infty} \sum_{m_2 \neq 0} \frac{A(m_1, m_2)}{|m_1 m_2|} \int_0^\infty \int_0^\infty W_{\text{Jacquet}} \left(\begin{pmatrix} y_1 y_2 m_1 |m_2| & & \\ & m_1 y_1 & \\ & & 1 \end{pmatrix}, \nu, \psi_{1,1} \right)$$

$$\times y_1^{s_1 - 1} y_2^{s_2 - 1} \frac{dy_1}{y_1} \frac{dy_2}{y_2}$$

$$= 4 \sum_{m_1=1}^{\infty} \sum_{m_2=1}^{\infty} \frac{A(m_1, m_2)}{m_1^{s_1} |m_2|^{s_2}} \int_0^\infty \int_0^\infty W_{\text{Jacquet}} \left(\begin{pmatrix} y_1 y_2 & & \\ & y_1 & \\ & & 1 \end{pmatrix}, \nu, \psi_{1,1} \right)$$

$$\times y_1^{s_1 - 1} y_2^{s_2 - 1} \frac{dy_1}{y_1} \frac{dy_2}{y_2}.$$

□

Proof of Proposition 6.6.5 By definition

$$\Lambda_1(s_1, s_2)$$

$$= \int_0^\infty \int_0^\infty \int_0^1 f_1 \left(\begin{pmatrix} 1 & 0 & u \\ 0 & 1 & 0 \\ 0 & 0 & 1 \end{pmatrix} \begin{pmatrix} y_1 y_2 & & \\ & y_1 & \\ & & 1 \end{pmatrix} \right) y_1^{s_1 - 1} y_2^{s_2 - 1} du \frac{dy_1}{y_1} \frac{dy_2}{y_2}$$

$$= \int_0^\infty \int_0^\infty \int_0^1 \tilde{f} \left(\begin{pmatrix} 1 & -u & 0 \\ 0 & 1 & 0 \\ 0 & 0 & 1 \end{pmatrix} \begin{pmatrix} y_1 y_2 y_2^{-1} & & \\ & y_2^{-1} & \\ & & 1 \end{pmatrix} \right) y_1^{s_1 - 1} y_2^{s_2 - 1} du \frac{dy_1}{y_1} \frac{dy_2}{y_2}$$

$$= \int_0^\infty \int_0^\infty \int_{-1}^0 \tilde{f} \left(\begin{pmatrix} 1 & u & 0 \\ 0 & 1 & 0 \\ 0 & 0 & 1 \end{pmatrix} \begin{pmatrix} y_1 y_2 & & \\ & y_1 & \\ & & 1 \end{pmatrix} \right) y_1^{s_1 - s_2} y_2^{s_1 - 1} du \frac{dy_1}{y_1} \frac{dy_2}{y_2},$$

□

where the last identity above is obtained after making the successive transformations $y_2 \mapsto y_1^{-1}$, $y_1 y_2 \mapsto y_2$, and $u \mapsto -u$. We now substitute the Whittaker expansion of Theorem 6.5.7 into the above. In this case the sum over $SL(2,\mathbb{Z})$ is not killed. Remarkably, however, the integral can be significantly simplified after several clever transformations. We have

$$\Lambda_1(s_1,s_2) = \sum_{\begin{pmatrix} a & b \\ c & d \end{pmatrix} \in U_2(\mathbb{Z})\backslash SL(2,\mathbb{Z})} \sum_{m_1=1}^{\infty} \sum_{m_2\neq 0} \frac{A(m_2,m_1)}{|m_1 m_2|} \int_0^\infty \int_0^\infty \int_{-1}^0$$
$$\times\, e^{2\pi i m_2 \Re\left(\frac{a(u+iy_2)+b}{c(u+iy_2)+d}\right)} \cdot W_{\text{Jacquet}}\left(\begin{pmatrix} \frac{y_1 y_2 m_1 |m_2|}{|c(u+iy_2)+d|} & & \\ & y_1 m_1 \cdot |c(u+iy_2)+d| & \\ & & 1 \end{pmatrix}, \nu, \psi_{1,1}\right)$$
$$\times\, y_1^{s_1-s_2} y_2^{s_1-1}\, du \frac{dy_1}{y_1}\frac{dy_2}{y_2}.$$

The term corresponding to $c=0$ will be killed in the above integral over u. So, we may assume $c\neq 0$. Note the identity: $(az+b)/(cz+d) = (a/c) - (1/c(cz+d))$. Hence,

$$\Re\left(\frac{a(u+iy_2)+b}{c(u+iy_2)+d}\right) = \Re\left(\frac{a}{c} - \frac{1}{c(c(u+iy_2)+d)}\right).$$

With this in mind, in the above integrals, make the transformations

$$cu+d \mapsto cu, \qquad d = cq + r(1\le r\le |c|,\ (r,c)=1).$$

Consequently

$$\Lambda_1(s_1,s_2) = \sum_{m_1=1}^{\infty} \sum_{m_2\neq 0} \sum_{q\in\mathbb{Z}} \sum_{c\neq 0} \sum_{\substack{r=1 \\ (r,c)=1}}^{|c|} e^{\frac{2\pi i m_2 \bar{r}}{c}} \frac{A(m_2,m_1)}{|m_1 m_2|}$$
$$\times \int_0^\infty \int_0^\infty \int_{q+\frac{r}{c}-1}^{q+\frac{r}{c}} e^{-\frac{2\pi i m_2 u}{|c(u+iy_2)|^2}}$$
$$\times\, W_{\text{Jacquet}}\left(\begin{pmatrix} \frac{y_1 y_2 m_1 |m_2|}{|c(u+iy_2)|} & & \\ & y_1 m_1 \cdot |c(u+iy_2)| & \\ & & 1 \end{pmatrix}, \nu, \psi_{1,1}\right)$$
$$\times\, y_1^{s_1-s_2} y_2^{s_1-1}\, du \frac{dy_1}{y_1}\frac{dy_2}{y_2}. \tag{6.6.6}$$

Note that for fixed r, c,

$$\sum_{q\in\mathbb{Z}} \int\limits_{q+\frac{r}{c}-1}^{q+\frac{r}{c}} = \int\limits_{-\infty}^{\infty}. \tag{6.6.7}$$

Next, we combine (6.6.6) and (6.6.7) and then make the successive transformations

$$y_1 \mapsto \frac{y_1}{|c(u+iy_2)|}, \qquad u \mapsto \frac{u}{c^2}, \qquad y_2 \mapsto \frac{y_2}{c^2}.$$

We obtain

$$\Lambda_1(s_1,s_2) = \sum_{m_1=1}^{\infty}\sum_{m_2\neq 0}\sum_{c\neq 0}\sum_{\substack{r=1\\(r,c)=1}}^{|c|} \frac{e^{\frac{2\pi i m_2 \bar{r}}{c}}}{|c|^{s_1+s_2}}\frac{A(m_2,m_1)}{|m_1m_2|}\int_0^\infty\int_0^\infty\int_{-\infty}^\infty e^{-\frac{2\pi i m_2 u}{|u+iy_2|^2}}$$
$$\times W_{\text{Jacquet}}\left(\begin{pmatrix} \frac{y_1y_2m_1|m_2|}{|u+iy_2|^2} & & \\ & y_1m_1 & \\ & & 1\end{pmatrix}, \nu, \psi_{1,1}\right) y_1^{s_1-s_2}y_2^{s_1-1}$$
$$\times |u+iy_2|^{s_2-s_1}\, du\, \frac{dy_1}{y_1}\frac{dy_2}{y_2}. \tag{6.6.8}$$

We now successively make the transformations

$$u\mapsto u\cdot y_2, \qquad y_2\mapsto y_2\cdot\frac{|m_2|}{u^2+1}, \qquad y_1\mapsto\frac{y_1}{m_1}$$

in (6.6.8). It follows that

$$\Lambda_1(s_1,s_2) = \sum_{m_1=1}^{\infty}\sum_{m_2\neq 0}\sum_{c\neq 0}\sum_{\substack{r=1\\(r,c)=1}}^{|c|} \frac{e^{\frac{2\pi i m_2 \bar{r}}{c}}}{|c|^{s_1+s_2}}\frac{A(m_2,m_1)}{|m_1m_2|}\int_0^\infty\int_0^\infty\int_{-\infty}^\infty e^{-\frac{2\pi i m_2 u}{y_2(u^2+1)}}$$
$$\times W_{\text{Jacquet}}\left(\begin{pmatrix} \frac{y_1m_1|m_2|}{y_2(u^2+1)} & & \\ & y_1m_1 & \\ & & 1\end{pmatrix}, \nu, \psi_{1,1}\right) y_1^{s_1-s_2}y_2^{s_2}$$
$$\times (u^2+1)^{(s_2-s_1)/2}\, du\, \frac{dy_1}{y_1}\frac{dy_2}{y_2}$$
$$= \sum_{m_1=1}^{\infty}\sum_{m_2\neq 0}\sum_{c\neq 0}\sum_{\substack{r=1\\(r,c)=1}}^{|c|} \frac{e^{\frac{2\pi i m_2 \bar{r}}{c}}}{|c|^{s_1+s_2}}\frac{A(m_2,m_1)}{m_1^{1+s_1-s_2}\,|m_2|^{1-s_2}}$$
$$\times\int_0^\infty\int_0^\infty\int_{-\infty}^\infty e^{-\frac{2\pi i u}{y_2}\cdot\frac{m_2}{|m_2|}} W_{\text{Jacquet}}\left(\begin{pmatrix} y_1y_2^{-1} & & \\ & y_1 & \\ & & 1\end{pmatrix}, \nu, \psi_{1,1}\right)$$
$$\times\frac{y_1^{s_1-s_2}y_2^{s_2}}{(u^2+1)^{(s_1+s_2)/2}}\, du\, \frac{dy_1}{y_1}\frac{dy_2}{y_2}. \tag{6.6.9}$$

Note that the Dirichlet series completely separates from the triple integral of the Whittaker function in (6.6.9). To complete the proof, we need two lemmas. The first lemma evaluates the Dirichlet series in (6.6.9) while the second evaluates the triple integral of the Whittaker function.

Lemma 6.6.10 *We have*

$$\sum_{m_1=1}^{\infty}\sum_{m_2\neq 0}\sum_{c\neq 0}\sum_{\substack{r=1\\(r,c)=1}}^{|c|}\frac{e^{\frac{2\pi i m_2 \bar{r}}{c}}}{|c|^{s_1+s_2}}\,\frac{A(m_2,m_1)}{m_1^{1+s_1-s_2}\,|m_2|^{1-s_2}}=\frac{4L_{\tilde{f}}(s_1)L_{\tilde{f}}(1-s_2)}{\zeta(s_1+s_2)}.$$

Lemma 6.6.11 *We have*

$$\int_0^\infty\int_0^\infty\int_{-\infty}^{\infty} e^{-\frac{2\pi i u}{y_2}\cdot\frac{m_2}{|m_2|}}\,W_{\text{Jacquet}}\left(\begin{pmatrix} y_1y_2^{-1} & & \\ & y_1 & \\ & & 1\end{pmatrix},\ \nu,\ \psi_{1,1}\right)$$
$$\times\frac{y_1^{s_1-s_2}\,y_2^{s_2}}{(u^2+1)^{(s_1+s_2)/2}}\,du\,\frac{dy_1}{y_1}\frac{dy_2}{y_2}$$
$$=\int_0^\infty\int_0^\infty W_{\text{Jacquet}}\left(\begin{pmatrix} y_1y_2 & & \\ & y_1 & \\ & & 1\end{pmatrix},\ \nu,\ \psi_{1,1}\right)$$
$$\times\, y_1^{s_1-s_2}\,y_2^{(s_1-s_2-1)/2}\,K_{\frac{s_1+s_2-1}{2}}(y_2)\,\frac{dy_1}{y_1}\frac{dy_2}{y_2}.$$

Proof To prove Lemma 6.6.10, we apply Proposition 3.1.7. Since every Maass form is even, we obtain

$$\sum_{m_1=1}^{\infty}\sum_{m_2\neq 0}\sum_{c\neq 0}\sum_{\substack{r=1\\(r,c)=1}}^{|c|}\frac{e^{(2\pi i m_2\bar{r})/c}}{|c|^{s_1+s_2}}\,\frac{A(m_2,m_1)}{m_1^{1+s_1-s_2}\,|m_2|^{1-s_2}}$$
$$=\frac{4}{\zeta(s_1+s_2)}\sum_{m_1=1}^{\infty}\sum_{m_2=1}^{\infty}\sum_{d|m_2}\frac{A(m_2,m_1)}{d^{s_1+s_2-1}m_1^{1+s_1-s_2}m_2^{1-s_2}}$$
$$=\frac{4}{\zeta(s_1+s_2)}\sum_{m_1=1}^{\infty}\sum_{m_2=1}^{\infty}\sum_{d=1}^{\infty}\frac{A(m_2d,m_1)}{m_1^{1+s_1-s_2}m_2^{1-s_2}d^{s_1}}$$
$$=\frac{4}{\zeta(s_1+s_2)}\sum_{m_1=1}^{\infty}\sum_{m_2=1}^{\infty}\sum_{d=1}^{\infty}\frac{A(m_2d,m_1)}{(m_1m_2)^{1-s_2}(m_1d)^{s_1}}$$
$$=\frac{4}{\zeta(s_1+s_2)}\sum_{m_2=1}^{\infty}\sum_{d=1}^{\infty}\sum_{m_1|(d,m_2)}\frac{A\big((m_2d/m_1^2),m_1\big)}{m_2^{1-s_2}d^{s_1}}$$

$$= \frac{4}{\zeta(s_1+s_2)} \sum_{m_2=1}^{\infty} \sum_{d=1}^{\infty} \frac{A(d,1)}{d^{s_1}} \cdot \frac{A(m_2,1)}{m_2^{1-s_2}}$$
$$= \frac{4L_{\tilde{f}}(s_1)L_{\tilde{f}}(1-s_2)}{\zeta(s_1+s_2)}.$$

Finally, to prove Lemma 6.6.11, we use the identity

$$\int_{-\infty}^{\infty} e^{\pm 2\pi i u/y_2}(u^2+1)^{\frac{-s_1-s_2}{2}}\, du$$
$$= \frac{2\pi^{\frac{s_1+s_2}{2}}}{\Gamma\left(\frac{s_1+s_2}{2}\right)} y_2^{\frac{1-s_1-s_2}{2}} K_{\frac{s_1+s_2-1}{2}}\left(2\pi y_2^{-1}\right).$$

The lemma immediately follows after applying the transformation $y_2 \mapsto y_2^{-1}$. This completes the proof of Proposition 6.6.5. □

Remark 6.6.12 Note that Bump's double Dirichlet series is generalized to $GL(n)$ in (Bump and Friedberg, 1990). This is a Rankin–Selberg construction involving two complex variables, one from an Eisenstein series, one of "Hecke" type. For $GL(3)$ it produces the Bump double Dirichlet series.

GL(n)pack functions The following **GL(n)pack** functions, described in the appendix, relate to the material in this chapter:

ApplyCasimirOperator	GetCasimirOperator
WeylGenerator	Whittaker
WhittakerGamma.	SpecialWeylGroup

7

The Gelbart–Jacquet lift

7.1 Converse theorem for $SL(3,\mathbb{Z})$

It was shown in Theorem 6.5.15 that the Godement–Jacquet L-function associated to a Maass form for $SL(3,\mathbb{Z})$ is entire and satisfies a simple functional equation. The development of a converse theorem for such L-functions (which generalizes Theorem 3.15.3) is really due to the efforts of Piatetski-Shapiro over several decades. In (Jacquet, Piatetski-Shapiro and Shalika, 1979) a Dirichlet series $\sum_{n=1}^{\infty} A(n,1)n^{-s}$ (with $A(1,1)=1$) is considered. It is assumed that this Dirchlet series is entire and bounded in vertical strips (EBV) and that the $A(m,n)$ satisfy the Hecke relations as given in Theorem 6.4.11. They proved that if all the twists

$$\sum_{n=1}^{\infty} A(n,1)\chi(n)n^{-s}, \qquad (\chi \text{ a Dirichlet character})$$

satisfy a suitable functional equation (very similar to the functional equation given in Theorem 6.5.15 with $\nu\in\mathbb{C}^2$) then

$$f(z) = \sum_{\gamma\in U_2(\mathbb{Z})\backslash SL(2,\mathbb{Z})} \sum_{m_1=1}^{\infty}\sum_{m_2\neq 0} \frac{A(m_1,m_2)}{m_1|m_2|} \times W_{\text{Jacquet}}\left(\begin{pmatrix} m_1|m_2| & & \\ & m_1 & \\ & & 1\end{pmatrix}\begin{pmatrix}\gamma & \\ & 1\end{pmatrix} z,\ \nu,\ \psi_{1,\frac{m_2}{|m_2|}}\right),$$

is, in fact, a Maass form for $SL(3,\mathbb{Z})$. This is the converse thoerem for $SL(3,\mathbb{Z})$, and provides a generalization of the converse theorem for $SL(2,\mathbb{Z})$ as given in Section 3.15.

The idea of introducing twisted functional equations to prove a converse theorem is due to Weil (1967). Cogdell and Piatetski-Shapiro have obtained a vast generalization of Weil's original converse theorem. In (Cogdell and Piatetski-Shapiro, 1994) it is shown that to prove an L-function is automorphic for $GL(n)$, it is necessary to twist by $GL(n-1)$ while in (Cogdell and Piatetski-Shapiro, 1999) this is improved to twists by $GL(n-2)$. In (Cogdell and Piatetski-Shapiro, 2001) there is an improvement in a different direction which involves restricting the ramification of the twisting. These papers have proved of fundamental importance in establishing cases of Langlands conjectures. A non-adelic version of the $GL(3)$ converse theorem was obtained by Miller and Schmid (2004). We shall present a new proof found recently by Goldfeld and Thillainatesan.

Interlude on Dirichlet characters A Dirichlet character (mod q) is a character of the cyclic group $(\mathbb{Z}/q\mathbb{Z})^*$. If χ is not trivial then $\chi(n) = 0$ if $(n, q) > 1$, and, otherwise is a $\phi(q)$th root of unity. It is a periodic function of n with period q. If q is the least period, then χ is said to be primitive. The Dirichlet characters (mod q), both primitive and imprimitive, form a basis for the functions on $(\mathbb{Z}/q\mathbb{Z})^*$. For Dirichlet characters χ (mod q) and $(n, q) = 1$, we have the following well-known identity (see (Davenport, 1967))

$$\tau(\bar{\chi})\cdot\chi(n) = \sum_{\ell=1}^{q} \bar{\chi}(\ell)e^{2\pi i \ell n/q}, \qquad \tau(\chi) = \sum_{\ell=1}^{q} \chi(\ell)e^{2\pi i \ell/q}, \tag{7.1.1}$$

where $\tau(\chi)$ is the Gauss sum. This allows us to represent $\chi(n)$ as a linear combination of qth roots of unity. Note that in (7.1.1), the condition $(n, q) = 1$ can be dropped if χ is primitive. We may also represent each qth root of unity as a linear combination of characters by the formula

$$\frac{1}{\phi(q)} \sum_{\chi \,(\mathrm{mod}\, q)} \bar{\chi}(\ell)\tau(\chi) = e^{2\pi i \ell/q}. \tag{7.1.2}$$

Theorem 7.1.3 (**$SL(3,\mathbb{Z})$ Converse theorem**) *For integers $m_1, m_2 \neq 0$, let $A(m_1, m_2) \in \mathbb{C}$ satisfy: $A(m_1, m_2) = A(-m_1, m_2) = A(m_1, -m_2)$, $A(1, 1) = 1$, and also the multiplicativity relations given in Theorem 6.4.11. Assume that for every primitive Dirichlet character χ (mod q), the Dirichlet series*

$$L_\chi(s) := \sum_{m=1}^{\infty} \frac{A(m,1)\chi(m)}{m^s}, \qquad \tilde{L}_\chi(s) := \sum_{m=1}^{\infty} \frac{A(1,m)\chi(m)}{m^s},$$

converge absolutely for $\Re(s)$ sufficiently large, and for fixed $\nu = (\nu_1, \nu_2) \in \mathbb{C}^2$, satisfy the functional equation

$$q^{\frac{3}{2}s} G_\nu(s+k) L_\chi(s) = i^{-k} \frac{\tau(\chi)^2}{\tau(\bar{\chi})\sqrt{q}} \cdot q^{\frac{3}{2}(1-s)} \tilde{G}_\nu(1+k-s) \tilde{L}_{\bar{\chi}}(1-s),$$

where $k = 0$ or 1 according as $\chi(-1)$ is $+1$ or -1, $G_\nu, \tilde{G}_\nu$ are products of three Gamma functions as in Theorem 6.5.15, and the functions $G_\nu(s)L(s)$, $\tilde{G}_\nu(s)\tilde{L}(s)$ are EBV. Then

$$\sum_{\gamma\in U_2(\mathbb{Z})\backslash SL(2,\mathbb{Z})} \sum_{m_1=1}^{\infty} \sum_{m_2\neq 0} \frac{A(m_1,m_2)}{|m_1m_2|} \times W_{\text{Jacquet}}\left(\begin{pmatrix} |m_1m_2| & & \\ & m_1 & \\ & & 1 \end{pmatrix}\begin{pmatrix} \gamma & \\ & 1 \end{pmatrix} z,\ \nu,\ \psi_{1,\frac{m_2}{|m_2|}}\right),$$

is a Maass form for $SL(3,\mathbb{Z})$.

Proof of the converse theorem, Step I Define

$$f(z) := \sum_{\gamma\in U_2(\mathbb{Z})\backslash SL(2,\mathbb{Z})} \sum_{m_1=1}^{\infty} \sum_{m_2\neq 0} \frac{A(m_1,m_2)}{|m_1m_2|} \times W_{\text{Jacquet}}\left(\begin{pmatrix} |m_1m_2| & & \\ & m_1 & \\ & & 1 \end{pmatrix}\begin{pmatrix} \gamma & \\ & 1 \end{pmatrix} z, (\nu_1,\nu_2), \psi_{1,\frac{m_2}{|m_2|}}\right), \tag{7.1.4}$$

and

$$\tilde{f}(z) := \sum_{\gamma\in U_2(\mathbb{Z})\backslash SL(2,\mathbb{Z})} \sum_{m_1=1}^{\infty} \sum_{m_2\neq 0} \frac{A(m_2,m_1)}{|m_1m_2|} \times W_{\text{Jacquet}}\left(\begin{pmatrix} |m_1m_2| & & \\ & m_1 & \\ & & 1 \end{pmatrix}\begin{pmatrix} \gamma & \\ & 1 \end{pmatrix} z, (\nu_2,\nu_1), \psi_{1,\frac{m_2}{|m_2|}}\right). \tag{7.1.5}$$

In the following, let

$$u = \begin{pmatrix} 1 & 0 & u_3 \\ 0 & 1 & u_1 \\ 0 & 0 & 1 \end{pmatrix}, \qquad w_2 = \begin{pmatrix} 0 & 0 & 1 \\ 1 & 0 & 0 \\ 0 & 1 & 0 \end{pmatrix}.$$

If f is automorphic then

$$f(z) = \tilde{f}\left(w_2 \cdot {}^t z^{-1}\right)$$

for all $z \in \mathfrak{h}^3$ and

$$\int_0^1\int_0^1 f(Auz)e^{-2\pi i k u_1}\,du_1du_3 = \int_0^1\int_0^1 \tilde{f}(w_2 \cdot {}^t(Auz)^{-1})e^{-2\pi i k u_1}\,du_1du_3 \tag{7.1.6}$$

for all $z \in \mathfrak{h}^3, k \in \mathbb{Z}, A = \begin{pmatrix} 1 & 0 & 0 \\ \alpha & 1 & 0 \\ 0 & 0 & 1 \end{pmatrix}$ with $\alpha \in \mathbb{Q}$. Our aim is to prove a converse to (7.1.6).

Basic Lemma 7.1.7 *Let $f, \tilde{f}$ be defined by (7.1.4), (7.1.5), respectively. Assume*

$$\int_0^1\int_0^1 f(Auz)e^{-2\pi i q u_1}\, du_1 du_3 = \int_0^1\int_0^1 \tilde{f}\big(w_2 \cdot {}^t(Auz)^{-1}\big)e^{-2\pi i q u_1} du_1 du_3$$

for all $z \in \mathfrak{h}^3,\ h, q \in \mathbb{Z},\ A = \begin{pmatrix} 1 & 0 & 0 \\ h/q & 1 & 0 \\ 0 & 0 & 1 \end{pmatrix}$, with $q \neq 0$. Then f is a Maass form of type ν for $SL(3,\mathbb{Z})$ and $\tilde{f}$ is its dual form.

Proof of Lemma 7.1.7 First of all, we claim that

$$f(pz) = f(z), \qquad \tilde{f}(pz) = \tilde{f}(z), \qquad \forall\, p \in \mathcal{P} = \begin{pmatrix} * & * & * \\ * & * & * \\ 0 & 0 & 1 \end{pmatrix} \subset SL(3,\mathbb{Z}). \tag{7.1.8}$$

This is due to the fact that $\mathcal{P}$ is generated by elements of the form

$$\begin{pmatrix} a & b & \\ c & d & \\ & & 1 \end{pmatrix}, \quad \begin{pmatrix} 1 & & r \\ & 1 & s \\ & & 1 \end{pmatrix}, \quad (\text{with } a,b,c,d,r,s \in \mathbb{Z},\ ad - bc = 1).$$

We first check that (7.1.8) holds for $p = \begin{pmatrix} a & b & \\ c & d & \\ & & 1 \end{pmatrix}$. This follows easily because

$$\begin{aligned} f(pz) &= \sum_{\gamma \in U_2(\mathbb{Z})\backslash SL(2,\mathbb{Z})} \sum_{m_1=1}^{\infty} \sum_{m_2 \neq 0} \frac{A(m_1,m_2)}{|m_1 m_2|} \\ &\quad \times W_{\text{Jacquet}}\left(\begin{pmatrix} |m_1 m_2| & & \\ & m_1 & \\ & & 1 \end{pmatrix} \begin{pmatrix} \gamma & \\ & 1 \end{pmatrix} \cdot pz,\ \nu,\ \psi_{1,\frac{m_2}{|m_2|}} \right) \\ &= f(z), \end{aligned}$$

since the sum over γ is permuted by p. A similar argument holds for $\tilde{f}$.

It only remains to show that (7.1.8) holds for $p = \begin{pmatrix} 1 & & r \\ & 1 & s \\ & & 1 \end{pmatrix}$. To show this, we just use the identity

$$\begin{pmatrix} |m_1m_2| & & \\ & m_1 & \\ & & 1 \end{pmatrix}\begin{pmatrix} a & b & \\ c & d & \\ & & 1 \end{pmatrix} \cdot \begin{pmatrix} 1 & & r \\ & 1 & s \\ & & 1 \end{pmatrix}$$
$$= \begin{pmatrix} 1 & & r' \\ & 1 & s' \\ & & 1 \end{pmatrix} \cdot \begin{pmatrix} |m_1m_2| & & \\ & m_1 & \\ & & 1 \end{pmatrix}\begin{pmatrix} a & b & \\ c & d & \\ & & 1 \end{pmatrix},$$

where

$$r' = |m_1m_2|(ar + bs), \qquad s' = m_1(cr + ds).$$

It follows that

$$f(pz) = \sum_{\gamma \in U_2(\mathbb{Z})\backslash SL(2,\mathbb{Z})} \sum_{m_1=1}^{\infty} \sum_{m_2 \neq 0} \frac{A(m_1, m_2)}{|m_1m_2|}$$
$$\times W_{\text{Jacquet}}\left(\begin{pmatrix} 1 & & r' \\ & 1 & s' \\ & & 1 \end{pmatrix} \cdot \begin{pmatrix} |m_1m_2| & & \\ & m_1 & \\ & & 1 \end{pmatrix}\begin{pmatrix} \gamma & \\ & 1 \end{pmatrix} z,\ \nu,\ \psi_{1, \frac{m_2}{|m_2|}}\right)$$
$$= f(z),$$

because

$$W_{\text{Jacquet}}\left(\begin{pmatrix} 1 & & r' \\ & 1 & s' \\ & & 1 \end{pmatrix} \cdot z, \nu, \psi\right) = W_{\text{Jacquet}}(z, \nu, \psi)$$

for all z, ν, ψ, if $r', s' \in \mathbb{Z}$. Again, a similar argument applies for $\tilde{f}$.

Lemma 7.1.9 *We have*

$$\tilde{f}\left(w_2 \cdot {}^t z^{-1}\right) = \tilde{f}\left(w_2 \cdot {}^t(pz)^{-1}\right)$$

for all $p = \begin{pmatrix} 1 & & r \\ & 1 & s \\ & & 1 \end{pmatrix}$, *with* $r, s \in \mathbb{Z}$.

Proof By (7.1.8), we may, without loss of generality replace w_2 by

$$w_2' = \begin{pmatrix} 1 & 0 & 0 \\ 0 & 0 & 1 \\ 0 & 1 & 0 \end{pmatrix}.$$

We compute:

$$w_2' \cdot {}^t p^{-1} \cdot w_2'^{-1} = w_2' \cdot \begin{pmatrix} 1 & & \\ & 1 & \\ -r & -s & 1 \end{pmatrix} \cdot w_2'^{-1} = \begin{pmatrix} 1 & & \\ -r & 1 & -s \\ & & 1 \end{pmatrix} \in \mathcal{P}.$$

Thus,

$$\begin{aligned} \tilde{f}\left(w_2' \cdot {}^t(pz)^{-1}\right) &= \tilde{f}\left(w_2' \cdot {}^t p^{-1} \cdot {}^t z^{-1}\right) = \tilde{f}\left(\left(w_2' \cdot {}^t p^{-1} \cdot w_2'^{-1}\right) \cdot w_2' \cdot {}^t z^{-1}\right) \\ &= \tilde{f}\left(\begin{pmatrix} 1 & & \\ -r & 1 & -s \\ & & 1 \end{pmatrix} \cdot w_2' \cdot {}^t z^{-1}\right) = \tilde{f}\left(w_2' \cdot {}^t z^{-1}\right). \end{aligned}$$

Here we have used the fact that $\tilde{f}$ is invariant under left multiplication by elements in $\mathcal{P}$. □

Now, if (7.1.6) holds for all $z \in \mathfrak{h}^3$ and all $h, q \in \mathbb{Z}$, $A = \begin{pmatrix} 1 & 0 & 0 \\ h/q & 1 & 0 \\ 0 & 0 & 1 \end{pmatrix}$, with $q \neq 0$, then on choosing $z = A^{-1}z'$, we must have

$$\begin{aligned} &\int_0^1 \int_0^1 f(AuA^{-1}z')e^{-2\pi i q u_1} du_1 du_3 \\ &\quad = \int_0^1 \int_0^1 \tilde{f}(w_2 \cdot {}^t(AuA^{-1}z')^{-1})e^{-2\pi i q u_1} du_1 du_3 \end{aligned}$$

for all $z' \in \mathfrak{h}^3$. Since

$$AuA^{-1} = \begin{pmatrix} 1 & 0 & u_3 \\ 0 & 1 & \frac{hu_3}{q} + u_1 \\ 0 & 0 & 1 \end{pmatrix},$$

it easily follows after a change of variables that

$$\int_0^1 \int_0^1 \left[f(uz) - \tilde{f}\left(w_2 \cdot {}^t(uz)^{-1}\right)\right] e^{-2\pi i q u_1} e^{-2\pi i h u_3}\, du_1 du_3 = 0, \qquad (7.1.10)$$

for all $z \in \mathfrak{h}^3$, and all $h, q \in \mathbb{Z}$, with $q \neq 0$.

We will next show that (7.1.10) holds for all $h \in \mathbb{Z}$ and $q = 0$. In fact, this case holds without any assumptions on f.

Lemma 7.1.11 *We have* $\int_0^1 \left[f(uz) - \tilde{f}(w_2 \cdot {}^t(uz)^{-1})\right] du_1 = 0$, *where* $u = \begin{pmatrix} 1 & 0 & 0 \\ 0 & 1 & u_1 \\ 0 & 0 & 1 \end{pmatrix}$.

Proof It follows from Theorem 6.5.7, after integrating in u_1, that

$$\int_0^1 [f(uz) - \tilde{f}(w_2 \cdot {}^t(uz)^{-1})] du_1$$

$$= \sum_{m_1=1}^{\infty} \sum_{m_2 \neq 0} \left[\frac{A(m_1, m_2)}{m_1 |m_2|} W_{\text{Jacquet}} \left(M \begin{pmatrix} 0 & 1 & 0 \\ 1 & 0 & 0 \\ 0 & 0 & 1 \end{pmatrix} z \right) \right.$$
$$\left. - \frac{A(m_2, m_1)}{m_1 |m_2|} \tilde{W}_{\text{Jacquet}} \left(M w_2 \cdot {}^t z^{-1} \right) \right],$$

where $M = \begin{pmatrix} m_1|m_2| & & \\ & m_1 & \\ & & 1 \end{pmatrix}$, and W_{Jacquet} is given as in Theorem 6.4.11. But

$$\tilde{W}_{\text{Jacquet}}(z) = W_{\text{Jacquet}} \left(w_3 \cdot {}^t z^{-1} \cdot w_3^{-1} \right), \qquad w_3 = \begin{pmatrix} & & 1 \\ & 1 & \\ 1 & & \end{pmatrix}.$$

The proof follows upon noting that

$$\tilde{W}_{\text{Jacquet}}(M w_2 \cdot {}^t z^{-1}) = W_{\text{Jacquet}}\left(w_3 \cdot M w_2 \cdot {}^t z^{-1} \cdot w_3^{-1}\right)$$
$$= W_{\text{Jacquet}} \left(M' \begin{pmatrix} 0 & 1 & 0 \\ 1 & 0 & 0 \\ 0 & 0 & 1 \end{pmatrix} z \right),$$

where M' is the same as M, but with m_1, m_2 interchanged. □

It follows from Lemma 7.1.9, that the function $f(z) - \tilde{f}(w_2 \cdot {}^t(z)^{-1})$ is invariant under left multiplication by the group $\begin{pmatrix} 1 & & * \\ & 1 & * \\ & & 1 \end{pmatrix}$. It, therefore, has a Fourier expansion in x_1, x_3. The identity (7.1.10) together with Lemma 7.1.11

tells us that every Fourier coefficient vanishes. Consequently

$$f(z) = \tilde{f}\big(\gamma \cdot {}^t(z)^{-1}\big) = \tilde{f}\big({}^t(\gamma z)^{-1}\big)$$

with $\gamma = w_2$, for all $z \in \mathfrak{h}^3$. Finally, we may complete the proof of Lemma 7.1.7. Note that $f(z) = \tilde{f}({}^t(\gamma z)^{-1})$ must hold for all $\gamma \in SL(3, \mathbb{Z})$ because it will hold with γ replaced by $p_1 \gamma p_2$ and $p_1, p_2 \in \mathcal{P}$. To see this note that $p_1 \in \mathcal{P}$ implies that p_1 is either of the form $\begin{pmatrix} \gamma & \\ & 1 \end{pmatrix}$ or of the form $\begin{pmatrix} \gamma & \\ & 1 \end{pmatrix} w_2 \begin{pmatrix} \gamma' & \\ & 1 \end{pmatrix}$ with $\gamma, \gamma' \in SL(2, \mathbb{Z})$. In the former case, the invariance due to multiplication by p_1 on the left follows from (7.1.8), while in the latter case we have already proved the invariance of w_2 on the left, so we are reduced to the first case. The invariance due to multiplication by p_2 on the right follows by letting $z \mapsto p_2^{-1} \cdot z$. So, we have shown that f is automorphic. □

Proof of the converse theorem, Step II Fix $A = \begin{pmatrix} 1 & 0 & 0 \\ h/q & 1 & 0 \\ 0 & 0 & 1 \end{pmatrix}$ with $h, q \in \mathbb{Z}$, and $q \neq 0$. Let $f, \tilde{f}$ be defined as in (7.1.4) and (7.1.5) and set

$$F(z, h, q) = \int_0^1 \int_0^1 f(Auz) e^{-2\pi i q u_1} \, du_1 du_3,$$

$$F_1(z, h, q) = \int_0^1 \int_0^1 \tilde{f}\big(w_2 \cdot {}^t(Auz)^{-1}\big) e^{-2\pi i q u_1} \, du_1 du_3,$$

where we recall that $u = \begin{pmatrix} 1 & & u_3 \\ & 1 & u_1 \\ & & 1 \end{pmatrix}$, $w_2 = \begin{pmatrix} & & 1 \\ & 1 & \\ 1 & & \end{pmatrix}$. If we can show that $F(z, h, q) = F_1(z, h, q)$ for all z, h, q, k then the basic Lemma 7.1.7 tells us that f is a Maass form for $SL(3, \mathbb{Z})$. It remains to show that the functional equations for $L_\chi(s)$ stated in Theorem 7.1.3 imply that $F(z, h, q) = F_1(z, h, q)$ for all z, h, q. We shall do this in two remaining steps. First we show that the functional equations for $L_\chi(s)$ imply that

$$\left(\frac{\partial}{\partial x_1}\right)^\ell \left(\frac{\partial}{\partial x_2}\right)^k F(z, h, q)\bigg|_{x_1 = x_2 = 0} = \left(\frac{\partial}{\partial x_1}\right)^\ell \left(\frac{\partial}{\partial x_2}\right)^k F_1(z, h, q)\bigg|_{x_1 = x_2 = 0}.$$

In the final step 3 of the proof, we extend the above result and prove the basic Lemma 7.1.7 in all cases.

The key point is that the functional equations for $L_\chi(s)$ imply that

$$\left(\frac{\partial}{\partial x_1}\right)^{\ell}\left(\frac{\partial}{\partial x_2}\right)^{k}\int_0^\infty\int_0^\infty F(z,h,q)\, y_1^{s_1-1}y_2^{s_2-1}\,\frac{dy_1dy_2}{y_1y_2}\Bigg|_{x_1=x_2=0}$$
$$=\left(\frac{\partial}{\partial x_1}\right)^{\ell}\left(\frac{\partial}{\partial x_2}\right)^{k}\int_0^\infty\int_0^\infty F_1(z,h,q)y_1^{s_1-1}y_2^{s_2-1}\,\frac{dy_1dy_2}{y_1y_2}\Bigg|_{x_1=x_2=0},$$

and then by taking inverse Mellin transforms we obtain the desired identity.

The above idea is exemplified in the following two lemmas.

Lemma 7.1.12 *Let $q,h\in\mathbb{Z}$ with $q\neq 0$. Define $\delta:=(h,q)$, $q_\delta:=q/\delta$, $h_\delta:=h/\delta$, and $\bar h_\delta\cdot h_\delta\equiv 1 \pmod{q_\delta}$. Then for integers $0\le \ell,k$, we have*

$$\left(\frac{\partial}{\partial x_1}\right)^{\ell}\left(\frac{\partial}{\partial x_2}\right)^{k}\int_0^\infty\int_0^\infty F(z,h,q)\, y_1^{s_1-1}y_2^{s_2-1}\,\frac{dy_1dy_2}{y_1y_2}\Bigg|_{x_1=x_2=0}$$
$$=\frac{(2\pi i q)^{\ell}}{q_\delta^{s_1-2s_2+1}\delta^{s_1}}\sum_{m_2\neq 0}\frac{A(\delta,m_2)}{|m|^{s_2}}\left(\frac{2\pi i m_2}{q_\delta^2}\right)^{k}e^{2\pi i m_2\bar h_\delta/q_\delta}$$
$$\times\int_0^\infty\int_0^\infty W_{\text{Jacquet}}(y,\nu,\psi_{1,1})y_1^{s_1-1}y_2^{s_2-1}\,\frac{dy_1}{y_1}\frac{dy_2}{y_2}.$$

Proof Note that

$$\begin{pmatrix} a & b\\ c & d\end{pmatrix}\cdot\begin{pmatrix} 1 & 0\\ h/q & 1\end{pmatrix}=\begin{pmatrix} a+bh/q & b\\ c+dh/q & d\end{pmatrix}=\begin{pmatrix} a' & b\\ c' & d\end{pmatrix}.$$

It follows from Theorem 6.5.7 that

$$\int_0^\infty\int_0^\infty F(z,h,q)\, y_1^{s_1-1}y_2^{s_2-1}\,\frac{dy_1dy_2}{y_1y_2}=\sum_{(c,d)=1}\sum_{m_1=1}^{\infty}\sum_{m_2\neq 0}\frac{A(m_1,m_2)}{m_1|m_2|}\int_0^\infty\int_0^\infty\int_0^1\int_0^1$$
$$\times e^{2\pi i m_1\left(c+\left(\frac{dh}{q}\right)\right)(x_3+u_3)}e^{2\pi i m_1 d(x_1+u_1)}e^{-2\pi i q u_1}e^{2\pi i m_2\Re\left(\frac{a'z_2+b}{c'z_2+d}\right)}$$
$$\times W_{\text{Jacquet}}\left(\begin{pmatrix}\frac{m_1|m_2|y_1y_2}{|c'z_2+d|} & & \\ & m_1y_1\cdot|c'z_2+d| & \\ & & 1\end{pmatrix},\nu,\psi_{1,\frac{m_2}{|m_2|}}\right)$$
$$\times y_1^{s_1-1}y_2^{s_2-1}\,\frac{dy_1}{y_1}\frac{dy_2}{y_2}du_1du_3.$$

The integrals over u_1, u_3 are zero unless $c = -dh/q,\ m_1 d = q$. Hence $c' = 0$, and $d = q/m_1$. Since $ad - bc = 1$ it immediately follows that $a(q/m_1) + b(h/m_1) = 1$. Consequently, $m_1 = (q,h) := \delta$, and letting $q_\delta = q/\delta,\ h_\delta = h/\delta,\ \bar{h}_\delta = h_\delta^{-1} \pmod{q_\delta}$, the above identity becomes:

$$\int_0^\infty\int_0^\infty F(z,h,q) y_1^{s_1-1} y_2^{s_2-1} \frac{dy_1 dy_2}{y_1 y_2}$$

$$= \sum_{m_2\neq 0} \frac{A(\delta, m_2)}{\delta\cdot|m_2|} e^{\frac{2\pi i m_2 \bar{h}_\delta}{q_\delta}} e^{2\pi i q x_1} e^{\frac{2\pi i m_2 x_2}{q_\delta^2}}$$

$$\times \int_0^\infty\int_0^\infty W_{\text{Jacquet}}\left(\begin{pmatrix} \frac{\delta|m_2|y_1y_2}{q_\delta} & & \\ & \delta y_1\cdot q_\delta & \\ & & 1\end{pmatrix}, \nu, \psi_{1,\frac{m_2}{|m_2|}}\right) y_1^{s_1-1} y_2^{s_2-1} \frac{dy_1}{y_1}\frac{dy_2}{y_2}.$$

The lemma follows after taking partial derivatives and then performing a simple variable change. □

Lemma 7.1.13 *Let $q, h \in \mathbb{Z}$ with $q \neq 0$. Then for $\ell, k \in \mathbb{Z}$ with $\ell \geq 0$, $k \in \{0,1\}$, we have*

$$\left(\frac{\partial}{\partial x_1}\right)^\ell \left(\frac{\partial}{\partial x_2}\right)^k \int_0^\infty\int_0^\infty F_1(z,h,q)\, y_1^{s_1-1} y_2^{s_2-1} \frac{dy_1}{y_1}\frac{dy_2}{y_2}\bigg|_{x_1=x_2=0}$$

$$= \frac{(2\pi i q)^{\ell+k} \pi^{\frac{s_1+s_2}{2}}}{q^{s_1+s_2}\Gamma\left(\frac{s_1+s_2}{2}\right)} \sum_{m_1|q}\sum_{m_2\neq 0} \frac{A(m_2,m_1)}{m_1^{2k+1-2s_2}|m_2|^{k+1-s_2}} \left(\frac{i m_2}{|m_2|}\right)^k$$

$$\times \sum_{\substack{r=1\\ \left(r,\frac{q}{m_1}\right)=1}}^{q/m_1} e^{2\pi i m_1 \frac{-hr+m_2\bar{r}}{q}} \int_0^\infty\int_0^\infty W_{\text{Jacquet}}(y,(\nu_2,\nu_1),\psi_{1,1})$$

$$\times K_{\frac{s_1+s_2-1-2k}{2}}(2\pi y_2) y_1^{2k+s_1-s_2} y^{\frac{2k+s_1-s_2-1}{2}} \frac{dy_1}{y_1}\frac{dy_2}{y_2}.$$

Proof Note that

$$w_2\cdot({}^t(Auz)^{-1})\cdot w_2^{-1} = \begin{pmatrix} 1 & -u_3-x_3 & -u_1-x_1+\frac{x_2(u_3+x_3)}{x_2^2+y_2^2} \\ & 1 & -h/q - \frac{x_2}{x_2^2+y_2^2} \\ & & 1\end{pmatrix}$$

$$\times \begin{pmatrix} \frac{y_1y_2}{\sqrt{x_2^2+y_2^2}} & & \\ & \frac{y_2}{x_2^2+y_2^2} & \\ & & 1\end{pmatrix} \bmod (O(3,\mathbb{R})\cdot\mathbb{R}^\times),$$

and

$$\int_0^\infty\!\!\int_0^\infty F_1(z,h,q)\, y_1^{s_1-1} y_2^{s_2-1}\,\frac{dy_1}{y_1}\frac{dy_2}{y_2}$$
$$= \int_0^\infty\!\!\int_0^\infty\!\!\int_0^1\!\!\int_0^1 \tilde{f}\left(w_2\cdot({}^t(Auz)^{-1})\cdot w_2^{-1}\right) e^{-2\pi i q u_1} du_1\, du_3 y_1^{s_1-1} y_2^{s_2-1}\frac{dy_1}{y_1}\frac{dy_2}{y_2}.$$

If we now make use of Theorem 6.5.7, in the above, we obtain

$$\int_0^\infty\!\!\int_0^\infty F_1(z,h,q)\, y_1^{s_1-1} y_2^{s_2-1}\,\frac{dy_1}{y_1}\frac{dy_2}{y_2}$$
$$= \sum_{(c,d)=1}\sum_{m_1=1}^{\infty}\sum_{m_2\neq 0}\frac{A(m_2,m_1)}{m_1|m_2|}\int_0^\infty\!\!\int_0^\infty\!\!\int_0^1\!\!\int_0^1 e^{2\pi i m_1 c\left[-u_1-x_1+\frac{x_2(u_3+x_3)}{x_2^2+y_2^2}\right]}$$
$$\times\, e^{-2\pi i m_1 d\left(\frac{h}{q}+\frac{x_2}{x_2^2+y_2^2}\right)} e^{2\pi i m_2 \Re\left(\frac{az_2'+b}{cz_2'+d}\right)} e^{-2\pi i q u_1}\, du_1$$
$$\times\, W_{\text{Jacquet}}\left(\begin{pmatrix}\frac{m_1|m_2|y_1y_2}{\sqrt{x_2^2+y_2^2}\cdot|cz_2'+d|} & & \\ & \frac{m_1 y_2|cz_2'+d|}{x_2^2+y_2^2} & \\ & & 1\end{pmatrix},\ (\nu_2,\nu_1),\ \psi_{1,1}\right)$$
$$\times\, du_3 y_1^{s_1-1} y_2^{s_2-1}\frac{dy_1}{y_1}\frac{dy_2}{y_2},$$

where

$$z_2' = -u_3 - x_3 + i y_1\sqrt{x_2^2+y_2^2}.$$

The integral in u_1 above, namely

$$\int_0^1 e^{-2\pi i m_1 c u_1} e^{-2\pi i q u_1}\, du_1$$

vanishes unless $m_1 c = -q$. Further, let $d = \ell c + r$ with $\ell \in \mathbb{Z}$ and $1 \le r < c$ where $(r,c) = 1$. If we make the change of variables $u_3 \mapsto u_3 + \ell + r m_1/q$, then we may re-express the above identity in the form

$$\int_0^\infty\!\!\int_0^\infty F_1(z,h,q)\, y_1^{s_1-1} y_2^{s_2-1}\,\frac{dy_1}{y_1}\frac{dy_2}{y_2}$$
$$= e^{2\pi i q x_1}\sum_{m_1|q}\sum_{m_2\neq 0}\frac{A(m_2,m_1)}{m_1|m_2|}\sum_{\ell\in\mathbb{Z}}\sum_{\substack{r=1\\ \left(r,\frac{q}{m_1}\right)=1}}^{q/m_1} e^{-\frac{2\pi i m_1 r h}{q}} e^{-\frac{2\pi i m_1 m_2 \bar{r}}{q}}$$

$$\times \int_0^\infty \int_0^\infty \int_{\ell+\frac{rm_1}{q}}^{\ell+\frac{rm_1}{q}+1} e^{-2\pi i q x_2\left(\frac{u_3+x_3}{x_2^2+y_2^2}\right)} e^{2\pi i m_2 \Re \frac{-1}{c^2 z_2'}}$$

$$\times W_{\text{Jacquet}}\left(\begin{pmatrix} \frac{m_1|m_2|y_1y_2}{\sqrt{x_2^2+y_2^2}\cdot|cz_2'|} & & \\ & \frac{m_1 y_2 |cz_2'|}{x_2^2+y_2^2} & \\ & & 1 \end{pmatrix}, (\nu_2, \nu_1), \psi_{1,1}\right)$$

$$\times du_3 y_1^{s_1-1} y_2^{s_2-1} \frac{dy_1}{y_1}\frac{dy_2}{y_2},$$

which after summing over $\ell \in \mathbb{Z}$, and, then successively making the transformations: $u_3 \mapsto u_3 - x_3$, $u_3 \mapsto u_3 \cdot y_1\sqrt{x_2^2+y_2^2}$, becomes

$$= e^{2\pi i q x_1} \sum_{m_1|q} \sum_{m_2\neq 0} \frac{A(m_2, m_1)}{m_1|m_2|} \sum_{\substack{r=1 \\ \left(r, \frac{q}{m_1}\right)=1}}^{q/m_1} e^{\frac{2\pi i m_1 r h}{q}} e^{\frac{2\pi i m_1 m_2 \bar{r}}{q}}$$

$$\times \int_0^\infty \int_0^\infty \int_{-\infty}^\infty e^{\frac{-2\pi i q x_2 y_1 u_3}{\sqrt{x_2^2+y_2^2}}} e^{\frac{2\pi i m_1^2 m_2 u_3}{q^2 y_1 \sqrt{x_2^2+y_2^2}\left(u_3^2+1\right)}}$$

$$\times W_{\text{Jacquet}}\left(\begin{pmatrix} \frac{m_1^2|m_2|y_2}{q\left(x_2^2+y_2^2\right)\sqrt{u_3^2+1}} & & \\ & \frac{q y_1 y_2 \sqrt{u_3^2+1}}{\sqrt{x_2^2+y_2^2}} & \\ & & 1 \end{pmatrix}, (\nu_2, \nu_1), \psi_{1,1}\right)$$

$$\times du_3\, y_1^{s_1} y_2^{s_2-1} \sqrt{x_2^2+y_2^2}\, \frac{dy_1}{y_1}\frac{dy_2}{y_2}.$$

If we now take partial derivatives with respect to x_1, x_2, set $x_1 = x_2 = 0$, and make the substitutions

$$y_1 \mapsto \frac{y_1}{q\sqrt{u_3^2+1}}, \qquad y_2 \mapsto \frac{m_1^2|m_2|y_2}{q\sqrt{u_3^2+1}},$$

it follows that

$$\left(\frac{\partial}{\partial x_1}\right)^\ell \left(\frac{\partial}{\partial x_2}\right)^k \int_0^\infty \int_0^\infty F_1(z, h, q)\, y_1^{s_1-1} y_2^{s_2-1} \frac{dy_1}{y_1}\frac{dy_2}{y_2}\Bigg|_{x_1=x_2=0}$$

$$= \frac{(2\pi i q)^\ell}{q^{s_1+s_2}} \sum_{m_1|q} \sum_{m_2\neq 0} \frac{A(m_2, m_1)}{m_1^{1-2s_2}|m_2|^{1-s_2}} \sum_{\substack{r=1 \\ \left(r, \frac{q}{m_1}\right)=1}}^{q/m_1} e^{\frac{2\pi i m_1 r h}{q}} e^{\frac{2\pi i m_1 m_2 \bar{r}}{q}}$$

$$\times \int_0^\infty \int_0^\infty \left(\frac{2\pi i q y_1 u_3}{m_1^2 |m_2| y_2}\right)^k W_{\text{Jacquet}}\left(\begin{pmatrix} y_2^{-1} & & \\ & y_1 & \\ & & 1 \end{pmatrix}, (\nu_2, \nu_1), \psi_{1,1}\right)$$

$$\times \int_{-\infty}^{\infty} e^{\frac{2\pi i u_3 m_2}{y_1 y_2 |m_2|}} \left(u_3^2+1\right)^{-(s_1+s_2)/2} du_3 \; y_1^{s_1} y_2^{s_2} \frac{dy_1}{y_1}\frac{dy_2}{y_2}.$$

Next, make the successive transformations: $y_2 \mapsto y_2^{-1}$, $y_2 \mapsto y_2 \cdot y_1$. We obtain

$$\frac{(2\pi i q)^\ell}{q^{s_1+s_2}} \sum_{m_1|q} \sum_{m_2 \neq 0} \frac{A(m_2, m_1)}{m_1^{1-2s_2} |m_2|^{1-s_2}} \sum_{\substack{r=1 \\ \left(r, \frac{q}{m_1}\right)=1}}^{q/m_1} e^{2\pi i m_1 \frac{(-hr+m_2\bar{r})}{q}}$$

$$\times \int_0^\infty \int_0^\infty \left(\frac{2\pi i q y_1^2 y_2}{m_1^2 |m_2|}\right)^k W_{\text{Jacquet}}\left(\begin{pmatrix} y_1 y_2 & & \\ & y_1 & \\ & & 1 \end{pmatrix}, (\nu_2, \nu_1), \psi_{1,1}\right)$$

$$\times \left(\int_{-\infty}^{\infty} e^{-2\pi i u_3 y_2 \frac{m_2}{|m_2|}} \left(u_3^2+1\right)^{-(s_1+s_2)/2} u_3^k \, du_3\right) y_1^{s_1-s_2} y_2^{-s_2} \frac{dy_1}{y_1}\frac{dy_2}{y_2}.$$

We may complete the proof of Lemma 7.1.13 by invoking the identities:

$$\int_{-\infty}^{\infty} e^{2\pi i u y_2} (u^2+1)^{-s} \, du = 2\frac{\pi^s}{\Gamma(s)} |y_2|^{s-\frac{1}{2}} K_{s-\frac{1}{2}}(2\pi |y_2|),$$

$$\int_{-\infty}^{\infty} e^{2\pi i u y_2} (u^2+1)^{-s} u \, du = 2\left(i\frac{y_2}{|y_2|}\right) \cdot \frac{\pi^s}{\Gamma(s)} |y_2|^{s-\frac{1}{2}} K_{s-\frac{3}{2}}(2\pi |y_2|).$$

□

Lemmas 7.1.12 and 7.1.13 involve additive twists of the Godement–Jacquet L-function which have very complicated functional equations. We would like to pass to twists by primitive Dirichlet characters which are much easier to deal with. This is the motivation for the next lemma.

Lemma 7.1.14 *Assume*

$$\left(\frac{\partial}{\partial x_1}\right)^\ell \left(\frac{\partial}{\partial x_2}\right)^k \int_0^\infty \int_0^\infty F(z, h, q)\, y_1^{s_1-1} y_2^{s_2-1} \frac{dy_1}{y_1}\frac{dy_2}{y_2}\Bigg|_{x_1=x_2=0}$$

$$= \left(\frac{\partial}{\partial x_1}\right)^\ell \left(\frac{\partial}{\partial x_2}\right)^k \int_0^\infty \int_0^\infty F_1(z, h, q)\, y_1^{s_1-1} y_2^{s_2-1} \frac{dy_1}{y_1}\frac{dy_2}{y_2}\Bigg|_{x_1=x_2=0} \quad (7.1.15)$$

for all $h, q, \ell \in \mathbb{Z}$ with $q \neq 0$, $\ell \geq 0$, and $k \in \{0, 1\}$. Define $\delta := (h, q)$ and $q_\delta := q/\delta$. Then

$$\frac{G_\nu(s_2)}{\tilde{G}_\nu(1 - s_2 + 2k)} \cdot \frac{\tau(\bar{\chi})q^{3s_2-1}}{\tau(\chi)\pi^{(s_1+s+2)/2}} \sum_{m_2 \neq 0} \frac{A(\delta, m_2)\,\chi(m_2)}{\delta^{s_1}|m_2|^{s_2}} \left(\frac{2\pi i m_2}{q_\delta^2}\right)^k$$
$$= (2\pi i q)^k \sum_{m_1|\delta} \sum_{m_2 \neq 0} \frac{A(m_2, m_1)\bar{\chi}(-m_1)}{m_1^{2k+1-2s_2}|m_2|^{k+1-s_2}} \left(\frac{i m_2}{|m_2|}\right)^k$$
$$\times \sum_{\substack{r=1 \\ (r,(q/m_1))=1}}^{q/m_1} \bar{\chi}(r) e^{\frac{2\pi i m_1 m_2 \bar{r}}{q}}, \tag{7.1.16}$$

for every character χ (mod q_δ), with $G_\nu, \tilde{G}_\nu$ as in Theorem 6.5.15. Further, if (7.1.16) holds for every character χ, each $k \in \{0, 1\}$, and all $s_1, s_2 \in \mathbb{C}$, then (7.1.15) holds for all $h, q, \ell \in \mathbb{Z}$, $k \in \{0, 1\}$, $\ell \geq 0$ and $q \neq 0$.

Note The identity (7.1.16) is a formal identity. It is understood that all series are well defined by analytic continuation. We present (7.1.16) and its proof in this form to simplify the exposition of the ideas.

Proof We have $h = h_\delta \cdot \delta$. Consequently if (7.1.15) holds, then

$$\left(\frac{\partial}{\partial x_1}\right)^\ell \left(\frac{\partial}{\partial x_2}\right)^k \int_0^\infty \int_0^\infty \sum_{h_\delta=1}^{q_\delta} \chi(h_\delta) F(z, h_\delta\delta, q) y_1^{s_1-1} y_2^{s_2-1} \frac{dy_1}{y_1}\frac{dy_2}{y_2}\bigg|_{x_1=x_2=0}$$
$$= \left(\frac{\partial}{\partial x_1}\right)^\ell \left(\frac{\partial}{\partial x_2}\right)^k \int_0^\infty \int_0^\infty \sum_{h_\delta=1}^{q_\delta} \chi(h_\delta) F_1(z, h_\delta\delta, q) y_1^{s_1-1} y_2^{s_2-1} \frac{dy_1}{y_1}\frac{dy_2}{y_2}\bigg|_{x_1=x_2=0}.$$

The result follows after several routine computations using Lemmas 7.1.12, 7.1.13, and formula (7.1.1). For example, on the left-hand side we will have from Lemma 7.1.12 the character sum

$$\sum_{h_\delta=1}^{q_\delta} \chi(h_\delta) e^{2\pi i m_2 \bar{h}_\delta/q_\delta} = \sum_{h_\delta=1}^{q_\delta} \bar{\chi}(h_\delta)\, e^{2\pi i m_2 h_\delta/q_\delta} = \tau(\bar{\chi}) \cdot \chi(m_2),$$

alternatively, on the right-hand side we will get from Lemma 7.1.13 (for $(r, q_\delta) = 1$), the character sum

$$\sum_{m_1|q} \sum_{h_\delta=1}^{q_\delta} \chi(h_\delta) e^{-2\pi i m_1 r \cdot h_\delta/q_\delta} = \sum_{m_1|q} \tau(\chi)\bar{\chi}(-m_1 r) = \sum_{m_1|\delta} \tau(\chi)\bar{\chi}(-m_1 r),$$

since $\chi(m_1) = 0$ if $(m_1, q_\delta) > 1$.

The last statement in Lemma 7.1.14 is a consequence of formula (7.1.2) which says that every qth root of unity can be expressed as a linear combination of Dirichlet characters (mod q). □

Let us explore Lemma 7.1.14 in the special case that $k = 0$, χ is an even primitive character (mod q) and $(h, q) = 1$ so that $\delta = 1$. In this case (using (7.1.1), Lemma 6.5.21 and (6.5.22)), the functional equation in Lemma 7.1.14 takes the simple form

$$q^{\frac{3}{2}s}\tilde{G}_\nu(s)L_\chi(s) = \frac{\tau(\chi)^2}{\tau(\bar{\chi})\sqrt{q}} \cdot q^{\frac{3}{2}(1-s)}\tilde{G}_\nu(1-s)\tilde{L}_{\bar{\chi}}(1-s),$$

where

$$L_\chi(s) = \sum_{n=1}^{\infty} \frac{A(n,1)\chi(n)}{n^s}, \qquad \tilde{L}_{\bar{\chi}}(s) = \sum_{n=1}^{\infty} \frac{A(1,n)\bar{\chi}(n)}{n^s},$$

and where $G_\nu(s)$, $\tilde{G}_\nu(s)$ are products of Gamma functions as defined in Theorem 6.5.15. Note that if we take $k = 1$ in the above situation, then the identity (7.1.16) holds automatically because each side will simply vanish since the sum over m_2 with positive and negative terms will cancel out. We may also consider odd primitive characters χ. In this case, the identity (7.1.16) automatically holds if $k = 0$ and we get a functional equation

$$q^{\frac{3}{2}s}\tilde{G}_\nu(s+1)L_\chi(s) = i\frac{\tau(\chi)^2}{\tau(\bar{\chi})\sqrt{q}} \cdot q^{\frac{3}{2}(1-s)}\tilde{G}_\nu(2-s)\tilde{L}_{\bar{\chi}}(1-s),$$

when $k = 1$. The two functional equations may be combined into a single functional equation

$$q^{\frac{3}{2}s}\tilde{G}_\nu(s+k)L_\chi(s) = i^k\frac{\tau(\chi)^2}{\tau(\bar{\chi})\sqrt{q}} \cdot q^{\frac{3}{2}(1-s)}\tilde{G}_\nu(1+k-s)\tilde{L}_{\bar{\chi}}(1-s), \tag{7.1.17}$$

where $k = 0$ if $\chi(-1) = 1$ and $k = 1$ if $\chi(-1) = -1$.

These are the simplest instances of the type of twisted functional equations needed to obtain the converse theorem. Once one knows these functional equations, then all the other required functional equations follow by the usual techniques (modified for the $GL(3)$ situation) for constructing functional equations with imprimitive characters. We shall not pursue this further as the technical complications become extremely messy.

Proof of the converse theorem, Step III We have proved that if $L_\chi(s)$ satisfies the functional equation (7.1.17) for all primitive Dirichlet characters

χ then

$$\frac{\partial^\ell}{\partial x_1^\ell}\frac{\partial^k}{\partial x_2^k}F(z,h,q)\Big|_{x_1=x_2=0} = \frac{\partial^\ell}{\partial x_1^\ell}\frac{\partial^k}{\partial x_2^k}F_1(z,h,q)\Big|_{x_1=x_2=0}, \tag{7.1.18}$$

for all $\ell = 0, 1, 2, 3, \ldots$, and $k = 0, 1$. Note that the proofs of Lemmas 7.1.12, 7.1.13 imply that $F(z) := F(z,h,q)$, $F_1(z) := F_1(z,h,q)$ are independent of x_3 and are functions of x_1, x_2, y_1, y_2 only. Consequently, if

$$z = \begin{pmatrix} 1 & x_2 & x_3 \\ & 1 & x_1 \\ & & 1 \end{pmatrix}\begin{pmatrix} y_1y_2 & & \\ & y_1 & \\ & & 1 \end{pmatrix},$$

then the function $F(z) - F_1(z)$ does not depend on x_3.

Since $F(z) - F_1(z)$ is a real analytic function, it has a power series expansion of the form

$$F(z) - F_1(z) = \sum_{i=0}^{\infty}\sum_{j=0}^{\infty} c_{i,j}(y_1, y_2)x_1^i x_2^j.$$

It immediately follows from (7.1.18) that

$$c_{\ell,k}(y_1, y_2) = 0, \qquad (\text{for } \ell = 0, 1, 2, 3, \ldots, \quad k = 0, 1). \tag{7.1.19}$$

Now, F, F_1 are eigenfunctions of the invariant differential operators on $GL(3,\mathbb{R})$. In particular, they are eigenfunctions of Δ_1 given in (6.1.1). Recall that

$$\Delta_1 = y_1^2\frac{\partial^2}{\partial y_1^2} + y_2^2\frac{\partial^2}{\partial y_2^2} - y_1y_2\frac{\partial^2}{\partial y_1\partial y_2} + y_1^2\left(x_2^2 + y_2^2\right)\frac{\partial^2}{\partial x_3^2} + y_1^2\frac{\partial^2}{\partial x_1^2} + y_2^2\frac{\partial^2}{\partial x_2^2} + 2y_1^2x_2\frac{\partial^2}{\partial x_1\partial x_3}.$$

We calculate the action of Δ_1 on $F - F_1$ and use this to prove $F(z) - F_1(z) \equiv 0$ by showing that $c_{i,j} = 0$ for all $i, j \geq 0$.

$$\Delta_1\left(x_1^i x_2^j\, c_{i,j}\right) = y_1^2x_1^ix_2^j\frac{\partial^2 c_{i,j}}{\partial y_1^2} + y_2^2x_1^ix_2^j\frac{\partial^2 c_{i,j}}{\partial y_2^2} - y_1y_2x_1^ix_2^j\frac{\partial^2 c_{i,j}}{\partial y_1\partial y_2} + y_1^2 i(i-1)x_1^{i-2}x_2^j c_{i,j} + y_2^2 j(j-1)x_1^ix_2^{j-2}c_{i,j},$$

$$\Delta_1\left(\sum_{i=0,\,j=0}^{\infty} x_1^ix_2^j c_{i,j}\right) = \sum_{i=0,\,j=0}^{\infty} x_1^ix_2^j\left(y_1^2\frac{\partial^2 c_{i,j}}{\partial y_1^2} + y_2^2\frac{\partial^2 c_{i,j}}{\partial y_2^2} - y_1y_2\frac{\partial^2 c_{i,j}}{\partial y_1\partial y_2}\right) + \sum_{i=2,\,j=0}^{\infty} y_1^2 i(i-1)x_1^{i-2}x_2^j c_{i,j} + \sum_{i=0,\,j=2}^{\infty} y_2^2 j(j-1)x_1^ix_2^{j-2}c_{i,j},$$

$$\Delta_1\left(\sum_{i=0,\, j=0}^{\infty} x_1^i x_2^j c_{i,j}\right) = \sum_{i=0,\, j=0}^{\infty} x_1^i x_2^j \left(y_1^2\frac{\partial^2 c_{i,j}}{\partial y_1^2} + y_2^2\frac{\partial^2 c_{i,j}}{\partial y_2^2} - y_1 y_2\frac{\partial^2 c_{i,j}}{\partial y_1 \partial y_2}\right)$$
$$+ \sum_{i=0,\, j=0}^{\infty} y_1^2(i+2)(i+1)x_1^i x_2^j c_{i+2,j}$$
$$+ \sum_{i=0,\, j=2}^{\infty} y_2^2(j+2)(j+1)x_1^i x_2^j c_{i,j+2}.$$

For fixed (i, j) we have

$$y_1^2\frac{\partial^2 c_{i,j}}{\partial y_1^2} + y_2^2\frac{\partial^2 c_{i,j}}{\partial y_2^2} - y_1 y_2\frac{\partial^2 c_{i,j}}{\partial y_1 \partial y_2} + y_1^2(i+2)(i+1)c_{i+2,j}$$
$$+ y_2^2(j+2)(j+1)c_{i,j+2} - \gamma_1 c_{i,j} = 0. \tag{7.1.20}$$

Now assume for fixed j, we have $c_{i,j} = 0$, this is certainly true for all i provided $j = 0$ or 1. Then from (7.1.20), it follows that $c_{i,j+2} = 0$. Further, from (7.1.19) and induction on j, we see that $c_{i,j} = 0$ for all i and j. This completes the proof of the converse Theorem 7.1.3.

7.2 Rankin–Selberg convolution for $GL(2)$

Let

$$f(z) = \sum_{n\neq 0} a(n)\sqrt{2\pi y}\cdot K_{\nu_f-\frac{1}{2}}(2\pi|n|y)\cdot e^{2\pi i n x}, \tag{7.2.1}$$

$$g(z) = \sum_{n\neq 0} b(n)\sqrt{2\pi y}\cdot K_{\nu_g-\frac{1}{2}}(2\pi|n|y)\cdot e^{2\pi i n x}, \tag{7.2.2}$$

be Maass forms of type ν_f, ν_g, respectively, for $SL(2,\mathbb{Z})$ as in Proposition 3.5.1. Both Rankin (1939) and Selberg (1940) independently found the meromorphic continuation and functional equation of the convolution L-function

$$L_{f\times g}(s) = \zeta(2s)\sum_{n=1}^{\infty}\frac{a(n)\overline{b(n)}}{n^s}. \tag{7.2.3}$$

They introduced, for the first time, the bold idea that $L_{f\times g}(s)$ can be constructed explicitly by taking an inner product of $f\cdot\bar{g}$ with an Eisenstein series. This beautiful construction has turned out to be extraordinarily important and has had many ramifications totally unforeseen by the original discoverers. The Rankin–Selberg convolution (for the case of Maass forms on $SL(2,\mathbb{Z})$) and its proof are given in the following theorem.

Theorem 7.2.4 (Rankin–Selberg convolution) *Let $f(z)$, $g(z)$ be Maass forms of type ν_f, ν_g, for $SL(2,\mathbb{Z})$ as in (7.2.1), (7.2.2). Then $L_{f\times g}(s)$, defined in (7.2.3), has a meromorphic continuation to all $s\in\mathbb{C}$ with at most a simple pole at $s=1$. Further, we have the functional equation*

$$\Lambda_{f\times g}(s)=\pi^{-2s}G_{\nu_f,\nu_g}(s)L_{f\times g}(s)=\Lambda_{f\times g}(1-s).$$

where $G_{\nu_f,\nu_g}(s)$ is the product of four Gamma factors

$$\Gamma\left(\frac{s+1-\nu_f-\nu_g}{2}\right)\Gamma\left(\frac{s+\nu_f-\nu_g}{2}\right)\Gamma\left(\frac{s-\nu_f+\nu_g}{2}\right)\Gamma\left(\frac{s-1+\nu_f+\nu_g}{2}\right).$$

If

$$L_f(s)=\prod_p\left(1-\frac{\alpha_p}{p^s}\right)^{-1}\left(1-\frac{\alpha_p'}{p^s}\right)^{-1}$$

and

$$L_g(s)=\prod_p\left(1-\frac{\beta_p}{p^s}\right)^{-1}\left(1-\frac{\beta_p'}{p^s}\right)^{-1},$$

then

$$L_{f\times g}(s)=\prod_p\left(1-\frac{\alpha_p\beta_p}{p^s}\right)^{-1}\left(1-\frac{\alpha_p\beta_p'}{p^s}\right)^{-1}\left(1-\frac{\alpha_p'\beta_p}{p^s}\right)^{-1}\left(1-\frac{\alpha_p'\beta_p'}{p^s}\right)^{-1}.$$

Proof Let $E(z,s)$ denote the Eisenstein series given in Definition 3.1.2. We compute, for $\Re(s)$ sufficiently large, the inner product

$$\begin{aligned}
\zeta(2s)\langle f\bar{g}, E(*,\bar{s})\rangle &= \zeta(2s)\iint\limits_{SL(2,\mathbb{Z})\backslash\mathfrak{h}^2} f(z)\overline{g(z)}\cdot\overline{E(z,\bar{s})}\,\frac{dxdy}{y^2}\\
&=\frac{\zeta(2s)}{2}\sum_{\gamma\in\Gamma_\infty\backslash SL(2,\mathbb{Z})}\ \iint\limits_{SL(2,\mathbb{Z})\backslash\mathfrak{h}^2} f(z)\overline{g(z)}\,I_s(\gamma z)\,\frac{dxdy}{y^2}\\
&=\frac{\zeta(2s)}{2}\iint\limits_{\Gamma_\infty\backslash\mathfrak{h}^2} f(z)\overline{g(z)}\,y^s\,\frac{dxdy}{y^2}\\
&=\frac{\zeta(2s)}{2}\int_0^\infty\int_0^1 f(z)\overline{g(z)}\,y^s\,\frac{dxdy}{y^2}\\
&=\pi\zeta(2s)\sum_{n\neq 0}a(n)\overline{b(n)}\\
&\quad\times\int_0^\infty K_{\nu_f-\frac12}(2\pi|n|y)K_{\nu_g-\frac12}(2\pi|n|y)\,y^s\,\frac{dy}{y}\\
&=(2\pi)^{1-s}L_{f\times g}(s)\cdot\int_0^\infty K_{\nu_f-\frac12}(y)K_{\nu_g-\frac12}(y)\,y^s\,\frac{dy}{y},
\end{aligned}$$

where

$$\int_0^\infty K_{\nu_f-\frac{1}{2}}(y)K_{\nu_g-\frac{1}{2}}(y)\,y^s\,\frac{dy}{y}$$
$$= 2^{s-3}\frac{\Gamma\left(\frac{s+1-\nu_f-\nu_g}{2}\right)\Gamma\left(\frac{s+\nu_f-\nu_g}{2}\right)\Gamma\left(\frac{s-\nu_f+\nu_g}{2}\right)\Gamma\left(\frac{s-1+\nu_f+\nu_g}{2}\right)}{\Gamma(s)}.$$

This computation gives the meromorphic continuation of $L_{f\times g}(s)$. By Theorem 3.1.10, the Eisenstein series $E(z,s)$ has a simple pole at $s=1$ with residue $3/\pi$. It follows that $L_{f\times g}(s)$ has a simple pole at $s=1$ if and only if $\langle f, g\rangle \neq 0$. Finally, the functional equation for $\Lambda_{f\times g}$ is a consequence of the functional equation for the Eisenstein series

$$E^*(z,s) = \pi^{-s}\Gamma(s)\zeta(2s)E(z,s) = E^*(z,1-s)$$

given in Theorem 3.1.10.

Finally, it remains to prove the Euler product representation for $L_{f\times g}(s)$. We shall actually prove a more general result (see (Bump, 1987)). Let $u, v : \mathbb{Z} \longrightarrow C$ be functions. Let z be a complex variable. Assume that

$$\sum_{n=0}^{\infty} u(n)z^n = (1-\alpha z)^{-1}(1-\alpha' z)^{-1},$$
$$\sum_{n=0}^{\infty} v(n)z^n = (1-\beta z)^{-1}(1-\beta' z)^{-1},$$

and for $|z|$ sufficiently small, the above series converge absolutely. We will show that

$$\sum_{n=0}^{\infty} u(n)v(n)z^n = \frac{1-\alpha\alpha'\beta\beta' z^2}{(1-\alpha\beta z)(1-\alpha'\beta z)(1-\alpha\beta' z)(1-\alpha'\beta' z)}. \qquad (7.2.5)$$

In fact, let us now show that (7.2.5) implies the Euler product representation of $L_{f\times g}$ in Theorem 7.2.4. Recalling (7.2.3), it follows that

$$L_{f\times g}(s) = \prod_p \frac{\left(\sum\limits_{n=0}^{\infty} a(p^n)\overline{b(p^n)}\,p^{-ns}\right)}{(1-p^{-2s})}. \qquad (7.2.6)$$

One may now easily check that our result follows from (7.2.6) and (7.2.5) if we choose

$$u(n) = a(p^n), \qquad v(n) = \overline{b(p^n)}, \qquad z = p^{-s}.$$

To prove (7.2.5), let us define

$$U(z) := \sum_{n=0}^{\infty} u(n)z^n, \qquad V(z) := \sum_{n=0}^{\infty} v(n)z^n,$$

and consider the integral

$$\frac{1}{2\pi i}\int_C U(z\cdot z_1)\,V\left(z_1^{-1}\right)\frac{dz_1}{z_1},$$

where C is the circle $|z_1| = \epsilon$. Here, $\epsilon > 0$, is sufficiently small so that the poles of $U(zz_1)$ are outside C and the poles of $V(z_1^{-1})$ are inside. It follows that

$$\frac{1}{2\pi i}\int_C U(z\cdot z_1)\,V\left(z_1^{-1}\right)\frac{dz_1}{z_1} = \sum_{n=0}^{\infty} u(n)v(n)z^n.$$

Thus, to prove (7.2.5), it is sufficient to show that

$$\frac{1}{2\pi i}\int_C (1-\alpha z z_1)^{-1}(1-\alpha' z z_1)^{-1}\left(1-\beta z_1^{-1}\right)^{-1}\left(1-\beta' z_1^{-1}\right)^{-1}\frac{dz_1}{z_1}$$
$$= \frac{1-\alpha\alpha'\beta\beta' z^2}{(1-\alpha\beta z)(1-\alpha'\beta z)(1-\alpha\beta' z)(1-\alpha'\beta' z)}.$$

This is easily proved, however, because the left-hand side is just equal to the sum of the residues at the poles: $z_1 = \beta$ and $z_1 = \beta'$, and summing these gives exactly the right-hand side of the above formula. □

This completes the proof of Theorem 7.2.4. □

7.3 Statement and proof of the Gelbart–Jacquet lift

The functional equation of a Maass form of type (ν_1, ν_2) for $SL(3,\mathbb{Z})$ is given in Theorem 6.5.15 and involves a product of three Gamma factors:

$$\Gamma\left(\frac{s+1-2\nu_1-\nu_2}{2}\right)\Gamma\left(\frac{s+\nu_1-\nu_2}{2}\right)\Gamma\left(\frac{s-1+\nu_1+2\nu_2}{2}\right).$$

If the Maass form is self dual then $\nu_1 = \nu_2 = \nu$, say, and the Gamma factors take the simpler form:

$$\Gamma\left(\frac{s+1-3\nu}{2}\right)\Gamma\left(\frac{s}{2}\right)\Gamma\left(\frac{s-1+3\nu}{2}\right).$$

In this case, if $L_\nu(s)$ denotes the L-function associated to the $SL(3,\mathbb{Z})$ Maass form, then its functional equation will be

$$\Lambda_\nu(s) := \pi^{-3s/2}\Gamma\left(\frac{s+1-3\nu}{2}\right)\Gamma\left(\frac{s}{2}\right)\Gamma\left(\frac{s-1+3\nu}{2}\right)L_\nu(s) = \Lambda_\nu(1-s). \tag{7.3.1}$$

This represents the simplest type of functional equation we may have on $SL(3,\mathbb{Z})$.

An intriguing question that arises in the mind of researchers first encountering this field is whether there might be any candidate L-functions occurring somewhere in the $GL(2)$ theory which have the functional equation (7.3.1). If one could prove that all the twists by Dirichlet characters of such a candidate L-function satisfy the functional equations and other conditions given in the converse Theorem 7.1.3, then the candidate L-function would have to be associated to a Maass form on $SL(3, \mathbb{Z})$. The fact that this phenomenon occurs is the substance of the Gelbart–Jacquet lift. In fact, if f is a Maass form of type ν_f for $SL(2, \mathbb{Z})$, then by Theorem 7.2.4, and the functional equation

$$\pi^{-s/2}\Gamma\left(\frac{s}{2}\right)\zeta(s) = \pi^{-(1-s)/2}\Gamma\left(\frac{1-s}{2}\right)\zeta(1-s)$$

of the Riemann zeta function, we see that

$$\pi^{-3s/2}\Gamma\left(\frac{s+1-2\nu_f}{2}\right)\Gamma\left(\frac{s}{2}\right)\Gamma\left(\frac{s-1+2\nu_f}{2}\right)\frac{L_{f\times f}(s)}{\zeta(s)}$$
$$= \pi^{-\frac{3}{2}(1-s)}\Gamma\left(\frac{2-s-2\nu_f}{2}\right)\Gamma\left(\frac{1-s}{2}\right)\Gamma\left(\frac{-s+2\nu_f}{2}\right)\frac{L_{f\times f}(1-s)}{\zeta(1-s)},$$

which exactly matches the functional equation of a self-dual Maass form on $SL(3, \mathbb{Z})$ of type $(2\nu_f/3, 2\nu_f/3)$. We now state and prove the Gelbart–Jacquet lift from $SL(2, \mathbb{Z})$ to $SL(3, \mathbb{Z})$.

Theorem 7.3.2 (Gelbart–Jacquet lift) *Let f be a Maass form of type ν_f for $SL(2, \mathbb{Z})$ with Fourier expansion (7.2.1). Assume that f is an eigenfunction of all the Hecke operators. Let $L_{f\times f}(s)$ be the convolution L-function as in (7.2.3). Then*

$$\frac{L_{f\times f}(s)}{\zeta(s)}$$

is the Godement–Jacquet L-function of a self-dual Maass form of type $(2\nu_f/3, 2\nu_f/3)$ for $SL(3, \mathbb{Z})$.

Proof Let us define, for $\Re(s)$ sufficiently large,

$$L(s) := \frac{L_{f\times f}(s)}{\zeta(s)} = \sum_{n=1}^{\infty} A(n, 1)n^{-s}. \tag{7.3.3}$$

We need to show that if we define $A(n, 1) = A(1, n)$ for all $n = 1, 2, 3, \ldots,$ then for $\Re(s)$ sufficiently large,

$$L(s) = \prod_p \left(1 - A(p, 1)p^{-s} + A(1, p)p^{-2s} - p^{-3s}\right)^{-1}, \tag{7.3.4}$$

and, in addition, we also must show that for every primitive Dirichlet character χ, the twisted L-function

$$L_\chi(s) := \sum_{n=1}^{\infty} A(n, 1)\chi(n)n^{-s}$$

satisfies the EBV condition and functional equation given in the converse Theorem 7.1.3. The proof will be accomplished in three independent steps. □

Proof of the Gelbart–Jacquet lift (Euler Product), Step I Consider the L-function, $L_f(s)$, associated to the Maass form f. Since f is an eigenfunction of all the Hecke operators we know by (3.13.2) that, for $\Re(s)$ sufficiently large, $L_f(s)$ has a degree two Euler product

$$L_f(s) = \prod_p \left(1 - \frac{\alpha_p}{p^s}\right)^{-1} \left(1 - \frac{\alpha'_p}{p^s}\right)^{-1}$$

where $\alpha_p \cdot \alpha'_p = 1$ for every rational prime p. □

Lemma 7.3.5 *For $\Re(s)$ sufficiently large, the L-function, $L(s)$, (defined in (7.3.3)) has a degree three Euler product*

$$L(s) = \prod_p \left(1 - \frac{\alpha_p^2}{p^s}\right)^{-1} \left(1 - \frac{\alpha_p \alpha'_p}{p^s}\right)^{-1} \left(1 - \frac{\alpha'^2_p}{p^s}\right)^{-1}.$$

Proof This follows immediately from the Euler product representation in Theorem 7.2.4 because the term $\prod_p (1 - (\alpha_p \alpha'_p / p^s))^{-1}$ just corresponds to $\zeta(s)$ since $\alpha \cdot \alpha' = 1$. □

Note that Lemma 7.3.5 immediately implies (7.3.4).

Proof of the Gelbart–Jacquet lift (Functional Equations), Step II In order to show that the candidate L-function

$$L(s) := \frac{L_{f \times f}(s)}{\zeta(s)} = \sum_{n=1}^{\infty} A(n, 1) n^{-s},$$

originally defined in (7.3.3) is actually the Godement–Jacquet L-function of a Maass form for $SL(3, \mathbb{Z})$ it is necessary to show that for every primitive Dirichlet character χ, the twisted L-function

$$L_\chi(s) = \sum_{n=1}^{\infty} A(n, 1) \chi(n) n^{-s},$$

satisfies the functional equation specified in the converse Theorem 7.1.3. Such functional equations, in some special cases of holomorphic modular forms, were first considered by Rankin (1939). This method was later generalized in (Li, 1975), (Atkin and Li, 1978), (Li, 1979). Also, Manin and Pančiškin (1977) obtained the required functional equations for the case of $\chi \pmod q$ where q is a prime power. An adelic version of the Rankin–Selberg method for $GL(2)$ was given in (Jacquet, 1972), but the requisite functional equations are not given in explicit form. We briefly sketch the method for obtaining such functional equations. □

Interlude on automorphic forms for $\Gamma_0(N)$ It is necessary, at this point, to introduce automorphic forms for the congruence subgroup

$$\Gamma_0(N) := \left\{ \begin{pmatrix} a & b \\ c & d \end{pmatrix} \in SL(2,\mathbb{Z}) \;\middle|\; c \equiv 0 \pmod{N} \right\}$$

with $N = 1, 2, 3, \ldots$ A function $F : \mathfrak{h}^2 \longrightarrow \mathbb{C}$ is said to be automorphic for $\Gamma_0(N)$, with character ψ, if it satisfies the automorphic condition $F(\gamma z) = \psi(\gamma)F(z)$ for all $\gamma \in \Gamma_0(N)$, $z \in \mathfrak{h}^2$, and for some character ψ of the group $\Gamma_0(N)$. Note that if χ is a Dirichlet character (mod N), then

$$\psi\left(\begin{pmatrix} a & b \\ c & d \end{pmatrix}\right) := \chi(d)$$

will always be a character of $\Gamma_0(N)$. To simplify notation, we will sometimes write $\chi\left(\begin{pmatrix} a & b \\ c & d \end{pmatrix}\right)$ to denote $\chi(d)$. The cusps of $\Gamma_0(N)$, denoted by Gothic letters $\mathfrak{a}, \mathfrak{b}, \mathfrak{c}, \ldots$, are defined to be elements of $\mathbb{Q} \cup \{i\infty\}$. Two cusps $\mathfrak{a}$, $\mathfrak{b}$, are termed equivalent if there exists $\gamma \in \Gamma_0(N)$ such that $\mathfrak{b} = \gamma\mathfrak{a}$. The stability group of a cusp $\mathfrak{a}$ is an infinite cyclic group, $\langle\gamma_{\mathfrak{a}}\rangle$, generated by some $\gamma_{\mathfrak{a}} \in \Gamma_0(N)$ which is defined by

$$\langle\gamma_{\mathfrak{a}}\rangle \;:=\; \left\{\gamma \in \Gamma_0(N) \;\middle|\; \gamma\mathfrak{a} = \mathfrak{a}\right\}.$$

Then there exists a scaling matrix $\sigma_{\mathfrak{a}} \in SL(2,\mathbb{Q})$ (which is determined up to right multiplication by a translation) by the conditions

$$\sigma_{\mathfrak{a}} i\infty = \mathfrak{a}, \qquad \sigma_{\mathfrak{a}}^{-1}\gamma_{\mathfrak{a}}\sigma_{\mathfrak{a}} = \begin{pmatrix} 1 & 1 \\ 0 & 1 \end{pmatrix}.$$

If $F(z)$ is automorphic for $\Gamma_0(N)$ and $\mathfrak{a}$ is any cusp, then $F(\sigma_{\mathfrak{a}} z)$ is invariant under the translation $z \mapsto z + 1$. It, therefore, has a Fourier expansion

$$F(\sigma_{\mathfrak{a}} z) = \sum_{n\in\mathbb{Z}} A_{\mathfrak{a}}(n, y)e^{2\pi i n x},$$

with $A_{\mathfrak{a}}(n, y) \in \mathbb{C}$ for all $n \in \mathbb{Z}$, $y > 0$. We say F is a cusp form if $A_{\mathfrak{a}}(m, y) = 0$ for all cusps $\mathfrak{a}$, every $m \le 0$, and all $y > 0$. If, in addition, F is an eigenfunction of the Laplacian $-y^2((\partial^2/\partial^2 x) + (\partial^2/\partial^2 y))$, then F is said to be a Maass form. If F does not come from a form of lower level then we say F is a newform. More precisely, if F is automorphic for $\Gamma_0(N)$ then for $r = 2, 3, 4, \ldots$, $F(rz)$

will be automorphic for $\Gamma_0(rN)$ because of the identity

$$\begin{pmatrix} r & 0 \\ 0 & 1 \end{pmatrix}\begin{pmatrix} a & b \\ c & d \end{pmatrix} = \begin{pmatrix} a & br \\ c/r & d \end{pmatrix}\begin{pmatrix} r & 0 \\ 0 & 1 \end{pmatrix}.$$

In this case $F(rz)$ is called an old form coming from F which lives on level N which is a lower level than rN.

The key ideas for obtaining meromorphic continuation and functional equations of the twisted L-functions $L_\chi(s)$ are based on the following two lemmas.

Lemma 7.3.6 *Let*

$$F(z) = \sum_{n\neq 0} A(n)\sqrt{2\pi y}\, K_{\nu-\frac{1}{2}}(2\pi|n|y)\, e^{2\pi i n x}$$

be a Maass form of type ν for $\Gamma_0(N)$ with trivial character. Let χ be a primitive character (mod q). *Then*

$$F_\chi(z) = \sum_{n\neq 0} A(n)\chi(n)\sqrt{2\pi y}\, K_{\nu-\frac{1}{2}}(2\pi|n|y)\, e^{2\pi i n x}$$

is a Maass form for $\Gamma_0(Nq^2)$ with character χ^2. If $N = 1$, then

$$F_\chi\left(\frac{-1}{q^2 z}\right) = \frac{\tau(\chi)}{\tau(\bar\chi)}\, F_{\bar\chi}(z).$$

Proof Since χ is primitive, the Gauss sum $\tau(\bar\chi)$, given by (7.1.1), cannot be zero. It also immediately follows from (7.1.1) that

$$F_\chi(z) = \tau(\bar\chi)^{-1}\sum_{\ell=1}^{q}\bar\chi(\ell)\cdot F\left(z+\frac{\ell}{q}\right).$$

Assume that $\gamma = \begin{pmatrix} a & b \\ c & d \end{pmatrix} \in \Gamma_0(Nq^2)$. We have the following matrix identity:

$$\begin{pmatrix} 1 & \ell/q \\ 0 & 1 \end{pmatrix}\begin{pmatrix} a & b \\ c & d \end{pmatrix} = \begin{pmatrix} a+\frac{\ell c}{q} & b-(ad-1)\frac{d\ell}{q}-\frac{cd^2\ell^2}{q^2} \\ c & d-\frac{cd^2\ell}{q} \end{pmatrix}\begin{pmatrix} 1 & d^2\ell/q \\ 0 & 1 \end{pmatrix}.$$

It follows that

$$\begin{aligned} F_\chi(\gamma z) &= \tau(\bar\chi)^{-1}\sum_{\ell=1}^{q}\bar\chi(\ell)\cdot F\left(z+\frac{d^2\ell}{q}\right)\\ &= \tau(\bar\chi)^{-1}\sum_{\ell=1}^{q}\bar\chi(\ell d^2)\chi(d)^2\cdot F\left(z+\frac{d^2\ell}{q}\right)\\ &= \chi(d)^2 F_\chi(z). \end{aligned}$$

This shows that F_χ is automorphic for $\Gamma_0(Nq^2)$ with character χ^2.

For the last part, suppose that $(\ell, q) = 1$. Then there exist integers r, s such that $rq - s\ell = 1$. Further, we have the matrix identity

$$\begin{pmatrix} 1 & \ell/q \\ 0 & 1 \end{pmatrix}\begin{pmatrix} 0 & -1 \\ q^2 & 0 \end{pmatrix} = \begin{pmatrix} 0 & -q \\ q & 0 \end{pmatrix}\begin{pmatrix} q & -s \\ -\ell & r \end{pmatrix}\begin{pmatrix} 1 & s/q \\ 0 & 1 \end{pmatrix}.$$

Note that $F\left(\begin{pmatrix} 0 & -q \\ q & 0 \end{pmatrix} z\right) = F\left(\begin{pmatrix} 0 & -1 \\ 1 & 0 \end{pmatrix} z\right) = F(z)$ for all z. Setting $s = \bar{\ell}$ where $\ell\bar{\ell} \equiv 1 \pmod{q}$, it follows from the prior matrix identity that

$$\begin{aligned}
\tau(\bar{\chi})F_{\chi}\left(\frac{-1}{q^2 z}\right) &= \sum_{\ell=1}^{q} \bar{\chi}(\ell) F\left(\begin{pmatrix} 1 & \ell/q \\ 0 & 1 \end{pmatrix}\begin{pmatrix} 0 & -1 \\ q^2 & 0 \end{pmatrix} z\right) \\
&= \sum_{\ell=1}^{q} \bar{\chi}(\ell) F\left(\begin{pmatrix} 1 & \bar{\ell}/q \\ 0 & 1 \end{pmatrix} z\right) \\
&= \sum_{\ell=1}^{q} \chi(\ell) F\left(\begin{pmatrix} 1 & \ell/q \\ 0 & 1 \end{pmatrix} z\right) \\
&= \tau(\chi) F_{\bar{\chi}}(z).
\end{aligned}$$

□

Lemma 7.3.7 *Let χ be an even primitive Dirichlet character* (mod q). *The Eisenstein series*

$$E(z, s, \chi) = \frac{1}{2} \sum_{\substack{(c,d)=1 \\ c \equiv 0 \pmod{q^2}}} \chi(d) \frac{y^s}{|cz + d|^{2s}}$$

is an automorphic form for $\Gamma_0(q^2)$ with character χ. For any fixed $z \in \mathfrak{h}^2$, the function

$$E^*(z, s, \chi) := \left(\frac{q^{\frac{5}{2}}}{\pi}\right)^s L(2s, \chi)\Gamma(s) E(z, s, \chi)$$

is entire in s and satisfies the functional equation

$$E^*(z, s, \chi) = \frac{\tau(\chi)}{\sqrt{q}} \cdot E^*\left(\frac{-1}{q^3 z}, 1 - s, \bar{\chi}\right).$$

Proof It is easy to see that $E(z, s, \chi)$ is automorphic when it is written in the form:

$$E(z, s, \chi) = \frac{1}{2} \sum_{\gamma \in \Gamma_\infty \backslash \Gamma_0(q^2)} \chi(\gamma) \Im(\gamma z)^s.$$

Let $w = u + iv$ with $v > 0$. The function $\sum_{m\in\mathbb{Z}} |m + w|^{-2s}$ is periodic in u and one may easily derive the Fourier expansion

$$\sum_{m\in\mathbb{Z}} |m+w|^{-2s} = 2\pi \frac{\Gamma(2s-1)}{\Gamma(s)^2} |2v|^{1-2s} + \frac{2\pi^s}{\Gamma(s)} \sum_{m\neq 0} |v|^{\frac{1}{2}-s} |m|^{s-\frac{1}{2}} K_{s-\frac{1}{2}}(2\pi |m| v) e^{2\pi i m u}.$$

It now follows from the definition of the Eisenstein series and the above Fourier expansion (after making the substitution $d = mcq^2 + r$) that

$$\begin{aligned}
&L(2s,\chi)E(z,s,\chi)\\
&= y^s L(2s,\chi) + \sum_{c=1}^{\infty}\sum_{d\in\mathbb{Z}} \chi(d)\cdot \frac{y^s}{|cq^2 z + d|^{2s}}\\
&= y^s L(2s,\chi) + \left(\frac{y}{q^4}\right)^s \sum_{c=1}^{\infty} |c|^{-2s} \sum_{r=1}^{cq^2} \chi(r) \sum_{m\in\mathbb{Z}} \left| m + z + \frac{r}{cq^2}\right|^{-2s}\\
&= y^s L(2s,\chi) + \left(\frac{y}{q^4}\right)^s \sum_{c=1}^{\infty} |c|^{-2s} \sum_{r=1}^{cq^2} \chi(r)\\
&\quad \times \Bigg[2\pi \frac{\Gamma(2s-1)}{\Gamma(s)^2} |2y|^{1-2s} + \frac{2\pi^s}{\Gamma(s)} \sum_{m\neq 0} \left|\frac{y}{m}\right|^{\frac{1}{2}-s}\\
&\qquad \times K_{s-\frac{1}{2}}(2\pi |m| y) e^{2\pi i m\left(x + \frac{r}{cq^2}\right)} \Bigg]\\
&= y^s L(2s,\chi) \;+\; \frac{2\pi^s \sqrt{y}}{q^{4s}\Gamma(s)} \sum_{m\neq 0} |m|^{s-\frac{1}{2}} K_{s-\frac{1}{2}}(2\pi |m| y) e^{2\pi i m x}\\
&\qquad \times \sum_{c=1}^{\infty} |c|^{-2s} \sum_{r=1}^{cq^2} \chi(r) e^{\frac{2\pi i m r}{cq^2}}.
\end{aligned} \tag{7.3.8}$$

Next, we show that

$$\sum_{r=1}^{cq^2} \chi(r) e^{2\pi i m r/cq^2} = \begin{cases} \tau(\chi)\bar{\chi}(\ell)\cdot qc & \text{if } m = \ell c q,\\ 0 & \text{otherwise.}\end{cases} \tag{7.3.9}$$

In fact, this follows easily because every r in the above sum is of the form $r = r_1 + tq$ with $1 \le r_1 \le q$ and $0 \le t < cq$. Hence, the sum takes the form

$$\sum_{r_1=1}^{q} \sum_{t=0}^{qc-1} \chi(r_1) e^{\frac{2\pi i r_1 m}{cq^2}} e^{\frac{2\pi i m t}{cq}}$$

and the inner sum over t above is zero unless $m = \ell cq$ for some $\ell \in \mathbb{Z}$. Combining (7.3.8) and (7.3.9), we obtain

$$L(2s,\chi)E(z,s,\chi) = y^s L(2s,\chi) + \tau(\chi)\frac{2\pi^s\sqrt{y}}{q^{3s-\frac{1}{2}}\Gamma(s)}\sum_{m\neq 0}\frac{\sigma_{2s-1}(m,\bar{\chi})}{|m|^{s-\frac{1}{2}}}K_{s-\frac{1}{2}}(2\pi|m|qy)e^{2\pi i m q x}, \tag{7.3.10}$$

where

$$\sigma_s(m,\chi) = \sum_{\substack{\ell|m\\ \ell\geq 1}} \chi(m/\ell)\cdot\left|\frac{m}{\ell}\right|^s.$$

Next, we compute the Fourier expansion around $z = 0$. We make use of the fact that $w_q = \begin{pmatrix} 0 & -1\\ q^2 & 0\end{pmatrix}$ is a normalizer for $\Gamma_0(q^2)$. This means that

$$w_q\Gamma_0(q^2)w_q^{-1} = \Gamma_0(q^2),$$

which can be easily seen from the calculation

$$\begin{pmatrix} 0 & -1\\ q^2 & 0\end{pmatrix}\begin{pmatrix} a & b\\ cq^2 & d\end{pmatrix}\begin{pmatrix} 0 & q^{-2}\\ -1 & 0\end{pmatrix} = \begin{pmatrix} d & -c\\ -bq^2 & a\end{pmatrix} \in \Gamma_0(q^2).$$

Thus, for every $\gamma \in \Gamma_\infty\backslash\Gamma_0(q^2)$, there exists a unique $\gamma' \in \Gamma_\infty\backslash\Gamma_0(q^2)$ such that $\gamma w_q = w_q\gamma'$. It follows, as in the computation for (7.3.8), that

$$\begin{aligned}
L(2s,\chi)E\left(\frac{-1}{q^2 z}, s, \chi\right) &= \frac{1}{2}L(2s,\chi)\sum_{\gamma\in\Gamma_\infty\backslash\Gamma_0(q^2)}\chi(\gamma)\,\Im(\gamma w_q z)^s\\
&= \frac{1}{2}L(2s,\chi)\sum_{\gamma\in\Gamma_\infty\backslash\Gamma_0(q^2)}\bar{\chi}(\gamma)\,\Im(w_q\gamma z)^s\\
&= \frac{L(2s,\chi)}{2}\sum_{\left(\begin{smallmatrix} a & b\\ c & d\end{smallmatrix}\right)\in\Gamma_\infty\backslash\Gamma_0(q^2)}\bar{\chi}(d)\,\Im\left(\frac{-1}{q^2\cdot(az+b)/(cz+d)}\right)^s\\
&= \frac{y^s}{q^{2s}}\sum_{a=1}^{\infty}\sum_{b\in\mathbb{Z}}\frac{\chi(a)}{|az+b|^{2s}} = \frac{y^s}{q^{2s}}\sum_{a=1}^{\infty}\frac{\chi(a)}{a^{2s}}\sum_{m\in\mathbb{Z}}\sum_{r=1}^{a}\left|z+m+\frac{r}{a}\right|^{-2s}\\
&= \frac{2\pi 2^{1-2s}y^{1-s}\Gamma(2s-1)}{q^{2s}\Gamma(s)^2}L(2s-1,\chi)\\
&\quad + \frac{2\pi^s\sqrt{y}}{q^{2s}\Gamma(s)}\sum_{m\neq 0}\frac{1}{|m|^{s-\frac{1}{2}}}\sigma_{2s-1}(m,\chi)K_{s-\frac{1}{2}}(2\pi|m|y)e^{2\pi i m x}.
\end{aligned}$$

The functional equation follows after comparing the above with (7.3.10). □

Finally, we return to the problem of determining the functional equation of

$$L_\chi(s) = \frac{L_{f\times f}(s,\chi)}{L(s,\chi)} = \frac{L_{f\times f_{\bar\chi}}(s)}{L(s,\chi)}$$

where f is a Maass form for $SL(2,\mathbb{Z})$. By Lemma 7.3.6, we know that f_χ is a Maass form for $\Gamma_0(q^2)$ with character χ^2. It follows, as in the proof of Theorem 7.2.4, that we may construct $L_{f\times f_{\bar\chi}}(s)$ as a Rankin–Selberg convolution

$$\langle f,\ f_{\bar\chi}\cdot E(*,s,\chi^2)\rangle,$$

proving that $L_{f\times f_{\bar\chi}}(s)$ has a meromorphic continuation to all $s\in\mathbb{C}$. If χ^2 is a primitive character, then we have the functional equation of Lemma 7.3.7. One may then easily show that $L_\chi(s)$ has exactly the functional equation given in the converse Theorem 7.1.3. This method will also work if χ^2 is not primitive, and we leave the details to the reader.

Proof of the Gelbart–Jacquet lift (EBV condition), Step III In Step II we have shown how to obtain the meromorphic continuation and functional equation for the twisted L-function

$$L_\chi(s) = \sum_{n=1}^\infty A(n,1)\chi(n)n^{-s} = \frac{L_{f\times f}(s,\chi)}{L(s,\chi)}.$$

It does not follow from these methods that $L_\chi(s)$ is entire, however. The problem is that we do not know that $L(s,\chi)$ divides $L_{f\times f}(s,\chi)$, and, for all we know, $L_\chi(s)$ could have poles at all the zeros of $L(s,\chi)$.

With a brilliant idea, Shimura (1975) was able to show that $L(s,\chi)$ always divides $L_{f\times f}(s,\chi)$, so that $L_\chi(s)$ is entire for all Dirichlet characters χ. We now explain Shimura's idea.

It follows from the methods used to prove Lemma 7.3.5 that

$$L(s,\chi)\sum_{n=1}^\infty \chi(n)a(n^2)n^{-s} = \sum_{n=1}^\infty \chi(n)a(n)^2 n^{-s}, \tag{7.3.11}$$

where $a(n)$ is the nth Fourier coefficient of f as in (7.2.1). Here, notice the difference from $a(n^2)$ to $a(n)^2$. Shimura showed that

$$\sum_{n=1}^\infty \chi(n)a(n^2)n^{-s}$$

is entire by considering the Rankin–Selberg convolution of f with a theta function

$$\theta_\chi(z) := \frac{1}{2}\sum_{n=-\infty}^{\infty} \bar\chi(n)n^\delta e^{2\pi i n^2 z},$$

where $\delta = 0$ or 1 according as $\chi(-1) = 1$ or -1. It is known (see (Shimura, 1973)) that the theta function is an automorphic form for $\Gamma_0(4q^2)$ where χ is a primitive Dirichlet character (mod q). The automorphic relation takes the explicit form

$$\theta_\chi(\gamma z) = \bar{\chi}(d)\left(\frac{-1}{d}\right)^\delta \left(\frac{c}{d}\right)\epsilon_d^{2\delta+1}(cz+d)^{\delta+\frac{1}{2}}\,\theta_\chi(z),$$

$$\left(\forall\,\gamma = \begin{pmatrix} a & b \\ c & d\end{pmatrix} \in \Gamma_0(4q^2)\right),$$

where $\left(\frac{c}{d}\right)$ is the Kronecker symbol and

$$\epsilon_d = \begin{cases} 1 & \text{if } d \equiv 1 \pmod 4 \\ i & \text{if } d \equiv 3 \pmod 4). \end{cases}$$

Note that θ_χ is automorphic of half-integral weight. The multiplier (factor of automorphy)

$$\omega(\gamma) := \left(\frac{-1}{d}\right)^\delta \left(\frac{c}{d}\right)\epsilon_d^{2\delta+1},$$

assures that we are always on the same branch of the square root $\sqrt{cz+d}$ function. In order to proceed with the Rankin–Selberg convolution, we need to construct an appropriate Eisenstein series which transforms with the same half-integral weight multiplier system. For example, we may construct

$$\tilde{E}(z,s,\chi) = \sum_{\substack{\gamma\in\Gamma_\infty\backslash\Gamma_0(4q^2)\\ \gamma=\left(\begin{smallmatrix} a & b\\ c & d\end{smallmatrix}\right)}} \overline{\omega(\gamma)}\,\bar{\chi}(d)(cz+d)^{-\delta-\frac{1}{2}}\Im(\gamma z)^s.$$

It is known that $\tilde{E}(z,s,\chi)$ has analytic continuation in $s \in \mathbb{C}$ with at most a simple pole at $s = 1 - ((2\delta+1)/4)$. The Rankin–Selberg unfolding method gives

$$\iint_{\Gamma_0(4q^2)\backslash\mathfrak{h}^2} f(z)\overline{\theta_\chi(z)}\,\tilde{E}(z,s,\chi)\,\frac{dx\,dy}{y^2} = \int_0^\infty\int_0^1 f(z)\overline{\theta_\chi(z)}\,y^s\,\frac{dx\,dy}{y^2}$$

$$= \sqrt{2\pi}\sum_{n=1}^\infty \frac{a(n^2)n^\delta\chi(n)}{(2\pi n^2)^{s-\frac{1}{2}}}\int_0^\infty K_{\nu_f-\frac{1}{2}}(y)\,y^{s-\frac{1}{2}}\,\frac{dy}{y}.$$

It now follows from this and (7.3.11) that

$$\frac{L_{f\times f}(s,\chi)}{L_f(s,\chi)}$$

has at most a simple pole at $s = \frac{3}{4} - \frac{\delta}{2}$. Actually, such a pole can only occur if χ is the trivial character, and in this case, we know from the Rankin–Selberg convolution in Step II that it cannot occur. The residue of the half-integral weight Eisenstein series is, up to a constant factor, $y^{\frac{1}{2}}\theta(z)$, where

$$\theta(z) = \sum_{n=-\infty}^{\infty} e^{2\pi i n^2 z}$$

denotes the classical theta function. As a consequence, one obtains the interesting result that

$$\iint_{\Gamma_0(4q^2)\backslash\mathfrak{h}^2} f(z)\, |\theta(z)|^2 y^{\frac{1}{2}} \frac{dxdy}{y^2} = 0.$$

Finally, from the growth properties of $\tilde{E}(z, s, \chi)$, one may obtain the EBV (entire and bounded in vertical strips of fixed width) conditions needed for the completion of the proof of the Gelbart–Jacquet lift. □

7.4 Rankin–Selberg convolution for $GL(3)$

The Rankin–Selberg convolution has been extended to automorphic forms on $GL(n)$ by Jacquet, Piatetski-Shapiro and Shalika (1983). A sketch of the method in the classical setting had been given in (Jacquet, 1981). We now work out, following (Friedberg, 1987b), the convolution for the special case of the group $SL(3, \mathbb{Z})$.

It is necessary to introduce the maximal parabolic Eisenstein series for $SL(3, \mathbb{Z})$. Let

$$\hat{\Gamma} = \begin{pmatrix} * & * & * \\ * & * & * \\ 0 & 0 & 1 \end{pmatrix} \in SL(3, \mathbb{Z}).$$

Lemma 7.4.1 *The cosets of $\hat{\Gamma}\backslash SL(3, \mathbb{Z})$ are in one-to-one correspondence with the relatively prime triples of integers via the map: $\hat{\Gamma}\gamma \to$ last row of γ.*

Proof One easily verifies that the map is well defined and injective. It is surjective because every relatively prime triple of integers can be completed to a matrix in $SL(3, \mathbb{Z})$. □

Lemma 7.4.2 *Let $z \in \mathfrak{h}^3$ have the representation*

$$z = \begin{pmatrix} 1 & x_2 & x_3 \\ & 1 & x_1 \\ & & 1 \end{pmatrix} \begin{pmatrix} y_1 y_2 & & \\ & y_1 & \\ & & 1 \end{pmatrix}.$$

Let $\gamma = \begin{pmatrix} * & * & * \\ * & * & * \\ a & b & c \end{pmatrix}$ *be a representative for the coset* $\hat{\Gamma}\backslash SL(3,\mathbb{Z})$. *Then*

$$\mathrm{Det}(\gamma z) = \frac{\mathrm{Det}(z)}{\left(y_1^2|az_2+b|^2+(ax_3+bx_1+c)^2\right)^{\frac{3}{2}}},$$

where $z_2 = x_2 + iy_2$, *and* $\mathrm{Det} : \mathfrak{h}^3 \to \mathbb{R}^+$ *denotes the determinant of a matrix, where the matrix is in Iwasawa canonical form as in Proposition 1.2.6.*

Proof Brute force computation. □

We now introduce the Eisenstein series

$$E(z,s) := \sum_{\gamma\in\hat{\Gamma}\backslash SL(3,\mathbb{Z})} \frac{\mathrm{Det}(\gamma z)^s}{2}.$$

Lemmas 7.4.1 and 7.4.2 then imply that the series above converges absolutely for $\Re(s) > 1$, and has the explicit representation

$$E(z,s) = \frac{1}{2}\sum_{\substack{a,b,c\in\mathbb{Z}\\(a,b,c)=1}} \frac{\left(y_1^2y_2\right)^s}{\left(y_1^2\left|az_2+b\right|^2+(ax_3+bx_1+c)^2\right)^{3s/2}}. \tag{7.4.3}$$

The most important properties of $E(z,s)$ are given in the next proposition.

Proposition 7.4.4 *The Eisenstein series $E(z,s)$ has a meromorphic continuation to all $s\in\mathbb{C}$ and satisfies the functional equation*

$$\pi^{-3s/2}\Gamma\left(\frac{3s}{2}\right)\zeta(3s)E(z,s) = \pi^{-(3-3s)/2}\Gamma\left(\frac{3-3s}{2}\right)\zeta(3-3s)\tilde{E}(z,1-s),$$

where $\tilde{E}(z,1-s) = E(w^t z^{-1} w, s)$ is the dual Eisenstein series, and w is the long element of the Weyl group, as in Proposition 6.3.1. Furthermore,

$$E^*(z,s) := \pi^{-3s/2}\Gamma\left(\frac{3s}{2}\right)\zeta(3s)E(z,s),$$

is holomorphic except for simple poles at $s = 0, 1$, with residues $-2/3$, $2/3$, respectively.

Proof The results will follow from the Fourier–Whittaker expansion in exactly the same way in which they were obtained for Eisenstein series on $SL(2,\mathbb{Z})$ in Theorems 3.1.8 and 3.1.10. We break the sum for $E^*(z,s)$, given in (7.4.3), into two pieces corresponding to $a = 0$ and $a \neq 0$. It follows that $E^*(z,s) = E_1^*(z,s) + E_2^*(z,s)$ where

$$E_1^*(z,s) = \pi^{-3s/2}\Gamma\left(\frac{3s}{2}\right)\sum_{b,c\in\mathbb{Z}^2-(0,0)} \frac{\left(y_1^2y_2\right)^s}{\left(y_1^2b^2+(bx_1+c)^2\right)^{3s/2}},$$

and

$$E_2^*(z,s) = 2\left(y_1^2 y_2\right)^s \sum_{a=1}^{\infty} \sum_{j,\ell=-\infty}^{\infty} \int_{-\infty}^{\infty}\int_{-\infty}^{\infty} \times \frac{\pi^{-3s/2}\Gamma(3s/2)\cdot e^{-2\pi i(ju+\ell v)}}{\left(y_1^2\left|az_2+u\right|^2+(ax_3+ux_1+v)^2\right)^{3s/2}}\, dudv. \tag{7.4.5}$$

But $E_1^*(z,s)$ is $\left(y_1^{\frac{1}{2}}y_2\right)^s$ times an $SL(2,\mathbb{Z})$ Eisenstein series with Fourier expansion given by Theorem 3.1.8 from which it follows that

$$\begin{aligned} E_1^*(z,s) = 2\left(y^{\frac{1}{2}} y_2\right)^s \Bigg\{ & y_1^{3s/2}\pi^{-3s/2}\Gamma\left(\frac{3s}{2}\right)\zeta(3s) \\ & + y_1^{1-(3s/2)}\pi^{(1-3s)/2}\Gamma\left(\frac{3s-1}{2}\right)\zeta(3s-1) \\ & + 2y_1^{\frac{1}{2}}\sum_{m_1\neq 0}|m_1|^{(3s-1)/2}\sigma_{1-3s}(|m_1|)K_{\frac{3s-1}{2}}(2\pi|m|y_1)e^{2\pi i m_1 x_1}\Bigg\}. \end{aligned} \tag{7.4.6}$$

Now $E^*(z,s)$ has a Fourier–Whittaker expansion of type

$$E^*(z,s) = \sum_{m_2\in\mathbb{Z}} \phi_{0,m_2}(z) + \sum_{\gamma\in\Gamma_\infty\backslash SL(2,\mathbb{Z})} \sum_{m_1\neq 0}\sum_{m_2\in\mathbb{Z}} \phi_{m_1,m_2}\left(\begin{pmatrix}\gamma & \\ & 1\end{pmatrix} z\right), \tag{7.4.7}$$

where

$$\phi_{m_1,m_2}(z) = \int_0^1\int_0^1\int_0^1 E^*\left(\begin{pmatrix}1 & \xi_2 & \xi_3\\ & 1 & \xi_1\\ & & 1\end{pmatrix} z\right) e^{-2\pi i(m_1\xi_1+m_2\xi_2)}\, d\xi_1\xi_2 d\xi_3.$$

We immediately observe that (7.4.6) only contributes to the Fourier expansion (7.4.7) when $m_2 = 0$. On the other hand, after some change of variables in (7.4.5), the contribution of $E_2^*(z,s)$ to the Fourier coefficient ϕ_{m_1,m_2} is given by

$$\begin{aligned} & 2\pi^{-3s/2}\Gamma\left(\frac{3s}{2}\right)\cdot\left(y_1^2y_2\right)^s \int_0^1\int_0^1\int_0^1 \sum_{a=1}^{\infty}\sum_{j,\ell=-\infty}^{\infty} \\ & \times e^{2\pi i\left(\ell a(x_3+\xi_3+\xi_2x_1)+(j-\ell(x_1+\xi_1))a(x_2+\xi_2)-m_1\xi_1-m_2\xi_2\right)} \\ & \times ay_2 \int_{-\infty}^{\infty}\int_{-\infty}^{\infty} \frac{e^{2\pi i(-\ell v+ay_2u(\ell(x_1+\xi_1)-j))}}{\left(y_1^2y_2^2a^2(u^2+1)+v^2\right)^{3s/2}}\, dudv\, d\xi_1 d\xi_2 d\xi_3. \end{aligned}$$

But this vanishes unless $\ell = 0$, and, in this case, $m_1 = 0$ also. Further, if $m_2 \neq 0$, we get a contribution only when $aj = m_2$, and when $m_2 = 0$, we must have $j = 0$. Consequently everything simplifies to

$$\delta_{m_1,0} \cdot 2\pi^{-3s/2}\Gamma\left(\frac{3s}{2}\right)\left(y_1^2 y_2\right)^s e^{2\pi i m_2 x_2}$$
$$\times \sum_{a|m_2} a y_2 \int_{-\infty}^{\infty}\int_{-\infty}^{\infty} \frac{e^{-2\pi i m_2 y_2 u}}{\left(y_1^2 y_2^2 a^2(u^2+1)+v^2\right)^{3s/2}}\, du\, dv,$$

when $m_2 \neq 0$, and to

$$\delta_{m_1,0} \cdot 2\pi^{-3s/2}\Gamma\left(\frac{3s}{2}\right)\left(y_1^2 y_2\right)^s$$
$$\times \sum_{a=1}^{\infty} a y_2 \int_{-\infty}^{\infty}\int_{-\infty}^{\infty} \left(y_1^2 y_2^2 a^2(u^2+1)+v^2\right)^{-3s/2}\, du\, dv,$$

when $m_2 = 0$. It follows that the Fourier coefficient $\phi_{m_1,m_2}(z)$ is given by

$$\begin{cases} 4\left(y^{\frac{1}{2}} y_2\right)^s y_1^{\frac{1}{2}} |m_1|^{\frac{3s-1}{2}} \sigma_{1-3s}(|m_1|) K_{\frac{3s-1}{2}}(2\pi |m_1| y_1) e^{2\pi i m_1 x_1}, & \\ \qquad\qquad\qquad\qquad\qquad\qquad\qquad \text{if } m_1 \neq 0, m_2 = 0, & \\ 4 y_1^{1-s} y_2^{1-(\frac{s}{2})} |m_2|^{\frac{3s}{2}-1} \sigma_{2-3s}(|m_2|) K_{\frac{3s}{2}-1}(2\pi |m_2| y_2) e^{2\pi i m_2 x_2}, & \\ \qquad\qquad\qquad\qquad\qquad\qquad\qquad \text{if } m_1 = 0, m_2 \neq 0, & \\ 2 y_1^{2s} y_2^{s} \pi^{\frac{-3s}{2}} \Gamma\left(\frac{3s}{2}\right)\zeta(3s) + 2 y_1^{1-s} y_2^{s} \pi^{\frac{1-3s}{2}} \Gamma\left(\frac{3s-1}{2}\right)\zeta(3s-1) & \\ \quad + 2 y_1^{1-s} y_2^{2-2s} \pi^{1-\frac{3s}{2}} \Gamma\left(\frac{3s}{2}-1\right)\zeta(3s-2), \qquad \text{if } m_1 = m_2 = 0. & \end{cases}$$

The meromorphic continuation and functional equation of $E(z,s)$ are an immediate consequence of the above explicit computation of the Fourier coefficients given in (7.4.7). □

The Rankin–Selberg convolution for $SL(3,\mathbb{Z})$ is very similar to the convolution for $SL(2,\mathbb{Z})$ given in Theorem 7.2.4. The idea is to take the inner product of two Maass forms with the Eisenstein series (7.4.3). Let

$$f(z) = \sum_{\gamma \in U_2(\mathbb{Z})\backslash SL(2,\mathbb{Z})} \sum_{m_1=1}^{\infty} \sum_{m_2 \neq 0} \frac{A(m_1,m_2)}{|m_1 m_2|} W_{\text{Jacquet}}\left(M\begin{pmatrix}\gamma & \\ & 1\end{pmatrix} z,\ \nu,\ \psi_{1,\frac{m_2}{|m_2|}}\right), \tag{7.4.8}$$

$$g(z) = \sum_{\gamma \in U_2(\mathbb{Z})\backslash SL(2,\mathbb{Z})} \sum_{m_1=1}^{\infty} \sum_{m_2 \neq 0} \frac{B(m_1,m_2)}{|m_1 m_2|} W_{\text{Jacquet}}\left(M\begin{pmatrix}\gamma & \\ & 1\end{pmatrix} z,\ \nu',\ \psi_{1,\frac{m_2}{|m_2|}}\right),$$

be Maass forms of type ν, ν', respectively, for $SL(3, \mathbb{Z})$. Here $M = \begin{pmatrix} m_1|m_2| & & \\ & m_1 & \\ & & 1 \end{pmatrix}$.

Theorem 7.4.9 (Rankin–Selberg convolution) *Let f, g be Maass forms for $SL(3, \mathbb{Z})$ of types ν, ν', as in (7.4.8). Then*

$$L_{f\times g}(s) := \zeta(3s) \sum_{m_1=1}^{\infty} \sum_{m_2=1}^{\infty} \frac{A(m_1, m_2)\,\overline{B(m_1, m_2)}}{m_1^{2s}\, m_2^{s}}$$

has a meromorphic continuation to all $s \in \mathbb{C}$ with at most a simple pole at $s = 1$. The residue is proportional to $\langle f, g\rangle$, the Petersson inner product of f with g.

Furthermore, if

$$G_{\nu,\nu'}(s) = \int_0^{\infty}\int_0^{\infty} W_{\text{Jacquet}}(y, \nu, \psi_{1,1}) \cdot \bar{W}_{\text{Jacquet}}(y, \nu', \psi_{1,1}) \left(y_1^2 y_2\right)^s \frac{dy_1 dy_2}{y_1^3 y_2^3},$$

then $L_{f\times g}$ satisfies the functional equation

$$\Lambda_{f\times g}(s) = \pi^{-3s/2}\Gamma\left(\frac{3s}{2}\right) G_{\nu,\nu'}(s) L_{f\times g}(s) \;=\; \Lambda_{\tilde{f}\times\tilde{G}}(1-s),$$

where $\tilde{f}, \tilde{g}$ denote the dual Maass forms as in Proposition 6.3.1.

Proof We compute the Rankin–Selberg inner product of $f \cdot \bar{g}$ with the Eisenstein series $E(z, s)$ given in (7.4.3). It follows that

$$\langle f\cdot\bar{g},\ E(*, \bar{s})\rangle = \int\limits_{SL(3,\mathbb{Z})\backslash\mathfrak{h}^3} f(z)\overline{g(z)}\cdot\overline{E(z, \bar{s})}\, d^*z,$$

with d^*z, the invariant measure as given in Proposition 1.5.3. If we now apply the usual unfolding trick, we obtain

$$\langle f\cdot\bar{g},\ E(*, \bar{s})\rangle = \frac{1}{2}\int\limits_{\hat{\Gamma}\backslash\mathfrak{h}^3} f(z)\overline{g(z)}\cdot\left(y_1^2 y_2\right)^s d^*z. \tag{7.4.10}$$

Now

$$\hat{\Gamma} = \begin{pmatrix} * & * & * \\ * & * & * \\ 0 & 0 & 1 \end{pmatrix} \subset SL(3, \mathbb{Z})$$

is generated by matrices of type

$$\begin{pmatrix} a & b & 0 \\ c & d & 0 \\ 0 & 0 & 1 \end{pmatrix}, \qquad \begin{pmatrix} 1 & 0 & e \\ 0 & 1 & f \\ 0 & 0 & 1 \end{pmatrix}$$

with $a, b, c, d, e, f \in \mathbb{Z}$, $ad - bc = 1$. Consequently, if we define $U_2(\mathbb{Z}) = \begin{pmatrix} 1 & * \\ & 1 \end{pmatrix} \subset SL(2, \mathbb{Z})$,

$$\bigcup_{\gamma \in U_2(\mathbb{Z})\backslash SL(2,\mathbb{Z})} \begin{pmatrix} \gamma & \\ & 1 \end{pmatrix} \cdot \hat{\Gamma}\backslash\mathfrak{h}^3 \cong U_3(\mathbb{Z})\backslash\mathfrak{h}^3, \tag{7.4.11}$$

where

$$U_3(\mathbb{Z}) = \begin{pmatrix} 1 & * & * \\ & 1 & * \\ & & 1 \end{pmatrix} \subset SL(3, \mathbb{Z}).$$

This result was also obtained in Lemma 5.3.13. Since the Fourier expansion of f (7.4.10) is given by a sum over $SL(2, \mathbb{Z})$, we may unfold further using (7.4.11) to obtain

$$\begin{aligned}
&\langle f \cdot \bar{g},\ E(*, \bar{s})\rangle \\
&\quad = \frac{1}{2} \int\limits_{U_3(\mathbb{Z})\backslash\mathfrak{h}^3} \sum_{m_1=1}^{\infty} \sum_{m_2 \neq 0} \frac{A(m_1, m_2)}{|m_1 m_2|} W_{\text{Jacquet}}(My,\ \nu,\ \psi_{1,1}) e^{2\pi i (m_1 x_1 + m_2 x_2)} \\
&\quad \times \sum_{\begin{pmatrix} a & b \\ c & d \end{pmatrix} \in U_2(\mathbb{Z})\backslash SL(2,\mathbb{Z})} \sum_{m_1'=1}^{\infty} \sum_{m_2' \neq 0} \overline{\frac{B(m_1', m_2')}{|m_1' m_2'|}} \\
&\qquad\qquad \times e^{-2\pi i \left[m_1'(cx_3 + dx_1) - m_2' \Re\left(\frac{az_2+b}{cz_2+d}\right)\right]} \\
&\qquad \times \bar{W}_{\text{Jacquet}}\left(M' \begin{pmatrix} \frac{y_1 y_2}{|cz_2+d|} & & \\ & y_1 \cdot |cz_2 + d| & \\ & & 1 \end{pmatrix},\ \nu',\ \psi_{1,1} \right) \\
&\qquad\qquad \times \left(y_1^2 y_2\right)^s d^*z,
\end{aligned}$$

after applying Theorem 6.5.7 to g. Further,

$$\int\limits_{U_3(\mathbb{Z})\backslash\mathfrak{h}^3} = \int_{x_1=0}^{1} \int_{x_2=0}^{1} \int_{x_3=0}^{1} \int_{y_1=0}^{\infty} \int_{y_2=0}^{\infty}.$$

It then follows by the standard techniques introduced in Section 6.5 that

$$\zeta(3s)\langle f\cdot\bar{g},\ E(*,\bar{s})\rangle \;=\; L_{f\times g}(s)\,G_{\nu,\nu'}(s).$$

The theorem is now a consequence of the functional equation of the Eisenstein series given in Proposition 7.4.4. □

We will end this section with an explicit computation of the Euler product for $L_{f\times g}$ assuming that f, g are both eigenfunctions of the Hecke operators and $L_f(s)$, $L_g(s)$ have Euler products as given in Definition 6.5.2.

Proposition 7.4.12 (Euler product) *Let f, g be Maass forms for $SL(3,\mathbb{Z})$ with Fourier coefficients $A(m_1,m_2)$, $B(m_1,m_2)$, respectively, as in (7.4.8). Let*

$$L_{f\times g}(s) := \zeta(3s)\sum_{m_1=1}^{\infty}\sum_{m_2=1}^{\infty}\frac{A(m_1,m_2)\,\overline{B(m_1,m_2)}}{m_1^{2s}\,m_2^{s}}$$

be the Rankin–Selberg convolution. Assume L_f, L_g have Euler products:

$$L_f(s)=\prod_p\prod_{i=1}^{3}\left(1-\frac{\alpha_{i,p}}{p^s}\right)^{-1},\qquad L_g(s)=\prod_p\prod_{j=1}^{3}\left(1-\frac{\beta_{j,p}}{p^s}\right)^{-1},$$

then

$$L_{f\times g}(s)=\prod_p\prod_{i=1}^{3}\prod_{j=1}^{3}\left(1-\frac{\alpha_{i,p}\,\overline{\beta}_{j,p}}{p^s}\right)^{-1}.$$

Proof The proof is based on a special case of Cauchy's identity (Weyl, 1939), (McDonald, 1979), (Bump, to appear), which takes the form:

$$\prod_{i=1}^{3}\prod_{j=1}^{3}(1-\alpha_i\beta_j x)^{-1} = \sum_{k_1=0}^{\infty}\sum_{k_2=0}^{\infty} S_{k_1,k_2}(\alpha_1,\alpha_2,\alpha_3)S_{k_1,k_2}(\beta_1,\beta_2,\beta_3)x^{k_1+2k_2}$$
$$\times\,(1-\alpha_1\alpha_2\alpha_3\beta_1\beta_2\beta_3x^3)^{-1}, \qquad (7.4.13)$$

where

$$S_{k_1,k_2}(x_1,x_2,x_3) := \frac{\begin{vmatrix} x_1^{k_1+k_2+2} & x_2^{k_1+k_2+2} & x_3^{k_1+k_2+2}\\ x_1^{k_1+1} & x_2^{k_1+1} & x_3^{k_1+1}\\ 1 & 1 & 1\end{vmatrix}}{\begin{vmatrix} x_1^2 & x_2^2 & x_3^2\\ x_1 & x_2 & x_3\\ 1 & 1 & 1\end{vmatrix}}$$

is the Schur polynomial given by a ratio of determinants. We shall defer the proof of Cauchy's identity for the moment. In the meantime, we will show that

$$A(p^{k_1}, p^{k_2}) = S_{k_1,k_2}(\alpha_{1,p}, \alpha_{2,p}, \alpha_{3,p}), \quad B(p^{k_1}, p^{k_2}) = S_{k_1,k_2}(\beta_{1,p}, \beta_{2,p}, \beta_{3,p}), \tag{7.4.14}$$

from which the proposition follows after choosing $\alpha_i = \alpha_{i,p}$, $\beta_j = \bar{\beta}_{i,p}$ for $(1 \leq i, j \leq 3)$ and $x = p^{-s}$ in (7.4.13).

To prove (7.4.14), we follow (Bump, 1984). We require the identity:

$$\sum_{k_1=0}^{\infty}\sum_{k_2=0}^{\infty} S_{k_1,k_2}(\alpha_1, \alpha_2, \alpha_3)p^{-k_1 s}p^{-k_2 s} \tag{7.4.15}$$
$$= (1 - p^{-2s})\prod_{i=1}^{3}(1 - \alpha_i p^{-s})^{-1} \prod_{1\leq i<j\leq 3} (1 - \alpha_i\alpha_j p^{-s})^{-1}.$$

This is proved by noting that the ratio of determinants in the definition of the Schur polynomial can be computed explicitly to yield

$$S_{k_1,k_2}(\alpha_1, \alpha_2, \alpha_3) = \frac{\alpha_1^{k_1+k_2+2}\left(\alpha_2^{k_1+1} - \alpha_3^{k_1+1}\right) + \alpha_2^{k_1+k_2+2}\left(\alpha_3^{k_1+1} - \alpha_1^{k_1+1}\right) + \alpha_3^{k_1+k_2+2}\left(\alpha_1^{k_1+1} - \alpha_2^{k_1+1}\right)}{(\alpha_2 - \alpha_1)(\alpha_3 - \alpha_2)(\alpha_1 - \alpha_3)}.$$

If we then multiply both sides of the above formula by $p^{-(k_1+k_2)s}$ and sum over k_1, k_2, then the identity (7.4.15) follows after some algebraic manipulations.

Now, by Section 6.5, we have

$$1-A(1, p)p^{-s}+A(p, 1)p^{-2s} - p^{-3s} = (1 - \alpha_1 p^{-s})(1 - \alpha_2 p^{-s})(1 - \alpha_3 p^{-s}),$$
$$1 - A(p, 1)p^{-s} + A(1, p)p^{-2s} - p^{-3s} = (1 - \alpha_2\alpha_3 p^{-s})(1 - \alpha_3\alpha_1 p^{-s}) \times (1 - \alpha_1\alpha_2 p^{-s}).$$

It follows that

$$\sum_{k_1=0}^{\infty}\sum_{k_2=0}^{\infty} A(p^{k_1}, p^{k_2})p^{-(k_1+k_2)s}$$
$$= (1 - p^{-2s})\sum_{k_1=0}^{\infty}\sum_{k_2=0}^{\infty}\sum_{k\leq\min(k_1,k_2)} A(p^{k_1-k}, p^{k_2-k})p^{-(k_1+k_2)s}$$
$$= (1 - p^{-2s})\sum_{k_1=0}^{\infty}\sum_{k_2=0}^{\infty} A(p^{k_1}, 1)A(1, p^{k_2})p^{-(k_1+k_2)s}$$
$$= (1 - p^{-2s})(1 - A(p, 1)p^{-s} + A(1, p)p^{-2s} - p^{-3s})^{-1} \times \left(1 - A(1, p)p^{-s} + A(p, 1)p^{-2s} - p^{-3s}\right)^{-1}$$

$$= (1-p^{-2s})(1-\alpha_1 p^{-s})^{-1}(1-\alpha_2 p^{-s})^{-1}(1-\alpha_3 p^{-s})^{-1}$$
$$\times (1-\alpha_2\alpha_3 p^{-s})^{-1}(1-\alpha_3\alpha_1 p^{-s})^{-1}(1-\alpha_1\alpha_2 p^{-s})^{-1}$$
$$= \sum_{k_1=0}^{\infty}\sum_{k_2=0}^{\infty} S_{k_1,k_2}(\alpha_1,\alpha_2,\alpha_3)p^{-(k_1+k_2)s}.$$

Finally, (7.4.14) is obtained by comparing coefficients on both sides.

It remains to prove Cauchy's identity (7.4.13). Note that Schur's polynomial is a ratio of determinants. The matrix in the denominator is a Vandermonde matrix, which is a matrix whose columns all have the form $(1, x_j, x_j^2, \ldots, x_j^{n-1})$ for some x_j and some $n > 2$. Now, if two of the x_js are equal, then two columns of the matrix are the same and the determinant is zero. This means, that as a polynomial, the determinant must have $(x_i - x_j)$ as a factor for all $1 \le i < j \le n$. Consequently, the determinant of the Vandermonde matrix must have

$$\prod_{1\le i<j\le n}(x_i - x_j) \tag{7.4.16}$$

as a factor. By comparing degrees one easily sees that the determinant of the Vandermonde matrix is the product (7.4.16) up to a constant factor. To show that the constant factor is 1, one may consider the main diagonal term $x_1^{n-1}x_2^{n-2}\cdots x_n^1 x_n^0$, which is exactly what one gets by taking the first positive terms in each factor of the product (7.4.16). This proves that

$$\begin{vmatrix} x_1^{n-1} & x_2^{n-1} & \cdots & x_n^{n-1} \\ x_1^{n-2} & x_2^{n-2} & \cdots & x_n^{n-2} \\ \vdots & \vdots & \cdots & \vdots \\ x_1 & x_2 & \cdots & x_n \\ 1 & 1 & \cdots & 1 \end{vmatrix} = \prod_{1\le i<j\le n}(x_i - x_j). \tag{7.4.17}$$

Next, we consider Cauchy's determinant which is defined to be the determinant of the $n \times n$ matrix whose i, jth entry is $1/(1 - x_i y_j)$ where x_i, y_j $(1 \le i, j \le n)$ are variables and $n \ge 2$. We write this matrix as

$$\left(\frac{1}{1 - x_i y_j}\right)_{1\le i,j\le n}.$$

Lemma 7.4.18 (Cauchy's determinant) *Let $n \ge 2$ and $x_i, y_i \in \mathbb{C}$ $(1 \le i \le n)$. Then*

$$\mathrm{Det}\left(\left(\frac{1}{1 - x_i y_j}\right)_{1\le i,j\le n}\right) = \frac{\prod\limits_{1\le i<j\le n}(x_i - x_j)\prod\limits_{1\le i<j\le n}(y_i - y_j)}{\prod\limits_{i=1}^{n}\prod\limits_{j=1}^{n}(1 - x_i y_j)}.$$

Proof We know the determinant is a rational function. The determinant will be zero if any two of the x_i or any two of the y_j are equal since, in this case, either two of the rows or columns of the matrix would be the same. Consequently, the numerator of the determinant must have factors $\prod(x_i - x_j)\prod(y_i - y_j)$. Clearly, the denominator of the determinant must have $(1 - x_i y_j)$ as a factor for every $1 \le i, j \le n$. This suggests that the determinant should be a constant multiple of

$$\frac{\prod\limits_{1\le i<j\le n}(x_i - x_j) \prod\limits_{1\le i<j\le n}(y_i - y_j)}{\prod\limits_{i=1}^{n}\prod\limits_{j=1}^{n}(1 - x_i y_j)}. \tag{7.4.19}$$

In fact, the numerator of (7.4.19) has degree $n(n-1)$ while the denominator has degree n^2 which is also the case for the determinant because all of its terms are products of n factors, each having one term in the denominator and none in the numerator with a net increase of n in the denominator. To show that the constant multiple must be 1, let $x_i = -y_i^{-1}$ for $i = 1, 2, \ldots, n$. In this case, (7.4.19) becomes

$$2^{-n} \cdot \frac{\prod\limits_{1\le i<j\le n}(y_i - y_j)^2}{\prod\limits_{1\le i<j\le n}(y_i + y_j)^2}.$$

We may now fix $y_2, y_3, \ldots, y_n$ and let $y_1 \to \infty$. The above expression collapses to

$$2^{-n} \cdot \frac{\prod\limits_{2\le i<j\le n}(y_i - y_j)^2}{\prod\limits_{2\le i<j\le n}(y_i + y_j)^2}.$$

Next, fix $y_3, \ldots, y_n$ and let $y_2 \to \infty$. Continue in this manner. The expression (7.4.19) turns into 2^{-n}. On the other hand, all upper triangular terms of the matrix become 0 after taking the above limits, while the diagonal terms are all $\frac{1}{2}$. So the value of Cauchy's determinant must also be 2^{-n}. This completes the proof of Lemma 7.4.18. □

We shall conclude this section by proving a more general version of Cauchy's identity (7.4.13). In order to do this, however, it is necessary to introduce a more general Schur polynomial. Let $n \ge 2$ and $k = (k_1, k_2, \ldots, k_{n-1})$ be a set of $n-1$ non-negative integers. We define

$$S_k(x_1, x_2, \ldots, x_n) = \frac{\begin{vmatrix} x_1^{k_1+k_2+\cdots+k_{n-1}+n-1} & \cdots & x_n^{k_1+k_2+\cdots+k_{n-1}+n-1} \\ x_1^{k_1+k_2+\cdots+k_{n-2}+n-2} & \cdots & x_n^{k_1+k_2+\cdots+k_{n-2}+n-2} \\ \vdots & \vdots & \vdots \\ x_1^{k_1+1} & \cdots & x_n^{k_1+1} \\ 1 & \cdots & 1 \end{vmatrix}}{\begin{vmatrix} x_1^{n-1} & x_2^{n-1} & \cdots & x_n^{n-1} \\ x_1^{n-2} & x_2^{n-2} & \cdots & x_n^{n-2} \\ \vdots & \vdots & \cdots & \vdots \\ x_1 & x_2 & \cdots & x_n \\ 1 & 1 & \cdots & 1 \end{vmatrix}},$$

where the denominator is the Vandermonde determinant which has the value given by (7.4.17). □

Proposition 7.4.20 (Cauchy's identity) *Let $n \ge 2$ and $\alpha_i, \beta_i \in \mathbb{C}$ for every $i = 1, \ldots, n$. For $x \in \mathbb{C}$ and $|x|$ sufficiently small, we have the identity*

$$\prod_{i=1}^{n}\prod_{j=1}^{n}(1-\alpha_i\beta_j x)^{-1} = \sum_{\substack{k_1=0 \\ k=(k_1,\ldots,k_{n-1})}}^{\infty} \cdots \sum_{k_{n-1}=0}^{\infty} \frac{S_k(\alpha_1,\ldots,\alpha_n)\, S_k(\beta_1,\ldots,\beta_n)\cdot x^{k_1+2k_2+\cdots+(n-1)k_{n-1}}}{1-\alpha\beta x^n},$$

where $\alpha = \prod_{i=1}^{n} \alpha_i$ and $\beta = \prod_{i=1}^{n} \beta_i$.

Proof We follow (Macdonald, 1979).

$$\begin{aligned} \mathrm{Det}\left(\left(\frac{1}{1-x_iy_j}\right)_{1\le i,j\le n}\right) &= \mathrm{Det}\left(\left(1 + x_iy_j + x_i^2y_j^2 + \cdots\right)_{1\le i,j\le n}\right) \\ &= \sum_{\ell_1=0}^{\infty}\cdots\sum_{\ell_n=0}^{\infty} \mathrm{Det}\left(\left(x_i^{\ell_j}y_j^{\ell_j}\right)_{1\le i,j\le n}\right) \\ &= \sum_{\substack{\ell_1=0 \\ \ell=(\ell_1,\ldots,\ell_n)}}^{\infty}\cdots\sum_{\ell_n=0} a_\ell((x_1,\ldots,x_n))y_1^{\ell_1}\cdots y_n^{\ell_n}, \end{aligned}$$

where

$$a_\ell((x_1,\ldots,x_n)) = \sum_{\sigma\in S_n} \epsilon(\sigma)\sigma\left(x_1^{\ell_1}\cdots x_n^{\ell_n}\right),$$

in which S_n is the group of permutations of $\{1,2,\ldots,n\}$, and $\epsilon(\sigma)$ is the sign of the permutation σ.

It is clear that the polynomial $a_\ell(x) := a_\ell((x_1,\ldots,x_n))$ satisfies

$$\sigma(a_\ell(x)) = \epsilon(\sigma)a_\ell(x)$$

for any $\sigma \in S_n$, and, therefore, vanishes unless $\ell_1,\ldots,\ell_n$ are all distinct. So, we may assume $\ell_1 > \ell_2 > \cdots > \ell_n \geq 0$. In view of the skew symmetry, we may write

$$(\ell_1,\ldots,\ell_n) = (\lambda_1+n-1,\ \lambda_2+n-2,\ \lambda_3+n-3,\ldots,\lambda_n).$$

It follows that

$$a_\ell(x) = a_{\lambda+\delta} = \operatorname{Det}\left(\left(x_i^{\lambda_j+n-j}\right)_{1\leq i,j\leq n}\right),$$

where $\lambda = (\lambda_1,\ldots,\lambda_n)$ and $\delta = (n-1,n-2,\ldots,1,0)$.

If the above computations are combined with Lemma 7.4.18 (Cauchy's determinant) we obtain the identity

$$\prod_{i=1}^n\prod_{j=1}^n(1-x_iy_j)^{-1} = \sum_\lambda \frac{a_{\lambda+\delta}(x)a_{\lambda+\delta}(y)}{\prod\limits_{1\leq i<j\leq n}(x_i-x_j)\prod\limits_{1\leq i<j\leq n}(y_i-y_j)}, \tag{7.4.21}$$

where the sum goes over all $\lambda = (\lambda_1,\ldots,\lambda_n) \in \mathbb{Z}^n$ with $\lambda_1 \geq \lambda_2 \geq \cdots \geq \lambda_n \geq 0$. To complete the proof, note that if we choose

$$\lambda = (k_1+\cdots+k_n,\ k_1+\cdots+k_{n-1},\ldots,k_1),$$

then

$$\frac{a_{\lambda+\delta}(x)}{\prod\limits_{1\leq i<j\leq n}(x_i-x_j)} = S_{k_2,k_3,\ldots,x_n}(x_1,x_2,\ldots,x_n).$$

Cauchy's identity (Proposition 7.4.20) immediately follows from this after renumbering $(k_2,k_3,\ldots,k_n)$ to $(k_1,k_2,\ldots,k_{n-1})$, and replacing x_i, y_j with $\alpha_i x,\ \beta_j x$, respectively in the identity (7.4.21). □

GL(n)pack functions: The following **GL(n)pack** functions, described in the appendix, relate to the material in this chapter:

HeckeMultiplicativeSplit SchurPolynomial.

8

Bounds for L-functions and Siegel zeros

8.1 The Selberg class

We have investigated carefully the theory of automorphic functions for the groups $SL(2, \mathbb{Z})$ and $SL(3, \mathbb{Z})$ as well as some of their subgroups. In these and all other known examples of Dirichlet series with arithmetic significance there has appeared certain expectations: a type of Riemann hypothesis will hold if the Dirichlet series has a functional equation and an Euler product or a converse theorem will hold allowing one to show that the Dirichlet series is, in fact, associated to an automorphic function on an arithmetic group. Following (Langlands, 1970), if

$$L(s) = \prod_p \prod_{i=1}^{n} \left(1 - \frac{\alpha_{p,i}}{p^s}\right)^{-1}$$

is the L-function associated to a Maass form on $GL(n)$, then one may consider

$$L(s, \vee^k) := \prod_p \prod_{1 \le i_1 \le i_2 \le \cdots \le i_k \le n} \left(1 - \left(\alpha_{p,i_1}\alpha_{p,i_2} \cdots \alpha_{p,i_k}\right) p^{-s}\right)^{-1},$$

the symmetric kth power L-function which is conjectured to be a Maass form on $GL(M)$ where

$$M = M(k, n) = \sum_{1 \le i_1 \le i_2 \le \cdots \le i_k \le n} 1.$$

We have shown this conjecture to hold in the case $n = 2$, $k = 2$ in Chapter 7. In this case, it is the Gelbart–Jacquet lift from $GL(2)$ to $GL(3)$. More recently, Kim and Shahidi (2002) proved this conjecture in the case of the symmetric cube lift from $GL(2)$ to $GL(4)$, i.e., when $n = 2$, $k = 3$, and Kim (2003) obtained the symmetric fourth power lift from $GL(2)$ to $GL(5)$. See also (Henniart, 2002).

Selberg (see (Selberg, 1991)) axiomatized certain expected properties of Dirichlet series and L-functions and introduced the **Selberg class** $\mathcal{S}$, consisting of all formal Dirichlet series (L-functions)

$$L(s) = \sum_{m=1}^{\infty} \frac{a_m}{m^s}, \qquad (a_1 = 1,\ a_m \in \mathbb{C} \text{ for } m = 2, 3, 4, \ldots)$$

which satisfiy the following axioms.

Axiom 8.1.1 (Analyticity) *The function $(s-1)^{\ell} L(s)$ is an entire function of finite order for some non-negative integer ℓ.*

Axiom 8.1.2 (Ramanujan hypothesis) *For fixed $\epsilon > 0$, we have $|a_m| \ll_{\epsilon} m^{\epsilon}$, for all $m = 1, 2, 3, \ldots$, where the constant implied by the $\ll$ symbol depends at most on ϵ.*

Axiom 8.1.3 (Functional equation) *Define $\Lambda(s) := A^s G(s) L(s)$, with $A > 0$, where $G(s) = \prod_{j=1}^{n} \Gamma(\lambda_j s + \mu_j)$ with $\lambda_j > 0, \Re(\mu_j) \geq 0$. Then we have the functional equation*

$$\Lambda(s) = \epsilon \cdot \overline{\Lambda(1-\bar{s})}, \qquad (|\epsilon| = 1).$$

Axiom 8.1.4 (Euler product) *We may express* $\log L(s)$ *by the Dirichlet series*

$$\log L(s) = \sum_{m=2}^{\infty} \frac{b_m}{m^s}$$

where $b_m = 0$ unless m is a positive power of a rational prime and $|b_m| \ll m^{\theta}$ for all $m = 2, 3, 4, \ldots$ with some fixed $\theta < \frac{1}{2}$.

Selberg made a number of interesting conjectures concerning Dirichlet series in $\mathcal{S}$. He also introduced the notion of primitive elements in $\mathcal{S}$ which cannot be factored into the product of two or more non-trivial members of $\mathcal{S}$. In (Murty, 1994), it is shown that Selberg's conjectures imply Artin's conjecture on the holomorphy of the L-series attached to finite-dimensional complex representations of $\mathrm{Gal}(\bar{\mathbb{Q}}/\mathbb{Q})$, and that the θ of Axiom 8.1.4 behaves like 0 on average. Here are Selberg's conjectures. Sums of type $\sum_p$ refer to sums over rational primes.

Conjecture 8.1.5 (Regularity of distribution) *Associated to each $L \in \mathcal{S}$, there is an integer $n_L \geq 0$, such that as $x \to \infty$,*

$$\sum_{p \leq x} \frac{|a_p|^2}{p} = n_L \log\log x + \mathcal{O}(1).$$

Conjecture 8.1.6 (Orthonormality) *If $L, L' \in \mathcal{S}$ are distinct and primitive, then $n_L = n_{L'} = 1$ and*

$$\sum_{p \le x} \frac{a_p \, a'_p}{p} = \mathcal{O}(1).$$

Conjecture 8.1.7 ($GL(1)$ twists) *Let χ be a primitive Dirichlet character and for $L(s) = \sum_{m=1}^{\infty} a_m/m^s \in \mathcal{S}$ define $L_\chi(s) = \sum_{m=1}^{\infty} a_m \, \chi(m)/m^s$ to be the twisted Dirichlet series. Then up to a finite Euler product, L_χ is also in $\mathcal{S}$.*

Conjecture 8.1.8 (Riemann hypothesis) *The Dirichlet series $L(s) \in \mathcal{S}$ have all their non-real zeros on the critical line $\Re(s) = \frac{1}{2}$. If $L(\beta) = 0$ with $\beta \in \mathbb{R}$, then $\beta = \frac{1}{2}$ or $\beta \le 0$.*

See (Kowalski, 2003), (Ramakrishnan and Wang, 2003) for further discussions of the following very interesting conjecture.

Conjecture 8.1.9 *For any $L(s) \in \mathcal{S}$, if it has a pole of order k at $s = 1$, then $\zeta(s)^k | L(s)$, i.e., $L(s) = \zeta(s)^k L_1(s)$, with $L_1(s) \in \mathcal{S}$.*

It follows from (Conrey and Ghosh, 1993) that Conjectures 8.1.5 and 8.1.6 imply Conjecture 8.1.9.

In order to classify the Dirichlet series $L(s)$ in the Selberg class, it is convenient to introduce the **degree** d_L of $L \in \mathcal{S}$ as

$$d_L = 2 \sum_{j=1}^{n} \lambda_j.$$

A fundamental conjecture asserts that the degree is always an integer. It has been shown by Richert (1957) that there are no elements in $\mathcal{S}$ with degree d satisfying $0 < d < 1$. Another proof was also found later by Conrey and Ghosh (1993) who also showed that the only elements of degree zero in $\mathcal{S}$ are the constant functions. Kaczorowski and Perelli (1999) determined the structure of the Selberg class for degree 1, showing that it contains only the Riemann zeta function and shifts of Dirichlet L-functions. Soundararajan (2004) found a simpler proof of this result. In another paper, Kaczorowski and Perelli (2002), showed that there are no elements of the Selberg class with degree $1 < d < 5/3$. It seems likely that the only elements in the Selberg class with degree $d \le 3$ are L-functions associated to automorphic functions on $GL(2)$, $GL(3)$.

8.2 Convexity bounds for the Selberg class

The maximum principle (see (Ahlfors, 1966)) states that if $f(z)$ is analytic on a closed bounded set E, then the maximum of $|f(z)|$ is taken on the boundary of E. A variation of the maximum principle is given in the following version of the Phragmén–Lindelöf theorem (Phragmén and Lindelöf, 1908) which has important applications to the theory of Dirichlet series.

Theorem 8.2.1 (Phragmén–Lindelöf) *Fix real numbers $\sigma_1 < \sigma_2$. Let $\phi(s)$ be holomorphic in the strip $\{s \mid \sigma_1 \le \Re(s) \le \sigma_2\}$. Suppose that $\phi(s)$ satisfies the bound*

$$|\phi(\sigma + it)| \le Ce^{|t|^\alpha}, \qquad \textit{(for some } \alpha, C > 0)$$

in this strip. Assume further that

$$|\phi(\sigma_1 + it)| \le B\,(1 + |t|)^{M_1}, \qquad |\phi(\sigma_2 + it)| \le B\,(1 + |t|)^{M_2},$$

for fixed constants $B > 0$, $M_1, M_2 \ge 0$, and for all $t \in \mathbb{R}$. Then

$$|\phi(\sigma + it)| \le B\,(1 + |t|)^{M(\sigma)}, \qquad \textit{with } M(\sigma) = \frac{M_1(\sigma_2 - \sigma) + M_2(\sigma - \sigma_1)}{\sigma_2 - \sigma_1},$$

for all $\sigma_1 \le \sigma \le \sigma_2$ and $t \in \mathbb{R}$.

Proof We first consider the simpler case that $M_1 = M_2 = 0$. Fix an integer $m > \alpha$ such that $m \equiv 2 \pmod 4$. Then for $\Im(s) = t \to +\infty$, we have $s/|s| \to i$, so that $(s/|s|)^m \to -1$. Consequently, there exists $T_0 > 0$ such that

$$\Re\left(\left(\frac{s}{|s|}\right)^m\right) < -\frac{1}{2}$$

for $t > T_0$.

Now fix $\epsilon > 0$, and define $\Phi_{\epsilon,m}(s) = e^{\epsilon s^m}\phi(s)$. Since

$$\Re(\epsilon s^m) = \epsilon \cdot \Re\left(\left(\frac{s}{|s|}\right)^m\right) \cdot |s|^m < -\frac{\epsilon}{2} \cdot |s|^m$$

it follows that for $T_\epsilon > T_0$, sufficiently large, that

$$|\Phi_{\epsilon,m}(s)| \le e^{-\frac{\epsilon}{2}|s|^m + |s|^\alpha} \cdot |\phi(s)| \le |\phi(s)| \le B,$$

for $\Re(s) = \sigma_1$, $\Re(s) = \sigma_2$, $\Im(s) = t \ge T_\epsilon$. Consequently, $|\Phi_{\epsilon,m}(s)| \le B$ on any line segment $\mathcal{L}_T$ (with $T \ge T_\epsilon$) where

$$\mathcal{L}_T = \left\{s \mid \Im(s) = T,\ \sigma_1 \le \Re(s) \le \sigma_2\right\}.$$

Therefore,

$$|\phi(s)| \le e^{\epsilon|s|^m} B \tag{8.2.2}$$

on $\mathcal{L}_T$. A similar argument can be given for the case $t < 0$ from which it follows that (8.2.2) also holds on any such $\mathcal{L}_{-T}$. By the maximum principle, we can now assert that (8.2.2) holds for any $\epsilon > 0$ on the entire strip with vertical sides consisting of the lines $\Re(s) = \sigma_1$, $\Re(s) = \sigma_2$. Letting $\epsilon \to 0$ in (8.2.2) establishes that

$$|\phi(s)| \le B$$

everywhere inside this strip which proves the theorem in the particular case considered.

In the general case, let

$$u(s) = e^{M(s)\log((-is)^\alpha)},$$

where the logarithm has its principal value. Then the function $u(s)$ is holomorphic for $\sigma_1 \le \Re(s) \le \sigma_2$, and $\Im(s) = t \ge 1$. If we write $M(s) = M(\sigma) + ibt$, then

$$\Re(M(s)\alpha\log(-is)) = \Re((M(\sigma)+ibt)\cdot\alpha\cdot\log(t - i\sigma)) = \alpha M(\sigma)\log t + \mathcal{O}(1).$$

Hence $|u(s)| = |t|^{M(\sigma)}e^{\mathcal{O}(1)}$, and, therefore, the function $\Phi(s) = \phi(s)/u(s)$ satisfies the same conditions as $\phi(s)$ did in the first part. Thus, $\Phi(s)$ is bounded in the strip and the theorem follows. □

The Phragmén–Lindelöf Theorem 8.2.1 allows one to obtain growth properties of holomorphic L-functions in the Selberg class. The key idea for doing this is that we know $L(s)$ is bounded in its region of absolute convergence $\Re(s) > 1 + \epsilon$, say. By the functional equation of Axiom 8.1.3 we may obtain a polynomial bound (in $|s|$) for the growth on the line $-\epsilon$. The Phragmén–Lindelöf theorem then gives us the growth of $L(s)$ in the critical strip $0 \le \Re(s) \le 1$. Growth properties obtained in this manner are called convexity bounds. We now establish such convexity bounds for the important subfamily of the Selberg class $\mathcal{S}$ where $\lambda_1 = \lambda_2 = \cdots = \lambda_n = \lambda$ and the μ_j $(j = 1, 2, \ldots, n)$ occur in conjugate pairs. Recall that these quantities occur in the Gamma factors of the functional equation of Axiom 8.1.3.

Restricting to this particular subfamily is not necessary. The reader may, with a modicum of effort, obtain convexity bounds in more general situations at the expense of more complicated notation. However, this subfamily includes all the examples treated in this book, so we will not consider other cases.

Theorem 8.2.3 (Convexity bound) *Let $L(s) \in \mathcal{S}$ have a pole of order ℓ at $s = 1$ and satisfy the functional equation*

$$\Lambda(s) = A^s \prod_{j=1}^{n} \Gamma(\lambda s + \mu_j) L(s) = \epsilon \cdot \overline{\Lambda(1 - \bar{s})}, \qquad (|\epsilon| = 1),$$

where $\lambda > 0$ and $\mu_j = \gamma_j + i\kappa_j$ (with $\gamma_j > 0, \kappa_j \in \mathbb{R}, j = 1, 2, \ldots, n$) occur in conjugate pairs. Then for every $\epsilon > 0$, there exists an effectively computable constant $C_\epsilon > 0$, such that

$$\left(\frac{s-1}{s+1}\right)^{\ell} |L(s)| \le C_\epsilon \left(A \cdot \prod_{j=1}^{n} (1 + |\lambda t + \kappa_j|)^{\lambda}\right)^{1-\sigma+\epsilon},$$

for all $s = \sigma + it$ with $0 \le \sigma \le 1$ and $t \in \mathbb{R}$.

Proof By Axiom 8.1.2 we know that for every fixed $\epsilon > 0$, the L-function $L(s)$ converges absolutely in the region $\Re(s) > 1 + \epsilon$, and, hence, is bounded in this region. Let $s = -\epsilon + it$. By the functional equation

$$|L(-\epsilon + it)| = \frac{A^{1+2\epsilon} |G(1 + \epsilon - it)| L(1 + \epsilon + it)}{|G(-\epsilon + it)|} \ll A^{1+2\epsilon} \frac{|G(1 + \epsilon - it)|}{|G(-\epsilon + it)|}.$$

In order to proceed further, we need to estimate the right-hand side of the above equation. We make use of Stirling's asymptotic formula

$$\Gamma(\sigma + it) = \sqrt{2\pi}\, (it)^{\sigma - \frac{1}{2}} e^{-\frac{\pi t}{2}} \left(\frac{|t|}{e}\right)^{it} \left\{1 + \mathcal{O}\left(\frac{1}{|t|}\right)\right\}, \qquad (8.2.4)$$

which is valid for fixed σ and $|t|$ sufficiently large. It follows that

$$\frac{|G(1+\epsilon - it)|}{|G(-\epsilon + it)|} = \frac{\prod_{j=1}^{n} |\Gamma(\lambda(1 + \epsilon - it) + \mu_j)|}{\prod_{j=1}^{n} |\Gamma(\lambda(-\epsilon + it) + \mu_j)|} \ll \left(\prod_{j=1}^{n} (1 + |\lambda t + \kappa_j|)\right)^{\lambda(1+2\epsilon)}.$$

We have thus established that $L(s)$ is bounded for $\Re(s) = 1 + \epsilon$ and is bounded by

$$A^{1+2\epsilon} \left(\prod_{j=1}^{n} (1 + |\lambda t + \kappa_j|)\right)^{\lambda(1+2\epsilon)}$$

for $\Re(s) = -\epsilon$. Before applying the Phragmén–Lindelöf Theorem 8.2.1, we first multiply $L(s)$ by $((s-1)/(s+1))^{\ell}$ to remove the multiple pole. The result then immediately follows from Theorem 8.2.1. □

A key ingredient in the proof of the convexity bound (Theorem 8.2.3) is Axiom 8.1.2 which states that the mth Dirichlet coefficient of an L-function in

$\mathcal{S}$ satisfies the Ramanujan bound of $\mathcal{O}(m^\epsilon)$. The ϵ here reappears as the ϵ (not necessarily the same ϵ) in the convexity bound. Molteni (2002), using an idea of Iwaniec (1990) (see also (Murty, 1994)) showed that Axiom 8.1.2 will automatically hold on average (which is all that is really needed for applications to convexity bounds) if certain symmetric square L-functions satisfy the expected growth properties.

8.3 Approximate functional equations

It is possible, in many cases, by the use of approximate functional equations, to obtain slight improvements on the convexity bound obtained in Theorem 8.2.3, i.e., one may replace the term

$$\left(A \cdot \prod_{j=1}^{n}(1 + |\lambda t + \kappa_j|)^\lambda\right)^\epsilon$$

by a power of the logarithm. Such considerations become important, for example, when estimating L-functions on the line $\Re(s) = 1$. Improvements of this type, based on the method of approximate functional equations, have a long history (see (Titchmarsh, 1986)).

In (Chandrasekharan and Narasimhan, 1962), the method of approximate functional equations is applied for the first time in a very general setting corresponding to the Selberg class. Their main interest, however, was in deducing the average order of the Dirichlet coefficients of an L-function rather than in obtaining bounds for the L-function in the critical strip. They made the surprising discovery that better error terms are available in the case where the Dirichlet series has positive coefficients. Lavrik (1966) obtained an explicit approximate functional equation for a very wide class of L-functions. See (Iwaniec and Kowalski, 2004) for a detailed exposition of more recent developments along these lines.

Following Lavrik (1966), (see also (Ivić, 1995), (Harcos, 2002), (Iwaniec and Kowalski, 2004)), we now derive a very general form of the approximate functional equation for $L \in \mathcal{S}$. We assume that $L(s)$ satisfies a functional equation of the type

$$\Lambda(s) = A^s \prod_{j=1}^{n} \Gamma(\lambda s + \mu_j) L(s) = \epsilon \cdot \overline{\Lambda(1 - \bar{s})}, \qquad (|\epsilon| = 1), \quad (8.3.1)$$

as given in Theorem 8.2.3. Define

$$\mathfrak{q}_w := A \cdot \prod_{j=1}^{n}(3 + |\lambda w + \mu_j|)^\lambda. \quad (8.3.2)$$

Theorem 8.3.3 (Approximate functional equation) *Let* $L(s) = \sum_{m=1}^{\infty} a_m/m^s \in \mathcal{S}$, *be entire and satisfy the functional equation (8.3.1). Define* $\mathfrak{q}_w$ *as in (8.3.2). Then there exists a smooth function* $F : (0, \infty) \to \mathbb{C}$ *such that for every* $w \in \mathbb{C}$ *with* $0 \le \Re(w) \le 1$, *we have*

$$L(w) = \sum_{m=1}^{\infty} \frac{a_m}{m^w} F\left(\frac{m}{\mathfrak{q}_w}\right) + \epsilon\lambda_w \sum_{m=1}^{\infty} \frac{\overline{a_m}}{m^{1-w}} \bar{F}\left(\frac{m}{\mathfrak{q}_{1-w}}\right),$$

where $\lambda_w = A^{1-w}G(1-w)/A^w G(w)$.

The function F *and its partial derivatives* $F^{(k)}$, $(k = 1, 2, \ldots)$ *satisfy, for any* $\sigma > 0$, *the following uniform growth estimates at* 0 *and* ∞:

$$F(x) = \begin{cases} 1 + \mathcal{O}_\sigma(x^\sigma) \\ \mathcal{O}_\sigma(x^{-\sigma}), \end{cases} \qquad F^{(k)}(x) = \mathcal{O}_\sigma(x^{-\sigma}).$$

The implied $\mathcal{O}_\sigma$*–constants depend only on* σ, k, n.

Remarks The approximate functional equation allows one to effectively compute and obtain bounds for special values of the L-function. It effectively reduces the computation to a short sum of $\ll \max(|q_w|, |q_{1-w}|)$ terms. One may also easily obtain a version of the approximate functional equation for L-functions with a pole at $s = 1$.

Proof Let $h(s)$ be a holomorphic function satisfying

$$h(s) = h(-s) = \overline{h(\bar{s})}, \qquad h(0) = 1, \tag{8.3.4}$$

and which is bounded in the vertical strip $-2 < \Re(s) < 2$. For every $w \in \mathbb{C}$, and $x > 0$, we define

$$H_w(x) := \frac{1}{2\pi i} \int_{2-i\infty}^{2+i\infty} \frac{G(s+w)}{G(w)} h(s) x^{-s} \frac{ds}{s}, \tag{8.3.5}$$

where

$$G(s) = \prod_{j=1}^{n} \Gamma(\lambda s + \mu_j),$$

as in Theorem 8.2.3. We assume that h has sufficient decay properties so that the integral in (8.3.5) converges absolutely.

As a first step in the proof of Theorem 8.3.3, we first derive an approximate functional equation in terms of the function H_w. For any small $\epsilon > 0$, and

$w \in \mathbb{C}$ with $0 \le \Re(w) \le 1$, consider the integral

$$I_L(w) := \frac{1}{2\pi i} \int_{1+\epsilon-i\infty}^{1+\epsilon+i\infty} \frac{A^{s+w} G(s+w) L(s+w)}{A^w G(w)} \, h(s) \, \frac{ds}{s}.$$

If we shift the line of integration to the left we pick up a residue of the pole of the integrand at $s = 0$. It follows, after applying the functional equation (8.3.1), and then transforming $s \to -s$, that

$$\begin{aligned} I_L(w) &= L(w) + \frac{1}{2\pi i} \int_{-1-\epsilon-i\infty}^{-1-\epsilon+i\infty} \frac{\epsilon\, A^{1-s-w} G(1-s-w) \overline{L(1-\bar{s}-\bar{w})}}{A^w G(w)} \, h(s) \, \frac{ds}{s} \\ &= L(w) - \frac{A^{1-w} G(1-w)}{2\pi i\, A^w G(w)} \\ &\qquad \times \int_{1+\epsilon-i\infty}^{1+\epsilon+i\infty} \frac{\epsilon\, A^{s+1-w} G(s+1-w) \tilde{L}(s+1-w)}{A^{1-w} G(1-w)} \, h(s) \, \frac{ds}{s}, \end{aligned}$$

or equivalently

$$L(w) = I_L(w) + \epsilon \cdot \frac{A^{1-w} G(1-w)}{A^w G(w)} \, I_{\tilde{L}}(1-w), \tag{8.3.6}$$

where $\tilde{L}(s) = \sum_{m=1}^{\infty} \overline{a_m}/m^s$ denotes the dual L-function. In (8.3.6) we may substitute the Dirichlet series for $L(s)$ and $\tilde{L}(s)$ and integrate term by term. It follows that

$$L(w) = \sum_{m=1}^{\infty} \frac{a_m}{m^w} H_w\left(\frac{m}{A}\right) + \epsilon \lambda_w \sum_{m=1}^{\infty} \frac{\overline{a_m}}{m^{1-w}} H_{1-w}\left(\frac{m}{A}\right). \tag{8.3.7}$$

To pass from (8.3.7) to the approximate functional equation in Theorem 8.3.3 it is necessary to choose a suitable test function h so that H_w takes the form of the function F in Theorem 8.3.3. To this end, we analyze H_w further using Stirling's approximation for the Gamma function. Let $w = u + iv$, $s = \sigma + it$. It follows from (8.2.4) that

$$\begin{aligned} \left| \frac{\Gamma(w+s)}{\Gamma(w)} \right| &\ll \frac{|w+s|^{u+\sigma-\frac{1}{2}}}{|w|^{u-\frac{1}{2}}} \cdot e^{\frac{\pi}{2}(|w|-|w+s|)} \\ &\ll (|w|+3)^{\sigma} e^{\frac{\pi}{2}|s|}. \end{aligned}$$

Consequently

$$\left|\frac{G(w+s)}{G(w)}\right| = \frac{\prod_{j=1}^{n} |\Gamma(\lambda(w+s)+\mu_j)|}{\prod_{j=1}^{n} |\Gamma(\lambda w+\mu_j)|} \ll \mathfrak{q}_w^{\sigma}\, e^{\frac{\pi}{2}n\lambda|s|}. \tag{8.3.8}$$

Next, following (8.3.5), we define

$$F(x) := H_w(\mathfrak{q}_w \cdot x) = \frac{1}{2\pi i}\int_{2-i\infty}^{2+i\infty} \frac{G(s+w)}{G(w)} h(s)(\mathfrak{q}_w x)^{-s}\,\frac{ds}{s}. \tag{8.3.9}$$

In (8.3.9), under the assumption that $h(s)$ has sufficient decay, we may shift the line of integration either to the left (picking up the residue 1 at the pole at $s=0$) or we may shift to the right. After shifting to an arbitrary line $\Re(s)=\sigma$ and differentiating k times with respect to x, we obtain from (8.3.9) that

$$\left(\frac{d}{dx}\right)^k F(x) = \delta_{\sigma,k} + \frac{(-1)^k}{2\pi i} \times \int_{\sigma-i\infty}^{\sigma+i\infty} \frac{G(s+w)}{G(w)} h(s)s(s+1)\cdots(s+k-1)\mathfrak{q}_w^{-s}x^{-s-k}\,\frac{ds}{s}, \tag{8.3.10}$$

where $\delta_{\sigma,k} = \begin{cases} 1 & \sigma<0,\ k=0,\\ 0 & \text{otherwise.}\end{cases}$

It now follows from (8.3.8) and (8.3.10) that

$$\left(\frac{d}{dx}\right)^k F(x) = \delta_{\sigma,k} + \mathcal{O}_{\sigma,k}\left(\int_{\sigma-i\infty}^{\sigma+i\infty} e^{\frac{\pi}{2}n|s|}\,(1+|s|)^k\cdot|h(s)|\cdot x^{-\sigma}\;|ds|\right).$$

To complete the proof of Theorem 8.3.3 it is enough to choose a test function h with sufficient decay properties so that the above integral converges absolutely. As an example, one may choose $h(s) = (\cos(\pi s/2))^{-2n}$. □

As an example of the approximate functional equation, consider a Dirichlet L-function $L(s,\chi)$ associated to a primitive Dirichlet character (mod q). Then the functional equation (see (Davenport, 1967)) takes the form

$$\Lambda(s,\chi) = \left(\frac{\pi}{q}\right)^{-(s+\mathfrak{a})/2}\Gamma\left(\frac{s+\mathfrak{a}}{2}\right)L(s,\chi) = \frac{\tau(\chi)}{i^{\mathfrak{a}}\sqrt{q}}\cdot\Lambda(1-s,\bar{\chi}),$$

where $\tau(\chi)$ is the Gauss sum, as in (7.1.1), and $\mathfrak{a} = \begin{cases} 0 & \text{if } \chi(-1)=1,\\ 1 & \text{if } \chi(-1)=-1.\end{cases}$

In this case, $\mathfrak{q}_w$ as defined in (8.3.2), satisfies

$$|\mathfrak{q}_w| \ll \sqrt{|q|} \cdot (1 + |w|).$$

It immediately follows from the approximate functional equation given in Theorem 8.3.3 that

$$|L(1, \chi)| \ll \log |q|. \tag{8.3.11}$$

8.4 Siegel zeros in the Selberg class

In the Selberg class $\mathcal{S}$ determined by Axioms 8.1.1–8.1.4, let us fix the integer $n \geq 1$, and we assume that we have a subfamily $\mathcal{S}_n$, all of whose elements have exactly n Gamma factors which are all of the same form. We shall now define Siegel zeros for this subfamily. Note, however, that by Selberg's Conjecture 8.1.8, we do not expect such Siegel zeros to exist.

Definition 8.4.1 (Siegel zero) *Fix a constant $c > 0$ and an integer $n \geq 1$. Let $L(s) \in \mathcal{S}_n$ satisfy the functional equation given in Axiom 8.1.3. Assume that $L(\beta) = 0$ for some real β satisfying*

$$1 - \frac{c}{\log(\lambda A + 1)} \leq \beta \leq 1,$$

where $\lambda = \max\limits_{1 \leq i \leq n} (\lambda_i + |\mu_i|)$. Then β is termed a Siegel zero for $L(s)$ relative to c.

Interlude on the history of Siegel zeros Let $D < 0$ denote the fundamental discriminant of an imaginary quadratic field $k = \mathbb{Q}(\sqrt{D})$. Then $D \equiv 1 \pmod 4$ and square-free, or of the form $D = 4m$ with $m \equiv 2$ or $3 \pmod 4$ and square-free. Define

$$h(D) = \# \left\{ \frac{\text{group of non-zero fractional ideals } \frac{\mathfrak{a}}{\mathfrak{b}}}{\text{group of principal ideals } (\alpha),\ \alpha \in k^\times} \right\}$$

to be the cardinality of the ideal class group of k. Gauss (1801) showed (using the language of binary quadratic forms) that $h(D)$ is always finite. He conjectured that

$$h(D) \to \infty \quad \text{as} \quad D \to -\infty,$$

which was first proved by Heilbronn (1934).

The *Disquisitiones* also contains a number of tables of binary quadratic forms (actually only even discriminants were considered) with small class numbers.

Gauss made the remarkable conjecture that his tables were complete. In modern parlance, we may rewrite Gauss' tables in the form (see also (Goldfeld, 2004)):

$h(D)$	1	2	3	4	5
# of fields	9	18	16	54	25
largest $\|D\|$	163	427	907	1555	2683

. (8.4.2)

The case of class number 1 is particularly interesting, because in this case, it can be shown that the imaginary quadratic field has the unique factorization property – *that every integer in the field can be uniquely factored into primes.* Note that unique factorization fails in $\mathbb{Q}(\sqrt{-5})$ because the integer 6 can be factored in two distinct ways, i.e.,

$$6 = 3 \cdot 2, \qquad 6 = (1 + \sqrt{-5}) \cdot (1 - \sqrt{-5}),$$

and each of $2, 3, 1 - \sqrt{-5}, 1 + \sqrt{-5}$, cannot be further factored, so are primes. For the case of class number 1, Gauss' conjecture takes the explicit form that there are only nine discriminants

$$-d = -3, -4, -7, -8, -11, -19, -43, -67, -163$$

where the imaginary quadratic field $\mathbb{Q}(\sqrt{-d})$ has class number 1.

The problem of finding an algorithm which would enable one to effectively determine all imaginary quadratic fields with a given class number h is now known as the Gauss class number problem, and it is in connection with this problem that Siegel zeros first arose. If such an effective algorithm did not exist, then a Dirichlet series associated to an imaginary quadratic field with small class number and large discriminant would have to have a Siegel zero. More concretely, it had been shown that if Gauss' tables (8.4.2) were not complete then a Siegel zero would have to exist and the Riemann hypothesis would be violated! This problem has a long and colorful history, (see (Goldfeld, 1985)), the first important milestones were obtained by Heegner (1952), Stark (1967, 1972), and Baker (1971) whose work led to the solution of the class number 1 and 2 problems. Finally Goldfeld, Gross, and Zagier (see (Goldfeld, 1976, 1985), (Gross and Zagier, 1986), (Iwaniec and Kowalski, 2004)) solved the Gauss class number problem completely. In (Watkins, 2004), the range of the complete (unconditional) solution for Gauss' class number prolem was extended to determining all imaginary quadratic fields with $h(d) \leq 100$. This was achieved after several months of computer computation.

Why are Siegel zeros so intimately related to the class numbers of imaginary quadratic fields? Although it may not seem so at first, the answer to this question is the substance of the following lemma.

Lemma 8.4.3 *Let $L \in \mathcal{S}$ have non-negative Dirichlet coefficients. Assume that $L(s)$ has a simple pole at $s = 1$ with residue R, and that $L(s)$ satisfies a growth condition on the line $\Re(s) = \frac{1}{2}$ of the form*

$$\left| L\left(\frac{1}{2} + it\right) \right| \le M(1 + |t|)^B \tag{8.4.4}$$

for some $M \ge 1$, $B \ge 0$, and all $t \in \mathbb{R}$. If $L(s)$ has no real zeros in the range $1 - (1/\log M) < s < 1$, then there exists an effective constant $c(B) > 0$ such that

$$R^{-1} \le c(B) \log M.$$

Proof Let $r > B$ be a fixed integer. We shall make use of the well-known integral transform

$$\frac{1}{2\pi i} \int_{2-i\infty}^{2+i\infty} \frac{x^s}{s(s+1)\cdots(s+r)}\, ds = \begin{cases} \frac{1}{r!}\left(1 - \frac{1}{x}\right)^r, & x > 1, \\ 0, & 0 < x \le 1, \end{cases}$$

which is proved by either shifting the line of integration to the left (if $x > 1$) or to the right (if $0 < x \le 1$), and then computing the sum of the residues with Cauchy's theorem. Now,

$$L(s) = \sum_{m=1}^{\infty} \frac{a(m)}{m^s}$$

with $a(1) = 1$, $a(m) \ge 0$, for $m = 2, 3, \ldots$ It follows that for all $x \ge 2$, and any $\frac{1}{2} < \beta < 1$,

$$1 \ll \frac{1}{2\pi i} \int_{2-i\infty}^{2+i\infty} \frac{L(s+\beta) x^s}{s(s+1)\cdots(s+r)}\, ds. \tag{8.4.5}$$

Here, as throughout this proof, the constant implied by the $\ll$–symbol is effective and depends at most on B. Now (8.4.4) implies that $L(s)$ has polynomial growth on the line $\Re(s) = \frac{1}{2}$. Further, $L(s)$ is bounded by $L(3)$ on the line $\Re(s) = 3$. By a convexity argument, one obtains that $L(\sigma + it) = \mathcal{O}(|t|^B)$ for all $\frac{1}{2} \le \sigma \le 3$, $t \ge 1$. It follows that one may shift the line of integration of the integral on the right-hand side of (8.4.5) to the line $\Re(s) = \frac{1}{2} - \beta < 0$, picking up residues at $s = 1 - \beta$, 0. Thus the integral becomes

$$\frac{R x^{1-\beta}}{(1-\beta)(2-\beta)\cdots(r+1-\beta)} + \frac{L(\beta)}{r!} + \mathcal{O}\big(M x^{\frac{1}{2}-\beta}\big).$$

If we now choose $x = M^C$, for a sufficiently large constant C, it follows from

(8.4.5) and the above residue computations that

$$1 \ll \frac{RM^{C(1-\beta)}}{1-\beta} + L(\beta). \tag{8.4.6}$$

The key point is that for real $s > 1$, the function $L(s)$ is positive. On the other hand, for $1 - (1/\log M) < s < 1$, the function $L(s)$ becomes negative since we have crossed the pole at $s = 1$ where there is a sign change and we have assumed that $L(s)$ has no zeros in the interval. One may then choose $1 - \beta = 1/\log M$ so that $L(\beta) \le 0$. Then (8.4.6) implies that $R^{-1} \ll \log M$. □

Let us now use Lemma 8.4.3 to relate the Siegel zero with the Gauss class number problem for imaginary quadratic fields. Let $D < 0$ be the fundamental discriminant of an imaginary quadratic field $k = \mathbb{Q}(\sqrt{D})$. The relation is through the zeta function

$$\zeta_k(s) = \zeta(s)L_\chi(s) = \sum_{m=1}^{\infty} \frac{a(m)}{m^s},$$

where $\chi(m) = (D/m)$ is the Kronecker symbol (primitive quadratic Dirichlet character of conductor D) where $a(1) = 1$, and

$$a(m) = \sum_{d|m} \chi(d) \ge 0 \tag{8.4.7}$$

for all $m = 1, 2, 3, \ldots$ It is a classical theorem of Dirichlet (see (Davenport, 1967) that $\zeta_k(s)$ has a simple pole at $s = 1$ with residue

$$R = \frac{\pi h(D)}{|D|^{\frac{1}{2}}} \tag{8.4.8}$$

provided $D < -4$. Further, by the convexity bound given in Theorem 8.2.3, $\zeta_k(s)$ satisfies the growth condition

$$\left|\zeta_k\left(\frac{1}{2} + it\right)\right| \ll |D|^{\frac{1}{2}+\epsilon} \cdot (1 + |t|)^{1+\epsilon}. \tag{8.4.9}$$

Actually, much stronger bounds are known, but we do not need them here. It follows from (8.4.7), (8.4.8), (8.4.9) that $\zeta_k(s)$ satisfies the conditions required in Lemma 8.4.3. It then follows from Lemma 8.4.3 that $\zeta_k(s)$ has a Siegel zero if the class number $h(D)$ is so small that

$$h(D) \ll \frac{|D|^{\frac{1}{2}}}{\log |D|}. \tag{8.4.10}$$

The above result was first published by Landau (1918), but Landau attributes this result to a lecture given by Hecke.

8.5 Siegel's theorem

It was shown in (8.4.10) that if the class number $h(D)$ of an imaginary quadratic field $k = \mathbb{Q}(\sqrt{D})$ is too small, then the zeta function $\zeta_k(s)$ of k must have a Siegel zero. In fact, since $\zeta_k(s)$ factors as $\zeta(s)L(s, \chi)$, and the Riemann zeta function $\zeta(s)$ does not vanish for $0 < s < 1$, it follows that the Dirichlet L-function $L(s, \chi)$ must have a Siegel zero. This result goes back to Landau (1918).

Fifteen years later there was further, rather surprising, progress.

- Deuring (1933) *proved that if the classical Riemann hypothesis is false then* $h(D) \geq 2$ *for* $-D$ *sufficiently large.*
- Mordell (1934) *showed if the Riemann hypothesis is false, then* $h(D) \to \infty$ *as* $-D \to \infty$.
- Heilbronn (1934) *proved that the falsity of the generalized Riemann hypothesis for Dirichlet L-functions implies that* $h(D) \to \infty$ *as* $-D \to \infty$.

When combined with the Landau–Hecke result (8.4.10) this gave an unconditional proof of Gauss' conjecture that the class number of an imaginary quadratic field goes to infinity with the discriminant. The surprising aspect of this chain of theorems is that first one assumes the Riemann hypothesis to establish a result and then one assumes that the Riemann hypothesis is false to obtain the exact same result! This is now called the Deuring–Heilbronn phenomenon, but has the defect of being totally ineffective. Siegel (1935) practically squeezed the last drop out of the Deuring–Heilbronn phenomenon. He proved the following theorem, which is the main subject matter of this section.

Theorem 8.5.1 (Siegel's theorem) *Let* $\mathbb{Q}(\sqrt{D})$ *be an imaginary quadratic field with fundamental discriminant* $D < 0$ *and class number* $h(D)$. *Then for every* $\epsilon > 0$, *there exists a constant* $c_\epsilon > 0$ *(which cannot be effectively computed) such that*

$$h(D) > c_\epsilon |D|^{\frac{1}{2}-\epsilon}.$$

Remarks Landau (1935) proved Theorem 8.5.1 with $\epsilon = \frac{1}{8}$ (also not effective). Siegel's theorem, with any $\epsilon > 0$, appeared in the same volume of Acta Arithmetica, but did not reference Landau's result at all!

Proof of Siegel's theorem We follow Goldfeld (1974). In view of Dirichlet's theorem (8.4.8), it is enough to prove that for every fixed $\epsilon > 0$ and for all real primitive quadratic Dirichlet characters $\chi \pmod D$, that

$$L(1, \chi) > c_\epsilon \, |D|^{-\epsilon}.$$

Note that since $h(D)$ is a positive rational integer, it follows from (8.4.8) that

$$L(1,\chi) \gg |D|^{-\frac{1}{2}}.$$

Let χ' be a real primitive quadratic Dirichlet character (mod D') for some other fundamental discriminant D'. Consider the zeta function

$$Z(s) = \zeta(s)L(s,\chi)L(s,\chi')L(s,\chi\chi'),$$

and let $R = L(1,\chi)L(1,\chi')L(1,\chi\chi')$ be the residue at $s = 1$. In view of the Euler product

$$Z(s) = \prod_p \left(1 - \frac{1}{p^s}\right)^{-1} \left(1 - \frac{\chi(p)}{p^s}\right)^{-1} \left(1 - \frac{\chi'(p)}{p^s}\right)^{-1} \left(1 - \frac{\chi(p)\chi'(p)}{p^s}\right)^{-1},$$

and the fact that χ, χ' can only take values among $\{-1, 0, +1\}$, one readily establishes that $Z(s)$ is a Dirichlet series whose first coefficient is 1 and whose other coefficients are non-negative.

Lemma 8.5.2 *For every* $\epsilon > 0$, *there exists* χ' (mod D') *and* $\beta \in \mathbb{R}$ *satisfying* $1 - \epsilon < \beta < 1$ *such that* $Z(\beta) \le 0$ *independent of what* χ (mod D) *may be.*

Proof If there are no zeros in $[1-\epsilon, 1]$ for any $L(s,\chi)$, then $Z(\beta) < 0$ if $1 - \epsilon < \beta < 1$ since $\zeta(\beta) < 0$ and all L-functions $L(s,\chi)$ will be positive in the interval. Here we use the fact that $L(s,\chi)$ is positive for $\Re(s) > 1$ and can only change sign in the interval $[1-\epsilon, 1]$ if the L-function vanishes in the interval.

On the other hand, if such real zeros do exist, let β be such a zero, with χ' the corresponding character. Then $Z(\beta) = 0$ independent of χ. □

Next, fix β as in Lemma 8.5.2. It follows, as in the proof of Lemma 8.4.3 that for $x \ge 1$,

$$\begin{aligned} 1 &\ll \frac{1}{2\pi i} \int_{2-i\infty}^{2+i\infty} Z(s+\beta) \cdot \frac{x^s}{s(s+1)(s+2)(s+3)(s+4)} ds \\ &= \frac{R \cdot x^{1-\beta}}{(1-\beta)(2-\beta)(3-\beta)(4-\beta)(5-\beta)} + \frac{Z(\beta)}{4!} + \mathcal{O}\left(\frac{|DD'|^{1+\epsilon} x^{-\beta}}{1-\beta}\right), \end{aligned}$$

after shifting the line of integration to $\Re(s) = -\beta$ and using convexity bounds of Theorem 8.2.3 for the growth of $Z(s)$. But Lemma 8.5.2 tells us that $Z(\beta) \le 0$. Therefore,

$$1 \ll R \cdot \frac{x^{1-\beta}}{1-\beta}$$

if $|DD'|^{2+\epsilon} \ll x$, since $R \gg 1/|DD'|$. Consequently, since

$$R \ll L(1, \chi)(\log|DD'|)(\log|D|),$$

by (8.3.11), we get

$$L(1, \chi) \gg \frac{|D|^{-(2+\epsilon)(1-\beta)}}{\log|D|},$$

where the implied constant in the $\gg$–sign depends only on χ', and, therefore, only on ϵ. This proves Siegel's theorem if $(2+\epsilon)(1-\beta) < \epsilon/2$ and $|D|$ is sufficiently large. □

Tatuzawa (1951) went a step beyond Siegel and proved that Siegel's Theorem 8.5.1 must hold with an effectively computable constant $c_\epsilon > 0$ for all $D < 0$, except for at most one exceptional discriminant D. This shows that the family of real Dirichlet L-functions can have at most one L-function with a Siegel zero. It can be shown (see (Davenport, 1967)) that complex Dirichlet L-functions cannot have Siegel zeros. Thus, the entire family of $GL(1)$ L-functions can have at most one exceptional L-function with a Siegel zero. The exceptional L-function must correspond to a real primitive Dirichlet character associated to a quadratic field.

8.6 The Siegel zero lemma

The following lemma plays a crucial role in all the recent work on the non-existence of Siegel zeros. It first appeared in (Goldfeld, Hoffstein and Lieman, 1994). While we have defined Siegel zeros relative to a constant $c > 0$, we shall suppress the constant in the following discussion because it can be easily computed and it is not really important to the flow of ideas.

Lemma 8.6.1 (Siegel zero lemma) *Let $L(s) \in \mathcal{S}$ have non-negative Dirichlet coefficients and satisfy the functional equation in Axiom 8.1.3. Assume $L'(s)/L(s)$ is negative for s real and > 1, and that $L(s)$ has a pole of order m at $s = 1$. Assume further that $\Lambda(s) = s^m(1-s)^m A^s G(s)L(s) = \overline{\Lambda(1-\bar{s})}$, where $G(s) = \prod_{j=1}^{n} \Gamma(\lambda_j s + \mu_j)$, as given in Axiom 8.1.3, is an entire function of order 1. Then $L(s)$ has at most m Siegel zeros.*

Proof Since $\Lambda(s)$ is an entire function of order 1 it has (see (Davenport, 1967)) a Hadamard factorization and can be represented in the form

$$\Lambda(s) = e^{a+bs} \prod_{\Lambda(\rho)=0} \left(1 - \frac{s}{\rho}\right) e^{s/\rho}, \tag{8.6.2}$$

for certain constants a, b. If $\Re(s) > 1$, we may take logarithmic derivatives in (8.6.2) to obtain

$$\frac{m}{s} + \frac{m}{s-1} + \log A + \frac{G'(s)}{G(s)} + \frac{L'(s)}{L(s)} = b + \sum_{\rho}\left(\frac{1}{s-\rho} + \frac{1}{\rho}\right).$$

By the functional equation

$$b + \sum_{\rho}\left(\frac{1}{s-\rho} + \frac{1}{\rho}\right) = -b - \sum_{\rho}\left(\frac{1}{1-s-\rho} + \frac{1}{\rho}\right),$$

so that

$$b = -\sum_{\rho}\frac{1}{\rho}.$$

It follows that

$$\frac{m}{s} + \frac{m}{s-1} + \log A + \frac{G'(s)}{G(s)} + \frac{L'(s)}{L(s)} = \sum_{\rho}\frac{1}{s-\rho}.$$

Now, by assumption, $L'(s)/L(s)$ is negative for $s \in \mathbb{R},\ s > 1$. Also, if we pair conjugate roots, then every term of $\sum(s-\rho)^{-1}$ is positive, so there exists an absolute effective constant $c_0 > 0$, such that

$$\sum_{j=1}^{r}\frac{1}{s-\beta_j} \leq \frac{m}{s-1} + c_0 \log M,$$

where r denotes the number of real zeros β_j of $L(s)$ in the interval $[1-(c/\log M), 1]$. Here the the constant c_0 can be computed from the integral representation

$$\log\Gamma(z) = \left(z - \frac{1}{2}\right)\log z - z + \frac{1}{2}\log(2\pi) + \int_0^{\infty}\left(\frac{1}{2} - \frac{t}{z} + \frac{t}{e^t - 1}\right)\frac{e^{-tz}}{t}dt,$$

for the Gamma function. If the constant c is chosen small enough, compared to the constant c_0, then a contradiction is obtained whenever $r \geq m+1$. □

8.7 Non-existence of Siegel zeros for Gelbart–Jacquet lifts

The existence of Siegel zeros for classical Dirichlet L-functions has been shown to be equivalent to the existence of primitive quadratic Dirchlet characters χ with the property that $L(1, \chi)$ takes on too small a value. This follows from

Lemma 8.4.3 when we choose $\zeta(s)L(s,\chi)$ as our Dirichlet series, which has all the required properties of Lemma 8.4.3 and has a simple pole at $s=1$ with residue $L(1,\chi)$. A real breakthrough was achieved in (Hoffstein and Lockhart, 1994) when they took the study of Siegel zeros outside the classical domain of Dirichlet L-functions and considered, for the first time, the question of whether such zeros could exist for L-functions associated to automorphic forms on $GL(3)$ occurring as symmetric square lifts (Gelbart–Jacquet lifts) from $GL(2)$. It now became possible to obtain lower bounds of special values of L-functions in the same way as the classical methods (using the Deuring–Heilbronn phenomenon) gave lower bounds for $L(1,\chi)$ with χ a real primitive Dirichlet character. This established a powerful new tool in modern analytic number theory.

In (Hoffstein and Lockhart, 1994), it was shown that if f is a Maass form for $SL(2,\mathbb{Z})$ which is an eigenfunction of the Hecke operators, and F is its symmetric square lift (Gelbart–Jacquet lift) to $SL(3,\mathbb{Z})$, then the lifted L-function,

$$L_F(s) = \frac{L_{f\times f}(s)}{\zeta(s)}, \tag{8.7.1}$$

given in Theorem 7.3.2 cannot have a Siegel zero. Actually, they proved a more general result valid for congruence subgroups of $SL(2,\mathbb{Z})$ and also considered Gelbart–Jacquet lifts of both holomorphic modular forms and non-holomorphic Maass forms. Their proof was based on a generalization of Siegel's Theorem 8.5.1 and, thereby, was not effective. In the appendix of their paper Goldfeld, Hoffstein and Lieman (1994) obtained an effective version of their theorem which was based on the Siegel-zero Lemma 8.6.1. This effective proof is the subject matter of this section.

The key idea in the proof of the non-existence of Siegel zeros for L-functions of type (8.7.1) is the construction of an auxiliary L-function which has non-negative Dirichlet coefficients and a multiple pole at $s=1$, and satisfies the requirements of Lemma 8.6.1. Accordingly, we introduce

$$Z(s) := \zeta(s)L_F(s)^2 L_{F\times F}(s). \tag{8.7.2}$$

Here, if $L_F(s) = \sum_{m=1}^{\infty} c(m)m^{-s}$, then $L_{F\times F}(s) = \zeta(3s)\sum_{m=1}^{\infty} |c(m)|^2 m^{-s}$ as we recall from Proposition 7.4.12. In terms of Euler products, if

$$L_f(s) = \prod_p \left(1 - \frac{\alpha_p}{p^s}\right)^{-1}\left(1 - \frac{\alpha'_p}{p^s}\right)^{-1},$$

with $\alpha_p \cdot \alpha'_p = 1$, and

$$L_F(s) = \prod_p \prod_p \left(1 - \frac{\alpha_p^2}{p^s}\right)^{-1} \left(1 - \frac{1}{p^s}\right)^{-1} \left(1 - \frac{{\alpha'_p}^2}{p^s}\right)^{-1},$$

then

$$\begin{aligned} L_{F\times F}(s) = \prod_p &\left(1 - \frac{\alpha_p^4}{p^s}\right)^{-1} \left(1 - \frac{\alpha_p^2}{p^s}\right)^{-1} \left(1 - \frac{1}{p^s}\right)^{-1} \left(1 - \frac{\alpha_p^2}{p^s}\right)^{-1} \\ &\times \left(1 - \frac{1}{p^s}\right)^{-1} \left(1 - \frac{{\alpha'_p}^2}{p^s}\right)^{-1} \left(1 - \frac{1}{p^s}\right)^{-1} \\ &\times \left(1 - \frac{{\alpha'_p}^2}{p^s}\right)^{-1} \left(1 - \frac{{\alpha'_p}^4}{p^s}\right)^{-1}. \end{aligned} \tag{8.7.3}$$

Recall that the symmetric square lift of F is given by

$$\begin{aligned} L_F(s, \vee^2) = \prod_p &\left(1 - \frac{\alpha_p^4}{p^s}\right)^{-1} \left(1 - \frac{\alpha_p^2}{p^s}\right)^{-1} \left(1 - \frac{1}{p^s}\right)^{-1} \\ &\times \left(1 - \frac{1}{p^s}\right)^{-1} \left(1 - \frac{{\alpha'_p}^2}{p^s}\right)^{-1} \left(1 - \frac{{\alpha'_p}^4}{p^s}\right)^{-1}. \end{aligned}$$

It follows that

$$L_{F\times F}(s) = L_F(s) L_F(s, \vee^2).$$

This actually factors further since

$$L_F(s, \vee^2) = \zeta(s) L_f(s, \vee^4).$$

Finally, we obtain

$$Z(s) = \zeta(s) L_F(s)^3 L_F(s, \vee^2). \tag{8.7.4}$$

Lemma 8.7.5 *Let $Z(s) = \sum_{m=1}^{\infty} a(m) m^{-s}$ be given by (8.7.2). Then $a(1) = 1$ and $a(m) \geq 0$ for $m = 2, 3, 4, \ldots$*

Proof The fact that $a(1) = 1$ is an immediate consequence of the Euler product for $Z(s)$. Now, the Euler product for $Z(s)$ takes the form

$$\prod_p \left(1 - \frac{\alpha_p^{\epsilon_1}\alpha_p'^{\,\epsilon_2}}{p^s}\right)^{-1}$$

where the product is taken over all sixteen possible pairs (ϵ_1, ϵ_2), and where ϵ_1, ϵ_2, independently run through the values 2, 0, 0, 2. If one takes logarithms, it follows that the pth term in the expansion of $\log Z(s)$ is

$$\sum_{\ell=1}^{\infty} \frac{(\alpha^{2\ell} + \alpha^{-2\ell} + 2)(\alpha'^{2\ell} + \alpha'^{-2\ell} + 2)}{\ell p^{\ell s}}.$$

Since α^2, α'^2, are either non-negative real numbers or lie on the unit circle, it follows that the above series has non-negative terms. Consequently, so does the series for $Z(s)$. □

Theorem 8.7.6 *Let f be a Maass form for $SL(2, \mathbb{Z})$ which is an eigenform for the Hecke operators. Let F be its symmetric square lift to $SL(3, \mathbb{Z})$. Then $L_F(s)$, given by (8.7.1), has no Siegel zero.*

Proof We make use of (8.7.4). It follows from the Euler product that for $s \in \mathbb{R}$, $s > 1$ that $Z'(s)/Z(s) < 0$. In (Bump and Ginzburg, 1992) it is shown that when f is not a lift from $GL(1)$, then $L_F(s, \vee^2)$ has a simple pole at $s = 1$, and any zero of $L_F(s)$ will be a zero of $Z(s)$ with order at least 3. Consequently, if $L(F, s)$ has a Siegel zero, then $Z(s) = \zeta(s)L_F(s)^3 L_F(s, \vee^2)$ will have 3 Siegel zeros. Since $Z(s)$ has a pole of order 2, we obtain a contradiction from the Siegel zero Lemma 8.6.1. □

In the classical case, the non-existence of Siegel zeros implies a lower bound for the class number of an imaginary quadratic field. One may ask what takes the place of class numbers in the $GL(3)$ setting. The answer to this question is given in Lemma 8.4.3 which says that we will obtain a lower bound for the residue (at $s = 1$) of the relevant L-function. We state an important and useful corollary to Theorem 8.7.6.

Corollary 8.7.7 *Let f be a Maass form for $SL(2, \mathbb{Z})$ of type ν. Then the Petersson inner product of f with itself satisfies*

$$\langle f, f \rangle \gg \frac{1}{\log(1 + |\nu|)}.$$

Proof Let F be the symmetric square lift of f. Then

$$L_{f\times f}(s) = \zeta(s)L_F(s),$$

and $L_{f\times f}(s)$ has no Siegel zero and satisfies the conditions of Lemma 8.4.3. This gives a lower bound for $L_F(1)$ of the type stated in Corollary 8.7.7. But, we know that $L_F(1) = c\cdot\langle f, f\rangle$ (for some constant c) by Theorem 7.2.4. This proves the corollary. □

Remark Corollary 8.7.7 can be generalized to the case of holomorphic modular forms and Maass forms for congruence subgroups of $SL(2,\mathbb{Z})$. It can also be further generalized to L-functions in the Selberg class satisfying Langlands' conjecture on the automorphicity of certain symmetric power L-functions.

8.8 Non-existence of Siegel zeros on $GL(n)$

We have already shown at the end of Section 8.5, that there is at most one classical Dirichlet L-function with a Siegel zero. The exceptional L-function, if it exists, will be associated to a real primitive Dirichlet character attached to a quadratic field. This is the situation for L-functions on $GL(1)$, and it has been known for a long time (Davenport, 1967).

With the breakthrough of Hoffstein and Lockhart (1994) it became possible, for the first time, to show the rarity of Siegel zeros of L-functions associated to Maass forms on $GL(n)$ with $n\ge 2$. We give a brief account of the known results and also specify the particular L-function used in conjunction with the Siegel zero Lemma 8.6.1 to obtain these results.

In accordance with Theorem 7.2.4 and Proposition 7.4.12, it is natural to formalize a more general version of the Rankin–Selberg convolution in terms of Euler products. Let

$$L_f(s) = \prod_p \prod_{i=1}^n \left(1-\frac{\alpha_{i,p}}{p^s}\right)^{-1}, \qquad L_g(s) = \prod_p \prod_{i=1}^n \left(1-\frac{\beta_{i,p}}{p^s}\right)^{-1},$$

be two L-functions in the Selberg class $\mathcal{S}$. We then define the Rankin–Selberg L-function, $L_{f\times g}$ by the new Euler product

$$L_{f\times g}(s) = \prod_p \prod_{i=1}^n \prod_{j=1}^n \left(1-\frac{\alpha_{i,p}\bar{\beta}_{j,p}}{p^s}\right)^{-1}. \tag{8.8.1}$$

In (Hoffstein and Ramakrishnan, 1995) it is shown that there are no Siegel zeros for L-functions associated to cusp forms (holomorphic or non-holomorphic Maass forms) on $GL(2)$. Their proof works over any number field. The idea of the proof is as follows. Let f be a cusp form on $GL(2)$ where

$L_f(s)$ has a Siegel zero. Let $L_f(s, \vee^2)$ denote the symmetric square lift as in Theorem 7.3.2, which is associated to a cusp form F on $GL(3)$.

Now, construct

$$\underbrace{\zeta(s)}_{GL(1)} \cdot \underbrace{L_f(s, \vee^2)}_{GL(3)} \cdot \underbrace{L_f(s)}_{GL(2)},$$

which corresponds to an automorphic form G on $GL(6)$. Let $L_{G\times G}(s)$, as in (8.8.1), denote the L-function of the Rankin–Selberg convolution of G with itself which will have non-negative Dirichlet coefficients. Then we have the identity

$$L_{G\times G}(s) = \zeta(s)L_{F\times F}(s)L_{f\times f}(s)L_F(s)^2L_f(s)^4L_f(s, \vee^3),$$

which can be verified by comparing Euler products. It follows from results of Bump and Ginzberg (1992), Bump, Ginzberg and Hoffstein (1996) and Shahidi (1989) that $L_{G\times G}(s)$ has a triple pole at $s = 1$. If $L_f(s)$ had a Siegel zero then $L_{G\times G}(s)$ would have to have four Siegel zeros which contradicts the Siegel zero Lemma 8.6.1.

We now consider the case of a non-self dual Maass form f on $GL(n)$ for $n \geq 3$. We follow (Hoffstein and Ramakrishnan 1995). Define

$$Z(s) = \zeta(s)L_f(s)L_{\tilde{f}}(s),$$

where $\tilde{f}$ denotes the Maass form dual to f. Let D(s) denote the Rankin–Selberg convolution of $Z(s)$ with itself. Then $D(s)$ will have a pole of order 3, but it will have two copies of $L_f(s)$ and an additional two copies of $L_{\tilde{f}}(s)$ as factors. So if $L_f(s)$ has a Siegel zero then $D(s)$ will have four Siegel zeros and a pole of order 3 at $s = 1$ which again contradicts the Siegel zero Lemma 8.6.1. This establishes that L-functions associated to non-self dual Maass forms on $GL(n)$, for $n \geq 3$, cannot have Siegel zeros. Note that this situation is analogous to the way one proves that complex Dirichlet L-functions cannot have Siegel zeros.

Finally, we consider the case of self dual Maass forms on $GL(n)$ with $n \geq 3$. We again follow Hoffstein and Ramakrishnan (1995) who proved that Siegel zeros cannot exist if one assumes Langlands' conjectures. Let f denote a self dual Maass form on $GL(n)$. Assume there exists some $g \neq f$ where g is not an Eisenstein series such that

$$L_{f\times f}(s) = L_g(s) \cdot D(s)$$

for some other Dirichlet series $D(s)$. Construct

$$Z(s) = \zeta(s)L_g(s)L_f(s),$$

and take the Rankin–Selberg convolution of Z with itself. Then this Rankin–Selberg convolution will have non-negative coefficients, a pole of order 3 at $s = 1$, and it will have $L_f(s)^4$ as a factor assuming everything else is analytic. If $L_f(s)$ had a Siegel zero, then this would contradict the Siegel zero Lemma 8.6.1.

For the case of $GL(3)$, Hoffstein and Ramakrishnan (1995) proved there are no Siegel zeros subject to a certain analyticity hypothesis. This hypothesis was subsequently proved in (Banks, 1997) which establishes that there are no Siegel zeros on $GL(3)$ except for the obvious cases.

9

The Godement–Jacquet L-function

9.1 Maass forms for $SL(n,\mathbb{Z})$

We briefly review Maass forms which were introduced in Section 5.1. For $n \geq 2$, the generalized upper half plane $\mathfrak{h}^n$ (see Definition 1.2.3) consists of all $n \times n$ matrices of the form $z = x \cdot y$ where

$$x = \begin{pmatrix} 1 & x_{1,2} & x_{1,3} & \cdots & & x_{1,n} \\ & 1 & x_{2,3} & \cdots & & x_{2,n} \\ & & \ddots & & & \vdots \\ & & & & 1 & x_{n-1,n} \\ & & & & & 1 \end{pmatrix}, \quad y = \begin{pmatrix} y_1 y_2 \cdots y_{n-1} & & & & \\ & y_1 y_2 \cdots y_{n-2} & & & \\ & & \ddots & & \\ & & & y_1 & \\ & & & & 1 \end{pmatrix},$$

with $x_{i,j} \in \mathbb{R}$ for $1 \leq i < j \leq n$ and $y_i > 0$ for $1 \leq i \leq n-1$.

Remark 9.1.1 It is particularly convenient to relabel the super diagonal elements of the unipotent matrix x so that $x_{1,2} = x_{n-1}, x_{2,3} = x_{n-2}, \quad \ldots, \quad x_{n-1,n} = x_1$, i.e.,

$$x = \begin{pmatrix} 1 & x_{n-1} & x_{1,3} & \cdots & x_{1,n} \\ & 1 & x_{n-2} & \cdots & x_{2,n} \\ & & \ddots & \ddots & \vdots \\ & & & 1 & x_1 \\ & & & & 1 \end{pmatrix}.$$

Henceforth, we will adhere to this notation.

Let U_n denote the group of $n \times n$ upper triangular matrices with 1s on the diagonal as in Section 5.2. Consider a Maass form $\phi(z)$ with $z \in SL(n,\mathbb{Z})\backslash\mathfrak{h}^n$

as defined in Definition 5.1.3. Then ϕ has a Fourier expansion of the form:

$$\phi(z) = \sum_{\gamma \in U_{n-1}(\mathbb{Z})\backslash SL(n-1,\mathbb{Z})} \sum_{m_1=1}^{\infty} \cdots \sum_{m_{n-2}=1}^{\infty} \sum_{m_{n-1}\neq 0} \tilde{\phi}_{(m_1,\ldots,m_{n-1})}\left(\begin{pmatrix} \gamma & \\ & 1 \end{pmatrix} z\right),$$

where the sum is independent of the choice of coset representatives γ,

$$\tilde{\phi}_{(m_1,\ldots,m_{n-1})}(z) := \int_0^1 \cdots \int_0^1 \phi(u \cdot z)\, e^{-2\pi i \left(m_1 u_1 + m_2 u_2 + \cdots + m_{n-1} u_{n-1}\right)}\, d^*u,$$

with $u \in U_n(\mathbb{R})$ given by (5.2.1) and

$$d^*u = du_1 \cdots du_{n-1} \prod_{1\le i<j+1\le n} du_{i,j}.$$

Note the change of notation in the Fourier expansion above to conform with Remark 9.1.1.

Now, we have shown that $\tilde{\phi}_{(m_1,\ldots,m_{n-1})}(z)$ is a Whittaker function. Further, $\tilde{\phi}_{(m_1,\ldots,m_{n-1})}(z)$ will inherit the growth properties of the Maass form ϕ and will satisfy the conditions of the multiplicity one theorem of Shalika (1974) which states that only the Jacquet Whittaker function (5.5.1) can occur in the Fourier expansion of a Maass form for $SL(n,\mathbb{Z})$ and that $\tilde{\phi}_{(m_1,\ldots,m_{n-1})}$ must be a constant multiple of the Jacquet Whittaker function. It follows from Theorem 5.3.2 and Proposition 5.5.2 that if ϕ is a Maass form of type $\nu = (\nu_1, \ldots, \nu_{n-1}) \in \mathbb{C}^{n-1}$ for $SL(n,\mathbb{Z})$ then

$$\phi(z) = \sum_{\gamma \in U_{n-1}(\mathbb{Z})\backslash SL(n-1,\mathbb{Z})} \sum_{m_1=1}^{\infty} \cdots \sum_{m_{n-2}=1}^{\infty} \sum_{m_{n-1}\neq 0} \frac{A(m_1,\ldots,m_{n-1})}{\prod_{k=1}^{n-1} |m_k|^{k(n-k)/2}}$$
$$\times W_{\text{Jacquet}}\left(M \cdot \begin{pmatrix} \gamma & \\ & 1 \end{pmatrix} z,\ \nu,\ \psi_{1,\ldots,1,\frac{m_{n-1}}{|m_{n-1}|}}\right), \tag{9.1.2}$$

where

$$M = \begin{pmatrix} m_1 \cdots m_{n-2} \cdot |m_{n-1}| & & & & \\ & \ddots & & & \\ & & m_1 m_2 & & \\ & & & m_1 & \\ & & & & 1 \end{pmatrix}, \qquad A(m_1,\ldots,m_{n-1}) \in \mathbb{C},$$

and

$$\psi_{1,\ldots,1,\epsilon}\left(\begin{pmatrix} 1 & u_{n-1} & & & * \\ & 1 & u_{n-2} & & \\ & & \ddots & \ddots & \\ & & & 1 & u_1 \\ & & & & 1 \end{pmatrix}\right) = e^{2\pi i\left(u_1+\cdots+u_{n-2}+\epsilon u_{n-1}\right)}.$$

The particular normalization $\frac{A(m_1,\ldots,m_{n-1})}{\prod\limits_{k=1}^{n-1}|m_k|^{k(n-k)/2}}$ is chosen so that later formulae are as simple as possible.

Lemma 9.1.3 (Fourier coefficients are bounded) *Let ϕ be a Maass form for $SL(n,\mathbb{Z})$ as in (9.1.2). Then for all non-zero integers $m_1,\ldots,m_{n-1}$,*

$$\frac{A(m_1,\ldots,m_{n-1})}{\prod\limits_{k=1}^{n-1}|m_k|^{k(n-k)/2}} = \mathcal{O}(1).$$

Proof Let $z\in\mathfrak{h}^n$. By the Fourier expansion, we know that

$$\frac{A(m_1,\ldots,m_{n-1})}{\prod\limits_{k=1}^{n-1}|m_k|^{k(n-k)/2}}\cdot W_{\text{Jacquet}}\left(My,\,\nu,\,\psi_{1,\ldots,1,\frac{m_{n-1}}{|m_{n-1}|}}\right)$$

$$= \int_0^1\cdots\int_0^1 \phi(z)\,e^{-2\pi i\left[m_1x_1+\cdots+m_{n-1}x_{n-1}\right]}\,d^*x,$$

where

$$d^*x = dx_1\cdots dx_{n-1}\prod_{1\le i<j+1\le n} dx_{i,j}.$$

Choosing $y_1=|m_1|^{-1}c_1,\ y_2=|m_2|^{-1}c_2,\ldots,y_{n-1}=|m_{n-1}|^{-1}c_{n-1}$, for suitable $c_1,\ldots,c_{n-1}$, and noting that ϕ is bounded implies that

$$\frac{A(m_1,\ldots,m_{n-1})}{\prod\limits_{k=1}^{n-1}|m_k|^{k(n-k)/2}} = \mathcal{O}(1).$$

□

9.2 The dual and symmetric Maass forms

Let $\phi(z)$ be a Maass form for $SL(n,\mathbb{Z})$ as in (9.1.2). We shall now define, $\tilde{\phi}(z)$, the dual Maass form associated to ϕ which plays an important role in automorphic form theory.

Proposition 9.2.1 *Let $\phi(z)$ be a Maass form of type $(\nu_1, \ldots, \nu_{n-1}) \in \mathbb{C}^{n-1}$ as in (9.1.2). Then (for $\lfloor x \rfloor$ the largest integer $\le x$)*

$$\tilde{\phi}(z) := \phi(w \cdot {}^t(z^{-1}) \cdot w), \qquad w = \begin{pmatrix} & & & (-1)^{[n/2]} \\ & & 1 & \\ & \cdot^{\cdot^{\cdot}} & & \\ 1 & & & \end{pmatrix}$$

is a Maass form of type $(\nu_{n-1}, \ldots, \nu_1)$ for $SL(n, \mathbb{Z})$. The Maass form $\tilde{\phi}$ is called the dual Maass form. If $A(m_1, \ldots, m_{n-1})$ is the $(m_1, \ldots, m_{n-1})$-Fourier coefficient of ϕ then $A(m_{n-1}, \ldots, m_1)$ is the corresponding Fourier coefficient of $\tilde{\phi}$.

Proof First, for every $\gamma \in SL(n, \mathbb{Z})$,

$$\tilde{\phi}(\gamma z) = \phi\big(w \cdot {}^t((\gamma z)^{-1}) \cdot w\big) = \phi\big(\gamma' w \cdot {}^t(z^{-1}) \cdot w\big) = \tilde{\phi}(z)$$

since $\gamma' = w \cdot {}^t(\gamma^{-1}) \cdot w \in SL(n, \mathbb{Z})$. Thus $\tilde{\phi}$ satisfies the automorphic condition (1) of Definition 5.1.3 of a Maass form.

Next, note that if

$$z = \begin{pmatrix} 1 & x_{n-1} & x_{1,3} & \cdots & x_{1,n} \\ & 1 & x_{n-2} & \cdots & x_{2,n} \\ & & \ddots & \ddots & \vdots \\ & & & 1 & x_1 \\ & & & & 1 \end{pmatrix} \begin{pmatrix} y_1 y_2 \cdots y_{n-1} & & & & \\ & y_1 y_2 \cdots y_{n-2} & & & \\ & & \ddots & & \\ & & & y_1 & \\ & & & & 1 \end{pmatrix},$$

then

$$w \cdot {}^t (z^{-1}) \cdot w^{-1} = \begin{pmatrix} 1 & \delta x_1 & & & \\ & 1 & -x_2 & & * \\ & & \ddots & \ddots & \\ & & & 1 & -x_{n-1} \\ & & & & 1 \end{pmatrix} \begin{pmatrix} y_1 y_2 \cdots y_{n-1} & & & & \\ & y_2 y_3 \cdots y_{n-2} & & & \\ & & \ddots & & \\ & & & y_{n-1} & \\ & & & & 1 \end{pmatrix} \tag{9.2.2}$$

where $\delta = (-1)^{\lfloor n/2 \rfloor + 1}$. One may then show that

$$\int\limits_{(SL(n,\mathbb{Z}) \cap U) \backslash U} \tilde{\phi}(z)\, du = 0,$$

for every upper triangular subgroup U of the form

$$U = \left\{ \begin{pmatrix} I_{r_1} & & & * \\ & I_{r_2} & & \\ & & \ddots & \\ & & & I_{r_b} \end{pmatrix} \right\},$$

with $r_1 + r_2 + \cdots r_b = n$. Here I_r denotes the $r \times r$ identity matrix. Thus $\tilde{\phi}$ satisfies the cuspidality condition (3) of Definition 5.1.3.

Now

$$I_{\nu_1,\ldots,\nu_{n-1}}(z) = I_{\nu_{n-1},\ldots,\nu_1}(w \cdot {}^t(z^{-1}) \cdot w^{-1})$$

since the involution $z \to w \cdot {}^t(z^{-1}) \cdot w^{-1}$ interchanges y_j and y_{n-j} for $j = 1, 2, \ldots, n-1$. This shows that $\tilde{\phi}$ is a Maass form of type $(\nu_{n-1}, \ldots, \nu_1)$.

Finally, if we integrate

$$\int_0^1 \cdots \int_0^1 \tilde{\phi}(z)\, e^{-2\pi i\left[m_1 x_1 + \cdots + m_{n-1} x_{n-1}\right]}\, d^*x$$

to pick off the $(m_1, \ldots, m_{n-1})$-Fourier coefficient, then because x_j and x_{n-j} are interchanged (for $j = 1, \ldots, n-1$) we will actually get $A(m_{n-1}, \ldots, m_1)$. □

In the $SL(2, \mathbb{Z})$ theory, the notions of even and odd Maass forms (see Section 3.9) played an important role. If $a(n)$ is the nth Fourier coefficient of an $SL(2, \mathbb{Z})$ Maass form then $a(n) = \pm a(-n)$ depending on whether the Maass form is even or odd. We shall see that a similar phenomenon holds in the case of $SL(n, \mathbb{Z})$ when n is even. On the other hand, if n is odd then all Maass forms are actually even (see Section 6.3 for the example of $n = 3$).

Consider a diagonal matrix δ of the form

$$\delta := \begin{pmatrix} \delta_1 \cdots \delta_{n-1} & & & & \\ & \ddots & & & \\ & & \delta_1\delta_2 & & \\ & & & \delta_1 & \\ & & & & 1 \end{pmatrix}$$

where $\delta_j \in \{+1, -1\}$ for $j = 1, \ldots, n-1$. We define an operator T_δ which maps Maass forms to Maass forms, and is given by

$$T_\delta \phi(z) := \phi(\delta z \delta) = \phi(\delta z),$$

since ϕ, as a function on $\mathfrak{h}^n$, is right-invariant under multiplication by $O(n, \mathbb{R})$. Note that

$$T_\delta\, \phi\left(\begin{pmatrix} 1 & x_{n-1} & & & \\ & 1 & x_{n-2} & & * \\ & & \ddots & \ddots & \\ & & & 1 & x_1 \\ & & & & 1 \end{pmatrix}\begin{pmatrix} y_1 y_2 \cdots y_{n-1} & & & & \\ & y_1 y_2 \cdots y_{n-2} & & & \\ & & \ddots & & \\ & & & y_1 & \\ & & & & 1 \end{pmatrix}\right) \tag{9.2.3}$$

$$= \phi\left(\begin{pmatrix} 1 & \delta_{n-1}x_{n-1} & & & \\ & 1 & \delta_{n-2}x_{n-2} & & * \\ & & \ddots & \ddots & \\ & & & 1 & \delta_1 x_1 \\ & & & & 1 \end{pmatrix}\begin{pmatrix} y_1 y_2 \cdots y_{n-1} & & & & \\ & y_1 y_2 \cdots y_{n-2} & & & \\ & & \ddots & & \\ & & & y_1 & \\ & & & & 1 \end{pmatrix}\right).$$

Clearly $(T_\delta)^2$ is the identity transformation, so the eigenvalues of T_δ can only be ± 1.

Definition 9.2.4 *A Maass form ϕ of type $\nu = (\nu_1, \ldots, \nu_{n-1}) \in \mathbb{C}^{n-1}$ for $SL(n, \mathbb{Z})$ is said to be symmetric if $T_\delta \phi = \pm\phi$ for all T_δ as in (9.2.3).*

Remark Note that every Maass form is a linear combination of symmetric 1s.

Proposition 9.2.5 *Assume $n \geq 2$ is an odd integer. Then every Maass form ϕ for $SL(n, \mathbb{Z})$ is even, i.e., $T_\delta \phi = \phi$, for all T_δ of the form (9.2.3). Furthermore, if $A(m_1, \ldots, m_{n-1})$ denotes the Fourier coefficient in the expansion (9.1.2) then*

$$A(m_1, \ldots, m_{n-2}, m_{n-1}) = A(m_1, \ldots, m_{n-2}, \; -m_{n-1}),$$

for all $m_j \geq 1$ (with $j = 1, 2, \ldots, n-1$).

Proof Any Maass form ϕ is invariant under left multiplication by elements in $SL(n, \mathbb{Z})$ of the form

$$\begin{pmatrix} \epsilon_1 & & & \\ & \epsilon_2 & & \\ & & \ddots & \\ & & & \epsilon_n \end{pmatrix}$$

with $\epsilon_j \in \{+1, -1\}$, for $j = 1, 2, \ldots, n$ and $\prod_{j=1}^{n} \epsilon_j = 1$. It is also invariant by the central element

$$\begin{pmatrix} -1 & & & \\ & -1 & & \\ & & \ddots & \\ & & & -1 \end{pmatrix}$$

which has determinant -1. Since these elements generate all possible T_δ, this proves $T_\delta \phi = \phi$ for all T_δ.

Next, let

$$T_{\delta_0} = \begin{pmatrix} -1 & & & \\ & 1 & & \\ & & \ddots & \\ & & & 1 \end{pmatrix}.$$

Then since $\delta_0 z \delta_0$ transforms $x_{n-1} \to -x_{n-1}$ and fixes x_j with $1 \le j \le n-2$, it follows that the integral

$$\int_0^1 \cdots \int_0^1 T_{\delta_0}\phi(z)\, e^{-2\pi i \left[m_1 x_1 + \cdots + m_{n-1} x_{n-1}\right]}\, d^*x$$

picks off the $A(m_1, \ldots, m_{n-2}, -m_{n-1})$ coefficient of ϕ, and this must be the same as $A(m_1, \ldots, m_{n-2}, m_{n-1})$ because $T_{\delta_0}\phi = \phi$. □

We next show that the theory of symmetric Maass forms for $SL(n, \mathbb{Z})$ (with n even) is very similar to the $SL(2, \mathbb{Z})$ theory. Basically, there are only two types of such Maass forms, even and odd Maass forms.

Proposition 9.2.6 *Assume $n \ge 2$ is an even integer. Let ϕ be a symmetric Maass form for $SL(n, \mathbb{Z})$ with Fourier coefficients $A(m_1, \ldots, m_{n-1})$ as in the expansion (9.1.2). Fix*

$$T_{\delta_0} = \begin{pmatrix} -1 & & & \\ & 1 & & \\ & & \ddots & \\ & & & 1 \end{pmatrix}.$$

Then for all $m_j \ge 1$ (with $j = 1, 2, \ldots, n-1$), we have

$$A(m_1, \ldots, m_{n-2},\ m_{n-1}) = \pm A(m_1, \ldots, m_{n-2},\ -m_{n-1}),$$

according as $T_{\delta_0}\phi = \pm\phi$. The Maass form ϕ is said to be even or odd accordingly.

Proof The diagonal elements with ± 1 entries are generated by such elements with determinant 1 and the additional special element T_{δ_0}. Since ϕ is invariant under left multiplication by $SL(n,\mathbb{Z})$ it follows that ϕ is symmetric if and only if $T_\delta \phi = \pm\phi$.

Now $\delta_0 z \delta_0$ transforms $x_{n-1} \to -x_{n-1}$ and fixes x_j with $1 \le j \le n-2$, it follows that the integral

$$\int_0^1 \cdots \int_0^1 T_{\delta_0}\phi(z)\, e^{-2\pi i\left[m_1 x_1 + \cdots + m_{n-1}x_{n-1}\right]}\, d^*x$$

picks off the $A(m_1, \ldots, m_{n-2}, -m_{n-1})$ coefficient of ϕ, and this must be the same as $\pm A(m_1, \ldots, m_{n-2}, m_{n-1})$ because $T_{\delta_0}\phi = \pm\phi$. □

9.3 Hecke operators for $SL(n, \mathbb{Z})$

We recall the general definition of Hecke operators given in Definition 3.10.5. Consider a group G that acts continuously on a topological space X. Let Γ be a discrete subgroup of G. For every g in $C_G(\Gamma)$, the commensurator of Γ in G, (i.e., $(g^{-1}\Gamma g) \cap \Gamma$ has finite index in both Γ and $g^{-1}\Gamma g$) we have a decomposition of a double coset into disjoint right cosets of the form

$$\Gamma g \Gamma = \bigcup_i \Gamma \alpha_i. \tag{9.3.1}$$

For each such g, the Hecke operator $T_g : \mathcal{L}^2(\Gamma\backslash X) \to \mathcal{L}^2(\Gamma\backslash X)$ is defined by

$$T_g f(x) = \sum_i f(\alpha_i x),$$

where $f \in \mathcal{L}^2(\Gamma\backslash X)$, $x \in X$, and α_i are given by (9.3.1). The Hecke ring consists of all formal sums

$$\sum_k c_k T_{g_k}$$

with integer coefficients c_k and g_k in a semigroup Δ as in Definition 3.10.8. Since two double cosets are either identical or totally disjoint, it follows that unions of double cosets are associated to elements in the Hecke ring. Finally, we recall Theorem 3.10.10 which states that the Hecke ring is commutative if there exists an antiautomorphism $g \mapsto g^*$ (i.e., $(gh)^* = h^* g^*$) for which $\Gamma^* = \Gamma$ and $(\Gamma g \Gamma)^* = \Gamma g \Gamma$ for every $g \in \Delta$.

We now consider, for $n \ge 2$, the general case

$$G = GL(n,\mathbb{R}), \qquad \Gamma = SL(n,\mathbb{Z}), \qquad X = GL(n,\mathbb{R})/(O(n,\mathbb{R})\cdot\mathbb{R}^\times) = \mathfrak{h}^n.$$

For every n-tuple of positive integers $(m_0, m_1, \ldots, m_{n-1})$, the matrix

$$\begin{pmatrix} m_0 \cdots m_{n-1} & & & \\ & \ddots & & \\ & & m_0 m_1 & \\ & & & m_0 \end{pmatrix} \in C_G(\Gamma),$$

the commensurator of Γ in G (defined in (3.10.2)). We define Δ to be the semigroup generated by all such matrices. As in the case of $SL(2, \mathbb{Z})$, we have the antiautomorphism

$$g \mapsto {}^t g, \qquad g \in \Delta,$$

where ${}^t g$ denotes the transpose of the matrix g. It is again clear that the conditions of Theorem 3.10.10 are satisfied so that the Hecke ring is commutative.

The following lemma is analogous to Lemma 3.12.1, which came up in the $SL(2, \mathbb{Z})$ situation.

Lemma 9.3.2 *Fix a positive integer $N \geq 1$. Define the set*

$$S_N := \left\{ \begin{pmatrix} c_1 & c_{1,2} & \cdots & c_{1,n} \\ & c_2 & \cdots & c_{2,n} \\ & & \ddots & \vdots \\ & & & c_n \end{pmatrix} \;\middle|\; \begin{matrix} c_\ell \geq 1 \; (\ell=1,2,\ldots,n) \\ \prod_{\ell=1}^{n} c_\ell = N \\ 0 \leq c_{i,\ell} < c_\ell \; (1 \leq i < \ell \leq n) \end{matrix} \right\}.$$

Then one has the disjoint partition

$$\bigcup_{m_0^n m_1^{n-1} \cdots m_{n-1} = N} \Gamma \begin{pmatrix} m_0 \cdots m_{n-1} & & & \\ & \ddots & & \\ & & m_0 m_1 & \\ & & & m_0 \end{pmatrix} \Gamma = \bigcup_{\alpha \in S_N} \Gamma \alpha. \qquad (9.3.3)$$

Proof First of all we claim the decomposition is disjoint. If not, there exists

$$\begin{pmatrix} \gamma_{1,1} & \gamma_{1,2} & \cdots & \gamma_{1,n} \\ \gamma_{2,1} & \gamma_{2,2} & \cdots & \gamma_{2,n} \\ \vdots & \vdots & & \vdots \\ \gamma_{n,1} & \gamma_{n,2} & \cdots & \gamma_{n,n} \end{pmatrix} \in \Gamma$$

such that

$$\begin{pmatrix} \gamma_{1,1} & \gamma_{1,2} & \cdots & \gamma_{1,n} \\ \gamma_{2,1} & \gamma_{2,2} & \cdots & \gamma_{2,n} \\ \vdots & \vdots & & \vdots \\ \gamma_{n,1} & \gamma_{n,2} & \cdots & \gamma_{n,n} \end{pmatrix} \cdot \begin{pmatrix} c_1 & c_{1,2} & \cdots & c_{1,n} \\ & c_2 & \cdots & c_{2,n} \\ & & \ddots & \vdots \\ & & & c_n \end{pmatrix} = \begin{pmatrix} c'_1 & c'_{1,2} & \cdots & c'_{1,n} \\ & c'_2 & \cdots & c'_{2,n} \\ & & \ddots & \vdots \\ & & & c'_n \end{pmatrix}. \tag{9.3.4}$$

This implies that $\gamma_{i,j} = 0$ for $1 \le j < i \le n$. Consequently, $\gamma_{\ell,\ell} \cdot c_\ell = c'_\ell$, for $1 \le \ell \le n$. But $\prod_{\ell=1}^{n} \gamma_{\ell,\ell} = 1$ and $c_\ell, c'_\ell \ge 1$ $(1 \le \ell \le n)$. It easily follows that

$$\gamma_{1,1} = \gamma_{2,2} = \cdots = \gamma_{n,n} = 1.$$

Note that the above shows that $c_\ell = c'_\ell$ $(1 \le \ell \le n)$. Therefore, (9.3.4) takes the form

$$\begin{pmatrix} 1 & \gamma_{1,2} & \cdots & \gamma_{1,n} \\ & 1 & \cdots & \gamma_{2,n} \\ & & \ddots & \vdots \\ & & & 1 \end{pmatrix} \cdot \begin{pmatrix} c_1 & c_{1,2} & \cdots & c_{1,n} \\ & c_2 & \cdots & c_{2,n} \\ & & \ddots & \vdots \\ & & & c_n \end{pmatrix} = \begin{pmatrix} c_1 & c'_{1,2} & \cdots & c'_{1,n} \\ & c_2 & \cdots & c'_{2,n} \\ & & \ddots & \vdots \\ & & & c_n \end{pmatrix}.$$

Since $0 \le c_{i,\ell},\ c'_{i,\ell} < c_\ell$ for $1 \le i < \ell \le n$, one concludes that $\gamma_{i,\ell} = 0$ for $1 \le i < \ell \le n$, and the decomposition is disjoint as claimed.

Now, by Theorem 3.11.2, every element on the right-hand side of (9.3.3) can be put into Smith normal form, so must occur as an element on the left-hand side of (9.3.3). Similarly, by Theorem 3.11.1, every element on the left-hand side of (9.3.3) can be put into Hermite normal form, so must occur as an element on the right-hand side of (9.3.3). This proves the equality of the two sides of (9.3.3). □

By analogy with the $SL(2,\mathbb{Z})$ situation (see (3.12.3)), it follows that for every integer $N \ge 1$, we have a Hecke operator T_N acting on the space of square integrable automorphic forms $f(z)$ with $z \in \mathfrak{h}^n$. The action is given by the formula

$$T_N f(z) = \frac{1}{N^{n-1/2}} \sum_{\substack{\prod_{\ell=1}^{n} c_\ell = N \\ 0 \le c_{i,\ell} < c_\ell \ (1 \le i < \ell \le n)}} f\left(\begin{pmatrix} c_1 & c_{1,2} & \cdots & c_{1,n} \\ & c_2 & \cdots & c_{2,n} \\ & & \ddots & \vdots \\ & & & c_n \end{pmatrix} \cdot z \right) \tag{9.3.5}$$

Note the normalizing factor of $1/N^{(n-1)/2}$ which was chosen to simplify later formulae. Clearly, T_1 is just the identity operator.

The ℂ-vector space $\mathcal{L}^2(\Gamma\backslash\mathfrak{h}^n)$ has a natural inner product, denoted $\langle , \rangle$, and defined by

$$\langle f, g\rangle = \int\limits_{\Gamma\backslash\mathfrak{h}^n} f(z)\overline{g(z)}\, d^*z,$$

for all $f, g \in \mathcal{L}^2(\Gamma\backslash\mathfrak{h}^n)$, where d^*z denotes the left invariant measure given in Proposition 1.5.3.

In the case of $SL(2,\mathbb{Z})$, we showed in Theorem 3.12.4 that the Hecke operators are self-adjoint with respect to the Petersson inner product. For $SL(n,\mathbb{Z})$ with $n \geq 3$, it is no longer true that the Hecke operators are self-adjoint. What happens is that the adjoint operator is again a Hecke operator and, therefore, the Hecke operator commutes with its adjoint which means that it is a normal operator.

Theorem 9.3.6 (Hecke operators are normal operators) *Consider the Hecke operators T_N, $(N = 1, 2, \ldots)$ defined in (9.3.5). Let T_N^* be the adjoint operator which satisfies*

$$\langle T_N f,\ g\rangle = \big\langle f,\ T_N^* g\big\rangle$$

for all $f, g \in \mathcal{L}^2(\Gamma\backslash\mathfrak{h}^n)$. Then T_N^* is another Hecke operator which commutes with T_N so that T_N is a normal operator. Explicitly, T_N^* is associated to the following union of double cosets:

$$\bigcup_{m_0^n m_1^{n-1}\cdots m_{n-1}=N} \Gamma \begin{pmatrix} N\cdot m_0^{-1} & & & \\ & N\cdot(m_0m_1)^{-1} & & \\ & & \ddots & \\ & & & N\cdot(m_0\cdots m_{n-1})^{-1} \end{pmatrix} \Gamma. \tag{9.3.7}$$

Proof It follows from (9.3.3), and also from the fact that transposition is an antiautomorphism (as in the proof of Theorem 3.10.10), that

$$\bigcup_{m_0^n m_1^{n-1}\cdots m_{n-1}=N} \Gamma \begin{pmatrix} m_0\cdots m_{n-1} & & & \\ & \ddots & & \\ & & m_0m_1 & \\ & & & m_0 \end{pmatrix} \Gamma = \bigcup_{\alpha\in S_N} \Gamma\alpha = \bigcup_{\alpha\in S_N} \alpha\Gamma. \tag{9.3.8}$$

Since the action of the Hecke operator is independent of the choice of right coset decomposition, we obtain

$$\begin{aligned}
\langle T_N f, g\rangle &= \frac{1}{N^{(n-1)/2}} \iint_{\Gamma\backslash\mathfrak{h}^n} \sum_{\alpha\in S_N} f(\alpha z)\,\overline{g(z)}\, d^*z \\
&= \frac{1}{N^{(n-1)/2}} \iint_{\Gamma\backslash\mathfrak{h}^n} f(z) \sum_{\alpha\in S_N} \overline{g(\alpha^{-1}z)}\, d^*z \qquad (9.3.9)\\
&= \frac{1}{N^{(n-1)/2}} \iint_{\Gamma\backslash\mathfrak{h}^n} f(z) \sum_{\alpha\in S_N} \overline{g\left(\begin{pmatrix} N & & & \\ & \ddots & & \\ & & N & \\ & & & N \end{pmatrix} \alpha^{-1}\cdot z\right)}\, d^*z,
\end{aligned}$$

after making the change of variables $z \to \alpha^{-1}z$. Multiplying by the diagonal matrix (with Ns on the diagonal) above does not change anything because g is well defined on $\mathfrak{h}^n$.

Now, it follows from (9.3.8) that for

$$\omega = \begin{pmatrix} & & & 1 \\ & & -1 & \\ & 1 & & \\ & \iddots & & \\ 1 & & & \end{pmatrix},$$

we have

$$\begin{aligned}
&\bigcup_{m_0^n m_1^{n-1}\cdots m_{n-1}=N} \Gamma\cdot\omega \begin{pmatrix} m_0\cdots m_{n-1} & & & \\ & \ddots & & \\ & & m_0 m_1 & \\ & & & m_0 \end{pmatrix}^{-1} \cdot\omega^{-1}\,\Gamma \\
&= \bigcup_{m_0^n m_1^{n-1}\cdots m_{n-1}=N} \Gamma\cdot \begin{pmatrix} m_0^{-1} & & & \\ & (m_0m_1)^{-1} & & \\ & & \ddots & \\ & & & (m_0\cdots m_{n-1})^{-1} \end{pmatrix} \cdot\Gamma.
\end{aligned} \qquad (9.3.10)$$

Finally, if we multiply both sides of (9.3.10) by the diagonal matrix $\begin{pmatrix} N & & \\ & \ddots & \\ & & N \end{pmatrix}$, with $N = m_0^n m_1^{n-1}\cdots m_{n-1}$, it follows that the adjoint Hecke operator defined by (9.3.9) is, in fact, associated to the union of double cosets given in (9.3.7). This completes the proof. □

The Hecke operators commute with the $GL(n, \mathbb{R})$-invariant differential operators, and they also commute with the operators T_δ given in (9.2.3). It follows by standard methods in functional analysis, that we may simultaneously diagonalize the space $\mathcal{L}^2(SL(n, \mathbb{Z})\backslash\mathfrak{h}^n)$ by all these operators. We shall be interested in studying Maass forms which are eigenfunctions of the full Hecke ring of all such operators. The following theorem is analogous to Theorem 3.12.8 which came up in the $SL(2, \mathbb{Z})$ situation.

Theorem 9.3.11 (Multiplicativity of the Fourier coefficients) *Consider*

$$\phi(z) = \sum_{\gamma \in U_{n-1}(\mathbb{Z})\backslash SL(n-1,\mathbb{Z})} \sum_{m_1=1}^{\infty} \cdots \sum_{m_{n-2}=1}^{\infty} \sum_{m_{n-1}\neq 0} \frac{A(m_1, \ldots, m_{n-1})}{\prod_{k=1}^{n-1} |m_k|^{k(n-k)/2}}$$

$$\times W_{\text{Jacquet}}\left(\begin{pmatrix} m_1 \cdots |m_{n-1}| & & & \\ & \ddots & & \\ & & m_1 & \\ & & & 1 \end{pmatrix} \cdot \begin{pmatrix} \gamma & \\ & 1 \end{pmatrix} z, \nu, \psi_{1,\ldots,1,\frac{m_{n-1}}{|m_{n-1}|}}\right),$$

a Maass form for $SL(n, \mathbb{Z})$, as in (9.1.2). Assume that ϕ is an eigenfunction of the full Hecke ring. If $A(1, \ldots, 1) = 0$, then ϕ vanishes identically. Assume $\phi \neq 0$ and it is normalized so that $A(1, \ldots, 1) = 1$. Then

$$T_m\phi = A(m, 1, \ldots, 1) \cdot \phi, \qquad \forall\, m = 1, 2, \ldots$$

Furthermore, we have the following multiplicativity relations

$$A\big(m_1 m_1', \ldots, m_{n-1} m_{n-1}'\big) = A(m_1, \ldots, m_{n-1}) \cdot A\big(m_1', \ldots, m_{n-1}'\big),$$

if $(m_1 \cdots m_{n-1},\ m_1' \cdots m_{n-1}') = 1$, and

$$A(m, 1, \ldots, 1)A(m_1, \ldots, m_{n-1}) = \sum_{\substack{\prod_{\ell=1}^{n} c_\ell = m \\ c_1|m_1, c_2|m_2, \ldots, c_{n-1}|m_{n-1}}} A\left(\frac{m_1 c_n}{c_1}, \frac{m_2 c_1}{c_2}, \ldots, \frac{m_{n-1} c_{n-2}}{c_{n-1}}\right)$$

Addendum *Let $\tilde{\phi}$ denote the Maass form dual to ϕ. As in Proposition 9.2.1, the $(m_1, \ldots, m_{n-1})$ Fourier coefficient of $\tilde{\phi}$ is $A(m_{n-1}, \ldots, m_1)$. If ϕ is an eigenform of the full Hecke ring then*

$$A(m_{n-1}, \ldots, m_1) = \overline{A(m_1, \ldots, m_{n-1})},$$

i.e., it is the complex conjugate of the Fourier coefficient of ϕ.

Proof Let $z = x \cdot y$ with

$$x = \begin{pmatrix} 1 & x_{1,2} & x_{1,3} & \cdots & x_{1,n} \\ & 1 & x_{2,3} & \cdots & x_{2,n} \\ & & \ddots & \ddots & \vdots \\ & & & 1 & x_{n-1,n} \\ & & & & 1 \end{pmatrix}, \quad y = \begin{pmatrix} y_1 y_2 \cdots y_{n-1} & & & & \\ & y_1 y_2 \cdots y_{n-2} & & & \\ & & \ddots & & \\ & & & y_1 & \\ & & & & 1 \end{pmatrix}.$$

In view of Theorem 5.3.2, Remark 9.1.1, and formula (9.1.2), we may write for, $m_1, \ldots, m_{n-1} \geq 1$,

$$\int_0^1 \cdots \int_0^1 \phi(z)\, e^{-2\pi i\left(m_1 x_{n-1,n} + m_2 x_{n-2,n-1} + \cdots + m_{n-1} x_{1,2}\right)} d^*x$$
$$= \frac{A(m_1, \ldots, m_{n-1})}{\prod\limits_{k=1}^{n-1} m_k^{k(n-k)/2}} \cdot W_{\text{Jacquet}}\left(\begin{pmatrix} m_1 y_1 \cdots m_{n-1} y_{n-1} & & & \\ & \ddots & & \\ & & m_1 y_1 & \\ & & & 1 \end{pmatrix}, \nu, \psi_{1,\ldots,1} \right), \tag{9.3.12}$$

where $d^*x = \prod\limits_{1 \leq i < j \leq n-1} dx_{i,j}$. While the notation adopted in Remark 9.1.1 is useful in most cases, it makes the Hecke operator computations extremely gruesome, so we temporarily return to our earlier notation for x.

If ϕ is an eigenfunction of the Hecke operator T_m defined by (9.3.5), then we have $T_m f(z) = \lambda_m f(z)$ for some eigenvalue λ_m. We can compute λ_m directly using a variation of (9.3.12). We begin by considering for $m = 1, 2, \ldots,$

$$\lambda_m \frac{A(m_1, \ldots, m_{n-1})}{\prod\limits_{k=1}^{n-1} m_k^{k(n-k)/2}} \cdot W_{\text{Jacquet}}\left(\begin{pmatrix} m_1 y_1 \cdots m_{n-1} y_{n-1} & & & \\ & \ddots & & \\ & & m_1 y_1 & \\ & & & 1 \end{pmatrix}, \nu, \psi_{1,\ldots,1} \right)$$
$$= \frac{1}{m^{n(n-1)/2}} \int_0^m \cdots \int_0^m T_m\phi(z)\, e^{-2\pi i\left(m_1 x_{n-1,n} + m_2 x_{n-2,n-1} + \cdots + m_{n-1} x_{1,2}\right)} d^*x$$

$$= \frac{1}{m^{(n+1)(n-1)/2}} \sum_{\substack{\prod_{\ell=1}^{n} c_\ell = m \\ 0 \le c_{i,\ell} < c_\ell \ (1 \le i < \ell \le n)}} \int_0^m \cdots \int_0^m$$

$$\times \phi\left(\begin{pmatrix} c_1 & c_{1,2} & \cdots & c_{1,n} \\ & c_2 & \cdots & c_{2,n} \\ & & \ddots & \vdots \\ & & & c_n \end{pmatrix} \begin{pmatrix} 1 & x_{1,2} & x_{1,3} & \cdots & x_{1,n} \\ & 1 & x_{2,3} & \cdots & x_{2,n} \\ & & \ddots & \ddots & \vdots \\ & & & 1 & x_{n-1,n} \\ & & & & 1 \end{pmatrix} \cdot y\right)$$

$$\times e^{-2\pi i\left(m_1 x_{n-1,n} + m_2 x_{n-2,n-1} + \cdots + m_{n-1} x_{1,2}\right)} \, d^*x. \qquad (9.3.13)$$

Next, if we let

$$\begin{pmatrix} c_1 & c_{1,2} & \cdots & c_{1,n} \\ & c_2 & \cdots & c_{2,n} \\ & & \ddots & \vdots \\ & & & c_n \end{pmatrix} \begin{pmatrix} 1 & x_{1,2} & x_{1,3} & \cdots & x_{1,n} \\ & 1 & x_{2,3} & \cdots & x_{2,n} \\ & & \ddots & \ddots & \vdots \\ & & & 1 & x_{n-1,n} \\ & & & & 1 \end{pmatrix}$$

$$= \begin{pmatrix} 1 & x'_{1,2} & x'_{1,3} & \cdots & x'_{1,n} \\ & 1 & x'_{2,3} & \cdots & x'_{2,n} \\ & & \ddots & \ddots & \vdots \\ & & & 1 & x'_{n-1,n} \\ & & & & 1 \end{pmatrix} \begin{pmatrix} c_1 & & & \\ & c_2 & & \\ & & \ddots & \\ & & & c_n \end{pmatrix},$$

then we may solve for $x'_{i,j}$ $(1 \le i < j \le n)$, and obtain

$$x'_{i,j} = \frac{1}{c_j} \sum_{k=i}^{j} c_{i,k} x_{k,j}, \qquad (9.3.14)$$

with the understanding that $c_{i,i} = c_i$ and $x_{i,i} = 1$ for $i = 1, 2, \ldots, n$. Note the special case:

$$x'_{i,i+1} = \frac{c_i x_{i,i+1} + c_{i,i+1}}{c_{i+1}} \qquad (i = 1, 2, \ldots, n-1), \qquad (9.3.15)$$

of (9.3.14).

With the computations (9.3.14), (9.3.15), the right-hand side of the identity (9.3.13) becomes

$$\frac{1}{m^{(n+1)(n-1)/2}} \sum_{\substack{\prod_{\ell=1}^{n} c_\ell = m \\ 0 \le c_{i,\ell} < c_\ell \ (1 \le i < \ell \le n)}} \prod_{1 \le i < j \le n} \int\limits_{\sum_{k=i+1}^{j} \frac{c_{i,k} x_{k,j}}{c_j}}^{\frac{c_i m}{c_j} + \sum_{k=i+1}^{j} \frac{c_{i,k} x_{k,j}}{c_j}}$$

$$\times \phi\left(\begin{pmatrix} 1 & x'_{1,2} & x'_{1,3} & \cdots & x'_{1,n} \\ & 1 & x'_{2,3} & \cdots & x'_{2,n} \\ & & \ddots & \ddots & \vdots \\ & & & 1 & x'_{n-1,n} \\ & & & & 1 \end{pmatrix} \begin{pmatrix} c_1 & & & \\ & c_2 & & \\ & & \ddots & \\ & & & c_n \end{pmatrix} \cdot y\right)$$

$$\times e^{2\pi i \sum_{r=1}^{n-1} \frac{c_{r,r+1} m_{n-r}}{c_r}} \cdot e^{-2\pi i \sum_{r=1}^{n-1} \frac{c_{r+1} m_{n-r}}{c_r} x'_{r,r+1}} \cdot \frac{c_j}{c_i}\, dx'_{i,j}.$$

In view of the periodicity of the above integrand, we may deduce that it takes the form:

$$\frac{1}{m^{(n+1)(n-1)/2}} \sum_{\substack{\prod_{\ell=1}^{n} c_\ell = m \\ 0 \le c_{i,\ell} < c_\ell (1 \le i < \ell \le n)}} \prod_{1 \le i < j \le n} \int_0^{\frac{c_i m}{c_j}}$$

$$\times \phi\left(\begin{pmatrix} 1 & x'_{1,2} & x'_{1,3} & \cdots & x'_{1,n} \\ & 1 & x'_{2,3} & \cdots & x'_{2,n} \\ & & \ddots & \ddots & \vdots \\ & & & 1 & x'_{n-1,n} \\ & & & & 1 \end{pmatrix} \begin{pmatrix} c_1 & & & \\ & c_2 & & \\ & & \ddots & \\ & & & c_n \end{pmatrix} \cdot y\right)$$

$$\times e^{2\pi i \sum_{r=1}^{n-1} \frac{c_{r,r+1} m_{n-r}}{c_r}} \cdot e^{-2\pi i \sum_{r=1}^{n-1} \frac{c_{r+1} m_{n-r}}{c_r} x'_{r,r+1}} \cdot \frac{c_j}{c_i}\, dx'_{i,j}.$$

But the above integral vanishes unless $c_r \mid c_{r+1} m_{n-r}$ for all $r = 1, 2, \ldots, n-1$. Furthermore, in this case we have

$$\sum_{0 \le c_{i,\ell} < c_\ell (1 \le i < \ell \le n)} e^{2\pi i \sum_{r=1}^{n-1} \frac{c_{r,r+1} m_{n-r}}{c_r}} = \begin{cases} \prod_{t=2}^{n} c_t^{t-1} & \text{if } c_r \mid m_{n-r} \ (1 \le r < n), \\ 0 & \text{otherwise.} \end{cases}$$

Consequently, it follows that our integral (9.3.13) may be written in the form:

$$\frac{1}{m^{(n+1)(n-1)/2}} \sum_{\prod_{\ell=1}^{n} c_\ell = m} \prod_{t=2}^{n} c_t^{t-1} \prod_{1\le i<j\le n} \int_0^{\frac{c_i m}{c_j}}$$

$$\times \phi\left(\begin{pmatrix} 1 & x'_{1,2} & x'_{1,3} & \cdots & x'_{1,n} \\ & 1 & x'_{2,3} & \cdots & x'_{2,n} \\ & & \ddots & \ddots & \vdots \\ & & & 1 & x'_{n-1,n} \\ & & & & 1 \end{pmatrix}\begin{pmatrix} c_1 & & & \\ & c_2 & & \\ & & \ddots & \\ & & & c_n \end{pmatrix} \cdot y\right)$$

$$\times e^{-2\pi i \sum\limits_{r=1}^{n-1} \frac{c_{r+1} m_{n-r}}{c_r} x'_{r,r+1}} \cdot \frac{c_j}{c_i}\, dx'_{i,j}.$$

Further, if $F : \mathbb{R} \to \mathbb{C}$ is a periodic function satisfying $F(x+1) = F(x)$ and M is a positive integer, then $\int_0^M F(x)\,dx = M \cdot \int_0^1 F(x)\,dx$. Consequently, the integral above is equal to

$$\frac{1}{m^{(n-1)/2}} \sum_{\prod_{\ell=1}^{n} c_\ell = m} \prod_{t=2}^{n} c_t^{t-1} \prod_{1\le i<j\le n} \int_0^{1}$$

$$\times \phi\left(\begin{pmatrix} 1 & x'_{1,2} & x'_{1,3} & \cdots & x'_{1,n} \\ & 1 & x'_{2,3} & \cdots & x'_{2,n} \\ & & \ddots & \ddots & \vdots \\ & & & 1 & x'_{n-1,n} \\ & & & & 1 \end{pmatrix}\begin{pmatrix} c_1 & & & \\ & c_2 & & \\ & & \ddots & \\ & & & c_n \end{pmatrix} \cdot y\right)$$

$$\times e^{-2\pi i \sum\limits_{r=1}^{n-1} \frac{c_{r+1} m_{n-r}}{c_r} x'_{r,r+1}}\, dx'_{i,j}. \qquad (9.3.16)$$

Finally, the multiple integral, $\prod\limits_{1\le i<j\le n} \int\limits_0^1 \cdots$, above can be evaluated with (9.3.12) and has the value

$$\frac{A\left(\frac{m_1 c_n}{c_{n-1}}, \frac{m_2 c_{n-1}}{c_{n-2}}, \ldots, \frac{m_{n-1} c_2}{c_1}\right)}{\prod\limits_{k=1}^{n-1}\left(\frac{c_{n+1-k} m_k}{c_{n-k}}\right)^{k(n-k)/2}}$$

$$\times W_{\text{Jacquet}}\left(\begin{pmatrix} m_1 y_1 \cdots m_{n-1} y_{n-1} c_n & & & \\ & \ddots & & \\ & & m_1 y_1 c_n & \\ & & & c_n \end{pmatrix}, \nu, \psi_{1,\ldots,1}\right),$$

from which it follows from (9.3.13) and (9.3.16) that

$$\lambda_n \frac{A(m_1,\ldots,m_{n-1})}{\prod\limits_{k=1}^{n-1} m_k^{k(n-k)/2}} \cdot W_{\text{Jacquet}}\left(\begin{pmatrix} m_1y_1\cdots m_{n-1}y_{n-1} & & & \\ & \ddots & & \\ & & m_1y_1 & \\ & & & 1\end{pmatrix}, \nu, \psi_{1,\ldots,1}\right)$$

$$= \frac{1}{m^{(n-1)/2}} \sum_{\prod\limits_{\ell=1}^{n} c_\ell = m} \prod_{t=2}^{n} c_t^{t-1} \cdot \frac{A\left(\frac{m_1c_n}{c_{n-1}}, \frac{m_2c_{n-1}}{c_{n-2}}, \ldots, \frac{m_{n-1}c_2}{c_1}\right)}{\prod\limits_{k=1}^{n-1}\left(\frac{c_{n+1-k}m_k}{c_{n-k}}\right)^{k(n-k)/2}}$$

$$\times W_{\text{Jacquet}}\left(\begin{pmatrix} m_1y_1\cdots m_{n-1}y_{n-1}c_n & & & \\ & \ddots & & \\ & & m_1y_1c_n & \\ & & & c_n\end{pmatrix}, \nu, \psi_{1,\ldots,1}\right).$$

Note that we may cancel the Whittaker functions on both sides of the above identity because the Whittaker functions are invariant under multiplication by scalar matrices. Further, one easily checks the identity

$$\frac{\prod\limits_{t=2}^{n} c_t^{t-1}}{\prod\limits_{k=1}^{n-1}\left(\frac{c_{n+1-k}}{c_{n-k}}\right)^{k(n-k)/2}} = (c_1\cdot c_2\cdots c_n)^{(n-1)/2} = m^{(n-1)/2}.$$

It immediately follows that

$$\lambda_m A(m_1,\ldots,m_{n-1}) = \sum_{\substack{\prod\limits_{\ell=1}^{n} c_\ell = m \\ c_{n-1}|m_1,\ c_{n-2}|m_2,\ldots,\ c_1|m_{n-1}}} A\left(\frac{m_1c_n}{c_{n-1}}, \frac{m_2c_{n-1}}{c_{n-2}}, \ldots, \frac{m_{n-1}c_2}{c_1}\right). \tag{9.3.17}$$

We now explore the consequences of the assumption that $A(1,\ldots,1) = 0$. It follows easily from (9.3.17) that $A(k, 1, \ldots, 1) = 0$ for all integers k, and then the left-hand side of (9.3.17) vanishes for all m, m_1 as long as $m_2 = m_3 = \cdots = m_{n-1} = 1$. By choosing $m_2 = \cdots = m_{n-1} = 1$, $m_1 = p$, $m = p$ one obtains $A(1, p, 1, \ldots, 1) = 0$. Arguing inductively, we may choose $m_2 = \cdots = m_{n-1} = 1$, $m_1 = p$, $m = p^\ell$ for $\ell = 1, 2, \ldots$ from which one can conclude that $A(p^\ell, p, 1, \ldots, 1) = 0$ for all $\ell = 0, 1, 2, \ldots$ One can continue to show that $A(p^{i_1}, p^{i_2}, 1, \ldots, 1) = 0$ for all non-negative integers i_1, i_2, and proceeding inductively one may show that

$$A\left(p^{i_1}, p^{i_2}, \ldots, p^{i_{n-1}}\right) = 0$$

for all non-negative integers $i_1, \ldots, i_{n-1}$. One may then proceed to products of two primes, products of three primes, etc. to eventually obtain that if $A(1, \ldots, 1) = 0$ then all coefficients $A(m_1, \ldots, m_{n-1})$ must vanish.

If $f \neq 0$ then we may assume it is normalized so that $A(1, \ldots, 1) = 1$. If we now choose $m_1 = m_2 = \cdots = m_{n-1} = 1$, it immediately follows from (9.3.17) that $\lambda_m = A(m, 1, \ldots, 1)$. Substituting this into (9.3.17), and changing indices (on the c_js), proves the identity

$$A(m, 1, \ldots, 1)A(m_1, \ldots, m_{n-1}) = \sum_{\substack{\prod_{\ell=1}^{n} c_\ell = m \\ c_1 | m_1,\ c_2 | m_2, \ldots,\ c_{n-1} | m_{n-1}}} A\left(\frac{m_1 c_n}{c_1}, \frac{m_2 c_1}{c_2}, \ldots, \frac{m_{n-1} c_{n-2}}{c_{n-1}}\right).$$

The rest of the proof of Theorem 9.3.11 follows easily.

To prove the addendum, consider the identity $\langle T\phi, \phi\rangle = \langle \phi, T^*\phi\rangle$, T is a Hecke operator and T^* is the adjoint operator, and $\langle\ \ \rangle$ denotes the Petersson inner product. The addendum follows from Theorem 9.3.6. □

9.4 The Godement–Jacquet L-function

Let

$$f(z) = \sum_{\gamma \in U_{n-1}(\mathbb{Z}) \backslash SL(n-1,\mathbb{Z})} \sum_{m_1=1}^{\infty} \cdots \sum_{m_{n-2}=1}^{\infty} \sum_{m_{n-1}\neq 0} \frac{A(m_1, \ldots, m_{n-1})}{\prod_{k=1}^{n-1} |m_k|^{k(n-k)/2}}$$
$$\times W_{\text{Jacquet}}\left(\begin{pmatrix} m_1 \cdots m_{n-2} \cdot |m_{n-1}| & & & \\ & \ddots & & \\ & & m_1 m_2 & & \\ & & & m_1 & \\ & & & & 1 \end{pmatrix} \cdot \begin{pmatrix} \gamma & \\ & 1 \end{pmatrix} z,\ \nu,\ \psi_{1,\ldots,1,\frac{m_{n-1}}{|m_{n-1}|}}\right),$$

be a non-zero Maass form for $SL(n, \mathbb{Z})$, normalized so that $A(1, \ldots, 1) = 1$, which is a simultaneous eigenfunction of all the Hecke operators as in Theorem 9.3.11. We want to build an L-function out of the Fourier coefficients of f. Lemma 9.1.3 tells us that we may form absolutely convergent Dirichlet series in a suitable half-plane.

The sum over $\gamma \in U_{n-1}(\mathbb{Z})\backslash SL(n-1,\mathbb{Z})$ in the Fourier expansion of f creates seemingly insurmountable complications, and it is not possible to simply set

$$x_{i,j} = 0 \ \ (1 \le i < j \le n),$$

and then take the $n-1$-fold Mellin transform in $y_1, y_2, \ldots, y_{n-1}$ which would be the exact analogue of what we did to create L-functions in the $SL(2,\mathbb{Z})$ situation. The ingenious construction of the L-functions and the proof of their functional equations was first obtained by Godement and Jacquet (1972).

By Theorem 9.3.11, the Fourier coefficients, $A(m_1, \ldots, m_{n-1})$, of f satisfy the multiplicativity relations

$$\begin{aligned} &A(m, 1, \ldots, 1)A(m_1, \ldots, m_{n-1}) \\ &= \sum_{\substack{\prod_{\ell=1}^{n} c_\ell = m \\ c_1|m_1,\ c_2|m_2, \ldots,\ c_{n-1}|m_{n-1}}} A\left(\frac{m_1 c_n}{c_1}, \frac{m_2 c_1}{c_2}, \ldots, \frac{m_{n-1}c_{n-2}}{c_{n-1}}\right). \end{aligned}$$

It follows that for all $k = 1, 2, \ldots$

$$\begin{aligned} A(p^k, 1, \ldots, 1)A(p, 1, \ldots, 1) &= A(p^{k+1}, 1, \ldots, 1) + A(p^{k-1}, p, 1, \ldots, 1), \\ A(p^k, 1, \ldots, 1)A(1, p, 1, \ldots, 1) &= A(p^k, p, 1, \ldots, 1) \\ &\quad + A(p^{k-1}, 1, p, 1, \ldots, 1), \\ A(p^k, 1, \ldots, 1)A(1, 1, p, 1, \ldots, 1) &= A(p^k, 1, p, 1, \ldots, 1) \\ &\quad + A(p^{k-1}, 1, 1, p, 1, \ldots, 1), \\ &\ \ \vdots \\ A(p^k, 1, \ldots, 1)A(1, \ldots, 1, p) &= A(p^k, 1, \ldots, 1, p) + A(p^{k-1}, 1, \ldots, 1), \end{aligned}$$

with the understanding that $A(1, \ldots, p^{-j}, \ldots, 1) = 0$ for any $j \ge 1$.

Therefore,

$$\begin{aligned} &\frac{1}{p^{ks}} \cdot \sum_{r=0}^{n-2} (-1)^r A(p^{k-r}, 1, \ldots, 1) A(\underbrace{1, \ldots, p}_{\text{position } r+1}, \ldots, 1) \\ &= \frac{A(p^{k+1}, 1, \ldots, 1) + (-1)^{n-2} A(p^{k-n+1}, 1, \ldots, 1)}{p^{ks}}. \end{aligned} \tag{9.4.1}$$

If we define

$$\phi_p(s) := \sum_{k=0}^{\infty} \frac{A(p^k, 1, \ldots, 1)}{p^{ks}},$$

then, after summing over k, equation (9.4.1) implies that

$$\phi_p(s) \cdot \left[\sum_{r=0}^{n-2} (-1)^r A(\underbrace{1, \ldots, p}_{\text{position } r+1}, \ldots, 1) p^{-rs} \right]$$
$$= \phi_p(s) p^s - p^s + (-1)^{n-2} \phi_p(s) p^{(-n+1)s}.$$

Solving for $\phi_p(s)$ yields

$$\phi_p(s) = \Big(1 - A(p, \ldots, 1) p^{-s} + A(1, p, \ldots, 1) p^{-2s} - \\ \cdots + (-1)^{n-1} A(1, \ldots, p) p^{(-n+1)s} + (-1)^n p^{-ns} \Big)^{-1}. \qquad (9.4.2)$$

In a manner completely analogous to the situation of $SL(2, \mathbb{Z})$, as in Definition 3.13.3, it is natural make the following definition.

Definition 9.4.3 *Let $s \in \mathbb{C}$ with $\Re(s) > (n + 1)/2$, and let $f(z)$ be a Maass form for $SL(n, \mathbb{Z})$, with $n \geq 2$, which is an eigenfunction of all the Hecke operators as in Theorem 9.3.11. We define the Godement–Jacquet L-function $L_f(s)$ (termed the L-function associated to f) by the absolutely convergent series*

$$L_f(s) = \sum_{m=1}^{\infty} A(m, 1, \ldots, 1) m^{-s} = \prod_p \phi_p(s),$$

with $\phi_p(s)$ given by (9.4.2).

Remark It is clear that the L-function associated to the dual Maass form $\tilde{f}$ takes the form

$$L_{\tilde{f}}(s) = \sum_{m=1}^{\infty} A(1, \ldots, 1, m) m^{-s}.$$

By analogy with the $GL(2)$ situation, we would like to construct the L-function $L_f(s)$ as a Mellin transform of the Maass form f. However, before taking the Mellin transform, it is necessary to kill the sum over $SL(n-1)$ in the Fourier Whittaker expansion (9.1.2). The procedure to do this uses an auxiliary integral which requires some preliminary preparation.

Set

$$M = \begin{pmatrix} m_1 \cdots m_{n-2} \cdot |m_{n-1}| & & & & \\ & \ddots & & & \\ & & m_1 m_2 & & \\ & & & m_1 & \\ & & & & 1 \end{pmatrix},$$

and $z = xy$ with

$$x = \begin{pmatrix} 1 & x_{n-1} & x_{1,3} & \cdots & x_{1,n} \\ & 1 & x_{n-2} & \cdots & x_{2,n} \\ & & \ddots & \ddots & \vdots \\ & & & 1 & x_1 \\ & & & & 1 \end{pmatrix},$$

$$y = \begin{pmatrix} y_1 y_2 \cdots y_{n-1} & & & & \\ & y_1 y_2 \cdots y_{n-2} & & & \\ & & \ddots & & \\ & & & y_1 & \\ & & & & 1 \end{pmatrix}.$$

A simple computation gives

$$M \cdot x = \begin{pmatrix} 1 & |m_{n-1}| x_{n-1} & & & \\ & & & * & \\ & \ddots & \ddots & & \\ & & 1 & m_2 x_2 & \\ & & & 1 & m_1 x_1 \\ & & & & 1 \end{pmatrix} \cdot M.$$

It follows from Definition 5.4.1 (2), that for any integers $\epsilon_1, \ldots, \epsilon_{n-1}$, the Jacquet Whittaker function satisfies

$$W_{\text{Jacquet}}\left(Mz,\ \nu,\ \psi_{\epsilon_1,\ldots,\epsilon_{n-1}}\right) = e^{2\pi i \left[m_1\epsilon_1 x_1 + \cdots + m_{n-2}\epsilon_{n-2}x_{n-2} + |m_{n-1}|\epsilon_{n-1}x_{n-1}\right]} \times W_{\text{Jacquet}}\left(M \cdot y,\ \nu,\ \psi_{\epsilon_1,\ldots,\epsilon_{n-1}}\right).$$

Further, for any $SL(n-1,\mathbb{Z})$ matrix

$$\gamma = \begin{pmatrix} a_{1,1} & \cdots & a_{1,n-1} \\ a_{2,1} & \cdots & a_{2,n-1} \\ \vdots & & \vdots \\ a_{n-1,1} & \cdots & a_{n-1,n-1} \end{pmatrix}, \tag{9.4.4}$$

we may put the $GL(n,\mathbb{R})$ matrix, $\begin{pmatrix} \gamma & \\ & 1 \end{pmatrix} \cdot z$, into Iwasawa form:

$$\begin{pmatrix} \gamma & \\ & 1 \end{pmatrix} \cdot z \equiv \begin{pmatrix} 1 & x_{n-1}^{\gamma} & x_{1,3}^{\gamma} & \cdots & x_{1,n}^{\gamma} \\ & 1 & x_{n-2}^{\gamma} & \cdots & x_{2,n}^{\gamma} \\ & & \ddots & \ddots & \vdots \\ & & & 1 & x_1^{\gamma} \\ & & & & 1 \end{pmatrix} \times \begin{pmatrix} y_1^{\gamma} y_2^{\gamma} \cdots y_{n-1}^{\gamma} & & & & \\ & y_1^{\gamma} y_2^{\gamma} \cdots y_{n-2}^{\gamma} & & & \\ & & \ddots & & \\ & & & y_1^{\gamma} & \\ & & & & 1 \end{pmatrix} \quad (\mathrm{mod}\ Z_n O(n,\mathbb{R})), \tag{9.4.5}$$

where $x_1^{\gamma} = a_{n-1,1}x_{1,n} + a_{n-1,2}x_{2,n} + \cdots + a_{n-1,n-1}x_1$.

It immediately follows from Proposition 5.5.2, (9.4.4) and (9.4.5) that we may write

$$\begin{aligned} W_{\mathrm{Jacquet}}&\left(M \cdot \begin{pmatrix} \gamma & \\ & 1 \end{pmatrix} \cdot z,\ \nu,\ \psi_{\epsilon_1,\ldots,\epsilon_{n-1}}\right) \\ &= e^{2\pi i\left[m_1\epsilon_1(a_{n-1,1}x_{1,n}+a_{n-1,2}x_{2,n}+\cdots+a_{n-1,n-1}x_1)+m_2\epsilon_2 x_2^{\gamma}+\cdots+|m_{n-1}|\epsilon_{n-1}x_{n-1}^{\gamma}\right]} \\ &\quad \times W_{\mathrm{Jacquet}}\left(M \cdot \begin{pmatrix} y_1^{\gamma}\cdots y_{n-1}^{\gamma} & & & \\ & \ddots & & \\ & & y_1^{\gamma} & \\ & & & 1 \end{pmatrix},\ \nu,\ \psi_{1,\ldots,1}\right). \end{aligned} \tag{9.4.6}$$

Finally, we obtain the following theorem which is the basis for the construction of the L-function $L_f(s)$ (given in Definition 9.4.3) as a Mellin transform.

Theorem 9.4.7 *Let $f(z)$ be a Maass form of type ν for $SL(n,\mathbb{Z})$ as in (9.1.2). Then we have the representation,*

$$\begin{aligned} f(z) = \sum_{\gamma\in U_{n-1}(\mathbb{Z})\backslash SL(n-1,\mathbb{Z})} \sum_{m_1=1}^{\infty} \cdots \sum_{m_{n-2}=1}^{\infty} \sum_{m_{n-1}\neq 0} &\frac{A(m_1,\ldots,m_{n-1})}{\prod_{k=1}^{n-1} |m_k|^{k(n-k)/2}} \\ &\times e^{2\pi i m_1(a_{n-1,1}x_{1,n}+\cdots+a_{n-1,n-1}x_1)}\, e^{2\pi i(m_2 x_2^{\gamma}+\cdots+m_{n-1}x_{n-1}^{\gamma})} \\ &\times W_{\mathrm{Jacquet}}(M\cdot y^{\gamma},\ \nu,\ \psi_{1,\ldots,1}), \end{aligned}$$

where γ is given by (9.4.4) and, x^γ, y^γ, are defined by: $\begin{pmatrix} \gamma & \\ & 1 \end{pmatrix} z \equiv x^\gamma \cdot y^\gamma$, *as in (9.4.5).*

Proof The proof follows from the Fourier expansion (9.1.2) and the identity (9.4.6). □

Note that Theorem 9.4.7 is a direct generalization of Theorem 6.5.7. The explicit realization of the Fourier expansion of the Maass form f given in Theorem 9.4.7 is of fundamental importance. It is the basis for the construction of the Godement–Jacquet L-function as a Mellin transform of a certain projection operator acting on the Maass form f.

Corollary 9.4.8 *Let $f(z)$ be a Maass form of type ν for $SL(n,\mathbb{Z})$ as in (9.1.2). Then*

$$\int_0^1 \cdots \int_0^1 f\left(\begin{pmatrix} 1 & 0 & \cdots & 0 & u_{1,n} \\ & \ddots & \ddots & \vdots & \vdots \\ & & 1 & 0 & u_{n-2,n} \\ & & & 1 & u_1 \\ & & & & 1 \end{pmatrix} \cdot z\right) e^{-2\pi i u_1}\, du_1 \prod_{j=1}^{n-2} du_{j,n}$$

$$= \sum_{\gamma \in U_{n-2}(\mathbb{Z})\backslash SL(n-2,\mathbb{Z})} \sum_{m_2=1}^{\infty} \cdots \sum_{m_{n-2}=1}^{\infty} \sum_{m_{n-1}\neq 0}$$

$$\times \frac{A(1, m_2, \ldots, m_{n-1})}{\prod_{k=2}^{n-1} |m_k|^{k(n-k)/2}} e^{2\pi i \left(x_1 + m_2 x_2^\gamma + \cdots + m_{n-1} x_{n-1}^\gamma\right)}$$

$$\times W_{\text{Jacquet}}\left(\begin{pmatrix} m_2 \cdots |m_{n-1}| y_1 y_2^\gamma \cdots y_{n-1}^\gamma & & & \\ & \ddots & & \\ & & m_2 y_1 y_2^\gamma & & \\ & & & y_1 & \\ & & & & 1 \end{pmatrix}, \nu, \psi_{1,\ldots,1}\right).$$

Proof We may use Theorem 9.4.7, to compute the integral:

$$\int_0^1 \cdots \int_0^1 f\left(\begin{pmatrix} 1 & & \cdots & & u_{1,n} \\ & \ddots & & & \vdots \\ & & 1 & & u_{n-2,n} \\ & & & 1 & u_1 \\ & & & & 1 \end{pmatrix} \cdot z\right) e^{-2\pi i u_1}\, du_1 \prod_{j=1}^{n-2} du_{j,n}.$$

The key point is that the integral

$$\int_0^1 \cdots \int_0^1 e^{2\pi i m_1 \left(a_{n-1,1}u_{1,n}+\cdots+a_{n-1,n-2}u_{n-2,n}+a_{n-1,n-1}u_1\right)}\, e^{-2\pi i u_1}\, du_1 \prod_{j=1}^{n-2} du_{j,n}$$

vanishes unless $m_1 a_{n-1,n-1} = 1$ (which implies $m_1 = 1, a_{n-1,n-1} = 1$), and

$$a_{n-1,1} = 0,\ a_{n-1,2} = 0, \qquad \cdots \qquad , a_{n-1,n-2} = 0.$$

The proof of Corollary 9.4.8 follows from this after noting that

$$\gamma = \begin{pmatrix} a_{1,1} & \cdots & a_{1,n-2} & a_{1,n-1} \\ a_{2,1} & \cdots & a_{2,n-2} & a_{2,n-1} \\ \vdots & \ddots & \vdots & \vdots \\ a_{n-2,1} & & a_{n-2,n-2} & a_{n-2,n-1} \\ 0 & \cdots & 0 & 1 \end{pmatrix} \in U_{n-1}(\mathbb{Z})\backslash SL(n-1,\mathbb{Z})$$

forces $a_{1,n-1} = a_{2,n-1} = \cdots = a_{n-2,n-1} = 0$, so that γ takes the form

$$\gamma = \begin{pmatrix} a_{1,1} & \cdots & a_{1,n-2} & 0 \\ a_{2,1} & \cdots & a_{2,n-2} & 0 \\ \vdots & \ddots & \vdots & \vdots \\ a_{n-2,1} & & a_{n-2,n-2} & 0 \\ 0 & \cdots & 0 & 1 \end{pmatrix} \in U_{n-2}(\mathbb{Z})\backslash SL(n-2,\mathbb{Z}).$$

□

Theorem 9.4.9 *Let $f(z)$ be a Maass form of type ν for $SL(n,\mathbb{Z})$ as in (9.1.2). Then for*

$$\hat{u} = \begin{pmatrix} 1 & 0 & u_{1,3} & \cdots & u_{1,n} \\ & 1 & u_{n-2} & \cdots & u_{2,n} \\ & & \ddots & \ddots & \vdots \\ & & & 1 & u_1 \\ & & & & 1 \end{pmatrix}, \qquad d^*\hat{u} = \prod_{j=1}^{n-2} u_j \prod_{2\le i<j-1\le n-1} du_{i,j},$$

we have

$$\int_0^1 \cdots \int_0^1 f(\hat{u}z)e^{-2\pi i(u_1+\cdots+u_{n-2})}\, d^*\hat{u}$$

$$= \sum_{m\neq 0} \frac{A(1,\ldots,1,m)}{|m|^{n-1/2}}\, e^{2\pi i m x_{n-1}}\, e^{2\pi i(x_1+\cdots+x_{n-2})}$$

$$\times W_{\text{Jacquet}}\left(\begin{pmatrix} |m| & & & \\ & 1 & & \\ & & \ddots & \\ & & & 1 \end{pmatrix}\cdot y;\ \nu,\ \psi_{1,\ldots,1}\right).$$

GL(n)pack functions The following **GL(n)pack** functions, described in the appendix, relate to the material in this chapter:

HeckeCoefficientSum HeckeOperator HeckePowerSum
HeckeEigenvalue HeckeMultiplicativeSplit.

10

Langlands Eisenstein series

The modern theory of Eisenstein series began with Maass (1949), who formally defined and studied the series (see Section 3.1)

$$E(z,s) = \frac{1}{2} \sum_{(c,d)=1} \frac{y^s}{|cz+d|^{2s}},$$

taking the viewpoint that it is an eigenfunction of the Laplacian. In (Roelcke, 1956), Eisenstein series for more general discrete groups commensurable with $SL(2,\mathbb{Z})$ were investigated. It was in (Selberg, 1956, 1963) that the spectral theory and the meromorphic continuation of Eisenstein series was fully worked out for $GL(2)$. The Selberg spectral decomposition given in Section 3.16 underscores the supreme importance of Eisenstein series in number theory. In (Selberg, 1960) (see also (Hejhal, 1983)) an extremely ingenious analytic method is introduced for obtaining the meromorphic continuation of Eisenstein series for higher rank groups, but it was not clear if the method would work for Langlands Eisenstein series twisted by Maass forms defined on lower rank groups.

The completion of this program in, perhaps, the most general context was attained by (Langlands, 1966, 1976). There were two main parts to Langlands theory.

- *The meromorphic continuation of Eisenstein series.*
- *The complete spectral decomposition of arithmetic quotients* $\Gamma\backslash G$ *where* G *is a reductive group and* Γ *is an arithmetic group.*

An excellent summary of Langlands theory of Eisenstein series was given in (Arthur, 1979). In the intervening years, two books have been published giving expositions of Langlands theory of Eisenstein series: (Osborne and Warner, 1981) and (Moeglin and Waldspurger, 1995) (see also (Jacquet, 1997)). A few years ago, a new proof of the meromorphic continuation of Langlands Eisenstein series has been attained by Bernstein (2002).

Another important direction in the theory of Eisenstein series was also initiated by Langlands who had the idea of studying automorphic L-functions by investigating the constant term (see (Langlands, 1971)) in the Fourier expansion of Eisenstein series. This theme was further developed in (Shahidi, 1981, 1988, 1990a, b, 1992), and is now called the Langlands–Shahidi method. This method has had a number of striking successes, one of the first being (Moeglin and Waldspurger, 1989) that the completed Rankin–Selberg L-function for $GL(n) \times GL(m)$ is holomorphic in the region $0 < \Re(s) < 1$. More recently, Gelbart and Shahidi (1988), Shahidi (1985, 1990a,b), Cogdell, Kim, Piatetski-Shapiro and Shahidi (2001), and Kim and Shahidi (2000) have led to many new examples of entire L-functions including the symmetric cube and fourth power lifts of $GL(2)$ Maass forms. The remarkable fact is that these particular symmetric power L-functions occur in the constant term of Eisenstein series which are associated to exceptional Lie groups! Unfortunately, since there are only a few exceptional Lie groups this puts a severe constraint on what one can expect to get by this method.

In this chapter we shall give an elementary exposition of Langlands Eisenstein series for the group $SL(n, \mathbb{Z})$. We only discuss meromorphic continuation that can be obtained from Fourier–Whittaker expansions or the Poisson summation formula. These methods work quite well for Eisenstein series that are *not* twisted by Maass forms of lower rank. We present a short elementary introduction to the Langlands–Shahidi method with an application to non-vanishing of L-functions on the line $\Re(s) = 1$. Finally, in Section 10.13, we give a simple proof (due to M. Thillainatesan) of the Langlands spectral decomposition for $GL(3, \mathbb{R})$.

10.1 Parabolic subgroups

We shall define the standard parabolic subgroups for $GL(n, \mathbb{R})$ (with $n \geq 2$) in an explicit manner, avoiding the more general abstract theory. Briefly, the standard parabolic subgroups are certain subgroups containing the standard Borel subgroup B which consists of all upper triangular non-singular matrices.

Each standard parabolic subgroup of $GL(n, \mathbb{R})$ is associated to a partition

$$n = n_1 + n_2 + \cdots + n_r,$$

where $1 \leq n_1, n_2, \ldots, n_r < n$ are integers.

Definition 10.1.1 *The standard parabolic subgroup associated to the partition $n = n_1 + n_2 + \cdots + n_r$ is denoted $P_{n_1,\dots,n_r}$, and is defined to be the group of all matrices of the form*

$$\begin{pmatrix} \mathfrak{m}_{n_1} & * & \cdots & * \\ 0 & \mathfrak{m}_{n_2} & \cdots & * \\ 0 & 0 & \cdots & * \\ \vdots & \vdots & \ddots & \vdots \\ 0 & 0 & \cdots & \mathfrak{m}_{n_r} \end{pmatrix},$$

where $\mathfrak{m}_{n_i} \in GL(n_i, \mathbb{R})$ for $1 \le i \le r$. The integer r is termed the rank of the parabolic subgroup $P_{n_1,\dots,n_r}$.

Example 10.1.2 (Parabolic subgroups of $GL(3, \mathbb{R})$) There are three standard parabolic subgroups of $GL(3, \mathbb{R})$ corresponding to the three partitions:

$$3 = 1 + 1 + 1, \qquad 3 = 1 + 2, \qquad 3 = 2 + 1.$$

Explicitly, we have

$$P_{1,1,1} = \left\{ \begin{pmatrix} * & * & * \\ 0 & * & * \\ 0 & 0 & * \end{pmatrix} \right\}, \qquad P_{1,2} = \left\{ \begin{pmatrix} * & * & * \\ 0 & * & * \\ 0 & * & * \end{pmatrix} \right\},$$

$$P_{2,1} = \left\{ \begin{pmatrix} * & * & * \\ * & * & * \\ 0 & 0 & * \end{pmatrix} \right\}.$$

Example 10.1.3 (Parabolic subgroups of $GL(4, \mathbb{R})$) There are seven standard parabolic subgroups of $GL(4, \mathbb{R})$ corresponding to the seven partitions:

$$4 = 1 + 1 + 1 + 1, \qquad 4 = 1 + 1 + 2, \qquad 4 = 1 + 2 + 1,$$
$$4 = 2 + 1 + 1, \qquad 4 = 1 + 3, \qquad 4 = 2 + 2, \qquad 4 = 3 + 1.$$

Explicitly, we have

$$P_{1,1,1,1} = \left\{ \begin{pmatrix} * & * & * & * \\ 0 & * & * & * \\ 0 & 0 & * & * \\ 0 & 0 & 0 & * \end{pmatrix} \right\}, \qquad P_{1,1,2} = \left\{ \begin{pmatrix} * & * & * & * \\ 0 & * & * & * \\ 0 & 0 & * & * \\ 0 & 0 & * & * \end{pmatrix} \right\},$$

$$P_{1,2,1} = \left\{ \begin{pmatrix} * & * & * & * \\ 0 & * & * & * \\ 0 & * & * & * \\ 0 & 0 & 0 & * \end{pmatrix} \right\}, \qquad P_{2,1,1} = \left\{ \begin{pmatrix} * & * & * & * \\ * & * & * & * \\ 0 & 0 & * & * \\ 0 & 0 & 0 & * \end{pmatrix} \right\},$$

$$P_{1,3} = \left\{ \begin{pmatrix} * & * & * & * \\ 0 & * & * & * \\ 0 & * & * & * \\ 0 & * & * & * \end{pmatrix} \right\}, \qquad P_{3,1} = \left\{ \begin{pmatrix} * & * & * & * \\ * & * & * & * \\ * & * & * & * \\ 0 & 0 & 0 & * \end{pmatrix} \right\},$$

$$P_{2,2} = \left\{ \begin{pmatrix} * & * & * & * \\ * & * & * & * \\ 0 & 0 & * & * \\ 0 & 0 & * & * \end{pmatrix} \right\}.$$

Definition 10.1.4 (Associate parabolics) *Fix integers $n \geq 2$, $1 < r \leq n$. Two standard parabolic subgroups $P_{n_1,\dots,n_r}$, $P_{n'_1,\dots,P_{n'_r}}$ of $GL(n, \mathbb{R})$, corresponding to the partitions,*

$$n = n_1 + \cdots + n_r = n'_1 + \cdots + n'_r,$$

are said to be associate ($P_{n_1,\dots,n_r} \sim P_{n'_1,\dots,n'_r}$) if the set of integers $\{n_1, \dots, n_r\}$ is a permutation of the set of integers $\{n'_1, \dots, n'_r\}$.

Definition 10.1.5 (Weyl group of associate parabolics) *Fix $n \geq 2$ and $1 < r \leq n$. Let $P = P_{n_1,\dots,n_r}$, $P' = P_{n'_1,\dots,n'_r}$ be two associate parabolic subgroups of $GL(n, \mathbb{R})$ corresponding to the partitions,*

$$n = n_1 + \cdots + n_r = n'_1 + \cdots + n'_r.$$

The Weyl group, denoted $\Omega(P, P')$, consists of all $\sigma \in S_r$ (permutation group on r symbols) such that $n'_i = n_{\sigma(i)}$ for all $i = 1, 2, \dots, r$. We shall also let $\Omega(P)$ denote $\Omega(P, P)$.

10.2 Langlands decomposition of parabolic subgroups

Let $n \geq 2$, and fix a partition $n = n_1 + n_2 + \cdots + n_r$ with $1 \leq n_1, n_2, \dots, n_r < n$. The parabolic subgroup (see Definition 10.1.1)

$$P_{n_1,\dots,n_r} = \left\{ \begin{pmatrix} \mathfrak{m}_{n_1} & * & \cdots & * \\ 0 & \mathfrak{m}_{n_2} & \cdots & * \\ 0 & 0 & \cdots & * \\ \vdots & \vdots & \ddots & \vdots \\ 0 & 0 & \cdots & \mathfrak{m}_{n_r} \end{pmatrix} \right\} \tag{10.2.1}$$

of $GL(n, \mathbb{R})$ can be factored as

$$P_{n_1,\dots,n_r} = N_{n_1,\dots,n_r} \cdot M_{n_1,\dots,n_r} \tag{10.2.2}$$

where

$$N_{n_1,\ldots,n_r} = \left\{ \begin{pmatrix} I_{n_1} & * & \cdots & * \\ 0 & I_{n_2} & \cdots & * \\ 0 & 0 & \cdots & * \\ \vdots & \vdots & \ddots & \vdots \\ 0 & 0 & \cdots & I_{n_r} \end{pmatrix} \right\} \qquad (I_k = k \times k \text{ Identity matrix}),$$

is the unipotent radical and

$$M_{n_1,\ldots,n_r} = \left\{ \begin{pmatrix} \mathfrak{m}_{n_1} & 0 & \cdots & 0 \\ 0 & \mathfrak{m}_{n_2} & \cdots & 0 \\ 0 & 0 & \cdots & 0 \\ \vdots & \vdots & \ddots & \vdots \\ 0 & 0 & \cdots & \mathfrak{m}_{n_r} \end{pmatrix} \right\} \qquad (\mathfrak{m}_k \in GL(k, \mathbb{R})),$$

is the so-called Levi component. The Levi component further decomposes into the direct product

$$M_{n_1,\ldots,n_r} = A_{n_1,\ldots,n_r} \cdot M'_{n_1,\ldots,n_r} \tag{10.2.3}$$

where $A_{n_1,\ldots,n_r}$ is the connected center of $M_{n_1,\ldots,n_r}$:

$$A_{n_1,\ldots,n_r} = \left\{ \begin{pmatrix} t_1 \cdot I_{n_1} & 0 & \cdots & 0 \\ 0 & t_2 \cdot I_{n_2} & \cdots & 0 \\ 0 & 0 & \cdots & 0 \\ \vdots & \vdots & \ddots & \vdots \\ 0 & 0 & \cdots & t_r \cdot I_{n_r} \end{pmatrix} \right\} \qquad (t_i \in \mathbb{R},\ t_i > 0),$$

and

$$M'_{n_1,\ldots,n_r} = \left\{ \begin{pmatrix} \mathfrak{m}'_{n_1} & 0 & \cdots & 0 \\ 0 & \mathfrak{m}'_{n_2} & \cdots & 0 \\ 0 & 0 & \cdots & 0 \\ \vdots & \vdots & \ddots & \vdots \\ 0 & 0 & \cdots & \mathfrak{m}'_{n_r} \end{pmatrix} \right\} \qquad \left(\det\left(\mathfrak{m}'_{n_i}\right) = \pm 1,\ i = 1, \ldots, r\right).$$

Definition 10.2.4 (Langlands decomposition) *The Langlands decomposition of a parabolic subgroup of the form (10.2.1) gives the factorizations (10.2.2), (10.2.3).*

Example 10.2.5 (Langlands decomposition on $GL(3, \mathbb{R})$) We explicitly write down the Langlands decomposition for the three parabolic subgroups of $GL(3,\mathbb{R})$.

$$P_{1,1,1}=\left\{\begin{pmatrix} * & * & * \\ 0 & * & * \\ 0 & 0 & * \end{pmatrix}\right\}$$
$$=\left\{\begin{pmatrix} 1 & * & * \\ 0 & 1 & * \\ 0 & 0 & 1 \end{pmatrix}\cdot\begin{pmatrix} t_1 & 0 & 0 \\ 0 & t_2 & 0 \\ 0 & 0 & t_3 \end{pmatrix}\cdot\begin{pmatrix} \pm1 & 0 & 0 \\ 0 & \pm1 & 0 \\ 0 & 0 & \pm1 \end{pmatrix}\right\},$$
$$(t_1,t_2,t_3>0)$$

$$P_{1,2}=\left\{\begin{pmatrix} * & * & * \\ 0 & * & * \\ 0 & * & * \end{pmatrix}\right\}$$
$$=\left\{\begin{pmatrix} 1 & * & * \\ 0 & 1 & 0 \\ 0 & 0 & 1 \end{pmatrix}\cdot\begin{pmatrix} t_1 & 0 & 0 \\ 0 & t_2 & 0 \\ 0 & 0 & t_2 \end{pmatrix}\cdot\begin{pmatrix} \pm1 & 0 & 0 \\ 0 & a & b \\ 0 & c & d \end{pmatrix}\right\},$$
$$(t_1,t_2>0,\ ad-bc=\pm1)$$

$$P_{2,1}=\left\{\begin{pmatrix} * & * & * \\ * & * & * \\ 0 & 0 & * \end{pmatrix}\right\}$$
$$=\left\{\begin{pmatrix} 1 & 0 & * \\ 0 & 1 & * \\ 0 & 0 & 1 \end{pmatrix}\cdot\begin{pmatrix} t_1 & 0 & 0 \\ 0 & t_1 & 0 \\ 0 & 0 & t_2 \end{pmatrix}\cdot\begin{pmatrix} a & b & 0 \\ c & d & 0 \\ 0 & 0 & \pm1 \end{pmatrix}\right\},$$
$$(t_1,t_2>0,\ ad-bc=\pm1).$$

Example 10.2.6 (Langlands decomposition for $P_{1,3,2}$) Finally, we shall give an example of the Langlands decomposition for the parabolic subgroup $P_{1,3,2}$ of $GL(6,\mathbb{R})$. We have

$$P_{1,3,2}=\left\{\begin{pmatrix} * & * & * & * & * & * \\ & * & * & * & * & * \\ & * & * & * & * & * \\ & * & * & * & * & * \\ & & & & * & * \\ & & & & * & * \end{pmatrix}\right\}$$
$$=\left\{\underbrace{\begin{pmatrix} 1 & * & * & * & * & * \\ & 1 & 0 & 0 & * & * \\ & 0 & 1 & 0 & * & * \\ & 0 & 0 & 1 & * & * \\ & & & & 1 & 0 \\ & & & & 0 & 1 \end{pmatrix}}_{\text{unipotent radical}}\right\}\cdot\left\{\underbrace{\begin{pmatrix} * & & & & & \\ & * & * & * & & \\ & * & * & * & & \\ & * & * & * & & \\ & & & & * & * \\ & & & & * & * \end{pmatrix}}_{\text{Levi component}}\right\},$$

and the further decomposition of the Levi component

$$\underbrace{\left\{\begin{pmatrix} * & & & & & \\ & * & * & * & & \\ & * & * & * & & \\ & * & * & * & & \\ & & & & * & * \\ & & & & * & * \end{pmatrix}\right\}}_{\text{Levi component}}$$

$$= \underbrace{\left\{\begin{pmatrix} t_1 & & & & & \\ & t_2 & 0 & 0 & & \\ & 0 & t_2 & 0 & & \\ & 0 & 0 & t_2 & & \\ & & & & t_3 & 0 \\ & & & & 0 & t_3 \end{pmatrix}\right\}}_{A_{1,3,2}} \cdot \underbrace{\left\{\begin{pmatrix} * & & & & & \\ & * & * & * & & \\ & * & * & * & & \\ & * & * & * & & \\ & & & & * & * \\ & & & & * & * \end{pmatrix}\right\}}_{M'_{1,3,2}},$$

where $t_1, t_2, t_3 > 0$ and the block matrices in $M'_{1,3,2}$ have determinant ± 1.

The parabolic subgroups of $GL(n, \mathbb{R})$ can be characterized as stabilizers of flags on $\mathbb{R}^n$. A flag of $\mathbb{R}^n$ is a sequence of subspaces:

$$\phi \subset V_1 \subset V_2 \cdots \subset V_r = \mathbb{R}^n$$

where $\subset$ denotes a proper subset, and ϕ is the empty set. The action of $GL(n, \mathbb{R})$ on a flag $(V_1, \ldots, V_r)$ is defined in the canonical way. That is if $g \in GL(n, \mathbb{R})$ then the action is given by $g(V_1, \ldots, V_r) = (gV_1, \ldots, gV_r)$, where the action of g on an element $(a_1, \ldots, a_n) \in \mathbb{R}^n$ is given by matrix multiplication $g \cdot {}^t(a_1, \ldots, a_n)$. The standard complete flag is $\phi \subset V_1 \subset \cdots \subset V_n = \mathbb{R}^n$ where $V_i = \mathbb{R}e_1 \oplus \cdots \oplus \mathbb{R}e_i$ and e_i is the vector (of length n) with a 1 in the ith position and zeros elsewhere. The stabilizer of a subflag $(V_{d_1}, \ldots, V_{d_r})$ (with $0 < d_1 < \cdots < d_r = n$) of the standard flag has the form

$$\begin{pmatrix} \mathfrak{m}_{11} & \mathfrak{m}_{12} & \mathfrak{m}_{13} & \cdots \\ 0 & \mathfrak{m}_{22} & \mathfrak{m}_{23} & \cdots \\ 0 & 0 & \mathfrak{m}_{33} & \cdots \\ \vdots & \vdots & \vdots & \ddots \end{pmatrix}, \tag{10.2.7}$$

where $\mathfrak{m}_{ii}$ is a square matrix of size $d_i - d_{i-1}$ for $i = 1, 2, \ldots, r$, where, by convention, $d_0 = 0$.

If a parabolic subgroup $P \leq GL(n, \mathbb{R})$ is the stabilizer of a flag $(V_1, \ldots, V_r)$, then for every $g \in P$, we must have $gV_i = V_i$ for $i = 1, \ldots, r$. It follows that g induces an automorphism g_i of V_i/V_{i-1} for every $1 \leq i \leq r$. Here, we set

V_0 to be the empty set. In the case that P takes the form (10.2.7), we have $g_i = \mathfrak{m}_{ii}$. The unipotent radical N_P of P is the subgroup of all $g \in P$ so that g_i is the identity on V_i/V_{i-1} for every i. For P of the form (10.2.7), $\mathfrak{m}_{ii}$ must be the identity matrix for every i. In a similar manner we may define the Levi component using flags. For each $1 \le i \le r$ choose a complementary subspace $X_i \subset \mathbb{R}^n$ so that $V_i = V_{i-1} \oplus X_i$. The Levi component M_P of P is defined to be the subgroup of P consisting of all $g \in P$ which stablize each X_i. In the case that P is of the form (10.2.7), the Levi component requires that $\mathfrak{m}_{ij} = 0$ for $1 \le i < j \le n$.

10.3 Bruhat decomposition

Let S_n denote the symmetric group of all permutations of n symbols. We have a homomorphism of S_n into $GL(n, \mathbb{R})$ whose image is the Weyl group W_n consisting of all $n \times n$ matrices which have exactly one 1 in each row and column, and zeros elsewhere.

Example 10.3.1 (Weyl group for $GL(3, \mathbb{R})$) The Weyl group W_3 is the group of six elements:

$$\begin{pmatrix} 1 & 0 & 0 \\ 0 & 1 & 0 \\ 0 & 0 & 1 \end{pmatrix}, \quad \begin{pmatrix} 1 & 0 & 0 \\ 0 & 0 & 1 \\ 0 & 1 & 0 \end{pmatrix}, \quad \begin{pmatrix} 0 & 1 & 0 \\ 1 & 0 & 0 \\ 0 & 0 & 1 \end{pmatrix},$$

$$\begin{pmatrix} 0 & 1 & 0 \\ 0 & 0 & 1 \\ 1 & 0 & 0 \end{pmatrix}, \quad \begin{pmatrix} 0 & 0 & 1 \\ 1 & 0 & 0 \\ 0 & 1 & 0 \end{pmatrix}, \quad \begin{pmatrix} 0 & 0 & 1 \\ 0 & 1 & 0 \\ 1 & 0 & 0 \end{pmatrix}.$$

Recall that the standard Borel subgroup B_n of $GL(n, \mathbb{R})$ is the group of invertible upper triangular matrices.

Proposition 10.3.2 (Bruhat decomposition) *For $n \ge 2$, we have*

$$GL(n, \mathbb{R}) = B_n W_n B_n.$$

Proof Let

$$g = \begin{pmatrix} g_{11} & g_{12} & \cdots & g_{1n} \\ g_{21} & g_{22} & \cdots & g_{2n} \\ \vdots & \vdots & \cdots & \vdots \\ g_{n1} & g_{n2} & \cdots & g_{nn} \end{pmatrix} \in GL(n, \mathbb{R}).$$

Let $g_{n\ell}$ denote the first non-zero entry in the bottom row of g. Then by right multiplication by some $b_1 \in B_n$, we can change this entry to 1 and make the rest of the bottom row 0. The matrix gb_1 now takes the form

$$gb_1 = \begin{pmatrix} g'_{11} & \cdots & g'_{1\ell} & \cdots & g'_{1n} \\ g'_{21} & \cdots & g'_{2\ell} & \cdots & g'_{2n} \\ \vdots & \cdots & \vdots & \cdots & \vdots \\ g'_{n-1,1} & \cdots & g'_{n-1,\ell} & \cdots & g'_{n-1,n} \\ 0 & \cdots & 1 & \cdots & 0 \end{pmatrix}.$$

If we now multiply gb_1 on the left by some matrix

$$b'_1 = \begin{pmatrix} b'_{11} & b'_{12} & b'_{13} & \cdots & b'_{1n} \\ 0 & b'_{22} & b'_{23} & \cdots & b'_{2n} \\ 0 & 0 & b'_{33} & \cdots & b'_{3n} \\ \vdots & \vdots & \vdots & \ddots & \vdots \\ 0 & 0 & 0 & \cdots & 1 \end{pmatrix} \in B_n,$$

it is easy to see that we may choose the $b'_{i,j}$ $(1 \le i < j \le n)$ so that $b'_1 g b_1$ takes the form

$$b'_1 g b_1 = \begin{pmatrix} g'_{11} & \cdots & 0 & \cdots & g'_{1n} \\ g'_{21} & \cdots & 0 & \cdots & g'_{2n} \\ \vdots & \cdots & \vdots & \cdots & \vdots \\ g'_{n-1,1} & \cdots & 0 & \cdots & g'_{n-1,n} \\ 0 & \cdots & 1 & \cdots & 0 \end{pmatrix}.$$

We call this clearing the (n)th row and (ℓ)th column.

We next consider the first non-zero entry in the $(n-1)$st row of $b'_1 g b_1$. Suppose it is g'_{n-1,ℓ_1}. We may again multiply $b'_1 g b_1$ on the right by some element $b_2 \in B_n$ so that we change this entry to 1 and make all other entries in the $(n-1)$st row 0. By left multiplication by some b'_2 we can make all the other entries in the (ℓ_1)st column 0 which results in clearing the $(n-1)$st row and (ℓ_1)st column.

Continuing in this manner, we obtain a set of n matrices

$$b'_1 g b_1, \quad b'_2 b'_1 g b_1 b_2, \quad b'_3 b'_2 b'_1 g b_1 b_2 b_3, \quad \ldots, \quad b'_n \cdots b'_3 b'_2 b'_1 g b_1 b_2 b_3 \cdots b_n$$

where the last entry must lie in W_n. □

We now seek a more explicit realization of the Bruhat decomposition. We follow (Friedberg, 1987a,c). It is necessary to introduce some more notation. Fix an integer $n \ge 2$. For every $\lambda = (\ell_1, \ldots, \ell_k) \in \mathbb{Z}^k$ (with $1 \le k \le n$) and

$g \in GL(n, \mathbb{R})$, define $M_\lambda(g)$ to be the $k \times k$ minor of g formed from the bottom k rows and the columns $\ell_1, \ldots, \ell_k$, indexed by the elements of λ. We may express $M_\lambda(g)$ using wedge products of $e_1, e_2, \ldots, e_n$, where e_i denotes the column vector of length n with a 1 at position i and zeros elsewhere. We have

$$M_\lambda(g)e_1 \wedge \cdots \wedge e_n = e_1 \wedge \cdots \wedge e_{n-k} \wedge \left(g \cdot e_{\ell_1}\right) \wedge \cdots \wedge \left(g \cdot e_{\ell_k}\right). \tag{10.3.3}$$

We shall also define

$$U_n = \begin{pmatrix} 1 & & * \\ & \ddots & \\ 0 & & 1 \end{pmatrix} \subset B_n, \tag{10.3.4}$$

to be the subgroup of upper triangular unipotent (1s on the diagonal) matrices.

Definition 10.3.5 *Let $w \in W_n$. We define $\omega \in S_n$ to be the permutation of the set $\{1, 2, \ldots, n\}$ associated to w, and defined by $we_i = e_{\omega^{-1}(i)}$ for all $1 \leq i \leq n$.*

Proposition 10.3.6 (Explicit Bruhat decomposition) *Every $g \in GL(n, \mathbb{R})$ $(n \geq 2)$ has a Bruhat decompostion $g = u_1 c w u_2$ with $u_1, u_2 \in U_n$, $w \in W_n$, and*

$$c = \begin{pmatrix} \epsilon/c_{n-1} & & & & \\ & c_{n-1}/c_{n-2} & & & \\ & & \ddots & & \\ & & & c_2/c_1 & \\ & & & & c_1 \end{pmatrix},$$

$$\epsilon = \det(w)\det(g), \quad c_i \neq 0 \, (1 \leq i < n).$$

Furthermore, for each $1 \leq i \leq n-1$,

$$|c_i| = \left|M_{(\omega(n), \omega(n-1), \ldots, \omega(n-i+1))}(g)\right|,$$

with ω as in Definition 10.3.5, and $M_{(\omega(n), \omega(n-1), \ldots, \omega(n-i+1))}(g)$ defined by (10.3.3).

Proof It easily follows from Proposition 10.3.2 that every $g \in GL(n, \mathbb{R})$ has a Bruhat decompostion $g = u_1 c w u_2$ with $u_1, u_2 \in U_n$, $w \in W_n$, and c a diagonal matrix. To determine c, we utilize (10.3.3) and write

$$M_\lambda(g)e_1 \wedge \cdots \wedge e_n = e_1 \wedge \cdots \wedge e_{n-i} \wedge ge_{\omega(n)} \wedge \cdots \wedge ge_{\omega(n-i+1)},$$

with $\lambda = (\omega(n), \omega(n-1), \ldots, \omega(n-i+1))$. Since $ue_i - e_i \in \text{Span}(e_1, \ldots, e_{i-1})$ for any $u \in U_n$, and $e_k \wedge e_k = 0$ for any $1 \leq k \leq n$, it follows that

$$M_\lambda(g) e_1 \wedge \cdots \wedge e_n = e_1 \wedge \cdots \wedge e_{n-i} \wedge cwu_2 \cdot e_{\omega(n)} \wedge \cdots \wedge cwu_2 \cdot e_{\omega(n-i+1)}. \tag{10.3.7}$$

We now write $u_2 = (\mu_{i,j})$ (where $\mu_{i,j}$ denotes the i, j entry of the matrix u_2). Note that for any $1 \leq \ell \leq n$,

$$u_2 e_\ell = \sum_{r=1}^{\ell} u_{r,\ell} e_r.$$

It follows from this and (10.3.7) that

$$\begin{aligned}
M_\lambda(g) e_1 \wedge \cdots \wedge e_n &= e_1 \wedge \cdots \wedge e_{n-i} \wedge \left(cw \sum_{r_1=1}^{\omega(n)} u_{r_1,\omega(n)} e_{r_1} \right) \\
&\quad \wedge \left(cw \sum_{r_2=1}^{\omega(n-1)} u_{r_2,\omega(n-1)} e_{r_2} \right) \wedge \cdots \wedge \left(cw \sum_{r_i=1}^{\omega(n-i+1)} u_{r_i,\omega(n-i+1)} e_{r_i} \right) \\
&= e_1 \wedge \cdots \wedge e_{n-i} \wedge \left(c \sum_{r_1=1}^{\omega(n)} u_{r_1,\omega(n)} e_{\omega^{-1}(r_1)} \right) \\
&\quad \wedge \left(c \sum_{r_2=1}^{\omega(n-1)} u_{r_2,\omega(n-1)} e_{\omega^{-1}(r_2)} \right) \wedge \cdots \wedge \left(c \sum_{r_i=1}^{\omega(n-i+1)} u_{r_i,\omega(n-i+1)} e_{\omega^{-1}(r_i)} \right) \\
&= e_1 \wedge \cdots \wedge e_{n-i} \wedge ce_n \wedge ce_{n-1} \wedge \cdots \wedge ce_{n-i+1} \\
&= c_i e_1 \wedge e_2 \wedge \cdots \wedge e_{n-i} \wedge e_n \wedge \cdots \wedge e_{n-i+1}.
\end{aligned}$$

□

10.4 Minimal, maximal, and general parabolic Eisenstein series

The minimal (smallest) standard parabolic subgroup for $GL(n, \mathbb{R})$ $(n \geq 2)$ is

$$\underbrace{P_{1,1,\ldots,1}}_{n \text{ ones}} = \begin{pmatrix} * & * & \cdots & * \\ & * & \cdots & * \\ & & \ddots & \vdots \\ & & & * \end{pmatrix}.$$

Set $\Gamma = SL(n, \mathbb{Z})$ and $P_{\min} = P_{1,1,\ldots,1} \cap \Gamma$.

Let $z \in \mathfrak{h}^n$ take the form

$$z = \begin{pmatrix} 1 & x_{1,2} & x_{1,3} & \cdots & x_{1,n} \\ & 1 & x_{2,3} & \cdots & x_{2,n} \\ & & \ddots & & \vdots \\ & & & 1 & x_{n-1,n} \\ & & & & 1 \end{pmatrix} \cdot \begin{pmatrix} y_1 y_2 \cdots y_{n-1} & & & & \\ & y_1 y_2 \cdots y_{n-2} & & & \\ & & \ddots & & \\ & & & y_1 & \\ & & & & 1 \end{pmatrix}.$$

Then the function (see also (5.1.1))

$$I_s(z) = \prod_{i=1}^{n-1} \prod_{j=1}^{n-1} y_i^{b_{i,j} s_j}$$

with

$$b_{i,j} = \begin{cases} ij & \text{if } i + j \leq n, \\ (n-i)(n-j) & \text{if } i + j \geq n, \end{cases}$$

and $s \in \mathbb{C}^{n-1}$ is invariant under transformations of the form

$$z \mapsto p \cdot z$$

with $p \in P_{\min}$. It follows that the sum

$$E_{P_{\min}}(z, s) := \sum_{\gamma \in P_{\min} \backslash \Gamma} I_s(\gamma z) \tag{10.4.1}$$

is well defined provided it converges absolutely.

Definition 10.4.2 *The series (10.4.1) is called the minimal parabolic Eisenstein series for* Γ.

Proposition 10.4.3 *The minimal parabolic Eisenstein series (10.4.1) converges absolutely and uniformly on compact subsets of* $\mathfrak{h}^n$ *to a* Γ *invariant function provided* $s = (s_1, \ldots, s_{n-1})$ *and* $\text{Re}(s_i)$ *is sufficiently large for every* $i = 1, 2, \ldots, n-1$.

Proof The fact that $E_{P_{\min}}(z, s)$ is invariant under Γ is easy to prove because the function is formed as a sum over a coset of the group Γ. For the absolute convergence, we follow Godement (see (Borel, 1966)). It is enough to show that for every point $z_0 \in \Gamma\backslash\mathfrak{h}^n$ and some (non-zero volume) compact subset C_{z_0} of $\Gamma\backslash\mathfrak{h}^n$ (with $z_0 \in C_{z_0}$), that the integral

$$\int_{C_{z_0}} \left| E_{P_{\min}}(z, s) \right| d^*z$$

converges. Here d^*z is the invariant measure given in Theorem 1.6.1. Without loss of generality, we may assume the s_i to be real. It follows that it is enough to show that the integral

$$\int_{C_{z_0}} \sum_{\gamma \in P_{\min}\backslash\Gamma} I_s(\gamma z)\, d^*z = \int_{(P_{\min}\backslash\Gamma)\cdot C_{z_0}} I_s(z)\, d^*z$$

converges. Now, it follows from Proposition 1.3.2 that there will be only finitely many $\gamma \in P_{\min}\backslash\Gamma$ such that $\gamma z_0 \in \Sigma_{\frac{\sqrt{3}}{2},\frac{1}{2}}$. By a continuity argument, one may deduce, for sufficiently small C_{z_0}, that there are only finitely many $\gamma \in P_{\min}\backslash\Gamma$ such that $\gamma z \in \Sigma_{\frac{\sqrt{3}}{2},\frac{1}{2}}$ for all $z \in C_{z_0}$. We immediately deduce that there exists some $a \geq \sqrt{3}/2$ such that

$$\gamma z \notin \Sigma_{a,\frac{1}{2}}$$

for all $\gamma \in P_{\min}\backslash\Gamma,\ z \in C_{z_0}$. It follows that

$$\int_{(P_{\min}\backslash\Gamma)\cdot C_{z_0}} I_s(z)\, d^*z \leq \int_0^1 \cdots \int_0^1 \int_0^a \cdots \int_0^a \prod_{i=1}^{n-1}\prod_{j=1}^{n-1} y_i^{b_{i,j}s_j} \prod_{1\leq i<j\leq n} dx_{i,j}$$
$$\times \prod_{k=1}^{n-1} y_k^{-k(n-k)-1}\, dy_k,$$

where the integrals from 0 to 1 are integrals for the variables $x_{i,j}(1 \leq i < j \leq n)$ and the integrals from 0 to a are integrals with respect to the variables $y_1, y_2, \ldots, y_{n-1}$. It is clear that the latter integral converges absolutely if the s_i are all sufficiently large. □

The two largest (maximal) parabolic subgroups of $\Gamma = SL(n,\mathbb{Z})$ are

$$P_{1,n-1} = \begin{pmatrix} * & * & \cdots & * \\ 0 & * & \cdots & * \\ \vdots & \vdots & \cdots & \vdots \\ 0 & * & \cdots & * \end{pmatrix}, \qquad P_{n-1,1} = \begin{pmatrix} * & \cdots & * & * \\ \vdots & \cdots & \vdots & \vdots \\ * & \cdots & * & * \\ 0 & \cdots & 0 & * \end{pmatrix}.$$

In a manner similar to the way we defined the minimal parabolic Eisenstein series (10.4.1), we would like to define maximal parabolic Eisenstein series, and more generally, Eisenstein series associated to any standard parabolic subgroup P of $GL(n,\mathbb{R})$. Since the function $I_s(z)$ is not usually invariant under translations $z \mapsto \gamma z$ (with $\gamma \in P \cap \Gamma$), we cannot simply sum over all left translations $\gamma \in P \cap \Gamma\backslash\Gamma$ as we did in (10.4.1). Nevertheless, we will show that for proper choice of s we can make $I_s(z)$ invariant under left multiplication $z \mapsto \gamma z$ with $\gamma \in P \cap \Gamma$.

The construction of such a P-invariant function $I_s(z)$ requires some preliminary preparation. We shall make use of the Langlands decomposition

$$P = NM = \left\{ \begin{pmatrix} I_{n_1} & * & \cdots & * \\ 0 & I_{n_2} & \cdots & * \\ 0 & 0 & \cdots & * \\ \vdots & \vdots & \ddots & \vdots \\ 0 & 0 & \cdots & I_{n_r} \end{pmatrix} \cdot \begin{pmatrix} \mathfrak{m}_{n_1} & 0 & \cdots & 0 \\ 0 & \mathfrak{m}_{n_2} & \cdots & 0 \\ 0 & 0 & \cdots & 0 \\ \vdots & \vdots & \ddots & \vdots \\ 0 & 0 & \cdots & \mathfrak{m}_{n_r} \end{pmatrix} \right\} \tag{10.4.4}$$

where $\mathfrak{m}_k \in GL(k, \mathbb{R})$.

Definition 10.4.5 *Let $P_{n_1,\ldots,n_r}$ be a parabolic subgroup of $GL(n, \mathbb{R})$, associated to the partition: $n = n_1 + \cdots + n_r$, with a Langlands decomposition (10.4.4). Let $s = (s_1, \ldots, s_r) \in \mathbb{C}^r$ satisfy $\sum_{i=1}^{r} n_i s_i = 0$. Let $K = O(n_1, \mathbb{R}) \times O(n_2, \mathbb{R}) \times \cdots \times O(n_r, \mathbb{R})$ be the direct product of orthogonal groups. We define the function, $I_s(*, P_{n_1,\ldots,n_r})$, which maps,*

$$P_{n_1,\ldots,n_r}/(K \cdot \mathbb{R}^\times) \to \mathbb{C},$$

by the formula

$$I_s(g, P_{n_1,\ldots,n_r}) = \prod_{i=1}^{r} \left|\mathrm{Det}\big(\mathfrak{m}_{n_i}(g)\big)\right|^{s_i},$$

for all

$$g = \begin{pmatrix} I_{n_1} & * & \cdots & * \\ 0 & I_{n_2} & \cdots & * \\ 0 & 0 & \cdots & * \\ \vdots & \vdots & \ddots & \vdots \\ 0 & 0 & \cdots & I_{n_r} \end{pmatrix} \cdot \begin{pmatrix} \mathfrak{m}_{n_1}(g) & 0 & \cdots & 0 \\ 0 & \mathfrak{m}_{n_2}(g) & \cdots & 0 \\ 0 & 0 & \cdots & 0 \\ \vdots & \vdots & \ddots & \vdots \\ 0 & 0 & \cdots & \mathfrak{m}_{n_r}(g) \end{pmatrix} \in P_{n_1,\ldots,n_r}.$$

Remarks Here $\mathfrak{m}_{n_i}$ is the whole group $GL(n_i, \mathbb{R})$ while $\mathfrak{m}_{n_i}(g)$ is a particular element in this group. The condition $\sum_{i=1}^{r} n_i s_i = 0$ insures that

$$I_s(\delta g, P_{n_1,\ldots,n_r}) = I_s(g, P_{n_1,\ldots,n_r})$$

for any matrix δ of the form $\delta = t \cdot I_n$ with $t \in \mathbb{R}^\times$, so that $I_s(*, P_{n_1,\ldots,n_r})$ is well defined on $P_{n_1,\ldots,n_r}/\mathbb{R}^\times$. It is also clear that $I_s(gk, P_{n_1,\ldots,n_r}) = I_s(g, P_{n_1,\ldots,n_r})$ for $k \in K$ since the determinant of an orthogonal matrix has absolute value 1.

It follows that I_s is well defined on the generalized upper half-plane $\mathfrak{h}^n$. This is because we may bring each matrix $m_{n_i}(g) \in GL(n_i, \mathbb{R})$ $(i = 1, \ldots, r)$ into diagonal form by right multiplication by an orthogonal matrix (see Proposition 1.2.6, the Iwasawa decomposition); and, then, with an additional multiplication by tI_n with $t \in \mathbb{R}^\times$ we may bring the entire matrix into Iwasawa form.

In particular, if

$$z = \begin{pmatrix} 1 & x_{1,2} & x_{1,3} & \cdots & x_{1,n} \\ & 1 & x_{2,3} & \cdots & x_{2,n} \\ & & \ddots & & \vdots \\ & & & 1 & x_{n-1,n} \\ & & & & 1 \end{pmatrix} \cdot \begin{pmatrix} y_1 y_2 \cdots y_{n-1} & & & & \\ & y_1 y_2 \cdots y_{n-2} & & & \\ & & \ddots & & \\ & & & y_1 & \\ & & & & 1 \end{pmatrix} \in \mathfrak{h}^n,$$

then $z \in P_{n_1,\ldots,n_r}$, since the standard Borel subgroup lies in every parabolic subgroup $P_{n_1,\ldots,n_r}$. It follows that

$$I_s(z, P_{n_1,\ldots,n_r}) = \left(\prod_{j_1=n-n_1+1}^{n} Y_{j_1} \right)^{s_1} \cdot \left(\prod_{j_2=n-n_1-n_2+1}^{n-n_1} Y_{j_2} \right)^{s_2}$$
$$\times \left(\prod_{j_3=n-n_1-n_2-n_3+1}^{n-n_1-n_2} Y_{j_3} \right)^{s_3} \cdots \left(\prod_{j_r=1}^{n_r} Y_{j_r} \right)^{s_r},$$

where we have defined $Y_1, Y_2, \ldots, Y_n$ by

$$\begin{pmatrix} Y_n & & & \\ & Y_{n-1} & & \\ & & \ddots & \\ & & & Y_1 \end{pmatrix} = \begin{pmatrix} y_1 y_2 \cdots y_{n-1} & & & & \\ & y_1 y_2 \cdots y_{n-2} & & & \\ & & \ddots & & \\ & & & y_1 & \\ & & & & 1 \end{pmatrix}.$$

One easily checks that $I_s(z, P_{n_1,\ldots,n_r})$ is precisely the standard function $I_{s'}(z)$ for suitable choice of s' depending on s.

Lemma 10.4.6 *Let $P_{n_1,\ldots,n_r}$ be a parabolic subgroup of $GL(n, \mathbb{R})$ with a Langlands decomposition (10.4.4). Let $s \in \mathbb{C}^r$ where $\sum_{i=1}^{r} n_i s_i = 0$. The function $I_s(*, P_{n_1,\ldots,n_r})$, as in Definition 10.4.5, satisfies*

$$I_s\big(\gamma z, P_{n_1,\ldots,n_r}\big) = I_s\big(z, P_{n_1,\ldots,n_r}\big)$$

for all $\gamma \in P_{n_1,\ldots,n_r} \cap SL(n, \mathbb{Z})$ and all $z \in \mathfrak{h}^n$.

Proof If $\gamma \in P_{n_1,\ldots,n_r} \cap SL(n, \mathbb{Z})$, then γ has a Langlands decomposition

$$\gamma = \begin{pmatrix} I_{n_1} & * & \cdots & * \\ 0 & I_{n_2} & \cdots & * \\ 0 & 0 & \cdots & * \\ \vdots & \vdots & \ddots & \vdots \\ 0 & 0 & \cdots & I_{n_r} \end{pmatrix} \cdot \begin{pmatrix} \mathfrak{m}_{n_1}(\gamma) & 0 & \cdots & 0 \\ 0 & \mathfrak{m}_{n_2}(\gamma) & \cdots & 0 \\ 0 & 0 & \cdots & 0 \\ \vdots & \vdots & \ddots & \vdots \\ 0 & 0 & \cdots & \mathfrak{m}_{n_r}(\gamma) \end{pmatrix}$$

where $\mathrm{Det}(\mathfrak{m}_{n_i}(\gamma)) = \pm 1$ for all $i = 1, \ldots, r$. The lemma immediately follows. □

Definition 10.4.7 *Let $P = P_{n_1,\ldots,n_r}$ be a parabolic subgroup of $GL(n, \mathbb{R})$ with a Langlands decomposition (10.4.4). Let $s \in \mathbb{C}^r$ where $\sum_{i=1}^{r} n_i s_i = 0$. We define the Eisenstein series associated to P, denoted $E_P(z, s)$, by the infinite series*

$$E_P(z, s) := \sum_{\gamma \in (P \cap \Gamma) \backslash \Gamma} I_s(\gamma z, P),$$

where $I_s(z, P)$ is given in Definition 10.4.5.

The absolute convergence of the Eisenstein series $E_P(z, s)$ for s in a suitable range follows from Proposition 10.4.3. This is because the function $I_s(z, P)$ is actually a special case of the I-function given in (5.1.1) and the set $P \cap \Gamma \backslash \Gamma$ has fewer elements than the set $P_{\min} \backslash \Gamma$ since $P_{\min} \subset P \cap \Gamma$.

Example 10.4.8 Maximal parabolic Eisenstein series for $SL(3, \mathbb{Z})$ Let $P_{1,2}, P_{2,1}$ denote the two maximal parabolic subgroups of $GL(3, \mathbb{R})$ given in Example 10.1.2. Then for $s = (s_1, s_2) \in \mathbb{C}^2$ satisfying $s_1 + 2s_2 = 0$ and

$$z = \begin{pmatrix} 1 & x_{1,2} & x_{1,3} \\ & 1 & x_{2,3} \\ & & 1 \end{pmatrix} \cdot \begin{pmatrix} y_1 y_2 & & \\ & y_1 & \\ & & 1 \end{pmatrix} \in \mathfrak{h}^3,$$

we have

$$I_s(z, P_{1,2}) = (y_1 y_2)^{s_1} \cdot y_1^{s_2} = \left(y_1^{\frac{1}{2}} y_2 \right)^{s_1}.$$

Similarly, for $s = (s_1, s_2) \in \mathbb{C}^2$ satisfying $2s_1 + s_2 = 0$, we have

$$I_s(z, P_{2,1}) = \left(y_1^2 y_2 \right)^{s_1}.$$

10.5 Eisenstein series twisted by Maass forms

Let $P = P_{n_1,\ldots,n_r}$ be a parabolic subgroup of $GL(n,\mathbb{R})$ which has a Langlands decomposition $P = NM$ as in (10.4.4).

We define a function:

$$\mathfrak{m}_P : P \to M,$$

by the formulae

$$\mathfrak{m}_P(g) = \begin{pmatrix} \mathfrak{m}_{n_1}(g) & 0 & \cdots & 0 \\ 0 & \mathfrak{m}_{n_2}(g) & \cdots & 0 \\ 0 & 0 & \cdots & 0 \\ \vdots & \vdots & \ddots & \vdots \\ 0 & 0 & \cdots & \mathfrak{m}_{n_r}(g) \end{pmatrix},$$

for all

$$g = \begin{pmatrix} I_{n_1} & * & \cdots & * \\ 0 & I_{n_2} & \cdots & * \\ 0 & 0 & \cdots & * \\ \vdots & \vdots & \ddots & \vdots \\ 0 & 0 & \cdots & I_{n_r} \end{pmatrix} \cdot \begin{pmatrix} \mathfrak{m}_{n_1}(g) & 0 & \cdots & 0 \\ 0 & \mathfrak{m}_{n_2}(g) & \cdots & 0 \\ 0 & 0 & \cdots & 0 \\ \vdots & \vdots & \ddots & \vdots \\ 0 & 0 & \cdots & \mathfrak{m}_{n_r}(g) \end{pmatrix} \in P. \tag{10.5.1}$$

Here, each $\mathfrak{m}_{n_i}(g) \in GL(n_i,\mathbb{R})$ $(i = 1,\ldots,r)$ as in (10.4.4). Note that every $z \in \mathfrak{h}^n$ is also an element of P so that $\mathfrak{m}_P(z)$ is well defined.

Let ϕ be a Maass form for the group M. Then ϕ is really a set of r Maass forms $\phi_1, \phi_2, \ldots, \phi_r$ where each ϕ_i is a Maass form for $GL(n_i,\mathbb{R})$ $(i = 1, 2, \ldots, r)$. For $g \in P$, of the form (10.5.1), we define

$$\phi\left(\mathfrak{m}_P(g)\right) = \prod_{i=1}^{r} \phi_i\left(\mathfrak{m}_{n_i}(g)\right). \tag{10.5.2}$$

We may now define the Langlands Eisenstein series twisted by Maass forms of lower rank.

Definition 10.5.3 *Let $P = P_{n_1,\ldots,n_r}$ be a standard parabolic subgroup of $GL(n,\mathbb{R})$ with Langlands decomposition: $P = NM$, as in (10.4.4). Let $s = (s_1,\ldots,s_r) \in \mathbb{C}^r$ satisfy*

$$\sum_{i=1}^{r} n_i s_i = 0,$$

and put $\Gamma = SL(n, \mathbb{Z})$. Then for ϕ a cusp form on M and $z \in \mathfrak{h}^n$, we define the Eisenstein series $E_P(z, s, \phi)$ by the infinite series

$$E_P(z, s, \phi) = \sum_{\gamma \in P \cap \Gamma \backslash \Gamma} \phi\left(\mathfrak{m}_P(\gamma z)\right) \cdot I_s(\gamma z, P),$$

where $I_s(z, P)$ is given by Definition 10.4.5 and ϕ is given by (10.5.2).

Example 10.5.4 ($SL(3, \mathbb{Z})$-Eisenstein series twisted by Maass forms) There are two classes of Eisenstein series twisted by Maass forms in the case of $SL(3, \mathbb{Z})$. Let ϕ be a Maass form for $SL(2, \mathbb{Z})$ with Fourier expansion

$$\phi\left(\begin{pmatrix} y & x & \\ & 1 & \\ & & 1 \end{pmatrix}\right) = \sum_{n \neq 0} a(n) \sqrt{y}\, K_r(2\pi |n| y) e^{2\pi i n x}, \qquad (x \in \mathbb{R},\ y > 0),$$

say, as in Proposition 3.5.1Note that, in order to conform to the notation (10.5.2), it is necessary to consider ϕ as being defined on 3×3 matrices. Let $s = (s_1, s_2)$ with $2s_1 + s_2 = 0$, and

$$z = \begin{pmatrix} 1 & x_{1,2} & x_{1,3} \\ & 1 & x_{2,3} \\ & & 1 \end{pmatrix} \cdot \begin{pmatrix} y_1 y_2 & & \\ & y_1 & \\ & & 1 \end{pmatrix} \in \mathfrak{h}^3.$$

The first class of such Eisenstein series is associated with the parabolic subgroup

$$P_{2,1} = \begin{pmatrix} * & * & * \\ * & * & * \\ 0 & 0 & * \end{pmatrix},$$

and consists of series of the form:

$$E_{P_{2,1}}(z, s, \phi) = \sum_{\gamma \in (P_{2,1} \cap SL(3,\mathbb{Z})) \backslash SL(3,\mathbb{Z})} I_s(\gamma z, P_{2,1}) \phi\left(\mathfrak{m}_{P_{2,1}}(\gamma z)\right). \qquad (10.5.5)$$

We seek a more explicit version of (10.5.5). By Example 10.4.8, we have

$$I_s(z, P_{2,1}) = \left(y_1^2 y_2\right)^{s_1} = \operatorname{Det}(z)^{s_1}.$$

It is also easy to see that

$$\mathfrak{m}_{P_{2,1}}(z) = \begin{pmatrix} y_2 & x_{1,2} & 0 \\ 0 & 1 & 0 \\ 0 & 0 & 1 \end{pmatrix} \begin{pmatrix} y_1 & 0 & 0 \\ 0 & y_1 & 0 \\ 0 & 0 & 1 \end{pmatrix}.$$

Then we have the explicit representation:

$$E_{P_{2,1}}(z, s, \phi) = \sum_{\gamma \in (P_{2,1} \cap SL(3,\mathbb{Z})) \backslash SL(3,\mathbb{Z})} \operatorname{Det}(\gamma z)^{s_1} \cdot \phi\left(\mathfrak{m}_{P_{2,1}}(\gamma z)\right). \qquad (10.5.6)$$

10.6 Fourier expansion of minimal parabolic Eisenstein series

Let $n \geq 2$ and consider $E_{P_{\min}}(z, s)$, the minimal parabolic Eisenstein series defined in (10.4.1). We shall first compute the Fourier coefficients of $E_{P_{\min}}(z, s)$ as in Theorem 5.3.2.Let $m = (m_1, m_2, \ldots, m_{n-1}) \in \mathbb{Z}^{n-1}$. Then the mth Fourier coefficient is

$$\mathcal{E}_m(z, s) = \int\limits_{U(\mathbb{Z})\backslash U(\mathbb{R})} E_{P_{\min}}(u \cdot z, s) \cdot \overline{\psi_m(u)}\, d^*u,$$

where

$$\psi_m(u) = e^{2\pi i (m_1 u_{1,2} + \cdots + m_{n-1} u_{n-1,n})},$$

and U denotes the group of upper triangular matrices with 1s on the diagonal.

The computation of this Fourier coefficient is based on the explicit Bruhat decomposition given in Proposition 10.3.6:

$$GL(n, \mathbb{R}) = \bigcup_{w \in W} G_w,$$

where W denotes the Weyl group and

$$G_w = UDwU = UwDU,$$

with U the group of upper triangular matrices with 1s on the diagonal (as above), and where D denotes the multiplicative group of diagonal matrices with non-zero determinant. Consider $P_{\min}(\mathbb{Z}) = P_{\min} \cap SL(n, \mathbb{Z})$.

The minimal parabolic Eisenstein series $E_{P_{\min}}(z, s)$ is constructed as a sum over the left quotient space $P_{\min}(\mathbb{Z})\backslash SL(n, \mathbb{Z})$. By the Bruhat decomposition, we may realize this left quotient space as a union:

$$P_{\min}(\mathbb{Z})\backslash SL(n, \mathbb{Z}) = \bigcup_{w \in W} P_{\min}(\mathbb{Z})\big\backslash (SL(n, \mathbb{Z}) \cap G_w).$$

It is natural then, for each $w \in W$, to study the left coset space

$$P_{\min}(\mathbb{Z})\big\backslash (SL(n, \mathbb{Z}) \cap G_w). \tag{10.6.1}$$

We would also like to take a further quotient of (10.6.1) on the right. In Lemma 10.6.3 we show that

$$\Gamma_w = \left(w^{-1} \cdot {}^t P_{\min}(\mathbb{Z}) \cdot w\right) \cap P_{\min}(\mathbb{Z})$$

acts properly on the right on (10.6.1), so that

$$P_{\min}(\mathbb{Z})\big\backslash (SL(n, \mathbb{Z}) \cap G_w)\big/\Gamma_w \tag{10.6.2}$$

is a well defined double coset space.

Lemma 10.6.3 *The group* $\Gamma_w = \left(w^{-1}\cdot {}^tP_{min}(\mathbb{Z})\cdot w\right)\cap P_{min}(\mathbb{Z})$ *acts properly on the right on the left coset space* $P_{min}(\mathbb{Z})\backslash(SL(n,\mathbb{Z})\cap G_w)$.

Proof For each $w \in W$, we introduce two additional spaces.

$$\begin{aligned} U_w &= (w^{-1}\cdot U\cdot w)\cap U \\ \bar{U}_w &= (w^{-1}\cdot {}^tU\cdot w)\cap U. \end{aligned} \tag{10.6.4}$$

To get a feel for these spaces, consider the example of $GL(3,\mathbb{R})$ where we have

$$U_{w_1} = \begin{pmatrix} 1 & * & * \\ & 1 & * \\ & & 1 \end{pmatrix}\cap U, \quad \bar{U}_{w_1} = \begin{pmatrix} 1 & & \\ * & 1 & \\ * & * & 1 \end{pmatrix}\cap U, \quad w_1 = \begin{pmatrix} 1 & & \\ & 1 & \\ & & 1 \end{pmatrix},$$

$$U_{w_2} = \begin{pmatrix} 1 & * & * \\ & 1 & \\ & * & 1 \end{pmatrix}\cap U, \quad \bar{U}_{w_2} = \begin{pmatrix} 1 & & \\ * & 1 & * \\ * & & 1 \end{pmatrix}\cap U, \quad w_2 = \begin{pmatrix} 1 & & \\ & & 1 \\ & 1 & \end{pmatrix},$$

$$U_{w_3} = \begin{pmatrix} 1 & & * \\ * & 1 & * \\ & & 1 \end{pmatrix}\cap U, \quad \bar{U}_{w_3} = \begin{pmatrix} 1 & * & \\ & 1 & \\ * & * & 1 \end{pmatrix}\cap U, \quad w_3 = \begin{pmatrix} & 1 & \\ 1 & & \\ & & 1 \end{pmatrix},$$

$$U_{w_4} = \begin{pmatrix} 1 & & \\ * & 1 & * \\ * & & 1 \end{pmatrix}\cap U, \quad \bar{U}_{w_4} = \begin{pmatrix} 1 & * & * \\ & 1 & \\ & * & 1 \end{pmatrix}\cap U, \quad w_4 = \begin{pmatrix} & 1 & \\ & & 1 \\ 1 & & \end{pmatrix},$$

$$U_{w_5} = \begin{pmatrix} 1 & * & \\ & 1 & \\ * & * & 1 \end{pmatrix}\cap U, \quad \bar{U}_{w_5} = \begin{pmatrix} 1 & & * \\ * & 1 & * \\ & & 1 \end{pmatrix}\cap U, \quad w_5 = \begin{pmatrix} & & 1 \\ 1 & & \\ & 1 & \end{pmatrix},$$

$$U_{w_6} = \begin{pmatrix} 1 & & \\ * & 1 & \\ * & * & 1 \end{pmatrix}\cap U, \quad \bar{U}_{w_6} = \begin{pmatrix} 1 & * & * \\ & 1 & * \\ & & 1 \end{pmatrix}\cap U, \quad w_6 = \begin{pmatrix} & & 1 \\ & 1 & \\ 1 & & \end{pmatrix}.$$

For example, $U_{w_5} = \begin{pmatrix} 1 & * & \\ & 1 & \\ & & 1 \end{pmatrix}$ and $\bar{U}_{w_5} = \begin{pmatrix} 1 & & * \\ & 1 & * \\ & & 1 \end{pmatrix}$. We may think of $\bar{U}_w$ as the space opposite to U_w in U. It is clear that for each $w \in W$, we have

$$U = U_w\cdot \bar{U}_w = \bar{U}_w\cdot U_w. \tag{10.6.5}$$

Returning to the proof of our lemma, it is plain that for every $w \in W$, the group Γ_w acts on $P_{\min}(\mathbb{Z})\backslash(SL(n,\mathbb{Z})\cap G_w)$ in the sense that right multiplication by Γ_w maps left cosets to left cosets. We only need to show that this action is proper, i.e., only the identity element acts trivially. To show this, suppose that

$$\gamma \in \Gamma_w = \left(w^{-1}\cdot {}^tP_{\min}(\mathbb{Z})\cdot w\right)\cap P_{\min}(\mathbb{Z}) = \bar{U}_w(\mathbb{Z})$$

fixes the left coset $P_{\min}(\mathbb{Z}) \cdot b_1 c w b_2$ where $c \in D$, and without loss of generality, $b_1 \in U_w$, $b_2 \in \bar{U}_w$.

Since

$$P_{\min}(\mathbb{Z}) \cdot b_1 c w b_2 \cdot \gamma = P_{\min}(\mathbb{Z}) \cdot b_1 c w b_2,$$

it follows that

$$b_2 \gamma b_2^{-1} \in U_w \cap \bar{U}_w = \{1\}.$$

This proves that γ must be the identity matrix and the action is proper. □

We now return to the computation of the Fourier coefficient $\mathcal{E}_m(z, s)$ initiated at the beginning of this section. In order to simplify presentation of formulae, we introduce the notation

$$\mathcal{G}_w = P_{\min}(\mathbb{Z}) \backslash (SL(n, \mathbb{Z}) \cap G_w),$$

with $G_w = UDwU$ as in the Bruhat decomposition in Proposition 10.3.6. The Bruhat decomposition tells us that

$$P_{\min}(\mathbb{Z}) \backslash SL(n, \mathbb{Z}) = \bigcup_{w \in W} \mathcal{G}_w.$$

We compute, using Lemma 10.6.3, (10.6.5), and the Bruhat decomposition above,

$$\begin{aligned}
\mathcal{E}_m(z, s) &= \int\limits_{U(\mathbb{Z}) \backslash U(\mathbb{R})} \sum_{\gamma \in P_{\min}(\mathbb{Z}) \backslash SL(n,\mathbb{Z})} I_s(\gamma u z)\, \overline{\psi_m(u)}\, d^*u \\
&= \sum_{w \in W_n} \sum_{c} \sum_{\substack{b_1 \in U_w(\mathbb{Q}),\ b_2 \in \bar{U}_w(\mathbb{Q}) \\ b_1 c w b_2 \in \mathcal{G}_w / \Gamma_w}} \sum_{\ell \in \Gamma_w} \int\limits_{U(\mathbb{Z}) \backslash U(\mathbb{R})} I_s(b_1 c w b_2 \ell u z)\, \overline{\psi_m(u)}\, d^*u \\
&= \sum_{w \in W_n} \sum_{c} \sum_{\substack{b_1 \in U_w(\mathbb{Q}),\ b_2 \in \bar{U}_w(\mathbb{Q}) \\ b_1 c w b_2 \in \mathcal{G}_w / \Gamma_w}} \sum_{\ell \in \Gamma_w = \bar{U}_w(\mathbb{Z})} \\
&\quad \times \int\limits_{U_w(\mathbb{Z}) \backslash U_w(\mathbb{R})} \int\limits_{\bar{U}_w(\mathbb{Z}) \backslash \bar{U}_w(\mathbb{R})} I_s(c w b_2 \ell u z)\, \overline{\psi_m(u)}\, d^*u \\
&= \sum_{w \in W_n} \sum_{c} \sum_{\substack{b_1 \in U_w(\mathbb{Q}),\ b_2 \in \bar{U}_w(\mathbb{Q}) \\ b_1 c w b_2 \in \mathcal{G}_w / \Gamma_w}} \psi_m(b_2) \cdot I_s(c) \\
&\quad \times \int\limits_{U_w(\mathbb{Z}) \backslash U_w(\mathbb{R})} \int\limits_{\bar{U}_w(\mathbb{R})} I_s(w u z)\, \overline{\psi_m(u)}\, d^*u.
\end{aligned}$$

The double integral above can be explicitly computed as follows. By (10.6.5), every $u \in U(\mathbb{R})$ can be written as $u = u_1 \cdot u_2$ with $u_1 \in U_w(\mathbb{R})$ and

$u_2 \in \bar{U}_w(\mathbb{R})$. Clearly $\psi_m(u_1u_2) = \psi_m(u_1)\psi_m(u_2)$. Since $wu_1 = u_1'w$ for some $u_1' \in U(\mathbb{R})$, we may, therefore, write

$$\begin{aligned}
&\int\limits_{U_w(\mathbb{Z})\backslash U_w(\mathbb{R})} \int\limits_{\bar{U}_w(\mathbb{R})} I_s(wuz)\, \overline{\psi_m(u)}\, d^*u \\
&= \int\limits_{U_w(\mathbb{Z})\backslash U_w(\mathbb{R})} \int\limits_{\bar{U}_w(\mathbb{R})} I_s\left(u_1' wu_2 z\right) \overline{\psi_m(u_1u_2)}\, d^*u_1\, d^*u_2. \\
&= \int\limits_{U_w(\mathbb{Z})\backslash U_w(\mathbb{R})} \overline{\psi_m(u_1)}\, d^*u_1 \cdot \int\limits_{\bar{U}_w(\mathbb{R})} I_s(wu_2z)\, \overline{\psi_m(u_2)}\, d^*u_2. \qquad (10.6.6)
\end{aligned}$$

Note that the last integral on the right-hand side of (10.6.6) will be a degenerate Whittaker function, as in Definition 5.10.1, if w is not the long element of the Weyl group.

It is instructive to illustrate the double integral (10.6.6) with an example. We shall consider $GL(4, \mathbb{R})$. In this case

$$U(\mathbb{R}) = \begin{pmatrix} 1 & * & * & * \\ & 1 & * & * \\ & & 1 & * \\ & & & 1 \end{pmatrix}.$$

For our example, we will let

$$w = \begin{pmatrix} & & 1 & \\ & 1 & & \\ & & & 1 \\ 1 & & & \end{pmatrix}.$$

In this case,

$$U_w(\mathbb{R}) = \begin{pmatrix} 1 & & & \\ & 1 & & * \\ & & 1 & * \\ & & & 1 \end{pmatrix}, \quad \bar{U}_w(\mathbb{R}) = \begin{pmatrix} 1 & * & * & * \\ & 1 & * & \\ & & 1 & \\ & & & 1 \end{pmatrix},$$

and the double integral (10.6.6) takes the form:

$$\int_0^1 e^{-2\pi i m_3 u_{3,4}}\, du_{3,4} \cdot \int_{-\infty}^{\infty}\int_{-\infty}^{\infty}\int_{-\infty}^{\infty}\int_{-\infty}^{\infty} I_s(wu_2z)\overline{\psi_m(u_2)}\, d^*u_2,$$

where

$$u_2 = \begin{pmatrix} 1 & u_{1,2} & u_{1,3} & u_{1,4} \\ & 1 & u_{2,3} & \\ & & 1 & \\ & & & 1 \end{pmatrix}, \qquad d^*u_2 = du_{1,2}du_{1,3}du_{1,4}du_{2,3}.$$

In the special case that $w = w_0$ is the long element

$$w_0 = \begin{pmatrix} & & & 1 \\ & & 1 & \\ & \cdot^{\cdot^{\cdot}} & & \\ 1 & & & \end{pmatrix},$$

we have $\bar{U}_{w_0}(\mathbb{R}) = U(\mathbb{R})$ and $U_{w_0}(\mathbb{R})$ is trivial. In this case, the double integral (10.6.6) is precisely the Jacquet Whittaker function given in Section 5.5. If the $(n-1)$-tuple $m = (m_1, m_2, \ldots, m_{n-1})$ satisfies $m_i \neq 0$ for all $1 \le i \le n-1$, i.e., the character ψ_m is non-degenerate, then the double integral (10.6.6) will vanish unless $w = w_0$ is the long element. This is because $U_w(\mathbb{Z})\backslash U_w(\mathbb{R})$ will be non-trivial and just a direct product of intervals $[0, 1]$, so that

$$\int\limits_{U_w(\mathbb{Z})\backslash U_w(\mathbb{R})} \overline{\psi_m(u_1)}\, d^*u_1 = 0.$$

In general, the integral $\int\limits_{U_w(\mathbb{Z})\backslash U_w(\mathbb{R})} \overline{\psi_m(u_1)}\, d^*u_1$ will be either 1 or 0, while the other integral

$$\int\limits_{\bar{U}_w(\mathbb{R})} I_s(wu_2z)\, \overline{\psi_m(u_2)}\, d^*u_2$$

will be a Whittaker function.

10.7 Meromorphic continuation and functional equation of maximal parabolic Eisenstein series

Let us fix (for this section) the notation

$$P = P_{n-1,1} = \begin{pmatrix} * & \cdots & * & * \\ \vdots & \cdots & \vdots & \vdots \\ * & \cdots & * & * \\ 0 & \cdots & 0 & 1 \end{pmatrix} \subset \Gamma,$$

to be the maximal parabolic subgroup of Γ whose elements have bottom row equal to $(0, \ldots, 0, 1)$. For $s \in \mathbb{C}$, $z \in \mathfrak{h}^n$, consider the maximal parabolic Eisenstein series

$$E_P^*(z, s) = \pi^{-ns/2}\,\Gamma\left(\frac{ns}{2}\right)\zeta(ns)E_P(z, s)$$

where

$$E_P(z, s) = \sum_{P\backslash\Gamma} \operatorname{Det}(\gamma z)^s, \qquad (\Re(s) > 2/n), \tag{10.7.1}$$

and where Det is the determinant function on $\mathfrak{h}^n$. While it is not yet clear that (10.7.1) converges absolutely for $\Re(s) > 2/n$, this will follow directly from (10.7.4). Note that in (10.7.1), we must put γz in canonical Iwasawa form (as in Proposition 1.2.6) before actually taking the determinant. The meromorphic continuation and functional equation of $E_P(z,s)$ can be obtained from the Poisson summation formula

$$\sum_{m\in\mathbb{Z}^n} f(m\cdot z) = \frac{1}{|\mathrm{Det}(z)|}\sum_{m\in\mathbb{Z}^n} \hat{f}(m\cdot({}^t z)^{-1}), \tag{10.7.2}$$

which holds for smooth functions $f : \mathbb{R}^n \to \mathbb{C}$ with sufficient decay at $\pm\infty$. Here

$$\tilde{f}((x_1,\ldots,x_n)) = \int_{-\infty}^{\infty}\cdots\int_{-\infty}^{\infty} f((t_1,\ldots,t_n))e^{-2\pi i(t_1x_1+\cdots+t_nx_n)}\,dt_1\cdots dt_n.$$

In order to be able to apply (10.7.2), it is necessary to rewrite the maximal parabolic Eisenstein series in the form of an Epstein zeta function. This is done as follows.

First note that if

$$\gamma = \begin{pmatrix} & * & \\ a_1 & \cdots & a_n \end{pmatrix}, \qquad \gamma' = \begin{pmatrix} & * & \\ a_1' & \cdots & a_n' \end{pmatrix} \in \Gamma,$$

then there exists $p \in P$ with $\gamma' = p\gamma$ if and only if $(a_1,\ldots,a_n) = (a_1',\ldots,a_n')$. Consequently, each coset of $P\backslash\Gamma$ is uniquely determined by n relatively prime integers $(a_1,\ldots,a_n)$.

Furthermore,

$$\gamma z = \begin{pmatrix} & * & \\ a_1 & \cdots & a_n \end{pmatrix} \cdot \begin{pmatrix} 1 & & x_{i,j} \\ & \ddots & \\ & & 1 \end{pmatrix}\begin{pmatrix} y_1\cdots y_{n-1} & & & \\ & \ddots & & \\ & & y_1 & \\ & & & 1 \end{pmatrix}$$
$$= \begin{pmatrix} & * & \\ b_1 & \cdots & b_n \end{pmatrix},$$

where

$$\begin{aligned} b_1 &= a_1y_1\cdots y_{n-1}\\ b_2 &= (a_1x_{1,2}+a_2)y_1\cdots y_{n-2}\\ &\;\vdots\\ b_n &= (a_1x_{1,n}+a_2x_{2,n}+\cdots+a_{n-1}x_{n-1,n}+a_n). \end{aligned} \tag{10.7.3}$$

On the other hand, by the Iwasawa decomposition (Proposition 1.2.6),

$$\gamma z = \tau \cdot k \cdot r I_n,$$

where $k \in O(n, \mathbb{R})$, $0 < r \in \mathbb{R}$, I_n denotes the $n \times n$ identity matrix, and τ is the canonical form for the Iwasawa decomposition. By comparing norms of the bottom rows, which amounts to the identity $(\gamma z) \cdot {}^t(\gamma z) = (\tau k r I_n) \cdot {}^t(\tau k r I_n) = \tau\, {}^t\tau \cdot r^2 I_n$, one obtains

$$b_1^2 + \cdots + b_n^2 = r^2.$$

Consequently

$$\begin{aligned}\mathrm{Det}(\gamma z) &= \mathrm{Det}(\tau) = |\mathrm{Det}(\gamma)|\, \mathrm{Det}(z) r^{-n} \\ &= \mathrm{Det}(z) \left[b_1^2 + \cdots + b_n^2\right]^{-n/2}.\end{aligned}$$

It immediately follows that

$$\zeta(ns) E_P(z, s) = \mathrm{Det}(z)^s \sum_{\substack{(a_1,\ldots,a_n) \in \mathbb{Z}^n \\ (a_1,\ldots,a_n) \neq (0,\ldots,0)}} \left[b_1^2 + \cdots + b_n^2\right]^{-ns/2}, \quad (10.7.4)$$

with b_i given by (10.7.3) for $i = 1, \ldots, n$. The right-hand side of (10.7.4) is termed an Epstein zeta function. We multiply by $\zeta(ns)$ on the left to convert the sum on the right to a sum over $(a_1, \ldots, a_n) \in \mathbb{Z}^n$, eliminating the relatively prime condition.

Next, we utilize the Poisson summation formula (10.7.2) to show that $E_P(z, s)$ has a meromorphic continuation and satisfies a functional equation.

Proposition 10.7.5 *The maximal parabolic Eisenstein series $E_P(z, s)$ defined in (10.7.1) has meromorphic continuation to all $s \in \mathbb{C}$ and satisfies the functional equation*

$$E_P^*(z, s) := \pi^{-ns/2} \Gamma\left(\frac{ns}{2}\right) \zeta(ns) E_P(z, s) = E_P^*({}^t z^{-1},\ 1 - s).$$

Further, $E_P^*(z, s)$ is holomorphic except for simple poles at $s = 0, 1$.

Proof Fix $u > 0$. For $x = (x_1, \ldots, x_n) \in \mathbb{R}^n$, define

$$f_u(x) := e^{-\pi (x_1^2 + \cdots + x_n^2) \cdot u}.$$

Then we have the Fourier transform $\hat{f}_u(x) = (1/u^{n/2}) f_u(x/u)$. We shall make use of the Poisson summation formula (10.7.2) with this choice of function f_u. It follows from (10.7.3), (10.7.4), and the integral representation of the Gamma function that

$$E_P^*(x,s) = \mathrm{Det}(z)^s \int_0^\infty \left[\sum_{(a_1,\ldots,a_n)\in\mathbb{Z}^n} f_u((a_1,\ldots,a_n)\cdot z) \;-\; f((0,\ldots,0)) \right] \times u^{ns/2}\,\frac{du}{u}.$$

The proposition follows by breaking the integral into two parts: $[0,1]$ and $[1,\infty]$, and then applying the Poisson summation formula (10.7.2) just as we did in the proof of the functional equation of the Riemann zeta function given on page 1. □

10.8 The L-function associated to a minimal parabolic Eisenstein series

To obtain the Fourier coefficients of the minimal parabolic Eisenstein series in a more precise form, we follow the method in Section 3.14. This method can also be used to obtain the meromorphic continuation and functional equation of the Eisenstein series. We compute the action of the Hecke operator (9.3.5) on the I-function. The I-function is an eigenfunction of all the Hecke operators, and the eigenvalue of the Hecke operator T_m will give us the $(m,1,\ldots,1)$th Fourier coefficient of the minimal parabolic Eisenstein series as in Theorem 9.3.11. Although Theorem 9.3.11 is stated for Maass forms, it can be easily generalized to Eisenstein series. Note that this method works up to a normalizing factor. We shall prove the following theorem.

Theorem 10.8.1 *Let $s=(s_1,\ldots,s_{n-1})\in\mathbb{C}^{n-1}$ with $n\geq 2$. Define $s-\frac{1}{n}$ to be $(s_1-\frac{1}{n},\ldots,s_{n-1}-\frac{1}{n})$. For $z\in\mathfrak{h}^n$, let $E(z,s)$ be the minimal parabolic Eisenstein series (10.4.1), for $\Gamma=SL(n,\mathbb{Z})$, with Fourier expansion*

$$E(z,s) = \mathrm{C}(z,s) + \sum_{\gamma\in U_{n-1}(\mathbb{Z})\backslash\Gamma} \sum_{m_1=1}^{\infty}\sum_{m_2=1}^{\infty}\cdots\sum_{m_{n-1}\neq 0} A((m_1,m_2,\ldots,m_{n-1}),s)$$
$$\times W_{\mathrm{Jacquet}}\left(\begin{pmatrix} |m_1m_2\cdots m_{n-1}| & & & \\ & \ddots & & \\ & & |m_1| & \\ & & & 1\end{pmatrix}\cdot\begin{pmatrix}\gamma & \\ & 1\end{pmatrix} z, s, \psi_{1,1,\ldots,\frac{m_{n-1}}{|m_{n-1}|}}\right)$$

as in Section 10.6. Here $\mathrm{C}(z,s)$ denotes the degenerate terms in the Fourier expansion associated to $(m_1,m_2,\ldots,m_{n-1})$ with $m_i=0$ for some $1\leq i\leq n-1$. We shall assume $E(z,s)$ is normalized (multiplied by a suitable *function of s) so that $A((1,1,\ldots,1),s) = 1$. Then for $m_1 = 1,2,3,\ldots$,*

$$A((m_1,1,1,\ldots,1),\ s) \ = \sum_{\substack{1\le c_1,c_2,\ldots,c_n\in\mathbb{Z}\\ \prod_{\ell=1}^{n} c_\ell = m_1}} I_{s-(1/n)}\left(\begin{pmatrix} c_1 & & & \\ & \ddots & & \\ & & c_{n-1} & \\ & & & c_n \end{pmatrix}\right).$$

Remarks It is easy to see that $A((m_1,1,1,\ldots,1),s) = A((m_1,1,1,\ldots,1), s')$ whenever

$$I_{s-(1/n)}(y) = I_{s'-(1/n)}(wy) \tag{10.8.2}$$

for some fixed w in the Weyl group of $GL(n,\mathbb{R})$ and all diagonal matrices y as in (5.9.1). This is due to the fact that the action of the Weyl group on the group of diagonal matrices just permutes the diagonal elements so that the sum over $1 \le c_1, c_2, \ldots, c_n \in \mathbb{Z}$ does not change. We shall show that $E(z,s)$ is actually an eigenfunction of the Hecke operators. The Hecke relations in Section 9.3 imply that if $A((p^k,1,1,\ldots,1),s)$ satisfies the functional equation $s \to s'$ for s, s' given by (10.8.2) and all primes p and all $k = 1,2,\ldots$, then $A((m_1,m_2,\ldots,m_{n-1}),s)$ must also satisfy the functional equations (10.8.2) for all $m_1, m_2, \ldots, m_{n-1} \in \mathbb{Z}^{n-1}$. It immediately follows from Theorem 5.9.8 that for $s_{j,k} = \sum_{i=0}^{j-1} (ns_{n-k+i} - 1)/2$, we have

$$\begin{aligned} E^*(z,s) &= \prod_{j=1}^{n-1} \prod_{j\le k\le n-1} \pi^{-\frac{1}{2}-s_{j,k}}\Gamma\left(\frac{1}{2}+s_{j,k}\right)\zeta(1+2s_{j,k})(E(z,s)-C(z,s)) \\ &= E^*(z,s') \end{aligned} \tag{10.8.3}$$

for all s, s' satisfying (10.8.2) It may also be shown that $C(z,s)$ satisfies these same functional equations.

Proof of Theorem 10.8.1 Recall the definition (see also (5.1.1)),

$$I_s(z) = \prod_{i=1}^{n-1}\prod_{j=1}^{n-1} y_i^{b_{i,j}s_j}$$

with

$$b_{i,j} = \begin{cases} ij & \text{if } i+j \le n, \\ (n-i)(n-j) & \text{if } i+j \ge n. \end{cases}$$

For $m_1 \geq 1$, we compute, using (9.3.5),

$$\begin{aligned}
T_{m_1} I_s(z) &= \frac{1}{m_1^{(n-1)/2}} \sum_{\substack{\prod_{\ell=1}^{n} c_\ell = m_1 \\ 0 \leq c_{i,\ell} < c_\ell \, (1 \leq i < \ell \leq n)}} I_s\left(\begin{pmatrix} c_1 & c_{1,2} & \cdots & c_{1,n} \\ & c_2 & \cdots & c_{2,n} \\ & & \ddots & \vdots \\ & & & c_n \end{pmatrix} \cdot z\right) \\
&= \frac{1}{m_1^{(n-1)/2}} \sum_{\substack{\prod_{\ell=1}^{n} c_\ell = m_1 \\ 0 \leq c_{i,\ell} < c_\ell \, (1 \leq i < \ell \leq n)}} I_s\left(\begin{pmatrix} c_1 & & & \\ & \ddots & & \\ & & c_{n-1} & \\ & & & c_n \end{pmatrix}\right) \cdot I_s(z) \\
&= \sum_{\prod_{\ell=1}^{n} c_\ell = m_1} \left(\prod_{i=1}^{n} c_i^{(2i-1-n)/2}\right) I_s\left(\begin{pmatrix} c_1 & & & \\ & \ddots & & \\ & & c_{n-1} & \\ & & & c_n \end{pmatrix}\right) \cdot I_s(z) \\
&= \sum_{\prod_{\ell=1}^{n} c_\ell = m_1} I_{s-(1/n)}\left(\begin{pmatrix} c_1 & & & \\ & \ddots & & \\ & & c_{n-1} & \\ & & & c_n \end{pmatrix}\right) \cdot I_s(z) \\
&= A((m_1, 1, \ldots, 1), s) \cdot I_s(z).
\end{aligned}$$

This computation shows that $I_s(z)$ is an eigenfunction of the Hecke operators T_{m_1} with eigenvalue $A((m_1, 1, \ldots, 1), s)$. We shall next show that this implies that $E(z, s)$ is also an eigenfunction of T_{m_1} with the same eigenvalue.

Let S_{m_1} denote the set of matrices

$$\begin{pmatrix} c_1 & c_{1,2} & \cdots & c_{1,n} \\ & c_2 & \cdots & c_{2,n} \\ & & \ddots & \vdots \\ & & & c_n \end{pmatrix}$$

where $\prod_{\ell=1}^{n} c_\ell = m_1$ and $0 \leq c_{i,\ell} < c_\ell$ $(1 \leq i < \ell \leq n)$.

Lemma 10.8.4 *For $m_1 = 1, 2, 3, \ldots,$ there exists a one-to-one correspondence between $S_{m_1} \times SL(n, \mathbb{Z})$ and $SL(n, \mathbb{Z}) \times S_{m_1}$.*

Proof It follows from the Hermite normal form, Theorem 3.11.1, that for any $\alpha' \in S_{m_1}$, $\gamma' \in SL(n, \mathbb{Z})$, there exists a unique $\alpha \in S_{m_1}$ and $\gamma \in SL(n, \mathbb{Z})$ such that $\gamma'\alpha' = \alpha\gamma$. The result follows easily from this. □

It is a consequence of Lemma 10.8.4 that for every $\alpha \in S_{m_1}$, $\gamma \in P_{\min}\backslash\Gamma$ there exists a unique $\alpha' \in S_{m_1}$, $\gamma' \in P_{\min}\backslash\Gamma$ such that $\alpha\gamma = \gamma'\alpha'$. It now follows from the definition of the action of the Hecke operator (9.3.5) and the definition of the minimal parabolic Eisenstein series (10.4.1) that for $m_1 = 1, 2, 3, \ldots$,

$$\begin{aligned} T_{m_1} E_{P_{\min}}(z, s) &= m_1^{-(n-1)/2} \sum_{\alpha \in S_{m_1}} \sum_{\gamma \in P_{\min}\backslash\Gamma} I_s(\gamma\alpha z) \\ &= m_1^{-(n-1)/2} \sum_{\gamma \in P_{\min}\backslash\Gamma} \sum_{\alpha \in S_{m_1}} I_s(\alpha\gamma z) \\ &= A((m_1, 1, 1, \ldots, 1), s) \cdot \sum_{\gamma \in P_{\min}\backslash\Gamma} I_s(\gamma z) \\ &= A((m_1, 1, 1, \ldots, 1), s) \cdot E_{P_{\min}}(z, s). \end{aligned}$$

This proves that $E_{P_{\min}}(z, s)$ is an eigenfunction of the Hecke operators. The proof of Theorem 10.8.1 is an immediate consequence of Theorem 9.3.11. Note that although we only proved Theorem 9.3.11 for Maass forms, it can be easily generalized to Eisenstein series. □

Let $n \geq 2$. For $v = (v_1, \ldots, v_{n-1}) \in \mathbb{C}^{n-1}$, $z \in \mathfrak{h}^n$, let $E_v(z) = E_{P_{\min}}(z, v)$ denote the minimal parabolic Eisenstein series (10.4.1). Finally, we are now in a position to compute the L-function associated to E_v, denoted $L_{E_v}(s)$, just as we did in Section 3.14 for the Eisenstein series on $SL(2, \mathbb{Z})$. It follows from Theorem 10.8.1 that

$$\begin{aligned} L_{E_v}(s) &= \sum_{m=1}^{\infty} m^{-s} \sum_{\substack{1 \leq c_1, c_2, \ldots, c_n \in \mathbb{Z} \\ \prod_{\ell=1}^{n} c_\ell = m}} I_{v-\frac{1}{n}}\left(\begin{pmatrix} \frac{c_1}{c_n} & & & \\ & \ddots & & \\ & & \frac{c_{n-1}}{c_n} & \\ & & & 1 \end{pmatrix}\right) \\ &= \sum_{m=1}^{\infty} m^{-s} \sum_{\substack{1 \leq c_1, \ldots, c_{n-1} \in \mathbb{Z} \\ \prod_{\ell=1}^{n-1} c_\ell \mid m}} I_{v-\frac{1}{n}}\left(\begin{pmatrix} \frac{c_1^2 c_2 \cdots c_{n-1}}{m} & & & \\ & \ddots & & \\ & & \frac{c_1 \cdots c_{n-2} c_{n-1}^2}{m} & \\ & & & 1 \end{pmatrix}\right) \\ &= \sum_{c_1=1}^{\infty} \cdots \sum_{c_{n-1}=1}^{\infty} \sum_{m=1}^{\infty} (m c_1 \cdots c_{n-1})^{-s}\, I_{v-\frac{1}{n}}\left(\begin{pmatrix} \frac{c_1}{m} & & & \\ & \ddots & & \\ & & \frac{c_{n-1}}{m} & \\ & & & 1 \end{pmatrix}\right). \end{aligned} \tag{10.8.5}$$

The above computation immediately implies the following theorem.

Theorem 10.8.6 (L-function associated to minimal parabolic Eisenstein series) *For $n \geq 2$, let $E_v(z) = E_{P_{min}}(z, v)$ denote the minimal parabolic Eisenstein series (10.4.1). Then there exist functions $\lambda_i : \mathbb{C}^{n-1} \to \mathbb{C}$, satisfying $\Re(\lambda_i(v)) = 0$ if $\Re(v_i) = \frac{1}{n}$ $(i = 1, \ldots, n-1)$, such that the L-function associated to E_v is just a product of shifted Riemann zeta functions of the form*

$$L_{E_v}(s) = \prod_{i=1}^{n} \zeta(s - \lambda_i(v)).$$

Furthermore, $L_{E_v}(s)$ satisfies the functional equation

$$G_{E_v}(s) L_{E_v}(s) = G_{\tilde{E}_v}(1-s) L_{\tilde{E}_v}(1-s),$$

where

$$G_{E_v}(s) = \prod_{i=1}^{n} \pi^{-\frac{s-\lambda_i(v)}{2}} \Gamma\left(\frac{s-\lambda_i(v)}{2}\right) = \pi^{-\frac{ns}{2}} \prod_{i=1}^{n} \Gamma\left(\frac{s-\lambda_i(v)}{2}\right),$$

and $\tilde{E}_v$ is the dual Eisenstein series as in Section 9.2.

We tabulate $L_{E_v}(s)$ for the cases $n = 2, 3, 4$, respectively:

$$\zeta\left(s + v - \frac{1}{2}\right) \zeta\left(s - v + \frac{1}{2}\right), \qquad (\text{if } n = 2)$$

$$\zeta\left(s + v_1 + 2v_2 - 1\right) \zeta\left(s - 2v_1 - v_2 + 1\right) \zeta(s + v_1 - v_2), \qquad (\text{if } n = 3)$$

$$\zeta\left(s + v_1 + 2v_2 + 3v_3 - \frac{3}{2}\right) \zeta\left(s - 3v_1 - 2v_2 - v_3 + \frac{3}{2}\right)$$
$$\times \zeta\left(s + v_1 - 2v_2 - v_3 + \frac{1}{2}\right) \zeta\left(s + v_1 + 2v_2 - v_3 - \frac{1}{2}\right), \qquad (\text{if } n = 4).$$

For example, in the last case ($n = 4$) we have

$$\lambda_1(v) = \frac{3}{2} - v_1 - 2v_2 - 3v_3,$$
$$\lambda_2(v) = 3v_1 + 2v_2 + v_3 - \frac{3}{2},$$
$$\lambda_3(v) = -v_1 + 2v_2 + v_3 - \frac{1}{2},$$
$$\lambda_4(v) = -v_1 - 2v_2 + v_3 + \frac{1}{2}.$$

Remark 10.8.7 The functions λ_i $(i = 1, \ldots, n)$ are uniquely determined by (10.8.5). It is easy to see that $\lambda_1(v) = ((n-1)/2) - v_1 - 2v_2 - \cdots - (n-1)v_{n-1}$. Theorem 10.8.6 is very important. It provides a template for all future functional equations. The reader can immediately check that when

$n = 3$ the functional equation of $L_{E_v}(s)$ given above is identical to the functional equation of a $GL(3)$-Maass form of type v obtained in Theorem 6.5.15. This is not an accident. In fact, the functional equation of $L_{E_v}(s)$ will always be identical to the functional equation of a symmetric Maass form of type v because the Whittaker functions in the Fourier expansion of the Maass form will match up exactly with the Whittaker functions of the Eisenstein series. The formal proof of the functional equation only uses the analytic properties of the Whittaker functions. The proof is entirely independent of the values of the arithmetic Fourier coefficients.

How to use Theorem 10.8.6 as a template for functional equations. Let us consider the example of twisting a Maass form by a primitive Dirichlet character $\chi \pmod q$. Its L-function will have the same functional equation as the L-function associated to the Eisenstein series E_v twisted by χ which will simply be

$$L_{E_v}(s, \chi) = \prod_{i=1}^{n-1} L(s - \lambda_i(v), \chi).$$

Now, the functional equation of the Dirichlet L-function $L(s, \chi)$ is (see (Davenport, 1974))

$$\Lambda(s, \chi) := \left(\frac{q}{\pi}\right)^{(s+\mathfrak{a}_\chi)/2} \Gamma\left(\frac{s + \mathfrak{a}_\chi}{2}\right) L(s, \chi) = \frac{\tau(\chi)}{i^{\mathfrak{a}_\chi}\sqrt{q}} \Lambda(1 - s, \bar{\chi}),$$

where $\tau(\chi)$ denotes the Gauss sum in (7.1.1). It immediately follows that $L_{E_v}(s, \chi)$ satisfies the functional equation

$$\begin{aligned}\Lambda_{E_v}(s, \chi) &:= \left(\frac{q}{\pi}\right)^{n(s+\mathfrak{a}_\chi)/2} \prod_{i=1}^{n} \Gamma\left(\frac{s + \mathfrak{a}_\chi - \lambda_i(\nu)}{2}\right) L_{E_v}(s, \chi)\\ &= \left(\frac{\tau(\chi)}{i^{\mathfrak{a}_\chi}\sqrt{q}}\right)^n \Lambda_{\tilde{E}_v}(1 - s, \bar{\chi}).\end{aligned}$$

This illustrates the method of obtaining functional equations of Maass forms of various types by carefully examining the template arising from the case of minimal parabolic Eisenstein series.

10.9 Fourier coefficients of Eisenstein series twisted by Maass forms

The method presented in Section 10.8 for obtaining the Fourier coefficients of minimal parabolic Eisenstein series easily extends to more general Eisenstein series which are associated to an arbitrary standard parabolic subgroup P. This is due to the fact that such an Eisenstein series as given in Definition 10.4.7 can actually be written in the form

$$E_P(z,s) = \sum_{\gamma\in(P\cap\Gamma)\backslash\Gamma} I_{s'}(\gamma z)$$

for suitable s' as explained in the discussion just prior to Lemma 10.4.6. It follows as in Theorem 10.8.1 that the normalized $(m_1, 1, 1, \ldots, 1)$th Fourier coefficient of $E_P(z,s)$ is just

$$\sum_{\substack{1\le c_1,c_2,\ldots,c_n\in\mathbb{Z}\\ \prod_{\ell=1}^{n} c_\ell = m_1}} I_{s'-\frac{1}{n}}\left(\begin{pmatrix} c_1 & & & \\ & \ddots & & \\ & & c_{n-1} & \\ & & & c_n \end{pmatrix}\right).$$

We now consider the more complex situation of Eisenstein series twisted by Maass forms of lower rank as in Section 10.5. Fix a parabolic subgroup $P = P_{n_1,\ldots,n_r}$ associated to the partition $n = n_1 + \cdots + n_r$, with $r \ge 2$. Let $\phi = (\phi_1, \ldots, \phi_r)$ be a set of r Maass forms where each ϕ_i is a Maass form of type $\lambda^i = (\lambda^i_1, \ldots, \lambda^i_{n_i-1}) \in \mathbb{C}^{n_i-1}$ $(i = 1, \ldots, r)$ for $SL(n_i, \mathbb{Z})$, as in Definition 5.1.3.

We shall show, in the next two propositions, that the Langlands Eisenstein series $E_P(z, s, \phi)$, given in Definition 10.5.3, is an eigenfunction of both the Hecke operators for $SL(n, \mathbb{Z})$, as in (9.3.5), and the invariant differential operators on $GL(n, \mathbb{R})$, as in Proposition 2.3.3 provided ϕ is a Hecke eigenform, i.e., each ϕ_i is an eigenfunction of the $SL(n_i, \mathbb{Z})$ Hecke operators. These properties allow one to obtain the meromorphic continuation and functional equation for the non-degenerate part of the Eisenstein series $E_P(z, s, \phi)$, just as we did previously in Section 10.8 for the minimal parabolic Eisenstein series.

Recall that a smooth function $F : \mathfrak{h}^n \to \mathbb{C}$ is of type $\nu \in \mathbb{C}^{n-1}$ if it is an eigenfunction of all the invariant differential operators (see Section 2.3) with the same eigenvalues as the function I_ν. In this regard, see also Definition 5.1.3.

Proposition 10.9.1 *For $r \ge 2$, let $P = P_{n_1,\ldots,n_r}$ be a standard parabolic subgroup associated to the partition $n = n_1 + \cdots + n_r$. Let $\phi = (\phi_1, \ldots, \phi_r)$ be a set of r Maass forms where each ϕ_i is a Maass form for $SL(n_i, \mathbb{Z})$ of type λ^i. Then for $s = (s_1, \ldots, s_r) \in \mathbb{C}^r$, the Eisenstein series $E_P(z, s, \phi)$ is an* *eigenfunction of type* $\lambda + s'$ *where* $s' \in \mathbb{C}^{n-1}$ *is such that*

$$I_{s'}(z) = \prod_{i=1}^{r} \text{Det}\big(\mathfrak{m}_{n_i}(z)\big)^{s_i}$$

for all $z \in \mathfrak{h}^n$, *and* $\lambda \in \mathbb{C}^{n-1}$ *is such that* $\prod_{i=1}^{r} \phi_i(\mathfrak{m}_{n_i}(z))$ *is of type* λ.

Proof Since the invariant differential operators commute with the action of $GL(n, \mathbb{R})$, it is enough to check that

$$\prod_{i=1}^{r} \phi_i\big(\mathfrak{m}_{n_i}(z)\big) \cdot \text{Det}\big(\mathfrak{m}_{n_i}(z)\big)^{s_i}, \tag{10.9.2}$$

the generating function of the Eisenstein series, is an eigenfunction. This will be the case because each Maass form ϕ_i is a linear combination of Whittaker functions $W_{\lambda^i}(z^i)$ of type λ^i with $z^i \in \mathfrak{h}^{n_i}$, where

$$W_{\lambda^i}(z^i) = \int_{U_{n_i}(\mathbb{R})} I_{\lambda^i}\big(w_{n_i} u z^i\big)\overline{\psi_i(u)}\, d^*u$$

is a Jacquet Whittaker function, for some character ψ_i of $U_{n_i}(\mathbb{R})$, as in (5.5.1). Here w_{n_i} is the long element of the Weyl group for $GL(n_i, \mathbb{R})$.

Now, for $w = \begin{pmatrix} w_{n_1} & & & \\ & w_{n_2} & & \\ & & \ddots & \\ & & & w_{n_r} \end{pmatrix}$,

$$\begin{aligned}
&\prod_{i=1}^{r} W_{\lambda^i}\big(\mathfrak{m}_{n_i}(z)\big)\text{Det}\big(\mathfrak{m}_{n_i}(z)\big)^{s_i} \\
&= \prod_{i=1}^{r} \int_{U_{n_i}(\mathbb{R})} I_{\lambda^i}\big(\mathfrak{m}_{n_i}(wuz)\big)\, \overline{\psi_i(u)} \cdot \text{Det}\big(\mathfrak{m}_{n_i}(z)\big)^{s_i}\, d^*u \\
&= \prod_{i=1}^{r} \int_{U_{n_i}(\mathbb{R})} I_{\lambda^i}\big(\mathfrak{m}_{n_i}(wuz)\big)\, \overline{\psi_i(u)} \cdot \text{Det}\big(\mathfrak{m}_{n_i}(wuz)\big)^{s_i}\, d^*u,
\end{aligned}$$

since $\text{Det}(wu) = 1$. It immediately follows that the above, and hence, (10.9.2) is an eigenfunction because it is obtained from $\prod_{\ell=1}^{n-1} y_\ell^{a_\ell}$ for suitable $a_1, a_2, \ldots, a_{n-1}$ by integrating a set of left translates. To show that $E_P(z, s, \phi)$ is an eigenfunction of type $\lambda + s'$ one uses the fact that $I_\lambda(z) \cdot I_{s'}(z) = I_{\lambda+s'}(z)$. □

Proposition 10.9.3 *With the notation of Proposition 10.9.1, let $E_P(z, s, \phi)$ be a Langlands Eisenstein series for $SL(n, \mathbb{Z})$. Let $s = (s_1, \ldots, s_r) \in \mathbb{C}^r$ satisfy $n_1 s_1 + \cdots + n_r s_r = 0$. Assume that ϕ is a Hecke eigenform, and let T_m denote the Hecke operator given in (9.3.5). Then for $m = 1, 2, 3, \ldots,$*

$$T_m\, E_P(z, s, \phi) = \lambda_m(s) \cdot E_P(z, s, \phi)$$

where

$$\lambda_m(s) = \sum_{\substack{1 \le C_1, C_2, \ldots, C_r \in \mathbb{Z} \\ C_1 C_2 \cdots C_r = m}} A_1(c_1) A_2(c_2) \cdots A_r(c_r) \cdot C_1^{s_1+\eta_1} C_2^{s_2+\eta_2} \cdots C_r^{s_r+\eta_r},$$

where $\eta_1 = 0$, and $\eta_i = n_1 + n_2 + \cdots + n_{i-1}$ for $i \ge 1$. In the above $A_i(C_i)$ denotes the eigenvalue of the $SL(n_i, \mathbb{Z})$ Hecke operator T_{C_i} acting on ϕ_i which may also be viewed as the $(C_i, \underbrace{1, 1, \ldots, 1}_{n_i - 1 \text{ terms}})$th Fourier coefficient of ϕ_i.

Proof For $m = 1, 2, \ldots,$ we compute, as in the proof of Theorem 10.8.1:

$$T_m \prod_{i=1}^{r} \phi_i\big(\mathfrak{m}_{n_i}(z)\big) \cdot \mathrm{Det}\big(\mathfrak{m}_{n_i}(z)\big)^{s_i}$$
$$= m^{-(n-1)/2} \sum_{\substack{\prod_{\ell=1}^{n} c_\ell = m \\ 0 \le c_{i,\ell} < c_\ell \ (1 \le i < \ell \le n)}} \prod_{i=1}^{r} \phi_i\big(\mathfrak{m}_{n_i}(cz)\big) \cdot \mathrm{Det}(\mathfrak{m}_{n_i}(cy))^{s_i} \qquad (10.9.4)$$

where

$$c = \begin{pmatrix} c_1 & c_{1,2} & \cdots & c_{1,n} \\ & c_2 & \cdots & c_{2,n} \\ & & \ddots & \vdots \\ & & & c_n \end{pmatrix}.$$

Note that

$$\mathfrak{m}_{n_i}(cz) = \mathfrak{m}_{n_i}(c) \cdot \mathfrak{m}_{n_i}(z).$$

Define $\eta_i = n_1 + n_2 + \cdots + n_{i-1}$ with the convention that $\eta_1 = 0$. Then for each $i = 1, 2, \ldots, r$, we have

$$\mathfrak{m}_{n_i}(cy) = \begin{pmatrix} c_{\eta_i+1} & c_{\eta_i+1,\eta_i+2} & \cdots & c_{\eta_i+1,\eta_{i+1}} \\ & c_{\eta_i+2} & \cdots & c_{\eta_i+2,\eta_{i+1}} \\ & & \ddots & \vdots \\ & & & c_{\eta_{i+1}} \end{pmatrix}$$
$$\times \begin{pmatrix} y_1 y_2 \cdots y_{n-\eta_i-1} & & & \\ & y_1 y_2 \cdots y_{n-\eta_i-2} & & \\ & & \ddots & \\ & & & y_1 y_2 \cdots y_{n-\eta_{i+1}} \end{pmatrix}.$$

To complete the proof, we rewrite the sum on the right-hand side of (10.9.4) so that for each $1 \le i \le r$, the sum turns into the local Hecke operator for $SL(n_i, \mathbb{R})$ acting on ϕ_i. This can be done as follows. Introduce r integers $1 \le C_1, C_2, \dots C_r$ satisfying $C_1C_2 \cdots C_r = \prod_{\ell=1}^{n} c_\ell = m$. Then (10.9.4) is equal to

$$m^{-(n-1)/2} \sum_{C_1C_2\cdots C_r=m} \prod_{i=1}^{r} \big(T_{C_i}\,\phi_i\big(\mathfrak{m}_{n_i}(z)\big)\big) \cdot C_i^{s_i+\eta_i} \cdot \mathrm{Det}\big(\mathfrak{m}_{n_i}(y)\big)^{s_i}$$

$$= m^{-(n-1)/2} \sum_{C_1C_2\cdots C_r=m} \prod_{i=1}^{r} A_i(C_i)\,\phi_i\big(\mathfrak{m}_{n_i}(z)\big) \cdot C_i^{s_i+\eta_i} \cdot \mathrm{Det}\big(\mathfrak{m}_{n_i}(y)\big)^{s_i},$$

where T_{C_i} denotes the Hecke operator on $SL(n_i, \mathbb{Z})$ and $\eta_i = n_1 + n_2 + \cdots + n_{i-1}$ if $i > 1$ while $\eta_1 = 0$. □

Propositions 10.9.1 and 10.9.3 allow one to show that the non-degenerate terms in the Whittaker expansion of general Langlands Eisenstein series for $SL(n, \mathbb{Z})$ with $n \ge 2$ have a meromorphic continuation and satisfy the same functional equation as the Whittaker functions that occur in the Fourier–Whittaker expansion.

10.10 The constant term

For $n, r \ge 2$, let $P = P_{n_1,\dots,n_r}$ be the standard parabolic subgroup of $GL(n, \mathbb{R})$ associated to the partition $n = n_1 + \cdots + n_r$. Let $\phi = (\phi_1, \dots, \phi_r)$ be a set of r Maass forms where for $i = 1, 2, \dots, r$, each ϕ_i is automorphic for $SL(n_i, \mathbb{Z})$. Let $s = (s_1, \dots, s_r) \in \mathbb{C}^r$ with $\sum_{i=1}^{r} n_i s_i = 0$. We are interested in determining the constant term (in the Fourier expansion) of the Langlands Eisenstein series

$$E_P(s, z, \phi) = \sum_{\gamma \in (P\cap\Gamma)\backslash\Gamma} \prod_{i=1}^{r} \phi_i\big(\mathfrak{m}_{n_i}(\gamma z)\big) \cdot \mathrm{Det}\big(\mathfrak{m}_{n_i}(\gamma z)\big)^{s_i} \qquad (10.10.1)$$

as in Definition 10.5.3. In (Langlands, 1966), the concept of the Fourier expansion along an arbitrary parabolic is introduced for the first time. This is a more general notion than the usual constant term and is required for the Langlands spectral decomposition. We shall now define this more general constant term.

Definition 10.10.2 (Constant term along an arbitrary parabolic) *Consider, $E_P(s, z, \phi)$, a Langlands Eisenstein series as in (10.10.1).Let P' be any parabolic subgroup of $GL(n, \mathbb{R})$ with Langlands decomposition $P' = N'M'$ as in (10.4.4). Then the constant term of E_P along the parabolic P' is given by*

$$\int_{N'(\mathbb{Z})\backslash N'(\mathbb{R})} E_P(\eta z, s, \phi)\, d^*\eta,$$

where $N'(\mathbb{Z}) = SL(n, \mathbb{Z}) \cap N'(\mathbb{R})$.

Theorem 10.10.3 (Langlands) *The constant term of E_P along a parabolic P' is zero if P has lower rank than P' or if P and P' have the same rank but are not associate (as in Definition 10.1.4).*

Proof We compute, for $P(\mathbb{Z}) = P \cap \Gamma$ with $\Gamma = SL(n, \mathbb{Z})$,

$$\begin{aligned}
&\int_{N'(\mathbb{Z})\backslash N'(\mathbb{R})} E_P(uz, s, \phi)\, d^*u \\
&\quad = \sum_{\gamma\in P(\mathbb{Z})\backslash\Gamma} \int_{N'(\mathbb{Z})\backslash N'(\mathbb{R})} \prod_{i=1}^{r} \phi_i\big(\mathfrak{m}_{n_i}(\gamma uz)\big) \cdot \mathrm{Det}\big(\mathfrak{m}_{n_i}(\gamma uz)\big)^{s_i}\, d^*u \\
&\quad = \sum_{\gamma\in P(\mathbb{Z})\backslash\Gamma/N'(\mathbb{Z})} \sum_{n'\in N'(\mathbb{Z})} \int_{N'(\mathbb{Z})\backslash N'(\mathbb{R})} \prod_{i=1}^{r} \phi_i\big(\mathfrak{m}_{n_i}(\gamma n' uz)\big) \\
&\qquad\qquad \times \mathrm{Det}\big(\mathfrak{m}_{n_i}(\gamma n' uz)\big)^{s_i}\, d^*u \\
&\quad = \sum_{\gamma\in P(\mathbb{Z})\backslash\Gamma/N'(\mathbb{Z})} \int_{((\gamma^{-1}P(\mathbb{Z})\gamma)\cap N'(\mathbb{Z}))\backslash N'(\mathbb{R})} \prod_{i=1}^{r} \phi_i\big(\mathfrak{m}_{n_i}(\gamma uz)\big) \\
&\qquad\qquad \times \mathrm{Det}\big(\mathfrak{m}_{n_i}(\gamma uz)\big)^{s_i}\, d^*u
\end{aligned} \tag{10.10.4}$$

because

$$P(\mathbb{Z})\gamma N'(\mathbb{Z})u_1 = P(\mathbb{Z})\gamma N'(\mathbb{Z})u_2$$

if and only if $u_1 u_2^{-1} \in N'(\mathbb{Z}) \cap (\gamma^{-1} P(\mathbb{Z})\gamma)$.

Now, by the Bruhat decomposition (see Propositions 10.3.2, 10.3.6), each

$$\gamma \in P(\mathbb{Z})\backslash\Gamma/N'(\mathbb{Z})$$

can be expressed in the form $\gamma = w\gamma'$ where w is in the Weyl group and $\gamma' \in B_n(\mathbb{Z})\backslash N'(\mathbb{Z})$. Making this replacement, a typical term in the sum on the right-hand side of (10.10.4) will be of the form

$$\int\limits_{((\gamma'^{-1}w^{-1}P(\mathbb{Z})w\gamma')\cap N'(\mathbb{Z}))\backslash N'(\mathbb{R})} \prod_{i=1}^{r} \phi_i\big(\mathfrak{m}_{n_i}(w\gamma' uz)\big)\cdot \mathrm{Det}\big(\mathfrak{m}_{n_i}(w\gamma' uz)\big)^{s_i}\, d^*u,$$

which after changing variables $u \mapsto \gamma'^{-1}u\gamma'$ becomes

$$\int\limits_{((w^{-1}P(\mathbb{Z})w)\cap N'(\mathbb{Z}))\backslash N'(\mathbb{R})} \prod_{i=1}^{r} \phi_i\big(\mathfrak{m}_{n_i}(wu\gamma' z)\big)\cdot \mathrm{Det}\big(\mathfrak{m}_{n_i}(wu\gamma' z)\big)^{s_i}\, d^*u.$$

Following Langlands, we define

$${}^0N = (M/A)\cap(wN'w^{-1}).$$

Then the above integral becomes

$$\int\limits_{((w^{-1}P(\mathbb{Z})w)\cap N'(\mathbb{Z}))\backslash(w^{-1}N'(\mathbb{R})w)\backslash N'(\mathbb{R})}\ \int\limits_{{}^0N(\mathbb{Z})\backslash {}^0N(\mathbb{R})} \prod_{i=1}^{r} \phi_i\big(\mathfrak{m}_{n_i}(u_1wuz)\big)$$
$$\times\, \mathrm{Det}\big(\mathfrak{m}_{n_i}(u_1wuz)\big)^{s_i}\, d^*u_1\, d^*u.$$

Since ϕ is a cusp form, the inner integral vanishes unless ${}^0N$ is the identity and this will only happen if P and P' are associate with $w \in \Omega(P, P')$. □

10.11 The constant term of $SL(3,\mathbb{Z})$ Eisenstein series twisted by $SL(2,\mathbb{Z})$-Maass forms

There are two maximal parabolic subgroups for $GL(3,\mathbb{R})$. They are $P_{1,2}$ and $P_{2,1}$, as in Example 10.1.2. The maximal parabolic Eisenstein series, associated to $P_{2,1}$, twisted by an $SL(2,\mathbb{Z})$ Maass form ϕ, as given in (10.5.5), takes the explicit form:

$$E_{P_{2,1}}(z,s,\phi) = \sum_{\gamma\in P_{21}\backslash\Gamma} \left(y_1^2y_2\right)^s \phi(z_2)\Big|_{\gamma}, \tag{10.11.1}$$

where $\Gamma = SL(3,\mathbb{Z})$, $z = \begin{pmatrix}1 & x_2 & x_3\\ & 1 & x_1\\ & & 1\end{pmatrix}\begin{pmatrix}y_1y_2 & & \\ & y_1 & \\ & & 1\end{pmatrix}$, with $z_2 = x_2 + iy_2$, and the slash operator $|_\gamma$ denotes the action of γ on z.

Proposition 10.11.2 *The constant terms of the Eisenstein series $E_{P_{2,1}}(z, s, \phi)$ given in (10.11.1) along the various parabolic subgroups take the form:*

$$\int_0^1\int_0^1\int_0^1 E_{P_{2,1}}\left(\begin{pmatrix}1 & u_2 & u_3\\ & 1 & u_1\\ & & 1\end{pmatrix} z, s, \phi\right) du_1 du_2 du_3 = 0,$$

$$\int_0^1\int_0^1 E_{P_{2,1}}\left(\begin{pmatrix}1 & & u_3\\ & 1 & u_1\\ & & 1\end{pmatrix} z, s, \phi\right) du_1 du_3 = 2\left(y_1^2 y_2\right)^s \phi(z_2),$$

$$\int_0^1\int_0^1 E_{P_{2,1}}\left(\begin{pmatrix}1 & u_2 & u_3\\ & 1 & \\ & & 1\end{pmatrix} z, s, \phi\right) du_2 du_3 = 2y_1^{1-s} y_2^{2-2s} \frac{\Lambda_\phi(\lambda-1)}{\Lambda_\phi(\lambda)}\,\phi(z_1),$$

where

$$\Lambda_\phi(s) = \pi^{-s}\Gamma\left(\frac{s+\epsilon+ir}{2}\right)\Gamma\left(\frac{s+\epsilon-ir}{2}\right)L_\phi(s),$$

(with $\epsilon = 0$ or 1 according as ϕ is even or odd), is the completed L-function of the Maass form ϕ (of type $\frac{1}{2}+ir$) as in Proposition 3.13.5, and where

$$\lambda = \begin{cases} 3s - \frac{1}{2} & \text{if } \phi \text{ is even} \\ 3s + \frac{1}{2} & \text{if } \phi \text{ is odd.} \end{cases}$$

Proof Omitted. □

10.12 An application of the theory of Eisenstein series to the non-vanishing of L-functions on the line $\Re(s) = 1$

The prime number theorem (Davenport, 1974) states that $\pi(x)$, the number of primes less than x, is asymptotic to $x/\log x$ as $x \to \infty$. A key ingredient to the analytic proof of the prime number theorem is the fact that the Riemann zeta function $\zeta(s)$ does not vanish on the line $\Re(s) = 1$. In (Jacquet and Shalika, 1976/77), a new proof of the non-vanishing of $\zeta(s)$ on the line $\Re(s) = 1$ was obtained. This proof had two very interesting features.

- *It made use of the theory of Eisenstein series.*
- *The proof could be vastly generalized.*

Recently, Sarnak (2004), Gelbart, Lapid and Sarnak (2004) obtained explicit zero–free regions for general automorphic L-functions by use of the Jacquet–Shalika method. We shall now present a short exposition of this method by considering the classical case of $GL(2)$ and then the not so classical case of $GL(3)$.

Non-vanishing of $\zeta(s)$ on $\Re(s) = 1$ Recall the Fourier expansion, given in Theorem 3.1.8, of the $SL(2, \mathbb{Z})$ Eisenstein series

$$E(z, s) = y^s + \phi(s)y^{1-s} + \frac{2\pi^s\sqrt{y}}{\Gamma(s)\zeta(2s)} \sum_{n\neq 0} \sigma_{1-2s}(n)|n|^{s-\frac{1}{2}} K_{s-\frac{1}{2}}(2\pi|n|y)e^{2\pi inx}$$

where

$$\phi(s) = \sqrt{\pi}\frac{\Gamma\left(s-\frac{1}{2}\right)}{\Gamma(s)}\frac{\zeta(2s-1)}{\zeta(2s)}, \qquad \sigma_s(n) = \sum_{\substack{d|n\\ d>0}} d^s,$$

and

$$K_s(y) = \frac{1}{2}\int_0^\infty e^{-\frac{1}{2}y(u+(1/u))}u^s\frac{du}{u}.$$

If $\zeta(1 + it_0) = 0$ for some $t_0 \in \mathbb{R}$, then one easily sees that if

$$E^*(z, s) = \pi^{-s}\Gamma(s)\zeta(2s)E(z, s) = E^*(z, 1 - s),$$

then $E^*(z, (1 + t_0)/2)$ must be a non-constant Maass form of type $\frac{1}{4} + t_0^2$. This is because the constant term of the Eisenstein series $E^*(z, s)$ will vanish when $s = (1 + t_0)/2$. It is easy to show that $E^*(z, (1 + t_0)/2)$ is non-constant, in particular non-zero, because the sum:

$$\sum_{n\neq 0} \sigma_{1-2s}(n)|n|^{s-\frac{1}{2}} K_{s-\frac{1}{2}}(2\pi|n|y)e^{2\pi inx}$$

will be non-zero high in the cusp. The key point is that an Eisenstein series can never be a cusp form because Eisenstein series are orthogonal to cusp forms, so the inner product

$$\left\langle E\left(*, \frac{1+t_0}{2}\right), E\left(*, \frac{1+t_0}{2}\right)\right\rangle$$

would have to be zero. This contradicts the fact that $E(*, (1 + t_0)/2)$ is not identically zero. Since we have obtained a contradiction, our original assumption: *that $\zeta(s)$ vanished on the line $\Re(s) = 1$*, must be false!

Non-vanishing of $GL(2)$ L-functions on the line $\Re(s) = 1$ The argument described above can be vastly generalized. We shall give another example based on Proposition 10.11.2. Let ϕ be a Maass form for $SL(2, \mathbb{Z})$ with associated L-function $L_\phi(s)$. Let $\Lambda_\phi(\lambda)$ denote the completed L-function as in Proposition 10.11.2, and define $E^*_{P_{2,1}}(z, s, \phi) = \Lambda(\lambda)E^*_{P_{2,1}}(z, s, \phi)$. If $L_\phi(s)$ vanishes on

$\Re(s) = 1$, then there exists a special value of s such that $E^*_{P_{2,1}}(z, s, \phi)$ will be a Maass form for $SL(3, \mathbb{Z})$, i.e., its constant term will vanish. One again obtains a contradiction by taking the inner product of E^* with itself.

10.13 Langlands spectral decomposition for $SL(3, \mathbb{Z})\backslash\mathfrak{h}^3$

We conclude this chapter with the Langlands spectral decomposition for the special case of $SL(3, \mathbb{Z})$. In order to state the main result succinctly, we introduce the following notation. Let u_j, $(j = 0, 1, 2, \ldots)$ denote a basis of normalized Maass forms for $SL(2, \mathbb{Z})$ with $z \in \mathfrak{h}^2$. Here u_0 is the constant function, and each u_j is normalized to have Petersson norm one. For each $j = 0, 1, 2, \ldots$, and $s \in \mathbb{C}$, define

$$E_j(z, s) := E_{P_{2,1}}(z, s, u_j)$$

as in (10.11.1). We also define for $s_1, s_2 \in \mathbb{C}$, the minimal parabolic Eisenstein series

$$E(z, s_1, s_2) := E_{P_{\min}}(z, (s_1, s_2))$$

as in (10.4.1). The following theorem generalizes Selberg's spectral decomposition given in Section 3.16. We follow the proof of M. Thillainatesan and thank her for allowing us to incorporate it here.

Theorem 10.13.1 (Langlands spectral decomposition) *Assume that $\phi \in \mathcal{L}^2(SL(3, \mathbb{Z})\backslash\mathfrak{h}^3)$ is orthogonal to the residues of all the Eisenstein series and is of sufficiently rapid decay that $\langle \phi, E\rangle$ converges absolutely for all the Eisenstein series E. Then the function*

$$\phi(z) - \frac{1}{(4\pi i)^2}\int_{(\frac{1}{3})}\int_{(\frac{1}{3})} \langle \phi, E(*, s_1, s_2)\rangle\, E(z, s_1, s_2)\, ds_1 ds_2$$
$$- \frac{1}{2\pi i}\sum_{j=0}^{\infty}\int_{(\frac{1}{2})} \langle \phi, E_j(*, s)\rangle\, E_j(z, s)\, ds,$$

is a cusp form for $SL(3, \mathbb{Z})$.

Proof Let $f(z)$ be an arbitrary automorphic form on $\mathfrak{h}^3$. We shall adopt the simplifying notation that $f_{0,0}$ denotes the constant term of f along the minimal parabolic, $f_{2,1}$ denotes the constant term of f along the maximal parabolic $P_{2,1}$, and $f_{1,2}$ denotes the constant term along $P_{1,2}$. In order to prove the Langlands spectral decomposition theorem for $SL(3, \mathbb{Z})$, it is necesssary to show that the following identities (10.13.2), (10.13.3), (10.13.4) hold for a Maass form

$\phi \in \mathcal{L}^2(SL(3,\mathbb{Z})\backslash\mathfrak{h}^3)$.

$$\phi_{00}(z) = \frac{1}{(4\pi i)^2} \iint\limits_{\Re(s_i)=\frac{1}{3}} \langle \phi, E(*, s_1, s_2)\rangle\, E_{00}(z, s_1, s_2)\, ds_1 ds_2. \tag{10.13.2}$$

Note that $E_{j,00}(z,s) = 0$ for $j \geq 1$ by Theorem 10.10.3.

$$\begin{aligned} \phi_{21}(z) = \frac{1}{(4\pi i)^2} \int_{(\frac{1}{3})}\int_{(\frac{1}{3})} \langle \phi, E(*, s_1, s_2)\rangle\, E_{21}(z, s_1, s_2)\, ds_1 ds_2 \\ - \frac{1}{2\pi i}\sum_{j=0}^{\infty} \int_{(\frac{1}{2})} \langle \phi, E_j(*, s)\rangle\, E_{j,21}(z, s)\, ds, \end{aligned} \tag{10.13.3}$$

$$\begin{aligned} \phi_{12}(z) = \frac{1}{(4\pi i)^2} \int_{(\frac{1}{3})}\int_{(\frac{1}{3})} \langle \phi, E(*, s_1, s_2)\rangle\, E_{12}(z, s_1, s_2)\, ds_1 ds_2 \\ - \frac{1}{2\pi i}\sum_{j=0}^{\infty} \int_{(\frac{1}{2})} \langle \phi, E_j(*, s)\rangle\, E_{j,12}(z, s)\, ds. \end{aligned} \tag{10.13.4}$$

The idea of the proof is to embed $\mathfrak{h}^2$ in $\mathfrak{h}^3$ and show that ϕ_{21} is invariant under $SL(2,\mathbb{Z})$ with this embedding. Then, we use a $GL(2)$ spectral decomposition of ϕ_{21}.

Following (Garrett, 2002), for $z \in \mathfrak{h}^3 = SL(3,\mathbb{R})/SO(3,\mathbb{R})$, we can write:

$$z = \begin{pmatrix} 1 & 0 & x_3 \\ 0 & 1 & x_1 \\ 0 & 0 & 1 \end{pmatrix} \begin{pmatrix} \multicolumn{2}{c}{z_2} & \begin{matrix}0\\0\end{matrix} \\ 0 \quad 0 & & 1 \end{pmatrix} \begin{pmatrix} 1/\sqrt{l} & 0 & 0 \\ 0 & 1/\sqrt{l} & 0 \\ 0 & 0 & l \end{pmatrix},$$

where $x_1, x_3, l \in \mathbb{R}$ and $l > 0$. Note that $z_2 \in \mathfrak{h}^2 = SL(2,\mathbb{R})/SO(2,\mathbb{R})$.

It can be shown that the constant term ϕ_{21}, as a function of z_2, is invariant under $\gamma \in SL(2,\mathbb{Z})$ and has constant term ϕ_{00}. So already the proof of the main theorem is reduced to showing (10.13.3) and (10.13.4).

In the proof of (10.13.3) and (10.13.4) we will need to know $\langle \phi, E\rangle$ for each of the Eisenstein series that we have defined. We begin by calculating $\langle \phi, E\rangle$ for the minimal parabolic Eisenstein series.

Lemma 10.13.5 *We have*

$$\langle \phi, E(*, \overline{s_1}, \overline{s_2})\rangle = \tilde{\phi}_{00}(2s_1 + s_2 - 2, s_1 + 2s_2 - 2),$$

where $\tilde{\phi}_{00}$ is the double Mellin transform of the constant term ϕ_{00}.

The proof of Lemma 10.13.5 is fairly straightforward and we will omit it. We will, however, give the calculation of $\langle \phi, E_j(*, \overline{s})\rangle$. It is part of our assumption that the integral converges absolutely.

$$\begin{aligned}\langle \phi, E_j(*, \overline{s})\rangle &= \int_{\Gamma\backslash\mathfrak{h}^3} \phi(z)\, \overline{E_j(z, \overline{s})}\, d^*z, \\ &= \int_{P_{21}\backslash\mathfrak{h}^3} \phi(z)\, y_1^{2s}\, y_2^{s}\, \overline{u_j(z_2)}\, d^*z.\end{aligned}$$

With the previous notation, l is given by $y_1^2 y_2 = l^3$. As in Section 1.6, we have the following change of coordinates.

$$P_{21}\backslash\mathfrak{h}^3 \equiv SL(2,\mathbb{Z})\backslash\mathfrak{h}^2 \times (\mathbb{R}/\mathbb{Z})^2 \times [0 < l < \infty],$$

and

$$d^*z = \frac{3}{2}\, l^{-3}\, d^*z_2\, dx_3\, dx_1\, \frac{dl}{l}.$$

Continuing the calculation gives

$$\begin{aligned}\langle \phi, E_j(*, \overline{s})\rangle &= \frac{3}{2}\int_0^\infty \int_0^1 \int_0^1 \int_{SL(2,\mathbb{Z})\backslash\mathfrak{h}^2} \phi(z)\, l^{3s-3}\, \overline{u_j(z_2)}\, d^*z_2\, dx_3\, dx_1\, \frac{dl}{l}, \\ &= \frac{3}{2}\int_0^\infty \int_{SL(2,\mathbb{Z})\backslash\mathfrak{h}^2} \phi_{21}(z_2, l)\, l^{3s-3}\, \overline{u_j(z_2)}\, d^*z_2\, \frac{dl}{l}, \\ &= \frac{3}{2}\int_0^\infty \langle \phi_{21}(*, l), u_j\rangle\, l^{3s-3}\, \frac{dl}{l}.\end{aligned}$$

The above calculation proves the following Lemma 10.13.6. When we use the notation of Garrett (2002), we will write $\phi_{21}(z_2, l)$ for $\phi_{21}(z)$.

Lemma 10.13.6 *For* $\Re(s) \gg 1$,

$$\langle \phi, E_j(*, \overline{s})\rangle = \frac{3}{2}\tilde{a}_j(3s - 3),$$

where $\tilde{a}_j(s)$ *is the Mellin transform of*

$$a_j(l) = \langle \phi_{21}(*, l), u_j\rangle.$$

As observed earlier, the constant term ϕ_{21} considered as a function of z_2, has a Selberg spectral expansion. We have given a proof in Section 3.16 of the Selberg ($SL(2,\mathbb{Z})$) spectral decomposition using Mellin transforms. Then

$$\phi_{21}(z_2, l) = \sum_{j=0}^{\infty} \langle \phi_{21}(*, l), u_j\rangle\, u_j(z_2) + \frac{1}{4\pi i}\int_{(\frac{1}{2})} \langle \phi_{21}(*, l), E(*, v)\rangle\, E(z_2, v)\, dv,$$

where $E(z, s)$ is the $GL(2)$ Eisenstein series.

Note that if we apply Mellin inversion to the results of Lemma 10.13.6 and by our assumptions move the line of integration to $(\frac{1}{2})$, we have the following.

$$a_j(l) = \langle \phi_{21}(*, l), u_j \rangle = \frac{1}{\pi i}\int_{(\frac{1}{2})} \langle \phi, E_j(*, \overline{s})\rangle\, l^{3-3s}\, ds. \tag{10.13.7}$$

Now we can put this together with the spectral decomposition of $\phi_{21}(z_2, l)$ to get

$$\phi_{21}(z_2, l) = \frac{1}{\pi i}\sum_{j=0}^{\infty}\int_{(\frac{1}{2})} \langle \phi, E_j(*, \overline{s})\rangle\, l^{3-3s}\, u_j(z_2)\, ds + \frac{1}{4\pi i}\int_{(\frac{1}{2})} \langle \phi_{21}(*, l), E(*, v)\rangle\, E(z_2, v)\, dv.$$

Make the change of variable $s \to 1 - s$ in the first integral. Then the equation above becomes:

$$\phi_{21}(z_2, l) = \frac{1}{\pi i}\sum_{j=0}^{\infty}\int_{(\frac{1}{2})} \langle \phi, E_j(*, s)\rangle\, l^{3s}\, u_j(z_2)\, ds + \frac{1}{4\pi i}\int_{(\frac{1}{2})} \langle \phi_{21}(*, l), E(*, v)\rangle\, E(z_2, v)\, dv. \tag{10.13.8}$$

We shall need to use Proposition 10.11.2 on the Fourier expansion of the maximal parabolic Eisenstein series. It follows that we can rewrite (10.13.8) in the form

$$\phi_{21}(z_2, l) = \frac{1}{2\pi i}\sum_{j=0}^{\infty}\int_{(\frac{1}{2})} \langle \phi, E_j(*, s)\rangle\, E_{j,21}(z_2, l, s)\, ds + \frac{1}{4\pi i}\int_{(\frac{1}{2})} \langle \phi_{21}(*, l), E(*, v)\rangle\, E(z_2, v)\, dv. \tag{10.13.9}$$

Let us now assume the following proposition, whose proof will be deferred to later.

Proposition 10.13.10 *We have*

$$\frac{1}{4\pi i}\int\limits_{(\frac{1}{2})} \langle \phi_{21}(*, l), E(*, v)\rangle\, E(z_2, v)\, dv = \frac{1}{(4\pi i)^2}\int\limits_{(\frac{1}{3})}\int\limits_{(\frac{1}{3})} \langle \phi, E(*, s_1, s_2)\rangle\, E_{21}(z_2, l, s_1, s_2)\, ds_1 ds_2.$$

Using the above proposition, we can rewrite (10.13.9) in the form

$$\phi_{21}(z) = \frac{1}{2\pi i}\sum_{j=0}^{\infty}\int\limits_{(\frac{1}{2})}\langle\phi, E_j(*, s)\rangle\, E_{j,21}(z, s)\, ds$$
$$+\frac{1}{(4\pi i)^2}\int\limits_{(\frac{1}{3})}\int\limits_{(\frac{1}{3})}\langle\phi, E(*, s_1, s_2)\rangle\, E_{21}(z, s_1, s_2)\, ds_1 ds_2.$$

This proves (10.13.3). This is almost the end because (10.13.4) is proved using (10.13.3) and a few properties of Eisenstein series. On $GL(3)$, there is an involution which preserves the Iwasawa form. The involution ι is defined by

$$^{\iota}z = w\, {}^t z^{-1}\, w, \qquad \text{where} \qquad w = \begin{pmatrix} & & 1\\ & 1 & \\ 1 & & \end{pmatrix}. \tag{10.13.11}$$

For any automorphic form $\phi(z)$, we can define another automorphic form, also on $SL(3, \mathbb{Z})$, denoted $\tilde{\phi}(z)$ given by

$$\tilde{\phi}(z) = \phi({}^{\iota}z). \tag{10.13.12}$$

We let $\tilde{E}_j(z, s) = E_j({}^{\iota}z, s)$. It is easy to show that $\tilde{E}(z, s_1, s_2) = E(z, s_2, s_1)$. It follows that $\phi_{21}({}^{\iota}z) = \tilde{\phi}_{12}(z)$ and $\phi_{12}({}^{\iota}z) = \tilde{\phi}_{21}(z)$. So, applying (10.13.3) to $\tilde{\phi}$ gives

$$\tilde{\phi}_{21}(z) = \frac{1}{2\pi i}\sum_{j=0}^{\infty}\int\limits_{(\frac{1}{2})}\langle\tilde{\phi}, E_j(*, s)\rangle\, E_{j,21}(z, s)\, ds$$
$$+\frac{1}{(4\pi i)^2}\int\limits_{(\frac{1}{3})}\int\limits_{(\frac{1}{3})}\langle\tilde{\phi}, E(*, s_1, s_2)\rangle\, E_{21}(z, s_1, s_2)\, ds_1 ds_2.$$

Since $\tilde{\phi}_{21}({}^{\iota}z) = \phi_{12}(z)$, it follows that:

$$\phi_{12}(z) := \frac{1}{2\pi i}\sum_{j=0}^{\infty}\int\limits_{(\frac{1}{2})}\langle\tilde{\phi}, E_j(*, s)\rangle\, E_{j,21}({}^{\iota}z, s)\, ds$$
$$+\frac{1}{(4\pi i)^2}\int\limits_{(\frac{1}{3})}\int\limits_{(\frac{1}{3})}\langle\tilde{\phi}, E(*, s_1, s_2)\rangle\, E_{21}({}^{\iota}z, s_1, s_2)\, ds_1 ds_2.$$

It is easy to see that $\langle\tilde{\phi}, \tilde{\psi}\rangle = \langle\phi, \psi\rangle$. It follows that $\langle\tilde{\phi}, E_j(*, s)\rangle = \langle\phi, \tilde{E}_j(*, s)\rangle$. Also, we have the following equalities:

$$\langle\tilde{\phi}, E(*, s_1, s_2)\rangle = \langle\phi, \tilde{E}(*, s_1, s_2)\rangle = \langle\phi, E(*, s_2, s_1)\rangle.$$

From these statements, we can continue the calculation of $\phi_{12}(z)$.

$$\phi_{12}(z) = \frac{1}{2\pi i}\sum_{j=0}^{\infty}\int_{(\frac{1}{2})}\langle\phi, \tilde{E}_j(*,s)\rangle\, \tilde{E}_{j,12}(z,s)\,ds + \frac{1}{(4\pi i)^2}\int_{(\frac{1}{3})}\int_{(\frac{1}{3})}\langle\phi, E(*,s_2,s_1)\rangle\, E_{12}(z,s_2,s_1)\,ds_1 ds_2.$$

Let $s \to 1-s$ in the first integral. Then

$$\phi_{12}(z) = \frac{1}{2\pi i}\sum_{j=0}^{\infty}\int_{(\frac{1}{2})}\langle\phi, \tilde{E}_j(*,1-s)\rangle\, \tilde{E}_{j,12}(z,1-s)\,ds + \frac{1}{(4\pi i)^2}\int_{(\frac{1}{3})}\int_{(\frac{1}{3})}\langle\phi, E(*,s_2,s_1)\rangle\, E_{12}(z,s_2,s_1)\,ds_1 ds_2. \quad (10.13.13)$$

Now, the maximal Eisenstein series has the functional equation:

$$\tilde{E}_j(z,1-s) = \theta(s)\,E_j(z,s),$$

where

$$\theta(s) = \frac{\Lambda_j(\lambda-1)}{\Lambda_j(\lambda)}.$$

The coefficients in $L_j(v)$ are real since they come from the Hecke form u_j, so we know that $\overline{\theta(s)} = \theta(\bar{s})$. Applying these observations to (10.13.13) gives

$$\phi_{12}(z) = \frac{1}{2\pi i}\sum_{j=0}^{\infty}\int_{(\frac{1}{2})}\theta(\bar{s})\langle\phi, E_j(*,s)\rangle\, \tilde{E}_{j,12}(z,1-s)\,ds + \frac{1}{(4\pi i)^2}\int_{(\frac{1}{3})}\int_{(\frac{1}{3})}\langle\phi, E(*,s_2,s_1)\rangle\, E_{12}(z,s_2,s_1)\,ds_1 ds_2,$$

$$\phi_{12}(z) = \frac{1}{2\pi i}\sum_{j=0}^{\infty}\int_{(\frac{1}{2})}\langle\phi, E_j(*,s)\rangle\, E_{j,12}(z,s)\,ds + \frac{1}{(4\pi i)^2}\int_{(\frac{1}{3})}\int_{(\frac{1}{3})}\langle\phi, E(*,s_1,s_2)\rangle\, E_{12}(z,s_1,s_2)\,ds_1 ds_2.$$

This proves (10.13.4), thus proving the main Theorem 10.13.1. □

Proof of Proposition 10.13.10 It still remains to prove Proposition 10.13.10. Let us define for $\Re(v_1) \gg 1$ and $\Re(v_2) \gg 1$,

$$E^*(z, v_1, v_2) = 2 \sum_{\gamma \in P_{21}\backslash\Gamma} \left(y_1^2 y_2\right)^{v_1} E(z_2, v_2)\Big|_{\gamma},$$

a series which is a rearrangement of the terms of the minimal parabolic Eisenstein series. We can see this as follows. For $\Re(s_1), \Re(s_2) \gg 1$,

$$\begin{aligned} E(z, s_1, s_2) &= \sum_{\gamma \in \Gamma_\infty\backslash\Gamma} y_1^{2s_1+s_2} y_2^{s_1+2s_2}\Big|_{\gamma}, \\ &= \sum_{\gamma \in \Gamma_\infty\backslash\Gamma} y_1^{2s_1+s_2} y_2^{s_1+(s_2/2)+3s_2/2}\Big|_{\gamma}, \\ &= \sum_{\beta \in \Gamma_\infty\backslash P_{21}} \sum_{\alpha \in P_{21}\backslash\Gamma} \left(y_1^2 y_2\right)^{s_1+(s_2/2)} y_2^{3s_2/2}\Big|_{\beta\alpha}. \end{aligned}$$

It was shown in Chapter 6 that the function $\det z = y_1^2 y_2$, is invariant under $\beta \in P_{21}$. So, $\det \beta z = \det z$.

Previously we had dropped the dependence of n in the notation, but to avoid confusion we need to introduce n in the notation. Thus $\Gamma(n) = SL(n, \mathbb{Z})$ and $\Gamma_\infty(n)$ is the set of upper-triangular matrices in $SL(n, \mathbb{Z})$ with 1s on the diagonal.

$$\begin{aligned} E(z, s_1, s_2) &= \sum_{\alpha \in P_{21}\backslash\Gamma(3)} \left(y_1^2 y_2\right)^{s_1+(s_2/2)} \sum_{\beta \in \Gamma_\infty(3)\backslash P_{21}} y_2^{3s_2/2}\Big|_{\beta}\Big|_{\alpha}, \\ &= 2 \sum_{\alpha \in P_{21}\backslash\Gamma(3)} \left(y_1^2 y_2\right)^{s_1+(s_2/2)} E\left(z_2, \frac{3s_2}{2}\right)\Big|_{\alpha}, \\ &= E^*\left(z, s_1 + \frac{s_2}{2}, \frac{3s_2}{2}\right). \end{aligned}$$

In the above we have used $\Gamma_\infty(3)\backslash P_{21} \equiv \Gamma_\infty(2)\backslash SL(2, \mathbb{Z})$, a fact which is easy to prove and also follows as a special case of Lemma 5.3.13.

Now, we will drop the dependence of the notation on n. To prove the proposition, we want to calculate the following inner product, in the region of absolute convergence of $E^*(z, v_1, v_2)$. In a manner similar to the calculation of Lemma 10.13.6, we have

$$\begin{aligned} \langle \phi, E^*(*, \overline{v_1}, \overline{v_2})\rangle &= \int_{\Gamma\backslash\mathfrak{h}^3} \phi(z)\, \overline{E^*(z, \overline{v_1}, \overline{v_2})}\, d^*z \\ &= \int_{\Gamma\backslash\mathfrak{h}^3} \phi(z)\, E^*(z, v_1, v_2)\, d^*z, \end{aligned}$$

$$= 3\int_0^\infty\int_0^1\int_0^1 \int\limits_{SL(2,\mathbb{Z})\backslash\mathfrak{h}^2} \phi(z) y_1^{2v_1} y_2^{v_1} l^{-3} E(z_2, v_2)$$
$$\times\, d^*z_2\, dx_3\, dx_1 \frac{dl}{l},$$
$$= 3\int_0^\infty \int\limits_{SL(2,\mathbb{Z})\backslash\mathfrak{h}^2} \phi_{21}(z_2, l)\, l^{3v_1-3}\, E(z_2, v_2)\, d^*z_2 \frac{dl}{l},$$
$$= 3\int_0^\infty \langle \phi_{21}(*, l), E(*, \overline{v_2})\rangle\, l^{3v_1-3} \frac{dl}{l}.$$

Let $a(l) = \langle \phi_{21}(*, l), E(*, \overline{v_2})\rangle$, then we have just shown that

$$\langle \phi, E^*(*, \overline{v_1}, \overline{v_2})\rangle = 3\,\tilde{a}(3v_1 - 3).$$

If we take the inverse Mellin transform of $\tilde{a}(v)$, then for some $\sigma \gg 0$ we have

$$a(l) = \frac{1}{2\pi i}\int_{(\sigma)} \tilde{a}(v) l^{-v}\, dv,$$
$$= \frac{3}{2\pi i}\int_{(\sigma')} \tilde{a}(3v_1 - 3)\, l^{3-3v_1}\, dv_1,$$
$$= \frac{1}{2\pi i}\int_{(\frac{1}{2})} \langle \phi, E^*(*, \overline{v_1}, \overline{v_2})\rangle\, l^{3-3v_1}\, dv_1.$$

In the last step we have used the assumption that ϕ is orthogonal to all the residues of the Eisenstein series to move the line to $\Re(v_1) = \frac{1}{2}$.

We have just shown that

$$\langle \phi_{21}(*, l), E(*, \overline{v_2})\rangle = \frac{1}{2\pi i}\int\limits_{(\frac{1}{2})} \langle \phi, E^*(*, \overline{v_1}, \overline{v_2})\rangle\, l^{3-3v_1}\, dv_1. \quad (10.13.14)$$

Recall that we want to show

$$\frac{1}{4\pi i}\int\limits_{(\frac{1}{2})} \langle \phi_{21}(*, l), E(*, v_2)\rangle\, E(z_2, v_2)\, dv_2$$
$$= \frac{1}{(4\pi i)^2}\int\limits_{(\frac{1}{3})}\int\limits_{(\frac{1}{3})} \langle \phi, E(*, s_1, s_2)\rangle\, E_{21}(z_2, l, s_1, s_2)\, ds_1 ds_2.$$

Using (10.13.14), we have that the left-hand side of the above is

$$
\begin{aligned}
LHS &= \frac{1}{4\pi i}\int_{(\frac{1}{2})} \langle \phi_{21}(*, l), E(*, v_2)\rangle\, E(z_2, v_2)\, dv_2,\\
&= \frac{1}{2(2\pi i)^2}\int_{(\frac{1}{2})}\int_{(\frac{1}{2})} \langle \phi, E^*(*, \overline{v_1}, v_2)\rangle\, l^{3(1-v_1)} E(z_2, v_2)\, dv_1 dv_2,\\
&= \frac{1}{2(2\pi i)^2}\int_{(\frac{1}{2})}\int_{(\frac{1}{2})} \langle \phi, E^*(*, v_1, v_2)\rangle\, l^{3v_1}\, E(z_2, v_2)\, dv_1 dv_2,
\end{aligned}
$$

where we have let $v_1 \to 1 - v_1$. Note that $l^3 = y_1^2 y_2$. So,

$$
LHS = \frac{1}{2(2\pi i)^2}\int_{(\frac{1}{2})}\int_{(\frac{1}{2})} \langle \phi, E^*(*, v_1, v_2)\rangle\, y_1^{2v_1} y_2^{v_1}\, E(z_2, v_2)\, dv_1 dv_2.
$$

Now let $v_1 = s_1 + (s_2/2)$ and $v_2 = 3s_2/2$. From the definition of $E^*(z, v_1, v_2)$, we get the following.

$$
LHS = \frac{3}{4(2\pi i)^2}\int_{(\frac{1}{3})}\int_{(\frac{1}{3})} \langle \phi, E(*, s_1, s_2)\rangle\, \left(y_1^2 y_2\right)^{s_1+(s_2/2)}\, E\left(z_2, \frac{3s_2}{2}\right) ds_1 ds_2. \tag{10.13.15}
$$

Using the fact that $\Phi_{21}(z_2, l)$ is invariant under $SL(2, \mathbb{Z})$ and so has a Fourier expansion, it is easy to show that for any form $\Phi(z)$,

$$
\Phi_{21}(z) = \sum_{n\in\mathbb{Z}} \Phi_{0,n}(z),
$$

where

$$
\Phi_{n_1,n_2}(z) = \int_0^1\int_0^1\int_0^1 \Phi\left(\begin{pmatrix} 1 & u_2 & u_3\\ 0 & 1 & u_1\\ 0 & 0 & 1\end{pmatrix} z\right) e(-n_1u_1 - n_2u_2)\, du_1 du_2 du_3.
$$

We will need some results from (Bump, 1984) and will use the notation from there. In (Bump, 1984), it is shown that the expansion of $E_{0,n}$ is the sum of three Whittaker functions indexed by $w \in V$, where V is a certain subset of the Weyl group W. (In the notation of (Bump, 1984), $V = \{w_1, w_2, w_4\}$.) So,

$$
E_{0,n}(z, s_1, s_2) = \frac{1}{2}\sum_{w\in V} a_w(n, s_1, s_2)\, W_{0,n}^{(s_1,s_2)}(z, w). \tag{10.13.16}
$$

Using the expansion of the $GL(2)$ Eisenstein series, it is easy to show that the factor $(y_1^2y_2)^{s_1+(s_2/2)}\, E\,(z_2, 3s_2/2)$ which appears in (10.13.15) is exactly:

$$
\left(y_1^2 y_2\right)^{s_1+(s_2/2)} E\left(z_2, \frac{3s_2}{2}\right) = \frac{1}{2}\sum_{n\in\mathbb{Z}} a_{w_2}(n, s_1, s_2)\, W_{0,n}^{(s_1,s_2)}(z, w_2).
$$

Now in (10.13.15), we will divide the integral into three parts and in each part let $(s_1, s_2) \to (u(s_1), u(s_2))$ for $u \in U$. We choose the subset U of W in such a way as to get the sum over u to give us $E_{0,n}(z, s_1, s_2)$. We can do this because the Whittaker functions have certain transformation properties. We will be interested in the transformation of $W_{0,n}^{(s_1,s_2)}(z, w_2)$ under the action of the Weyl group on (s_1, s_2); specifically for $w \in V$, there exists a $u \in U$ depending on w, such that

$$W_{0,n}^{(u(s_1),u(s_2))}(z, w_2) = B_w^u \, W_{0,n}^{(s_1,s_2)}(z, w). \tag{10.13.17}$$

Recall that $w \in V$ are those that appear in the expansion of E_{21}.

LHS

$$\begin{aligned}
&= \sum_{n\in\mathbb{Z}} \frac{3}{8(2\pi i)^2} \int_{(\frac{1}{3})}\int_{(\frac{1}{3})} \langle \phi, E(*, s_1, s_2)\rangle \, a_{w_2}(n, s_1, s_2) W_{0,n}^{(s_1,s_2)}(z, w_2)\, ds_1 ds_2,\\
&= \sum_{n\in\mathbb{Z}}\sum_{u\in U} \frac{1}{8(2\pi i)^2} \int\limits_{(\frac{1}{3})}\int\limits_{(\frac{1}{3})} \langle \phi, E(*, u(s_1), u(s_2))\rangle \, a_{w_2}(n, u(s_1), u(s_2))\\
&\qquad \times W_{0,n}^{(u(s_1),u(s_2))}(z, w_2)\, ds_1 ds_2,\\
&= \sum_{n\in\mathbb{Z}}\sum_{w\in V} \frac{1}{8(2\pi i)^2} \int\limits_{(\frac{1}{3})}\int\limits_{(\frac{1}{3})} \langle \phi, E(*, u(s_1), u(s_2))\rangle \, a_{w_2}(n, u(s_1), u(s_2))\\
&\qquad \times B_w^u \, W_{0,n}^{(s_1,s_2)}(z, w) ds_1 ds_2.
\end{aligned}$$

Also note that at $\Re(s_i) = \frac{1}{3}$, we have $\Re(u(s_i)) = \frac{1}{3}$.
From Bump again, we can show that for any $w \in W$,

$$E(z, w(s_1), w(s_2)) = \varphi_w(s_1, s_2) E(z, s_1, s_2)$$

for some φ_w independent of z. This is done as follows. First, let us define, as in (Bump, 1984), the function $G(z, s_1, s_2)$ in the following way.

$$G(z, s_1, s_2) = B(s_1, s_2)\, E(z, s_1, s_2),$$

where

$$\begin{aligned}
B(s_1, s_2) = \frac{\zeta(3s_1)\,\zeta(3s_2)\,\zeta(3s_1+3s_2-1)}{4\pi^{3s_1+3s_2-1/2}} \Gamma\left(\frac{3s_1}{2}\right) \Gamma\left(\frac{3s_2}{2}\right)&\\
\times \Gamma\left(\frac{3s_1+3s_2-1}{2}\right) E(z, s_1, s_2).&
\end{aligned}$$

Now from (Bump, 1984), we know that $G(z, s_1, s_2)$ is invariant under the action of the Weyl group on the pair (s_1, s_2). In particular, it follows,

$$B(w(s_1), w(s_2))\, E(z, w(s_1), w(s_2)) = B(s_1, s_2) E(z, s_1, s_2). \tag{10.13.18}$$

So, $\varphi_w(s_1, s_2) = \dfrac{B(s_1, s_2)}{B(w(s_1), w(s_2))}$. From the definition $\overline{B(s_1, s_2)} = B(\overline{s_1}, \overline{s_2})$. At $\Re(s_i) = \frac{1}{3}$, we have that $\overline{s_i} = \frac{2}{3} - s_i$ and $\Re(w(s_i)) = \frac{1}{3}$. It follows that at $\Re(s_i) = \frac{1}{3}$,

$$\overline{\varphi_w(s_1, s_2)} = \frac{B\left(\frac{2}{3} - s_1, \frac{2}{3} - s_2\right)}{B\left(\frac{2}{3} - w(s_1), \frac{2}{3} - w(s_2)\right)}.$$

Consequently,

$$LHS = \frac{1}{2(4\pi i)^2} \iint\limits_{\Re s_i = \frac{1}{3}} \langle \phi, E(*, s_1, s_2)\rangle \cdot \sum_{n\in\mathbb{Z}} \sum_{w\in V} \frac{B\left(\frac{2}{3} - s_1, \frac{2}{3} - s_2\right)}{B\left(\frac{2}{3} - u(s_1), \frac{2}{3} - u(s_2)\right)}$$
$$\times\, a_{w_2}(n, u(s_1), u(s_2)) B_w^u\, W_{0,n}^{(s_1,s_2)}(z, w)\, ds_1 ds_2.$$

To prove the proposition, it remains to show the following. □

Lemma 10.13.19 *The constant term $E_{21}(z, s_1, s_2)$, of the Eisenstein series $E(z, s_1, s_2)$ is*

$$E_{21}(z, s_1, s_2) = \frac{1}{2} \sum_{n\in\mathbb{Z}} \sum_{w\in V} \frac{B\left(\frac{2}{3} - s_1, \frac{2}{3} - s_2\right)}{B\left(\frac{2}{3} - u(s_1), \frac{2}{3} - u(s_2)\right)}$$
$$\times\, a_{w_2}(n, u(s_1), u(s_2)) B_w^u\, W_{0,n}^{(s_1,s_2)}(z, w).$$

Proof We have defined coefficients $a_w(n, s_1, s_2)$ in (10.13.16) such that

$$E_{21}(z, s_1, s_2) = \frac{1}{2} \sum_{n\in\mathbb{Z}} \sum_{w\in V} a_w(n, s_1, s_2)\, W_{0,n}^{(s_1,s_2)}(z, w). \qquad (10.13.20)$$

By the linear independence of these Whittaker functions, we are reduced to showing that

$$a_w(n, s_1, s_2) = \frac{B\left(\frac{2}{3} - s_1, \frac{2}{3} - s_2\right)}{B\left(\frac{2}{3} - u(s_1), \frac{2}{3} - u(s_2)\right)}\, a_{w_2}(n, u(s_1), u(s_2)) B_w^u, \qquad (10.13.21)$$

where u depends on w as noted in (10.13.17).

The proof of the above lemma will follow if we put all the definitions and transformation properties together. We can write equation (10.13.18) using the expansion (10.13.16). So for any $u \in U$, we have the equalities below.

$$\frac{B(s_1,s_2)}{B(u(s_1),u(s_2))}\,E_{21}(z,s_1,s_2) = E_{21}(z,u(s_1),u(s_2))$$

$$\sum_{v\in V}\frac{B(s_1,s_2)}{B(u(s_1),u(s_2))}\,a_v(n,s_1,s_2)\,W_{0,n}^{(s_1,s_2)}(z,v)$$

$$= \sum_{w\in V} a_w(n,u(s_1),u(s_2))W_{0,n}^{(u(s_1),u(s_2))}(z,w).$$

Now if we consider this last equality along with the transformation property

$$W_{0,n}^{(u(s_1),u(s_2))}(z,w_2) = B_w^u\,W_{0,n}^{(s_1,s_2)}(z,w)$$

and use the fact that the Whittaker functions are linearly independent, we get the equality:

$$\frac{B(s_1,s_2)}{B(u(s_1),u(s_2))}\,a_w(n,s_1,s_2)W_{0,n}^{(s_1,s_2)}(z,w)$$
$$= a_{w_2}(n,u(s_1),u(s_2))B_w^u\,W_{0,n}^{(s_1,s_2)}(z,w).$$

So,

$$\frac{B(s_1,s_2)}{B(u(s_1),u(s_2))}\,a_w(n,s_1,s_2) = a_{w_2}(n,u(s_1),u(s_2))B_w^u. \qquad (10.13.22)$$

Now (10.13.21) reduces to showing that

$$\frac{B\left(u(s_1),u(s_2)\right)}{B(s_1,s_2)} = \frac{B\left(\frac{2}{3}-s_1,\frac{2}{3}-s_2\right)}{B\left(\frac{2}{3}-u(s_1),\frac{2}{3}-u(s_2)\right)}. \qquad (10.13.23)$$

To prove this and to finish the proof of the lemma, we have to show that the product

$$B(s_1,s_2)\,B\left(\frac{2}{3}-s_1,\frac{2}{3}-s_2\right)$$

is invariant under the action of the Weyl group on (s_1,s_2). Since

$$B(s_1,s_2)\,B\left(\frac{2}{3}-s_1,\frac{2}{3}-s_2\right)$$

is a product of six gamma functions and six zeta functions and since the quantities

$$3s_1,\quad 3s_2,\quad 3s_1+3s_2-1,\quad 2-3s_1,\quad 2-3s_2,\quad 3-3s_1-3s_2$$

are permuted by the Weyl group, this completes the proof of Lemma 10.13.19 and Proposition 10.13.10. □

GL(n)pack functions The following **GL(n)pack** functions, described in the appendix, relate to the material in this chapter:

BruhatCVector	BruhatForm
BlockMatrix	KloostermanBruhatCell
KloostermanCompatibility	LanglandsForm
LanglandsIFun	LongElement
MPEisensteinGamma	MPEisensteinLambdas
MPEisensteinSeries	MPExteriorPowerGamma
MPExteriorPowerLFun	MPExteriorPowerGamma
MPSymmetricPowerLFun	ParabolicQ

11

Poincaré series and Kloosterman sums

Poincaré series and Kloosterman sums associated to the group $SL(3, \mathbb{Z})$ were introduced and studied in (Bump, Friedberg and Goldfeld, 1988) following the point of view of Selberg (1965). A very nice exposition of the $GL(2)$ theory is given in (Cogdell and Piatetski-Shapiro, 1990). The method was first generalized to $GL(n)$ in (Friedberg, 1987), (Stevens, 1987). In (Bump, Friedberg and Goldfeld, 1988) it is shown that the $SL(3, \mathbb{Z})$ Kloosterman sums are hyper Kloosterman sums associated to suitable algebraic varieties. Non-trivial bounds were obtained by using Hensel's lemma and Deligne's estimates for hyper-Kloosterman sums (Deligne, 1974) in (Larsen, 1988), and later (Dabrowski and Fisher, 1997) improved these bounds by also using methods from algebraic geometry following (Deligne, 1974). Sharp bounds for special types of Kloosterman sums were also obtained in (Friedberg, 1987a,c). In (Dabrowski, 1993), the theory of Kloosterman sums over Chevalley groups is developed. Important applications of the theory of $GL(n)$ Kloosterman sums were obtained in (Jacquet, 2004b) (see also (Ye, 1998)).

Another fundamental direction for research in the theory of Poincaré series and Kloosterman sums was motivated by the $GL(2)$ Kuznetsov trace formula, (see (Kuznecov, 1980) and also (Bruggeman, 1978)). Generalizations of the Kuznetsov trace formula to $GL(n)$, with $n \geq 3$ were obtained in (Friedberg, 1987), (Goldfeld, 1987), (Ye, 2000), but they have not yet proved useful for analytic number theory. The chapter concludes with a new version of the $GL(n)$ Kuznetsov trace formula derived by Xiaoqing Li.

11.1 Poincaré series for $SL(n, \mathbb{Z})$

Let $n \geq 2$ and let $s = (s_1, \ldots, s_{n-1}) \in \mathbb{C}^{n-1}$. Fix a character ψ of the upper triangular unipotent group $U_n(\mathbb{R})$ of $n \times n$ matrices with coefficients in $\mathbb{R}$ and

1s on the diagonal. Note that for fixed integers $m_1, m_2, \ldots, m_{n-1}$, the character ψ may be defined by

$$\psi(u) = \psi_{m_1,\ldots,m_{n-1}}(u) = e^{2\pi i(m_1u_1+m_2u_2+\cdots+m_{n-1}u_{n-1})} \tag{11.1.1}$$

where

$$u = \begin{pmatrix} 1 & u_{n-1} & & & * \\ & 1 & u_{n-2} & & \\ & & \ddots & \ddots & \\ & & & 1 & u_1 \\ & & & & 1 \end{pmatrix} \in U_n(\mathbb{R}).$$

The Poincaré series is constructed from the following functions:

- *The $I_s(z)$–function given in (5.1.1).*
- *A bounded function: $e_\psi : \mathfrak{h}^n \to \mathbb{C}$ which is characterized by the property:*

$$e_\psi(uz) = \psi(u)e_\psi(z)$$

for all $u \in U_n(\mathbb{R})$, and $z \in \mathfrak{h}^n$. Such functions are termed e-functions.

Definition 11.1.2 ***(Poincaré series)*** *For $n \geq 2$, let $s \in \mathbb{C}^{n-1}$, and fix an e-function $e_\psi(z)$ as above. Then the Poincaré series for $\Gamma = SL(n, \mathbb{Z})$, denoted $P(z, s, e_\psi)$, is formally given by*

$$P(z, s, e_\psi) := \sum_{\gamma \in U_n(\mathbb{Z})\backslash\Gamma} I_s(\gamma z)e_\psi(\gamma z).$$

Remark Since e_ψ is bounded on $\mathfrak{h}^n$, it easily follows from Proposition 10.4.3 that the infinite series above defining the Poincaré series converges absolutely and uniformly on compact subsets of $\mathfrak{h}^n$ provided $\Re(s_i)$ is sufficiently large for every $i = 1, 2, \ldots, n-1$.

The Poincaré series is characterized by a very useful property. The inner product of the Poincaré series with a Maass form will give a certain Mellin transform of an individual Fourier coefficient of the Maass form. This is made explicit in the following proposition.

Proposition 11.1.3 *Let ϕ be a Maass form for $SL(n, \mathbb{Z})$ with Fourier expansion given by*

$$\phi(z) = \sum_{\gamma \in U_{n-1}(\mathbb{Z})\backslash SL(n-1,\mathbb{Z})} \sum_{m_1=1}^{\infty} \cdots \sum_{m_{n-2}=1}^{\infty} \sum_{m_{n-1}\neq 0} \frac{A(m_1, \ldots, m_{n-1})}{\prod_{k=1}^{n-1} |m_k|^{k(n-k)/2}}$$
$$\times W_{\text{Jacquet}}\left(M \cdot \begin{pmatrix} \gamma & \\ & 1 \end{pmatrix} z,\ \nu,\ \psi_{1,\ldots,1,\frac{m_{n-1}}{|m_{n-1}|}}\right),$$

as in (9.1.2). Then $\langle \phi,\ P(*, s, e_\psi)\rangle$ *is equal to*

$$\frac{A(m_1, \ldots, m_{n-1})}{\prod\limits_{k=1}^{n-1} |m_k|^{k(n-k)/2}} \int\limits_0^\infty \cdots \int\limits_0^\infty W_{\text{Jacquet}}(My, \nu, \psi_{1,\ldots,1}) \cdot I_{\bar{s}}(y)\overline{e_\psi(y)} \prod_{i=1}^{n-1} \frac{dy_i}{y_i^{i(n-i)+1}}.$$

Proof By the usual unfolding argument, we have

$$\begin{aligned}
&\langle \phi,\ P(*, s, e_\psi)\rangle \\
&\quad = \int\limits_{\Gamma\backslash\mathfrak{h}^n} \phi(z)\, \overline{P(z, s, e_\psi)}\, d^*z \\
&\quad = \int\limits_{U_n(\mathbb{Z})\backslash\mathfrak{h}^n} \phi(xy)\overline{e_\psi(y)} \cdot \bar{\psi}_{m_1,\ldots,m_{n-1}}(x) \prod_{1\le j<k\le n} dx_{j,k} \prod_{i=1}^{n-1} \frac{dy_i}{y_i^{i(n-i)+1}},
\end{aligned}$$

where we have used the explicit measure d^*z given in Theorem 1.6.1. The inner dx-integral picks off the $(m_1, \ldots, m_{n-1})$th Fourier coefficient, and the result follows. □

11.2 Kloosterman sums

The Fourier expansion of the Poincaré series can be computed in the same way the Fourier expansion was computed for minimal parabolic Eisenstein series in Section 10.6. Note that if the e-function is the constant function, then the Poincaré series is just an Eisenstein series. The Fourier expansion involves certain Kloosterman sums which we shall forthwith define.

Let W_n denote the Weyl group of $GL(n, \mathbb{R})$. In Section 10.3 we established the Bruhat decomposition

$$GL(n, \mathbb{R}) = \bigcup_{w\in W_n} G_w, \qquad (\text{where } G_w = U_n w D_n U_n),$$

and D_n denotes the subgroup of diagonal matrices in $GL(n, \mathbb{R})$. We let $U_n(\mathbb{Z})$ denote the matrices in U_n with integer coefficients. Let $c_1, \ldots, c_{n-1}$ be non-zero integers, and set

$$c = \begin{pmatrix} 1/c_{n-1} & & & & \\ & c_{n-1}/c_{n-2} & & & \\ & & \ddots & & \\ & & & c_2/c_1 & \\ & & & & c_1 \end{pmatrix} \tag{11.2.1}$$

We adopt the notation: $\Gamma = SL(n, \mathbb{Z})$ and $\Gamma_w = (w^{-1} \cdot {}^t U_n(\mathbb{Z}) \cdot w) \cap U_n(\mathbb{Z})$.

Definition 11.2.2 (Kloosterman sums) *Fix $w \in W_n$, $c \in D_n$ as in (11.2.1), and ψ, ψ' characters of U_n as in (11.1.1). The $SL(n, \mathbb{Z})$ Kloosterman sum is defined by*

$$S_w(\psi, \psi', c) := \sum_{\substack{\gamma \in U_n(\mathbb{Z})\backslash\Gamma\cap G_w/\Gamma_w \\ \gamma = b_1 c w b_2}} \psi(b_1)\,\psi'(b_2),$$

provided this sum is well defined, (i.e., it is independent of the choice of Bruhat decomposition for γ). Otherwise, the Kloosterman sum is defined to be zero.

Example 11.2.3 ($SL(2, \mathbb{Z})$) **(Kloosterman sum)** In this example,

$$c = \begin{pmatrix} c_1^{-1} & 0 \\ 0 & c_1 \end{pmatrix}, \qquad w = \begin{pmatrix} 0 & -1 \\ 1 & 0 \end{pmatrix}, \qquad \Gamma_w = \begin{pmatrix} 1 & * \\ & 1 \end{pmatrix} = U_2(\mathbb{Z}).$$

The Kloosterman sum is based on the Bruhat decomposition. It is convenient to set $b_1 = \begin{pmatrix} 1 & b_1'/c_1 \\ & 1 \end{pmatrix}$, $b_2 = \begin{pmatrix} 1 & b_2'/c_1 \\ & 1 \end{pmatrix}$. In this case, the Bruhat decomposition says that every $\gamma \in SL(2, \mathbb{Z})$, $\gamma \neq \begin{pmatrix} \pm 1 & * \\ 0 & \pm 1 \end{pmatrix}$ can be written in the form

$$\begin{aligned} \gamma &= \begin{pmatrix} 1 & b_1'/c_1 \\ & 1 \end{pmatrix} \begin{pmatrix} c_1^{-1} & \\ & c_1 \end{pmatrix} \begin{pmatrix} & -1 \\ 1 & \end{pmatrix} \begin{pmatrix} 1 & b_2'/c_1 \\ & 1 \end{pmatrix} \\ &= \begin{pmatrix} b_1' & (b_1' b_2' - 1)/c_1 \\ c_1 & b_2' \end{pmatrix} \end{aligned}$$

where

$$b_1' b_2' \equiv 1 \pmod{c_1}.$$

The $SL(2, \mathbb{Z})$ Kloosterman sum takes the form

$$S_w(\psi, \psi', c) = \sum_{\substack{b_1' \pmod{c_1} \\ b_1'\overline{b_1'} \equiv 1 \pmod{c_1}}} \psi\left(\begin{pmatrix} 1 & b_1'/c_1 \\ 0 & 1 \end{pmatrix}\right) \psi'\left(\begin{pmatrix} 1 & \overline{b_1'}/c_1 \\ 0 & 1 \end{pmatrix}\right). \tag{11.2.4}$$

If we let $M, N \in \mathbb{Z}$ such that

$$\psi\left(\begin{pmatrix} 1 & u \\ 0 & 1 \end{pmatrix}\right) = e^{2\pi i M u}, \qquad \psi'\left(\begin{pmatrix} 1 & u \\ 0 & 1 \end{pmatrix}\right) = e^{2\pi i N u},$$

then the Kloosterman sum in (11.2.4) can be written in the more traditional form

$$S_w(\psi,\psi',c) := S(M,N,c_1) = \sum_{\substack{\beta \pmod{c_1} \\ \beta\bar{\beta}\equiv 1 \pmod{c_1}}} e^{2\pi i\frac{M\beta+N\bar{\beta}}{c_1}}. \tag{11.2.5}$$

Example 11.2.6 (**$SL(3,\mathbb{Z})$ long element Kloosterman sum**) This example is taken from (Bump, Friedberg and Goldfeld, 1988). Let

$$c=\begin{pmatrix}1/c_2 & & \\ & c_2/c_1 & \\ & & c_1\end{pmatrix},\ w=\begin{pmatrix} & & -1\\ & 1 & \\ 1 & & \end{pmatrix},\ \Gamma_w=\begin{pmatrix}1 & * & *\\ & 1 & *\\ & & 1\end{pmatrix}=U_3(\mathbb{Z}).$$

Here w is the so-called long element in the Weyl group. Set

$$b_1=\begin{pmatrix}1 & \alpha_2 & \alpha_3\\ & 1 & \alpha_1\\ & & 1\end{pmatrix}\in U_3(\mathbb{Q}),\qquad b_2=\begin{pmatrix}1 & \beta_2 & \beta_3\\ & 1 & \beta_1\\ & & 1\end{pmatrix}\in U_3(\mathbb{Q}).$$

We are interested in determining when an element

$$\gamma=\begin{pmatrix}a_{1,1} & a_{1,2} & a_{1,3}\\ a_{2,1} & a_{2,2} & a_{2,3}\\ a_{3,1} & a_{3,2} & a_{3,3}\end{pmatrix}\in SL(3,\mathbb{Z})$$

has the Bruhat decomposition

$$\gamma = b_1 c w b_2. \tag{11.2.7}$$

It follows from Proposition 10.3.6 that (11.2.7) holds if

$$c_1 = a_{3,1}\neq 0,\qquad c_2 = a_{2,2}a_{3,1}-a_{2,1}a_{3,2}\neq 0. \tag{11.2.8}$$

With this choice of c_1, c_2, the identity (11.2.7) implies that γ must be equal to

$$\begin{pmatrix} a_{3,1}\alpha_3 & a_{2,2}\alpha_2-\frac{a_{2,1}a_{3,2}\alpha_2}{c_1}+a_{3,1}\alpha_3\beta_2 & a_{2,2}\alpha_2\beta_1-\frac{1}{c_2}-\frac{a_{2,1}a_{3,2}\alpha_2\beta_1}{c_1}+c_1\alpha_3\beta_3\\ c_1\alpha_1 & a_{2,2}-\frac{a_{2,1}a_{3,2}}{c_1}+c_1\alpha_1\beta_2 & a_{2,2}\beta_1-\frac{a_{2,1}a_{3,2}\beta_1}{c_1}+c_1\alpha_1\beta_3\\ c_1 & c_1\beta_2 & c_1\beta_3\end{pmatrix}.$$

One may use this to systematically solve for the coefficients of b_1 and b_2. In this manner, one obtains

$$b_1=\begin{pmatrix}1 & c_1a_{1,2}-\frac{a_{1,1}a_{3,2}}{c_2} & \frac{a_{1,1}}{c_1}\\ & 1 & \frac{a_{2,1}}{c_1}\\ & & 1\end{pmatrix},$$

$$b_2=\begin{pmatrix}1 & \frac{a_{3,2}}{c_1} & \frac{a_{3,3}}{c_1}\\ & 1 & c_1a_{2,3}-\frac{a_{2,1}a_{3,3}}{c_2}\\ & & 1\end{pmatrix}. \tag{11.2.9}$$

For $M = (M_1, M_2) \in \mathbb{Z}^2$, $N = (N_1, N_2) \in \mathbb{Z}^2$, let us define two characters ψ_M, ψ_N by

$$\psi_M(u) = e^{2\pi i(M_1u_1+M_2u_2)}, \qquad \psi_N(u) = e^{2\pi i(N_1u_1+N_2u_2)},$$

where $u = \begin{pmatrix} 1 & u_2 & u_3 \\ & 1 & u_1 \\ & & 1 \end{pmatrix} \in U_3(\mathbb{R})$.

It follows from (11.2.7), (11.2.8), (11.2.9) that the long element Kloosterman sum takes the form

$$S_w(\psi_M, \psi_N, c) = \sum_{\substack{a_{2,1},a_{3,2} \pmod{c_1} \\ c_1a_{2,3}-a_{2,1}a_{3,3} \pmod{c_2} \\ c_1a_{1,2}-a_{1,1}a_{3,2} \pmod{c_2}}} e^{2\pi i\left[\frac{M_1a_{2,1}}{c_1}+\frac{M_2(c_1a_{1,2}-a_{1,1}a_{3,2})}{c_2}\right]} \cdot e^{2\pi i\left[\frac{N_1(c_1a_{2,3}-a_{2,1}a_{3,3})}{c_2}+\frac{N_2a_{3,2}}{c_1}\right]},$$

subject to the constraint that there exist integers $a_{1,2}, a_{1,3}, a_{2,2}$ such that

$$c_1(a_{1,2}a_{2,3}-a_{1,3}a_{2,2})+a_{3,2}\big(a_{1,3}a_{2,1} - a_{1,1}a_{2,3}\big) + a_{3,3}\big(a_{1,1}a_{2,2} - a_{1,2}a_{2,1}\big)=1,$$

i.e., that $\mathrm{Det}(\gamma) = 1$.

It is important to determine the compatibility conditions for which Kloosterman sums are well defined and non-zero. The following proposition is taken from (Friedberg, 1987).

Proposition 11.2.10 (Compatibility condition) *Fix $n \geq 2$. The Kloosterman sum $S_w(\psi, \psi', c)$ given in Definition 11.2.2 is well defined if and only if*

$$\psi(cwuw^{-1}) = \psi'(u)$$

for all $u \in (w^{-1}U_n(\mathbb{R})w) \cap U_n(\mathbb{R})$. Hence the Kloosterman sum is non-zero only in this case.

Proof See Lemma 10.6.3. □

Definition 11.2.11 (Kloosterman zeta function) *Fix $n \geq 2$. Let ψ, ψ' be two characters of $U_n(\mathbb{R})$, c of the form (11.2.1), and $S_w(\psi, \psi', c)$ the Kloosterman sum as in Definition 11.2.2. Let $s = (s_1, \ldots, s_{n-1}) \in \mathbb{C}^{n-1}$ with $\Re(s_i)$ sufficiently large for all $i = 1, \ldots, n-1$. The Kloosterman zeta function, $Z(\psi, \psi', s)$ is defined by the absolutely convergent series*

$$Z(\psi, \psi', s) := \sum_{c_1=1}^{\infty} \cdots \sum_{c_{n-1}=1}^{\infty} \frac{S_w(\psi, \psi', c)}{c_1^{ns_1} \cdots c_{n-1}^{ns_{n-1}}}.$$

Remarks The Kloosterman zeta function was first introduced, for the case of $GL(2)$, in (Selberg, 1965) who showed the meromorphic continuation in s and the existence of infinitely many simple poles on the line $\Re(s) = \frac{1}{2}$ occurring at the eigenvalues of the Laplacian. Selberg used Weil's bound for $GL(2)$ Kloosterman sums to deduce that the Kloosterman zeta function converged absolutely in the region $\Re(s) > \frac{3}{4}$. He was able to immediately deduce from this the bound $\geq \frac{3}{16}$ for the lowest eigenvalue of the Laplacian (see (Selberg, 1965)).

For $\Re(s) > \frac{1}{2} + \epsilon$, it was shown in (Goldfeld and Sarnak, 1983) that the Kloosterman zeta function is

$$\mathcal{O}\left(\frac{|s|^{\frac{1}{2}}}{\Re(s) - \frac{1}{2}}\right).$$

Such results have not been obtained for higher-rank Kloosterman zeta functions. This would constitute an important research problem.

11.3 Plücker coordinates and the explicit evaluation of Kloosterman sums

Definition 11.3.1 (Plücker coordinates) *Let $g \in GL(n, \mathbb{R})$ with $n \geq 2$. The Plücker coordinates of g are the $n - 1$ row vectors $\rho_1, \ldots, \rho_{n-1}$, where, for each $1 \leq k \leq n - 1$, $\rho_k \in \mathbb{R}^{\binom{n}{k}}$ consists of every possible $k \times k$ minor formed with the bottom k rows of the matrix g.*

As an example, consider

$$g = \begin{pmatrix} g_{1,1} & g_{1,2} & g_{1,3} \\ g_{2,1} & g_{2,2} & g_{2,3} \\ g_{3,1} & g_{3,2} & g_{3,3} \end{pmatrix} \in GL(3, \mathbb{R}).$$

Then the Plücker coordinates are

$$\rho_1 = \{g_{3,1},\ g_{3,2},\ g_{3,3}\}$$

$$\rho_2 = \{g_{2,1}g_{3,2} - g_{3,1}g_{2,2},\ \ g_{2,1}g_{3,3} - g_{3,1}g_{2,3},\ \ g_{2,2}g_{3,3} - g_{3,2}g_{2,3}\}.$$

The key theorem needed for the evaluation of Kloosterman sums is the following (see (Friedberg, 1987)).

Theorem 11.3.2 *Let $n \geq 2$. If $g \in SL(n, \mathbb{R})$, a necessary and sufficient condition for the coset of g in $U_n(\mathbb{Z})\backslash SL(n, \mathbb{R})$ to contain a representative in $SL(n, \mathbb{Z})$ is that each of the rows ρ_k (with $1 \leq k \leq n - 1$) of Plücker coordinates of g consist of coprime integers.*

Proof It is convenient to have a notation for the minors that occur in the Plücker coordinates. Fix $n \geq 2$. For $1 \leq k \leq n-1$, let $\mathcal{L}_k$ denote the set of all k-element subsets of $\{1, 2, \ldots, n\}$ with the lexicographical ordering $<$, where

$$\{\ell_1, \ldots, \ell_k\} < \{\ell'_1, \ldots, \ell'_k\}$$

if $\ell_1 < \ell'_1$. If the first elements are equal, but only in that case, you look at the second and compare them, whichever is higher, and that is the "greater" set. One continues naturally in this manner obtaining an ordering which is analogous to alphabetical order. Then the set $\mathcal{L}_k$ can be used to index the minors that occur in the Plücker coordinate ρ_k. For example, if $\{\ell_1, \ldots, \ell_k\} \in \mathcal{L}_k$, then we may consider the minor formed with the columns $\ell_1, \ldots, \ell_k$.

For each k with $1 \leq k < n-1$, we may list the elements of $\mathcal{L}_{k+1}$ (in some order) as follows:

$$\mathcal{L}_{k+1} = \left\{\lambda_{k,1}, \ldots, \lambda_{k,\binom{n}{k+1}}\right\}$$

with

$$\lambda_{k,i} = \{\ell_{k,i,1}, \ldots, \ell_{k,i,k+1}\}, \qquad \ell_{k,i,1} < \ell_{k,i,2} < \cdots < \ell_{k,i,k+1}.$$

This allows us to index the coordinates of $\mathbb{R}^{2^n-2}$ by $\mathcal{L}_1 \cup \cdots \cup \mathcal{L}_{n-1}$. Next, define $\mathcal{V} \subseteq \mathbb{R}^{2^n-2}$ to be the affine algebraic set of all

$$\left\{v_1, \ldots, v_n, v_{1,2}, v_{1,3}, \ldots, v_{n-1,n}, \quad \ldots \quad , v_{2,3,\ldots,n}\right\} \subseteq \mathbb{R}^{2^n-2} \tag{11.3.3}$$

satisfying

$$\begin{pmatrix} v_{\lambda_{k,1}} \\ v_{\lambda_{k,2}} \\ \vdots \\ v_{\lambda_{k,\binom{n}{k+1}}} \end{pmatrix} \in \text{Image}(T), \tag{11.3.4}$$

for all $1 \leq k < n-1$, where $T = (t_{i,j})$ (with $1 \leq i \leq \binom{n}{k+1}$, $1 \leq j \leq n$) is the linear transformation

$$t_{i,j} = \begin{cases} (-1)^{r-1} v_{\lambda_{k,i}-\{j\}} & \text{if } j = \ell_{k,i,r} \in \lambda_{k,i}, \\ 0 & \text{otherwise.} \end{cases} \tag{11.3.5}$$

Here $\lambda_{k,i}$ is a set of k integers and $\lambda_{k,i} - \{j\}$ means to delete the element j from the set $\lambda_{k,i}$. Note also that the linear transformation T maps $\mathbb{R}^n$ to $\mathbb{R}^n$ by simple left matrix multiplication of T on an n-dimensional column vector in $\mathbb{R}^n$. Thus, given coordinates v_λ (with $\lambda \in \mathcal{L}_k$) we see that (11.3.4), (11.3.5) gives a condition which allows us to see which v_λ (with $\lambda \in \mathcal{L}_{k+1}$) occur.

For example, if $n = 3$, then V is given by the set of all $\{v_1, v_2, v_3, v_{1,2}, v_{1,3}, v_{2,3}\}$ subject to the constraint that

$$\begin{pmatrix} v_{1,2} \\ v_{1,3} \\ v_{2,3} \end{pmatrix} \in \text{Image}\left(\begin{pmatrix} v_2 & -v_1 & 0 \\ v_3 & 0 & -v_1 \\ 0 & v_3 & -v_2 \end{pmatrix}\right),$$

i.e., such that

$$v_3 v_{1,2} - v_2 v_{1,3} + v_1 v_{2,3} = 0.$$

We now consider the map $M : GL(n, \mathbb{R}) \to \mathbb{R}^{2^n-2}$ given by

$$M(g) := \{M_\lambda(g) \mid \lambda \in \mathcal{L}_1 \cup \cdots \cup \mathcal{L}_{n-1}\},$$

where for $\lambda = \{\ell_1, \ldots, \ell_k\}$, the function $M_\lambda(g)$ denotes the $k \times k$ minor formed from the bottom k rows and columns $\ell_1, \ldots, \ell_k$.

The first step in the proof of Theorem 11.3.2 is to show that the above map $M : GL(n, \mathbb{R}) \to \mathbb{R}^{2^n-2}$ induces a bijection from $U_n(\mathbb{R})\backslash GL(n, \mathbb{R})$ to

$$V_1 := \left\{v \in V \;\middle|\; v_\lambda \neq 0 \text{ for some } \lambda \in \mathcal{L}_{n-1}\right\}.$$

To see this, note that the map M factors through $U_n(\mathbb{R})\backslash GL(n, \mathbb{R})$. Furthermore, condition (11.3.4) holds for $M(g)$ since $M_\lambda(g)$ can be computed for $\lambda \in \mathcal{L}_k$ in terms of $M_{\lambda'}(g)$ with $\lambda' \in \mathcal{L}_{k-1}$ for all $1 < k < n$, by expanding the determinant along the top row. Since $\text{Det}(g) \neq 0$ it follows that M maps into V_1. Now given $v \in V_1$, the condition (11.3.4) guarantees that there exists an $(n-1) \times n$ matrix with appropriate minors. Since $v_\lambda \neq 0$ for some $\lambda \in \mathcal{L}_{n-1}$, this can be completed to a matrix $g \in GL(n, \mathbb{R})$ with image v. Consequently, the map M is onto V_1. It remains to show that the map $M : U_n(\mathbb{R})\backslash GL(n, \mathbb{R}) \to V_1$ is injective.

To see this, note first that by Proposition 10.3.6, if

$$M_\lambda(g) = \begin{cases} 1 & \text{if } \lambda \in \left\{\lambda_1^{(1)}, \ldots, \lambda_{n-1}^{(1)}\right\} \\ 0 & \text{otherwise,} \end{cases}$$

where $\lambda_i^{(1)} = \{n, n-1, \ldots, n-i+1\}$ for $1 \le i \le n-1$, then g must lie in $U_n(\mathbb{R})$. Suppose that $M(g) = M(h)$. Then for any $\lambda = \{\ell_1, \ldots, \ell_k\} \in \mathcal{L}_k$ we have by Proposition 10.3.6 that

$$\begin{aligned} &g^{-1}(e_1) \wedge \cdots \wedge g^{-1}(e_{n-k}) \wedge e_{\ell_1} \wedge \cdots \wedge e_{\ell_k} \\ &\quad = h^{-1}(e_1) \wedge \cdots \wedge h^{-1}(e_{n-k}) \wedge e_{\ell_1} \wedge \cdots \wedge e_{\ell_k}. \end{aligned}$$

Hence

$$\begin{aligned} M_\lambda(gh^{-1})e_1 \wedge \cdots \wedge e_n &= g^{-1}e_1 \wedge \cdots \wedge g^{-1}e_{n-k} \wedge h^{-1}e_{\ell_1} \wedge \cdots \wedge h^{-1}e_{\ell_k} \\ &= h^{-1}e_1 \wedge \cdots \wedge h^{-1}e_{n-k} \wedge h^{-1}e_{\ell_1} \wedge \cdots \wedge h^{-1}e_{\ell_k} \\ &= e_1 \wedge \cdots \wedge e_{n-k} \wedge e_{\ell_1} \wedge \cdots \wedge e_{\ell_k}. \end{aligned}$$

It follows that

$$M_\lambda(gh^{-1}) = \begin{cases} 1 & \text{if } \lambda = \lambda_k^{(1)}, \\ 0 & \text{if } \lambda \neq \lambda_k^{(1)}, \end{cases}$$

and, therefore, $U_n(\mathbb{R})g = U_n(\mathbb{R})h$.

We now complete the proof of Theorem 11.3.2. Let

$$V_2 = \left\{ v \in V \;\middle|\; v_\lambda \in \mathbb{Z} \text{ for all } \lambda,\ \gcd(v_\lambda \mid \lambda \in \mathcal{L}_k) = 1 \text{ for } 1 \le k \le n-1 \right\}.$$

It is enough to show that the map $M : U_n(\mathbb{Z})\backslash SL(n,\mathbb{Z}) \to V_2$ is surjective. It follows from the previous arguments that given $v \in V_2$, there exists a matrix $g \in GL(n,\mathbb{R})$ with the desired minors. We must show that the coset $U_n(\mathbb{R})g$ contains an integral element. Let $p_\lambda \in \mathbb{Z}$ satisfy

$$\sum_{\lambda \in \mathcal{L}_k} p_\lambda M_\lambda(g) = 1, \qquad (1 \le k < n).$$

It follows from this and Proposition 10.3.6 that

$$\sum_{\substack{\lambda \in \mathcal{L}_k \\ \lambda = \{\ell_1,\ldots,\ell_k\}}} p_\lambda g\left(e_{\ell_1}\right) \wedge \cdots \wedge g\left(e_{\ell_k}\right) = \sum_{\substack{\lambda \in \mathcal{L}_k \\ \lambda = \{\ell_1,\ldots,\ell_k\}}} b_\lambda e_{\ell_1} \wedge \cdots \wedge e_{\ell_k}, \tag{11.3.6}$$

where $b_\lambda \in \mathbb{R}$ and $b_{n-k+1,\ldots,n} = 1$.

Next, define a matrix h determined by the identities

$$e_1 \wedge \cdots \wedge e_{n-1} \wedge h(e_i) = e_1 \wedge \cdots \wedge e_{n-1} \wedge g(e_i), \tag{11.3.7}$$

and

$$\begin{aligned} &e_1 \wedge \cdots \wedge e_{j-1} \wedge h(e_i) \wedge e_{j+1} \wedge \cdots \wedge e_n \qquad\qquad (11.3.8) \\ &\quad = e_1 \wedge \cdots \wedge e_{j-1} \wedge h(e_i) \wedge \left(\sum_{\lambda \in \mathcal{L}_{n-j}} p_\lambda\, g\left(e_{\ell_1}\right) \wedge \cdots \wedge g\left(e_{\ell_{n-j}}\right) \right), \end{aligned}$$

for all $1 \le i \le n,\ 1 \le j \le n-1$. Then h is an $n \times n$ matrix with integer coefficients since each entry is a sum of minors of g multiplied by the integers p_λ. It remains to show that $h \in SL(n,\mathbb{Z})$ and $M(g) = M(h)$, so it suffices to show that $hg^{-1} \in U_n(\mathbb{R})$.

Now (11.3.7) implies that

$$h(e_i) = g(e_i) + \sum_{j=1}^{n-1} a_{ji} e_j \tag{11.3.9}$$

for $a_{ji} \in \mathbb{R}$. If we substitute this in (11.3.8) with $j < n$, it then follows from (11.3.6) that

$$a_{ji} e_1 \wedge \cdots \wedge e_n = \sum_{\substack{\lambda \in \mathcal{L}_{n-j} \\ \lambda \neq \{j+1,\ldots,n\}}} b_\lambda \, e_1 \wedge \cdots \wedge e_{j-i} \wedge g(e_i) \wedge e_{\ell_1} \wedge \cdots \wedge e_{\ell_{n-j}}.$$

Define $\mu = (\mu_{ij})$ by $hg^{-1} = I + \mu$. Since $(I + \mu)g = h$, it follows from (11.3.9) that

$$a_{ji} e_1 \wedge \cdots \wedge e_n = e_1 \wedge \cdots \wedge e_{j-i} \wedge \mu g(e_i) \wedge e_{j+1} \wedge \cdots \wedge e_n \tag{11.3.10}$$

for $1 \le i \le n,\ 1 \le j \le n-1$. By comparing (11.3.9), (11.3.10), we obtain

$$\sum_{k=1}^{n} g_{ki} \mu_{jk} = \sum_{\substack{\lambda \in \mathcal{L}_{n-j} \\ \lambda = \{j,\ldots,\hat{m},\ldots,n\}, m > j}} b_\lambda \, g_{mi} (\pm 1)_\lambda,$$

where $g(e_i) = \sum g_{ji} e_j$, and $\hat{m}$ means that the term m is omitted. In particular, $\mu_{jk} = 0$ when $j \ge k$, which is what was required. □

The Plücker relations are defined by a set of quadratic forms in the bottom row based minor determinants of any $n \times n$ matrix in $GL(n, \mathbb{R})$. These forms have coefficients ± 1. They must vanish if the values assigned to symbols representing the minor determinants come from any square matrix. The number of Plücker relations grows rapidly with n, because each $j \times j$ sub-matrix, with elements chosen from the bottom j rows and any j columns, also gives rise to a set of relationships of the given type. In dimension 2 there are no relationships and in dimension 3 just one, the Cramer's rule relationship $v_1 v_{23} - v_2 v_{13} + v_3 v_{12} = 0$. (Here v_λ, where λ is an ordered subset of $\{1, 2, \ldots, n\}$, is used to represent the minor determinant based on the bottom $|\lambda|$ rows and the columns indexed by the elements of λ.)

Example 11.3.11 ($GL(4)$ **Plücker relations**) I would like to thank Kevin Broughan for this example in which ten relations for dimension $n = 4$ are derived.

(i) The simplest relation is obtained by expanding the matrix using the bottom row-based minors of size $n - 1$ along the bottom row. By Cramer's rule we obtain

$$(-1)^{1+1} v_1 v_{234} + (-1)^{1+2} v_2 v_{134} + (-1)^{1+3} v_3 v_{124} + (-1)^{1+4} v_4 v_{123} = 0.$$

(ii) Now let the rows of a fixed but arbitrary 4×4 matrix be represented by the vectors a_1, a_2, a_3, a_4. Then because for all vectors v, $v \wedge v = 0$:

$$0 = (a_3 \wedge a_4) \wedge (a_3 \wedge a_4) = \alpha\, e_1 \wedge e_2 \wedge e_3 \wedge e_4,$$

where the e_i are the standard unit vectors and α is a real constant with value $\nu_{12}\nu_{34} - \nu_{24}\nu_{13} + \nu_{14}\nu_{23}$, which is therefore 0.

(iii) The relation $\nu_{24}\nu_{123} + \nu_{12}\nu_{234} - \nu_{23}\nu_{124} = 0$ will now be derived. First expand the wedge product of the bottom two rows:

$$\begin{aligned} a_3 \wedge a_4 &= \nu_{12}e_1 \wedge e_2 + \nu_{13}e_1 \wedge e_3 + \nu_{14}e_1 \wedge e_4 \\ &\quad + \nu_{23}e_2 \wedge e_3 + \nu_{24}e_2 \wedge e_4 + \nu_{34}e_3 \wedge e_4. \end{aligned}$$

Then

$$\begin{aligned} a_3 \wedge a_4 &= e_2 \wedge (-\nu_{12}e_1 + \nu_{23}e_3 + \nu_{24}e_4) \\ &\quad + \nu_{13}e_1 \wedge e_3 + \nu_{14}e_1 \wedge e_4 + \nu_{34}e_3 \wedge e_4 \\ &= e_2 \wedge \omega + \eta, \end{aligned}$$

say, where ω is a 1-form and η a 2-form with e_2 not appearing. Then

$$a_2 \wedge a_3 \wedge a_4 = \nu_{123}e_1 \wedge e_2 \wedge e_3 + \nu_{234}e_2 \wedge e_3 \wedge e_4 + \nu_{124}e_1 \wedge e_2 \wedge e_4$$

so

$$\begin{aligned} a_2 \wedge a_3 \wedge a_4 \wedge \omega &= (\nu_{24}\nu_{123} + \nu_{12}\nu_{234} - \nu_{23}\nu_{124})e_1 \wedge e_2 \wedge e_3 \wedge e_4 + 0 \\ &= \lambda\, e_1 \wedge e_2 \wedge e_3 \wedge e_4. \end{aligned}$$

Since each term of η has two of $\{e_1, e_3, e_4\}$ and each term of ω one of this set we can write

$$\eta \wedge \omega = (\nu_{12}\nu_{34} - \nu_{24}\nu_{13} + \nu_{14}\nu_{23})e_1 \wedge e_3 \wedge e_4 = 0$$

by the relation derived in (ii) above.

But then

$$\begin{aligned} a_2 \wedge a_3 \wedge a_4 \wedge \omega &= a_2 \wedge (a_3 \wedge a_4) \wedge \omega \\ &= a_2 \wedge (e_2 \wedge \omega + \eta) \wedge \omega \\ &= a_2 \wedge (\eta \wedge \omega) = 0, \end{aligned}$$

so, therefore, $\lambda = 0$, and we get $\nu_{24}\nu_{123} + \nu_{12}\nu_{234} - \nu_{23}\nu_{124} = 0$.

Three similar relationships are derived by factoring out in turn each of the unit vectors e_1, e_3 and e_4.

(iv) The four remaining relations are obtained by applying the dimension 3 relations to each of the four subsets of $\{1, 2, 3, 4\}$ of column numbers of size 3.

Example 11.3.12 (The $w = \begin{pmatrix} & \pm 1 \\ I_{n-1} & \end{pmatrix}$ Kloosterman sum) This example was first worked out in (Friedberg, 1987). For $w = \begin{pmatrix} & \pm 1 \\ I_{n-1} & \end{pmatrix}$, we have

$$U_w := (w^{-1}U_n(\mathbb{R})w) \cap U_n(\mathbb{R}) = \begin{pmatrix} 1 & & * & 0 \\ & \ddots & & \vdots \\ & & & 0 \\ & & & 1 \end{pmatrix}.$$

It follows from Proposition 11.2.10 (brute force computation) that the Kloosterman sum is non-zero only when $c_1|c_2,\ c_2|c_3, \ldots, c_{n-2}|c_{n-1}$.

Next, we show that the map $M : GL(n, \mathbb{R}) \to \mathbb{R}^{n^2-2}$ arising in the proof of Theorem 11.3.2, composed with projection, gives a bijection between the sets

$$\left\{ \gamma \in U_n(\mathbb{Z})\backslash SL(n, \mathbb{Z}) \cap G_w/\Gamma_w \;\middle|\; \mathrm{diag}(\gamma) = c \right\}$$

and

$$\mathcal{S} = \left\{ (v_n, v_{n-1,n}, \ldots, v_{2,3,\ldots,n}) \in \mathbb{Z}^{n-1} \;\middle|\; v_n \pmod{c_1},\ v_{n-1,n} \pmod{c_2}, \ldots, \right.$$
$$\left. v_{2,3,\ldots,n} \pmod{c_{n-1}},\ (v_{n-j+1,\ldots,n},\ c_j/c_{j-1}) = 1 \text{ for all } 1 \le j \le n-1 \right\},$$

with $c_0 = 1$.

To see the bijection, note that the image of M composed with projection is contained in $\mathcal{S}$, since $\mathrm{Det}(\gamma) = 1$ implies that the relative primality conditions must hold. Further, the map is one-to-one because the information in $\mathcal{S}$ determines all minors of the form

$$M_{n-k,\ldots,\widehat{n-i},\ldots,n}(\gamma), \qquad i < k \pmod{c_k}$$

by induction on k for

$$M_{n-k,\ldots,\widehat{n-i},\ldots,n}(\gamma) = \frac{c_k}{c_{k-1}} \cdot M_{n-k+1,\ldots,\widehat{n-i},\ldots,n}(\gamma).$$

Since the remaining $M_\lambda(\gamma)$ are determined by Proposition 10.3.6 it follows that the map is one-to-one. As for surjectivity, it suffices to show that the point in V arising from a point of $\mathcal{S}$ is actually in V_2. This follows by induction since

$$\gcd\left\{ v_{n-k,\ldots,\widehat{n-i},\ldots,n} \;\middle|\; 0 \le i \le k-1 \right\} = \frac{c_k}{c_{k-1}}.$$

The above bijection can be used to prove

Theorem 11.3.13 *Let $n \geq 2$ and define $M = (M_1, \ldots, M_{n-1})$, and $N = (N_1, \ldots, N_{n-1})$ to be two $(n-1)$–tuples of integers. Let ψ_M, ψ_N be two characters of $U_n(\mathbb{R})$, as in (11.1.1), determined by the conditions*

$$\psi_M(u) = e^{2\pi i(M_1 u_1 + \cdots + u_{n-1} M_{n-1})}, \qquad \psi_N(u) = e^{2\pi i(N_1 u_1 + \cdots + u_{n-1} N_{n-1})},$$

for all $u \in U_n(\mathbb{R})$. Let c be given by (11.2.1), and assume $w = \begin{pmatrix} & \pm 1 \\ I_{n-1} & \end{pmatrix}$. Suppose that $c_1 | c_2 | \cdots | c_{n-1}$ and

$$N_i = \frac{M_{i+1} c_{n-i} c_{n-i-2}}{c_{n-i-1}^2} \tag{11.3.14}$$

for $i = 1, 2, \ldots, n-2$ with $c_0 = 1$. Then

$$S(\psi_M, \psi_N, c) = \sum_{\substack{x_i \pmod{c_i} \\ (x_i,\, c_i/c_{i-1})=1}} e^{2\pi i\left(\frac{M_1 \overline{x_{n-1}}}{c_{n-1}/c_{n-2}} + \frac{M_2 x_{n-1} \overline{x_{n-2}}}{c_{n-2}/c_{n-3}} + \cdots + \frac{M_{n-1} x_2 \overline{x_1}}{c_1/c_0} + \frac{N_{n-1} x_1}{c_1}\right)},$$

where $x_i \bar{x}_i \equiv 1 \pmod{c_i/c_{i-1}}$ for $1 \leq i \leq n-1$.

Remark The compatibility condition of Proposition 11.2.10 implies the divisibility criterion $c_1 | c_2 | \cdots | c_{n-1}$ together with the identitities (11.3.14).

11.4 Properties of Kloosterman sums

We now proceed to determine some of the more important properties of the Kloosterman sums.

Proposition 11.4.1 *Let $n \geq 2$ and define $M = (M_1, \ldots, M_{n-1})$, and $N = (N_1, \ldots, N_{n-1})$ to be two $(n-1)$–tuples of integers. Let ψ_M, ψ_N be two characters of $U_n(\mathbb{R})$, as in (11.1.1), determined by the conditions*

$$\psi_M(u) = e^{2\pi i(M_1 u_1 + \cdots + u_{n-1} M_{n-1})}, \qquad \psi_N(u) = e^{2\pi i(N_1 u_1 + \cdots + u_{n-1} N_{n-1})},$$

for all $u \in U_n(\mathbb{R})$. Let w be in the Weyl group of $GL(n, \mathbb{Z})$, let c be given by (11.2.1), and assume that the Kloosterman sum $S_w(\psi_M, \psi_N, c)$ is well defined. Then its value depends only on

$$M_i \pmod{c_i}, \qquad \text{and} \qquad N_i \pmod{c_{n-i}},$$

for $i = 1, 2, \ldots, n-1$.

Proof It is clear that the proposition holds for the long element Kloosterman sum. For non-zero integers $c_1, \ldots, c_{n-1}$, let $U(c_1, \ldots, c_{n-1})$ denote the group

$$U(c_1, \ldots, c_{n-1}) = \left\{ u \in U_n(\mathbb{R}) \;\middle|\; c_i u_i \in \mathbb{Z} \right\},$$

where u is as in (11.1.1). Then, by the use of Cramer's rule, one may show that any $\gamma \in SL(n, \mathbb{Z}) \cap G_w$ has Bruhat decompositions $\gamma = b_1 c w b_2 = b_1' c w b_2'$

where

$$b_2 \in U(c_{n-1}, \ldots, c_1), \qquad b_1' \in U(c_1, \ldots, c_{n-1}).$$

By choosing the appropriate one of the above decompositions when evaluating the Kloosterman sum, the result follows. □

Proposition 11.4.2 (Multiplicativity condition) *Fix $n \geq 2$, and let ψ_M, ψ_N be characters of $U_n(\mathbb{R})$ as in Proposition 11.4.1. Let*

$$c = \begin{pmatrix} 1/c_{n-1} & & & & \\ & c_{n-1}/c_{n-2} & & & \\ & & \ddots & & \\ & & & c_2/c_1 & \\ & & & & c_1 \end{pmatrix},$$

$$c' = \begin{pmatrix} 1/c'_{n-1} & & & & \\ & c'_{n-1}/c'_{n-2} & & & \\ & & \ddots & & \\ & & & c'_2/c'_1 & \\ & & & & c'_1 \end{pmatrix}$$

with positive integers $c_1, c_1', \ldots, c_{n-1}, c_{n-1}'$. Let w be in the Weyl group of $GL(n, \mathbb{R})$. Then there exist characters $\psi_{N'}, \psi_{N''}$ of $U_n(\mathbb{R})$ such that

$$S_w(\psi_M, \psi_N, cc') = S_w(\psi_M, \psi_{N'}, c) \cdot S_w(\psi_M, \psi_{N''}, c').$$

Proof This was first proved for $n = 3$ in (Bump, Friedberg and Goldfeld, 1988). See (Friedberg, 1987), (Stevens, 1987) for the proof. □

It was shown in (Friedberg, 1987) that the Kloosterman sums $S_w(\psi_M, \psi_N, c)$ are non-zero only if w is of the form

$$w = \begin{pmatrix} & & I_{i_1} \\ & I_{i_2} & \\ \cdot^{\cdot^{\cdot}} & & \\ I_{i_\ell} & & \end{pmatrix}$$

where the I_k are $k \times k$ identity matrices and $i_1 + \cdots + i_\ell = n$. It was further shown in (Friedberg, 1987) that if w_n is the long element and, in addition, $c_1, \ldots, c_{n-1}$ are pairwise coprime, then the long element Kloosterman sum $S_{w_n}(\psi_M, \psi_N, c)$ factors into $GL(2)$ Kloosterman sums.

In the special case that $c_1, c_2, \ldots, c_{n-1}$ are all suitable powers of a fixed prime p, then the Kloosterman sum $S_w(\psi_M, \psi_N, c)$, with c given by (11.2.1), will be associated to an algebraic variety defined over the finite field $\mathbb{F}_p$ of p

elements. By use of a deep theorem of Deligne (1977), on bounds for hyper Kloosterman sums of type

$$\sum_{\substack{x_1\cdots x_n=1\\ x_i\in\mathbb{F}_p,(i=1,\ldots,n)}} e^{2\pi i\left(\frac{x_1+\cdots+x_n}{p}\right)},$$

it is proven in (Friedberg, 1987) (see also Theorem 11.3.13) that for

$$w=\begin{pmatrix} & 1\\ I_{n-1} & \end{pmatrix},$$

then for all $1\le j\le n-1$,

$$|S_w(\psi_M,\psi_N,c)| = \mathcal{O}\left(c_j^{((n-1)^2/2j)+\epsilon}\prod_{\substack{p|c_{n-1}\\ p\nmid c_j}} p^{(n-j-1)(n-j-2)/2}\right).$$

11.5 Fourier expansion of Poincaré series

Let $n\ge 2$, $z\in\mathfrak{h}^n$, $s\in\mathbb{C}^{n-1}$, and, ψ a character of $U_n(\mathbb{R})$ as in (11.1.1). Consider an e-function, $e_\psi:\mathfrak{h}^n\to\mathbb{C}$, which is a bounded function characterized by the property that

$$e_\psi(uz)=\psi(u)e_\psi(z)$$

for all $u\in U_n(\mathbb{R})$, and $z\in\mathfrak{h}^n$.

Let

$$P(z,s,e_\psi):=\sum_{\gamma\in U_n(\mathbb{Z})\backslash\Gamma} I_s(\gamma z)e_\psi(\gamma z)$$

be a Poincaré series for $SL(n,\mathbb{Z})$ as in Definition 11.1.2. The main goal of this section is an explicit computation of the $N=(N_1,\ldots,N_{n-1})\in\mathbb{Z}^{n-1}$ Fourier coefficient

$$\int\limits_{U_n(\mathbb{Z})\backslash U_n(\mathbb{R})} P(uz,s,e_\psi)\overline{\psi_N(u)}\,d^*u, \tag{11.5.1}$$

with

$$u=\begin{pmatrix} 1 & u_{n-1} & & u_{i,j} & \\ & 1 & u_{n-2} & & \\ & & \ddots & \ddots & \\ & & & 1 & u_1\\ & & & & 1\end{pmatrix}\in U_n(\mathbb{R}),$$

$d^*u=\prod_{1\le i<j\le n} du_{i,j}$, and $\psi_N(u)=e^{2\pi i(u_1N_1+\cdots+u_{n-1}N_{n-1})}$.

The computation of (11.5.1) requires some additional notation.

Definition 11.5.2 *Fix $n \geq 2$. We define the group V_n of diagonal matrices:*

$$V_n := \left\{ v = \begin{pmatrix} v_1 & & \\ & \ddots & \\ & & v_n \end{pmatrix} \;\middle|\; v_i = \pm 1 \forall 1 \leq i \leq n,\ \mathrm{Det}(v) = \prod_{i=1}^{n} v_i = 1 \right\}.$$

Definition 11.5.3 *Fix $n \geq 2$. Let $\psi : U_n(\mathbb{R}) \to \mathbb{C}$ be a character of $U_n(\mathbb{R})$ as in (11.1.1). Then for any $v \in V_n$, we define a new character $\psi^v : U_n(\mathbb{R}) \to \mathbb{C}$ by the condition*

$$\psi^v(u) := \psi(v^{-1}uv) = \psi(vuv),$$

for all $u \in U_n(\mathbb{R})$.

We are now ready to state and prove the main theorem of this section which was first proved in (Bump, Friedberg and Goldfeld, 1988) for $n = 3$, and then more generally in (Friedberg, 1987).

Theorem 11.5.4 (Fourier coefficient of Poincaré series) *Fix $n \geq 2$, and let $M = (M_1, \ldots, M_{n-1})$, $N = (N_1, \ldots, N_{n-1}) \in \mathbb{Z}^{n-1}$. Then, for $\psi = \psi_M$, and $\Re(s_i)$ sufficiently large ($i = 1, \ldots, n-1$), the Fourier coefficient (11.5.1) is given by*

$$\sum_{w \in W_n} \sum_{v \in V_n} \sum_{c_1=1}^{\infty} \cdots \sum_{c_{n-1}=1}^{\infty} \frac{S_w\big(\psi_M, \psi_N^v, c\big)\, J_w\big(z; \psi_M, \psi_N^v, c\big)}{c_1^{ns_1} \cdots c_{n-1}^{ns_{n-1}}},$$

where $S_w(\psi_M, \psi_N^v, c)$ is the Kloosterman sum as in Definition 11.2.2, and

$$J_w\big(z; \psi_M, \psi_N^v, c\big) = \int_{\bar{U}_w(\mathbb{R})} I_s(wuz)\, e_{\psi_M}(wcuz)\, \overline{\psi_N^v(u)}\, d^*u,$$

with c given by (11.2.1) and $\bar{U}_w(\mathbb{R}) = (w^{-1}U_n(\mathbb{R})w) \cap U_n(\mathbb{R})$ as in (10.6.4).

Proof It follows from (11.5.1) and the definition of the Poincaré series that

$$\int_{U_n(\mathbb{Z})\backslash U_n(\mathbb{R})} P(uz, s, e_{\psi_M})\overline{\psi_N(u)} d^*u = \int_{U_n(\mathbb{Z})\backslash U_n(\mathbb{R})} \sum_{\gamma \in U_n(\mathbb{Z})\backslash\Gamma} I_s(\gamma uz) e_{\psi_M}(\gamma uz)\overline{\psi_N(u)} d^*u.$$

The sum over $\gamma \in U_n(\mathbb{Z})\backslash\Gamma$ on the right above may be rewritten using the Bruhat decomposition (see Section 11.2) after taking an additional quotient by

V_n as in Definition 11.5.2. In this manner, one obtains

$$\sum_{\gamma\in U_n(\mathbb{Z})\backslash\Gamma} = \sum_{w\in W_n}\sum_{v\in V_n}\sum_{\gamma\in U_n(\mathbb{Z})\backslash(\Gamma\cap G_w)/V_n}.$$

Choose a set of representatives R_w for $U_n(\mathbb{Z})\backslash(\Gamma \cap G_w)/V_n\Gamma_w$ with

$$\Gamma_w = \left(w^{-1}\cdot {}^tU_n(\mathbb{Z})\cdot w\right)\cap U_n(\mathbb{Z}),$$

as before. By Proposition 10.3.6, for every $\gamma \in R_w$, we may choose a Bruhat decomposition

$$\gamma = b_1cwb_2, \qquad b_1, b_2 \in U_n(\mathbb{Q}),$$

and c as in (11.2.1). In an anologous manner, we may rewrite

$$\int_{U_n(\mathbb{Z})\backslash U_n(\mathbb{R})} = \int_{U_w(\mathbb{Z})\backslash U_w(\mathbb{R})}\int_{\bar{U}_w(\mathbb{Z})\backslash\bar{U}_w(\mathbb{R})},$$

with $U_w, \bar{U}_w$ as in (10.6.4), (10.6.5). The result follows after several computations such as those preceding and including (10.6.6). □

11.6 Kuznetsov's trace formula for $SL(n, \mathbb{Z})$

Kuznetsov's trace formula for $SL(2, \mathbb{Z})$ was established by Kuznetsov and first published by him in (Kuznecov, 1980). The Kuznetsov trace formula was also written up by Bruggeman (1978). Later on, it was systematically developed and heavily used by Iwaniec and his collaborators to:

- *extend Bombieri's mean value theorem on primes to large moduli (see (Bombieri, Friedlander and Iwaniec, 1987));*
- *achieve subconvexity bounds of L functions (see (Iwaniec and Kowalski, 2004) and the references there, for example).*

Another spectacular application was due to Motohashi (1997) who proved the asymptotic formula for the fourth moment of the Riemann zeta function by making use of Kuznetsov's formula on $SL(2, \mathbb{Z})$. In this section, we will generalize this valuable tool to the case of $SL(n, \mathbb{Z})$ with $n \geq 2$.

The proof of the Kuznetsov trace formula which we give here is a generalization of Zagier (unpublished notes) on Kuznetsov's trace formula on $SL(2, \mathbb{Z})$. It was worked out by Xiaoqing Li, and I would like to thank her for allowing me to include it in this section.

Throughout Section 11.6, we identify, for $n \geqslant 2$, the generalized Siegel upper half-plane

$$\mathfrak{h}^n = SL(n, \mathbb{R})/SO(n, \mathbb{R}).$$

Warning *Note that this does not conform to the definition of $\mathfrak{h}^n$ as given in Definition 1.2.3, although it is a very close approximation. We shall only adopt this notation temporarily in this section because of the particular applications in mind, and the fact that the Selberg transform may take a very simple form in some cases.*

For $z, z' \in \mathfrak{h}^n$ and every $k \in C_c^\infty(SO(n, \mathbb{R})\backslash SL(n, \mathbb{R})/SO(n, \mathbb{R}))$, a space which contains infinitely differentiable compactly supported $SO(n, \mathbb{R})$ bi-invariant functions on $SL(n, \mathbb{R})$, we define an automorphic kernel

$$K(z, z') = \sum_{\gamma \in SL(n,\mathbb{Z})} k(z^{-1}\gamma z'). \tag{11.6.1}$$

Clearly

$$K(z, z') \in \mathcal{L}^2(SL(n, \mathbb{Z})\backslash \mathfrak{h}^n). \tag{11.6.2}$$

For $M = (M_1, \ldots, M_{n-1}),\ N = (N_1, \ldots, N_{n-1}) \in \mathbb{Z}^{n-1}$, where none of M_i and N_i, $1 \leqslant i \leqslant n-1$ are zero, we call them non-degenerate, we may compute the (M, N)th Fourier coefficients of $K(z, z')$ in two ways. One way is to use the Bruhat decomposition of $GL(n, \mathbb{R})$, and the other is to use the spectral decomposition of $\mathcal{L}^2(SL(n, \mathbb{Z})\backslash \mathfrak{h}^n)$. After these two computations are made, we end up with an identity which is termed the pre-Kuznetsov trace formula.

To start, define

$$P_M(y, z') := \int_{U_n(\mathbb{Z})\backslash U_n(\mathbb{R})} K(xy, z')\psi_M(x)\, d^*x. \tag{11.6.3}$$

Here

$$x = \begin{pmatrix} 1 & x_{1,2} & x_{1,3} & \cdots & x_{1,n} \\ & 1 & x_{2,3} & \cdots & x_{2,n} \\ & & \ddots & & \vdots \\ & & & 1 & x_{n-1,n} \\ & & & & 1 \end{pmatrix}, \quad x_i := x_{n-i,n-i+1}\ (i = 1, \ldots, n-1),$$

$$y = \begin{pmatrix} y_1 y_2 \dots y_{n-1} & & & & \\ & y_1 y_2 \dots y_{n-2} & & & \\ & & \ddots & & \\ & & & y_1 & \\ & & & & 1 \end{pmatrix} \cdot \left(y_1^{n-1} y_2^{n-2} \dots y_{n-2}^2 y_{n-1}\right)^{-1/n}, \tag{11.6.4}$$

$$\psi_M(x) = e^{2\pi i (M_1 x_1 + \cdots M_{n-1} x_{n-1})}, \tag{11.6.5}$$

and

$$d^*x = \prod_{1 \le i < j \le n} dx_{i,j}. \tag{11.6.6}$$

Lemma 11.6.7 *The function $P_M(y, z')$ is a Poincaré series in z'.*

Proof We rewrite $P_M(y, z')$ as

$$\begin{aligned} P_M(y, z') &= \int\limits_{U_n(\mathbb{Z})\backslash U_n(\mathbb{R})} K(xy, z')\psi_M(x)\, d^*x \\ &= \sum_{\gamma \in SL(n,\mathbb{Z})} \int\limits_{U_n(\mathbb{Z})\backslash U_n(\mathbb{R})} k(z^{-1}\gamma z')\psi_M(x)\, d^*x. \end{aligned}$$

Define

$$e_{\psi_M}(y, z') := \int\limits_{U_n(\mathbb{Z})\backslash U_n(\mathbb{R})} k(z^{-1} z')\psi_M(x)\, d^*x. \tag{11.6.8}$$

Then for $u \in U_n(\mathbb{R})$, we have

$$\begin{aligned} e_{\psi_M}(y, uz') &= \int\limits_{U_n(\mathbb{Z})\backslash U_n(\mathbb{R})} k(z^{-1}uz')\psi_M(x)\, d^*x \\ &= \int\limits_{U_n(\mathbb{Z})\backslash U_n(\mathbb{R})} k(z^{-1}z')\psi_M(ux)\, d^*x \\ &= \psi_M(u) \int\limits_{U_n(\mathbb{Z})\backslash U_n(\mathbb{R})} k(z^{-1}z')\psi_M(x)\, d^*x \\ &= \psi_M(u) e_{\psi_M}(y, z'), \end{aligned}$$

so that $e_{\psi_M}(y, z')$ is an e-function. It is also clear that $P_M(y, z') \in \mathcal{L}^2(SL(n, \mathbb{Z})\backslash \mathfrak{h}^n)$ as a function of z'. Hence, $P_M(y, z')$ is a Poincaré series as in Definition 11.1.2 with $s = (0, 0, \dots, 0)$. □

It follows from Theorem 11.5.4 and Lemma 11.6.7 that

$$\int\limits_{U_n(\mathbb{Z})\backslash U_n(\mathbb{R})} P_M(y, x'y')\overline{\psi_N(x')}\, d^*x' \tag{11.6.9}$$

$$= \sum_{w\in W_n}\sum_{v\in V_n}\sum_{c_1=1}^{\infty}\cdots\sum_{c_{n-1}=1}^{\infty} S_w\big(\psi_M, \psi_N^v; c\big)\, J_w\big(y, y'; \psi_M, \psi_N^v, c\big),$$

where, after applying (11.6.8),

$$J_w\big(y, y'; \psi_M, \psi_N^v, c\big) = \int\limits_{\bar{U}_w} e_{\psi_M}(y, wcuy')\overline{\psi_N^v(u)}\, d^*u$$

$$= \int\limits_{\bar{U}_w}\ \int\limits_{U_n(\mathbb{Z})\backslash U_n(\mathbb{R})} k(z^{-1}\cdot wcuy')\, \psi_M(x)\, \overline{\psi_N^v(u)}\, d^*x\, d^*u, \tag{11.6.10}$$

and $S_w(\psi_M, \psi_N^v; c)$ is the Kloosterman sum as in Definition 11.2.2.

The other way of computing the Fourier coefficient (11.6.9) is to make use of the Langlands spectral decomposition (Langlands, 1966, 1976) (see also Theorem 10.13.1). Langlands spectral decomposition states that

$$\mathcal{L}^2(SL(n,\mathbb{Z})\backslash\mathfrak{h}^n) = \mathcal{L}^2_{\text{discrete}}(SL(n,\mathbb{Z})\backslash\mathfrak{h}^n) \oplus \mathcal{L}^2_{\text{cont}}(SL(n,\mathbb{Z})\backslash\mathfrak{h}^n)$$

where

$$\mathcal{L}^2_{\text{discrete}}(SL(n,\mathbb{Z})\backslash\mathfrak{h}^n) = \mathcal{L}^2_{\text{cusp}}(SL(n,\mathbb{Z})\backslash\mathfrak{h}^n) \oplus \mathcal{L}^2_{\text{residue}}(SL(n,\mathbb{Z})\backslash\mathfrak{h}^n).$$

Here $\mathcal{L}^2_{\text{cusp}}$ denotes that the space of Maass forms, $\mathcal{L}^2_{\text{residue}}$ consists of iterated residues of Eisenstein series twisted by Maass forms, and $\mathcal{L}^2_{\text{cont}}$ is the space spanned by integrals of Langlands Eisenstein series.

For each $k \in C_c^\infty(SO(n,\mathbb{R})\backslash SL(n,\mathbb{R})/SO(n,\mathbb{R}))$, we can define an integral operator L_k which acts on $f \in \mathcal{L}^2(SL(n,\mathbb{Z})\backslash\mathfrak{h}^n)$, by

$$(L_k f)(z) := \int\limits_{\mathfrak{h}^n} f(w)k(z^{-1}w)\, d^*w$$

$$= \sum_{\gamma\in\Gamma}\ \int\limits_{SL(n,\mathbb{Z})\backslash\mathfrak{h}^n} f(\gamma w)k(z^{-1}\gamma w)\, d^*w$$

$$= \int\limits_{SL(n,\mathbb{Z})\backslash\mathfrak{h}^n} f(w)K(z, w)\, d^*w.$$

Then L_k commutes with all the invariant differential operators. Furthermore, we have

Lemma 11.6.11 (Selberg) *Fix $n \geq 2$. Let ϕ be a Maass form of type $\nu \in \mathbb{C}^{n-1}$ for $SL(n, \mathbb{Z})$ as in Definition 5.1.3. Then*

$$(L_k\phi)(z) = \hat{k}(\nu)\phi(z),$$

where $\hat{k}(\nu)$ depends only on k and ν. More precisely,

$$\hat{k}(\nu) = (L_k I_\nu)(I) = \int_{\mathfrak{h}^n} k(w) I_\nu(w) dw,$$

where I is the $n \times n$ identity matrix and I_ν is given in Definition 2.4.1.

Usually, $\hat{k}(\nu)$, as in Lemma 11.6.11, is called the Selberg transform of k and

$$\bar{k}(y) = \int_{U_n(\mathbb{R})} k(xy)\, d^*x$$

is called the Harish transform of k. One can think of the Selberg transform as the Mellin transform of its Harish transform. The Selberg transform has an inversion formula due to Harish-Chandra, Bhanu-Murthy, Gangolli, etc., see (Terras, 1988) and the references there.

For simplicity, we introduce the parameters a_k for $1 \leq k \leq n-1$, which are defined by

$$\frac{j(n-j)}{2} + \sum_{k=1}^{n-j} \frac{a_k}{2} = \sum_{i=1}^{n-1} b_{ji}\nu_i$$

for any $1 \leqslant j \leqslant n-1$, where b_{ji}, ν_i are defined in (5.1.1).

It is easy to check that

$$I_\nu(z) = I_a(z) := \prod_{j=1}^{n-1} y_j^{\left[(j(n-j)/2) + \sum_{k=1}^{n-j} a_k/2\right]}.$$

For $z \in GL(n, \mathbb{R})$,

$$h_a(z) := \int_{SO(n,\mathbb{R})} p_a(k^t z k)\, dk,$$

with

$$p_a(z) = \prod_{i=1}^{n-1} |z_i|^{\frac{1}{2}(\bar{a}_{n-i}/2) - (\bar{a}_{n-i+1}/2) - 1)},$$

where z_i is the $i \times i$ upper left-hand corner in z for $1 \leqslant i \leqslant n-1$, and where we normalize the measure dk such that

$$\int_{SO(n,\mathbb{R})} dk = 1.$$

Proposition 11.6.12 (Inversion formula) *For $k \in \mathcal{L}^2(SO(n,\mathbb{R})\backslash\mathfrak{h}^n)$, we have*

$$\begin{aligned} k(z) &= w_n \int\limits_{\substack{\Re(a_i)=0\\ 1\leqslant i\leqslant n-1}} \int\limits_{\mathfrak{h}^n} \frac{k(w) I_a(w) h_a(z^t z)}{|c_n(a)|^2} dw da_1 \ldots da_{n-1} \\ &= w_n \int\limits_{\substack{\Re(a_i)=0\\ 1\leqslant i\leqslant n-1}} \frac{\hat{k}(a) h_a(z^t z)}{|c_n(a)|^2}\, da_1 \ldots da_{n-1}, \end{aligned}$$

where w_n is some constant depending on n, and

$$|c_n(a)|^{-2} = \prod_{1\leqslant i\leqslant j\leqslant n-1} \frac{\pi}{4}|a_i - a_{j+1}| \cdot \tanh\left(\frac{\pi}{4}|a_i - a_{j+1}|\right),$$

with

$$\sum_{i=1}^{n} a_i = 0.$$

Proof See page 88 in (Terras, 1988). Our model $\mathfrak{h}^n$ can be identified with her model $\mathcal{SP}_n$ of $n \times n$ positive definite matrices of determinant 1 by considering the map: $\mathfrak{h}^n \to \mathcal{SP}_n : z \to z^t z$. □

Now, we continue with the derivation of Kuznetsov's trace formula. Let ϕ_j, $j \geqslant 1$ be an orthonormal basis for the cuspidal spectrum and Ω_j, $j \geqslant 1$ be an orthonormal basis for the residual spectrum. For $k \in C_c^\infty(SO(n,\mathbb{R})\backslash\mathfrak{h}^n)$ we have the following spectral expansion of the kernel function $K(z, z')$:

$$\begin{aligned} K(z, z') &= \sum_{j\geqslant 1} \hat{k}(\nu_j)\phi_j(z')\overline{\phi_j(z)} + \sum_{j\geqslant 1} \hat{k}(\Omega_j)\Omega_j(z')\overline{\Omega_j(z)} \\ &\quad + \sum_{\text{Eisenstein series } E} c_E \int \hat{k}(s) E(z', s)\overline{E(z, s)}\, ds, \end{aligned}$$

where c_E are constants depending on E.

Next, we compute the Fourier coefficients of the Poincaré series $P_M(y, z')$ by the above spectral expansion of K. It follows, as in the proof of Lemma 9.1.3 that

$$
\begin{aligned}
&\int\limits_{U_n(\mathbb{Z})\backslash U_n(\mathbb{R})} P_M(y, x'y')\overline{\psi_N(x')}\, d^*x' \\
&= \sum_{j\geqslant 1} \hat{k}(\nu_j) \frac{A_j(N)\,\overline{A_j(M)}}{\prod\limits_{i,j=1}^{n-1} |N_i|^{i(n-i)/2}|M_j|^{j(n-j)/2}} W_{\text{Jacquet}}(N^*y', \nu_j)\overline{W_{\text{Jacquet}}(M^*y, \nu_j)} \\
&\qquad + \sum_{\text{Eisenstein series } E} c_E \int \hat{k}(s) \frac{\varepsilon(N, s)}{\prod\limits_{i=1}^{n-1} |N_i|^{i(n-i)/2}} \frac{\overline{\varepsilon(M, s)}}{\prod\limits_{i=1}^{n-1} |M_i|^{i(n-i)/2}} \\
&\qquad \times W_{\text{Jacquet}}(N^*y', s)\overline{W_{\text{Jacquet}}(M^*y, s)}\, ds, \qquad (11.6.13)
\end{aligned}
$$

where for each ϕ_j in the cuspidal spectrum, every Eisenstein series E in the continuous spectrum, and all $N = (N_1, \ldots, N_{n-1}) \in \mathbb{Z}^{n-1}$ with $N_1 \cdots N_{n-1} \neq 0$, we have

$$
N^* := \begin{pmatrix} N_1 \cdots |N_{n-1}| & & & & \\ & \ddots & & & \\ & & N_1 N_2 & & \\ & & & N_1 & \\ & & & & 1 \end{pmatrix},
$$

$$
\frac{A_j(N)}{\prod\limits_{i=1}^{n-1} |N_i|^{i(n-i)/2}} W_{\text{Jacquet}}(N^*y', \nu_j) = \int\limits_{U_n(\mathbb{Z})\backslash U_n(\mathbb{R})} \phi_j(xy')\overline{\psi_N(x)}\, d^*x,
$$

and, correspondingly,

$$
\frac{\varepsilon(N, s)}{\prod\limits_{i=1}^{n-1} |N_i|^{i(n-i)/2}} W_{\text{Jacquet}}(N^*y', s) = \int\limits_{U_n(\mathbb{Z})\backslash U_n(\mathbb{R})} E(xy', s)\overline{\psi_N(x)}\, d^*x.
$$

The contribution from the residual spectrum is 0 due to the lack of non-degenerate Whittaker models for the residual spectrum.

Remark To simplify notation, we have supressed the character $\psi_{1,\ldots,1,\frac{N_{n-1}}{|N_{n-1}|}}$ in the Jacquet Whittaker functions above.

By comparing (11.6.9) and (11.6.13), we immediately obtain the following pre-Kuznetsov trace formula.

Proposition 11.6.14 (Pre-Kuznetsov trace formula) *Fix $n \geq 2$, and let $k \in C_c^\infty(SO(n,\mathbb{R})\backslash\mathfrak{h}^n)$, $M = (M_1, \ldots, M_{n-1}) \in \mathbb{Z}^{n-1}$, $N = (N_1, \ldots, N_{n-1}) \in \mathbb{Z}^{n-1}$ be nondegenerate. Then*

$$\sum_{j\geqslant 1} \hat{k}(\nu_j) \frac{A_j(N)\,\overline{A_j(M)}}{\prod\limits_{i,j=1}^{n-1} |N_i|^{i(n-i)/2}|M_j|^{j(n-j)/2}} W_{\text{Jacquet}}(N^* y', \nu_j)\overline{W_{\text{Jacquet}}(M^* y, \nu_j)}$$

$$+ \sum_{\text{Eisenstein series } E} c_E \int \hat{k}(s) \frac{\varepsilon(N,s)}{\prod\limits_{i=1}^{n-1} |N_i|^{i(n-i)/2}} \frac{\overline{\varepsilon(M,s)}}{\prod\limits_{i=1}^{n-1} |M_i|^{i(n-i)/2}}$$

$$\times W_{\text{Jacquet}}(N^* y', s)\overline{W_{\text{Jacquet}}(M^* y, s)}\, ds$$

$$= \sum_{w\in W_n} \sum_{v\in V_n} \sum_{c_1=1}^{\infty} \cdots \sum_{c_{n-1}=1}^{\infty} S_w\big(\psi_M, \psi_N^v; c\big) J_w\big(y, y'; \psi_M, \psi_N^v, c\big).$$

It is useful to be able to remove Jacquet's Whittaker functions in the above formula. This can be done by employing Stade's formula (see (Stade, 2002)) for the Mellin transform of products of Whittaker functions on $GL(n)$. In order to present Stade's formula succinctly, we introduce parameters α_k and β_k for $1 \leqslant k \leqslant n-1$, which are defined by

$$\frac{j(n-j)}{2} + \sum_{k=1}^{n-j} \alpha_k = \sum_{i=1}^{n-1} b_{ji}\nu_i \tag{11.6.15}$$

for any $1 \leqslant j \leqslant n-1$, where b_{ji}, ν_i are defined in (5.1.1). We also set

$$\alpha_n = -\sum_{i=1}^{n-1} \alpha_i. \tag{11.6.16}$$

Similar relations hold for β_k and ν_i'.

Proposition 11.6.17 (Stade's formula) *For $\Re(s) \geqslant 1$ we have*

$$2 \int\limits_{(\mathbb{R}^+)^{n-1}} W_{\text{Jacquet}}(y,\nu)\, W_{\text{Jacquet}}(y,\nu') \prod_{j=1}^{n-1} \pi^{(n-j)s} y_j^{(n-j)(s-j)} \frac{dy_j}{y_j}$$

$$= \Gamma\left(\frac{ns}{2}\right)^{-1} \prod_{j=1}^{n} \prod_{k=1}^{n} \Gamma\left(\frac{s+\alpha_j+\beta_k}{2}\right),$$

where α_j and ν_i are related by (11.6.15), (11.6.16), and similarly for β_j and ν_i'.

Remark Stade proves the above formula for $\Re(s)$ large, but one may obtain $|\Re\alpha_j| < \frac{1}{2}$ by Proposition 12.1.9, and similarly for β_k, so the right-hand side of the formula is analytic for $\Re s \geqslant 1$. On the other hand Jacquet (2004a) shows the integral on the left is absolutely convergent for $\Re(s) \geqslant 1$, so by analytic continuation Stade's formula holds for $\Re s \geqslant 1$.

Definition 11.6.18 *Fix $n \geq 2$, $N = (N_1, \ldots, N_{n-1}) \in \mathbb{Z}^{n-1}$. For $j = 1, 2, \ldots$, let*

$$\rho_j(N) = \frac{A_j(N)}{\Gamma(n/2)\prod\limits_{i=1}^{n-1} |N_i|^{i(n-i)/2}} \prod_{j=1}^{n}\prod_{k=1}^{n} \Gamma\left(\frac{1+\alpha_j+\overline{\alpha_k}}{2}\right),$$

$$\eta_s(N) = \frac{\varepsilon_s(N)}{\Gamma(n/2)\prod\limits_{i=1}^{n-1} |N_i|^{i(n-i)/2}} \prod_{j=1}^{n}\prod_{k=1}^{n} \Gamma\left(\frac{1+\alpha_j+\overline{\alpha_k}}{2}\right)$$

be normalized Nth *Fourier coefficients of $SL(n, \mathbb{Z})$ Maass cusp forms and Eisenstein series, respectively.*

We are now ready to state and prove the Kuznetsov trace formula.

Theorem 11.6.19 (Kuznetsov trace formula) *Fix $n \geq 2$, and choose $k \in C_c^\infty(SO(n, \mathbb{R})\backslash\mathfrak{h}^n)$, $M = (M_1, \ldots, M_{n-1}) \in \mathbb{Z}^{n-1}$, $N = (N_1, \ldots, N_{n-1}) \in \mathbb{Z}^{n-1}$ satisfying $M_1 N_1 \cdots M_{n-1} N_{n-1} \neq 0$. Then*

$$\sum_{j\geqslant 1} \hat{k}(\nu_j)\rho_j(N)\bar{\rho}_j(M) + \sum_{\text{Eisenstein series } E} c_E \int \hat{k}(s)\eta_s(N)\bar{\eta}_s(M)\, ds$$

$$= \sum_{w\in W_n}\sum_{v\in V_n}\sum_{c_1=1}^{\infty}\cdots\sum_{c_{n-1}=1}^{\infty} S_w\left(\psi_M, \psi_N^v; c\right) H_w\left(\psi_N^v, c\right),$$

where $\hat{k}$ is given in Lemma 11.6.11, W_n is the Weyl group of $GL(n, \mathbb{R})$, V_n is as in Definition 11.5.2, ψ_N^v is as in Definition 11.5.3, $S_w(\psi_M, \psi_N^v; c)$ is the Kloosterman sum as in Definition 11.2.2, c_E is as in (11.6.13), and ρ_j, η_s are given in Definition 11.6.18.

Furthermore,

$$H_w\left(\psi_N^v, c\right) = \int\limits_{(\mathbb{R}^+)^{n-1}} \int\limits_{U_n(\mathbb{Z})\backslash U_n(\mathbb{R})} \int\limits_{\bar{U}_w} k\left(t_M^{-1}x^{-1}wcut_N\right)\psi_M(x)\overline{\psi_N^v(u)}\, d^*u\, d^*x$$

$$\times \prod_{j=1}^{n-1} 2\pi^{(n-j)}\, t_j^{(n-j)(1-j)} \frac{dt_j}{t_j}$$

where $t_M = t(M^*)^{-1}$, $t_N = t(N^*)^{-1}$, *and*

$$M^* := \begin{pmatrix} M_1 \cdots |M_{n-1}| & & & \\ & \ddots & & \\ & & M_1 & \\ & & & 1 \end{pmatrix}, \quad N^* := \begin{pmatrix} N_1 \cdots |N_{n-1}| & & & \\ & \ddots & & \\ & & N_1 & \\ & & & 1 \end{pmatrix},$$

$$t := \begin{pmatrix} t_1 \cdots t_{n-1} & & & \\ & \ddots & & \\ & & t_1 & \\ & & & 1 \end{pmatrix}.$$

Proof In the pre-Kuznetsov trace formula, given in Proposition 11.6.14, choose y, y' so that $M^*y = N^*y' = t$. This may be accomplished by setting

$$t_i = M_i y_i = N_i y_i', \qquad (i = 1, 2, \ldots, n-1).$$

Next, multiply both sides of the pre-Kuznetsov trace formula by

$$\delta_n \prod_{j=1}^{n-1} \pi^{n-j} t_j^{(n-j)(1-j)} 2^{-j(n-j)} \frac{dt_j}{t_j}$$

and then integrate over $(\mathbb{R}^+)^{n-1}$. Theorem 11.6.19 immediately follows from Proposition 11.6.17. □

Concluding Remarks

(i) The test function k can be extended to a larger space, say, Harish-Chandra's Schwartz space, as long as the convergence of both sides is not a problem.

(ii) The nice feature of the above Kuznetsov type formula is that the residual spectrum does not appear, while its appearance is inevitable in the Selberg–Arthur trace formula. In this sense, the Kuznetsov formula is more handy to treat the cuspidal spectrum.

(iii) In the case of $GL(2)$, Kuznetsov also derived an inversion formula so that one can put any good test function on the Kloosterman sum side. Such an inversion formula on $GL(n)$ does not exist yet, although it is certainly a very interesting research problem.

GL(n)pack functions The following **GL(n)pack** functions, described in the appendix, relate to the material in this chapter:

KloostermanBruhatCell KloostermanCompatibility KloostermanSum PluckerCoordinates PluckerInverse PluckerRelations Whittaker.

12

Rankin–Selberg convolutions

The Rankin–Selberg convolution for L-functions associated to automorphic forms on $GL(2)$ was independently discovered by Rankin (1939) and Selberg (1940). The method was discussed in Section 7.2 in connection with the Gelbart–Jacquet lift and a generalization to $GL(3)$ was given in Section 7.4. A much more general interpretation of the original Rankin–Selberg convolution for $GL(2) \times GL(2)$ in the framework of adeles and automorphic representations was first obtained in (Jacquet, 1972). The theory was subsequently further generalized in (Jacquet and Shalika, 1981).

The Rankin–Selberg convolution for the case $GL(n) \times GL(n')$, $(1 \leq n < n')$, requires new ideas. Note that this includes $GL(1) \times GL(n')$ which is essentially the Godement–Jacquet L-function whose holomorphic continuation and functional equation was first obtained by Godement and Jacquet (1972). A sketch of the general Rankin–Selberg convolution for $GL(n) \times GL(n')$ in classical language was given in (Jacquet, 1981). The theory was further extended in the context of automorphic representations in (Jacquet, Piatetski-Shapiro and Shalika, 1983). In this chapter, we shall present an elementary and self contained account of both the meromorphic continuation and functional equation of Rankin–Selberg L-functions associated to $GL(n) \times GL(n')$. In particular, this will give the meromorphic continuation and functional equation of the Godement–Jacquet L-function.

The fact that Rankin–Selberg L-functions have Euler products is a consequence of the uniqueness of Whittaker functions for local fields. The non-Archimedean theory was worked out in (Jacquet, Piatetski-Shapiro and Shalika, 1983) while the Archimedean local factors were computed in (Jacquet and Shalika, 1981, 1990).

Explicit computations of the local factors in the $GL(n) \times GL(n')$ Rankin–Selberg convolution were obtained by Bump (1987). Bump started with a theorem of Shintani (1976) which relates the Fourier coefficients of Maass forms with Schur polynomials as was discussed in Section 7.4, and ties everything together with an old identity of Cauchy. Such formulae were originally conjectured by Langlands (1970) and generalizations were found by Kato (1978) and Casselman and Shalika (1980). We shall follow this approach as in (Bump, 1984, 1987, 2004). Alternatively, a basic reference for an adelic treatment of the Rankin–Selberg theory is Cogdell's, *analytic theory of L-functions for GL_n*, in (Bernstein and Gelbart, 2003).

The Rankin–Selberg convolution is one of the most important constructions in the theory of L-functions. Naturally it has had inumerable generalizations. The excellent survey paper of Bump (to appear) gives a panoramic overview of the entire subject. We give applications of the Rankin–Selberg convolution method towards the generalized Ramanujan and Selberg conjectures. In particular, the chapter concludes with the theorem of Luo, Rudnick and Sarnak (1995, 1999) which has been used to obtain the current best bounds for the Ramanujan and Selberg conjectures.

12.1 The $GL(n) \times GL(n)$ convolution

Fix an integer $n \geq 2$. Let f, g be Maass forms for $SL(n, \mathbb{Z})$ of type ν_f, $\nu_g \in \mathbb{C}^{n-1}$, respectively, with Fourier expansions:

$$f(z) = \sum_{\gamma \in U_{n-1}(\mathbb{Z})\backslash SL(n-1,\mathbb{Z})} \sum_{m_1=1}^{\infty} \cdots \sum_{m_{n-2}=1}^{\infty} \sum_{m_{n-1}\neq 0} \frac{A(m_1, \ldots, m_{n-1})}{\prod_{k=1}^{n-1} |m_k|^{k(n-k)/2}}$$
$$\times W_{\text{Jacquet}}\left(M \cdot \begin{pmatrix} \gamma & \\ & 1 \end{pmatrix} z,\ \nu_f,\ \psi_{1,\ldots,1,m_{n-1}/|m_{n-1}|}\right),$$

$$g(z) = \sum_{\gamma \in U_{n-1}(\mathbb{Z})\backslash SL(n-1,\mathbb{Z})} \sum_{m_1=1}^{\infty} \cdots \sum_{m_{n-2}=1}^{\infty} \sum_{m_{n-1}\neq 0} \frac{B(m_1, \ldots, m_{n-1})}{\prod_{k=1}^{n-1} |m_k|^{k(n-k)/2}}$$
$$\times W_{\text{Jacquet}}\left(M \cdot \begin{pmatrix} \gamma & \\ & 1 \end{pmatrix} z,\ \nu_g,\ \psi_{1,\ldots,1,\, m_{n-1}/|m_{n-1}|}\right),$$
$$(12.1.1)$$

as in (9.1.2).

Definition 12.1.2 (Rankin–Selberg L-function) *For $n \ge 2$, let f, g be two Maass forms for $SL(n, \mathbb{Z})$ as in (12.1.1). Let $s \in \mathbb{C}$. Then the Rankin–Selberg L-function, denoted $L_{f\times g}(s)$, is defined by*

$$L_{f\times g}(s) = \zeta(ns) \sum_{m_1=1}^{\infty} \cdots \sum_{m_{n-1}=1}^{\infty} \frac{A(m_1, \ldots, m_{n-1}) \cdot \overline{B(m_1, \ldots, m_{n-1})}}{\left(m_1^{n-1} m_2^{n-2} \cdots m_{n-1}\right)^s},$$

which converges absolutely provided $\Re(s)$ is sufficiently large.

This definition conforms with the notation of Theorem 7.4.9.

Theorem 12.1.3 (Euler product) *Fix $n \ge 2$. Let f, g be two Maass forms for $SL(n, \mathbb{Z})$ as in (12.1.1) with Euler products*

$$L_f(s) = \sum_{m=1}^{\infty} \frac{A(m, 1, \ldots, 1)}{m^s} = \prod_p \prod_{i=1}^{n} (1 - \alpha_{p,i}\, p^{-s})^{-1},$$

$$L_g(s) = \sum_{m=1}^{\infty} \frac{B(m, 1, \ldots, 1)}{m^s} = \prod_p \prod_{i=1}^{n} (1 - \beta_{p,i}\, p^{-s})^{-1},$$

then $L_{f\times g}$ will have an Euler product of the form:

$$L_{f\times g}(s) = \prod_p \prod_{i=1}^{n} \prod_{j=1}^{n} (1 - \alpha_{p,i}\, \overline{\beta_{p,j}}\, p^{-s})^{-1}.$$

Proof The proof follows as in Section 7.4. The key point is that the identity (7.4.14) generalizes. In this case, we have

$$A\left(p^{k_1}, p^{k_2}, \ldots, p^{k_{n-1}}\right) = S_{k_1,\ldots,k_{n-1}}(\alpha_{p,1}, \alpha_{p,2}, \ldots, \alpha_{p,n}),$$

and Theorem 12.1.3 is a consequence of Cauchy's identity, Proposition 7.4.20.

□

Our main result is the following generalization of Theorems 7.2.4 and 7.4.9.

Theorem 12.1.4 (Functional equation) *For $n \ge 2$, let f, g be two Maass forms of types ν_f, ν_g for $SL(n, \mathbb{Z})$, as in (12.1.1), whose associated L-functions L_f, L_g, satisfy the functional equations:*

$$\Lambda_f(s) := \prod_{i=1}^{n} \pi^{\frac{-s+\lambda_i(\nu_f)}{2}}\, \Gamma\left(\frac{s - \lambda_i(\nu_f)}{2}\right) L_f(s) = \Lambda_{\tilde f}(1-s)$$

$$\Lambda_g(s) := \prod_{j=1}^{n} \pi^{\frac{-s+\lambda_j(\nu_g)}{2}}\, \Gamma\left(\frac{s - \lambda_j(\nu_g)}{2}\right) L_g(s) = \Lambda_{\tilde g}(1-s),$$

as in Theorem 10.8.6 and Remark 10.8.7 where $\tilde f, \tilde g$, are the dual Maass forms (see Section 9.2). Then the Rankin–Selberg L-function, $L_{f\times g}(s)$ (see Definition 12.1.2) has a meromorphic continuation to all $s \in \mathbb{C}$ with at most a simple *pole at $s = 1$ with residue proportional to $\langle f, g\rangle$, the Petersson inner product of f with g.*

Furthermore, $L_{f\times g}(s)$ satisfies the functional equation

$$\Lambda_{f\times g}(s) := \prod_{i=1}^{n}\prod_{j=1}^{n} \pi^{\frac{-s+\lambda_i(\nu_f)+\overline{\lambda_j(\nu_g)}}{2}}\, \Gamma\left(\frac{s-\lambda_i(\nu_f)-\overline{\lambda_j(\nu_g)}}{2}\right) L_{f\times g}(s)$$
$$= \Lambda_{\tilde{f}\times\tilde{g}}(1-s).$$

Remark It follows from (10.8.5) and Remark 10.8.7 that the powers of π occurring in the above functional equations take the much simpler form:

$$\prod_{i=1}^{n} \pi^{\frac{-s+\lambda_i(\nu)}{2}} = \pi^{-ns/2}, \qquad \prod_{i=1}^{n}\prod_{j=1}^{n} \pi^{\frac{-s+\lambda_i(\nu_f)+\overline{\lambda_j(\nu_g)}}{2}} = \pi^{-n^2 s/2}.$$

Proof This Rankin–Selberg convolution requires the maximal parabolic Eisenstein series $E_P(z, s)$, with $P = P_{n-1,1}$, as defined in (10.7.1). The proof of Theorem 12.1.4 is a generalization of the proof of Theorem 7.4.9. The idea is to compute the inner product of $f \cdot \bar{g}$ with $E_P(z, \bar{s})$. It follows that for $\Gamma = SL(n, \mathbb{Z})$,

$$\langle f\cdot\bar{g},\ E_P(*, \bar{s})\rangle = \int_{\Gamma\backslash\mathfrak{h}^n} f(z)\overline{g(z)}\cdot\overline{E_P(z,\bar{s})}\, d^*z,$$

with d^*z, the invariant measure as given in Proposition 1.5.3. If we now apply the unfolding trick, we obtain

$$\langle f\cdot\bar{g},\ E_P(*, \bar{s})\rangle = \int_{P\backslash\mathfrak{h}^n} f(z)\overline{g(z)}\cdot \mathrm{Det}(z)^s\, d^*z,$$

Now

$$P = \begin{pmatrix} * & \cdots & * & * \\ \vdots & \cdots & \vdots & \vdots \\ * & \cdots & * & * \\ 0 & \cdots & 0 & 1 \end{pmatrix} \subset \Gamma,$$

is generated by matrices of type

$$\begin{pmatrix} & & & 0 \\ & \boxed{\mathfrak{m}_{n-1}} & & \vdots \\ & & & 0 \\ 0 & \cdots & 0 & 1 \end{pmatrix}, \quad \begin{pmatrix} 0 & \cdots & 0 & r_1 \\ \vdots & \cdots & \vdots & \vdots \\ 0 & \cdots & 0 & r_{n-1} \\ 0 & \cdots & 0 & 1 \end{pmatrix},$$

where $\mathfrak{m}_{n-1} \subset SL(n-1, \mathbb{Z})$. Consequently

$$\bigcup_{\gamma \in U_{n-1}(\mathbb{Z})\backslash SL(n-1,\mathbb{Z})} \begin{pmatrix} \gamma & \\ & 1 \end{pmatrix} \cdot P\backslash\mathfrak{h}^n \cong U_n(\mathbb{Z})\backslash\mathfrak{h}^n, \tag{12.1.5}$$

a result obtained earlier in Lemma 5.3.13. Since the Fourier expansion of f given in (12.1.1) is written as a sum over $SL(n-1, \mathbb{Z})$, and $\text{Det}(z)$ is invariant under left multiplication by $\begin{pmatrix} \gamma & \\ & 1 \end{pmatrix}$ with $\gamma \in SL(n-1, \mathbb{Z})$, we may unfold further using (12.1.5) to obtain

$$\begin{aligned}
&\langle f \cdot \bar{g},\ E_P(*, \bar{s})\rangle \\
&= \int\limits_{U_n(\mathbb{Z})\backslash\mathfrak{h}^n} \sum_{m_1=1}^{\infty} \cdots \sum_{m_{n-2}=1}^{\infty} \sum_{m_{n-1}\neq 0} \frac{A(m_1, \ldots, m_{n-1})}{\prod\limits_{k=1}^{n-1} |m_k|^{k(n-k)/2}} \\
&\quad \times W_{\text{Jacquet}}\left(My, \nu_f, \psi_{1,\ldots,1}\right) e^{2\pi i(m_1x_1+\cdots+m_{n-1}x_{n-1})} \cdot \bar{g}(z)\, \text{Det}(z)^s\, d^*z \\
&= \int_0^{\infty} \cdots \int_0^{\infty} \sum_{m_1=1}^{\infty} \cdots \sum_{m_{n-2}=1}^{\infty} \sum_{m_{n-1}\neq 0} \frac{A(m_1, \ldots, m_{n-1})\overline{B(m_1, \ldots, m_{n-1})}}{\prod\limits_{k=1}^{n-1} |m_k|^{k(n-k)}} \\
&\quad \times W_{\text{Jacquet}}(My,\ \nu_f,\ \psi_{1,\ldots,1}) \cdot \bar{W}_{\text{Jacquet}}(My,\ \nu_g,\ \psi_{1,\ldots,1})\, \text{Det}(y)^s\, d^*y.
\end{aligned}$$

Recall from Proposition 1.5.3 that

$$d^*y = \prod_{k=1}^{n-1} y_k^{-k(n-k)} \frac{dy_k}{y_k}.$$

It follows that

$$\begin{aligned}
&\zeta(ns)\langle f \cdot \bar{g},\ E_P(*, \bar{s})\rangle \\
&= L_{f\times g}(s) \int_0^{\infty} \cdots \int_0^{\infty} W_{\text{Jacquet}}(y, \nu_f, \psi_{1,\ldots,1}) \bar{W}_{\text{Jacquet}}(y, \nu_g, \psi_{1,\ldots,1}) \text{Det}(y)^s d^*y \\
&= L_{f\times g}(s) G_{\nu_f,\nu_g}(s),
\end{aligned}$$

say.

The meromorphic continuation and functional equation of $L_{f\times g}(s)$ now follows from the meromorphic continuation and functional equation of $E_P(z, s)$, (given in Proposition 10.7.5) provided we know the meromorphic continuation and functional equation of $G_{\nu_f,\nu_g}(s)$, the Mellin transform of the product of Whittaker functions. This latter functional equation was obtained in a very explicit form by Stade (2001) which allows one to prove the functional equation stated in Theorem 12.1.4.

A much simpler way to obtain the explicit functional equation of $L_{f\times g}(s)$ is by using Remark 10.8.7. In this approach, the functional equation of the minimal parabolic Eisenstein series is used as a template. By formally taking the Rankin-Selberg convolution of the minimal parabolic Eisensteins series $E_{\nu_f}(s)$ and $E_{\nu_g}(s)$, one readily sees that

$$L_{E_{\nu_f}\times E_{\nu_g}}(s) = \zeta(ns)\prod_{i=1}^{n}\prod_{j=1}^{n}\zeta\big(s-\lambda_i(\nu_f)-\overline{\lambda_j(\nu_g)}\big).$$

The correct form of the functional equation can immediately be seen. Of course, one needs to deal with the difficult convergence problems when taking Rankin–Selberg convolutions which are not of rapid decay, but the techniques for dealing with this are now well known (Zagier, 1982), (Liemann, 1993). □

One of the original applications of the Rankin-Selberg method (see (Rankin, 1939), (Selberg, 1940)) was to obtain strong bounds for the Fourier coefficients of automorphic forms. If $A(m_1,\ldots,m_{n-1})$ denotes the Fourier coefficient of a Maass form on $SL(n,\mathbb{Z})$ as in (12.1.1), then we have already obtained the bound

$$|A(m_1,\ldots,m_{n-1})| \ll \prod_{k=1}^{n-1}|m_k|^{k(n-k)/2}$$

in Lemma 9.1.3. For example, this bound gives

$$A(m,1,\ldots,1) \ll m^{(n-1)/2},$$

which unfortunately grows exponentially with n. By using the Rankin–Selberg convolution for $SL(n,\mathbb{Z})$ we can eliminate the dependence on n in the exponent. The following bound was first obtained in a much more general setting in (Jacquet and Shalika, 1981).

Proposition 12.1.6 *For $n\ge 2$, let $A(m_1,\ldots,m_{n-1})$ denote the Fourier coefficient of a Maass form on $SL(n,\mathbb{Z})$ as in (12.1.1). Then there exists a constant $C>0$ such*

$$|A(m,1,\ldots,1)| \le C\cdot|m|^{\frac{1}{2}}.$$

Proof In view of the addendum to Theorem 9.3.11, it is enough to prove that $|A(1,\ldots,1,m)| \le C\cdot|m|^{\frac{1}{2}}$. For $\sigma>0$, consider the identity

$$\frac{1}{2\pi i}\int_{\sigma-i\infty}^{\sigma+i\infty}\frac{x^s}{s^\ell}\,ds = \begin{cases}\frac{1}{(\ell-1)!}(\log x)^{\ell-1} & \text{if } x>1,\\ 0 & \text{if } 0<x\le 1,\end{cases}$$

which is easily proved by shifting the line of integration to the left and applying Cauchy's residue theorem. In view of the fact that the Rankin–Selberg L-function, $L_{f\times f}(s)$, specified in Definition 12.1.2 has positive coefficients and $|A(1,\ldots,1,m)|^2$ occurs as a coefficient of m^{-s}, one obtains for fixed σ, ℓ sufficiently large and all $|m|$ sufficiently large that

$$|A(1,\ldots,1,m)|^2 \ll \frac{1}{2\pi i}\int_{\sigma-i\infty}^{\sigma+i\infty} L_{f\times f}(s)\,\frac{(2m)^s}{s^\ell}\,ds. \qquad (12.1.7)$$

Here one chooses σ sufficiently large so that $L_{f\times f}(\sigma+it)$ converges absolutely for all $t\in\mathbb{R}$. Also, one chooses ℓ sufficiently large so that $|L_{f\times f}(s)| \ll |s|^{\ell-2}$ for $\Re(s)>0$. This can be done, because by standard methods, the functional equation of theorem 12.1.4 implies that $L_{f\times f}(s)$ has polynomial growth properties in the strip $\Re(s)>0$. Proposition 12.1.6 can be proved by shifting the line of integration in (12.1.7) to the left and noting that $L_{f\times f}(s)$ has a simple pole at $s=1$ with residue $\mathcal{R}$, say. By Cauchy's theorem, it follows that $|A(1,\ldots,m)|^2 \ll \mathcal{R}\cdot 2m + \mathcal{O}(1)$. □

Remark 12.1.8 Note that the method for proving Proposition 12.1.6 also shows that $A(m_1,m_2,\ldots,m_{n-1})$ behaves like a constant on average. If we write

$$L_{f\times f}(s) = \sum_{m=1}^{\infty}\frac{b(m)}{m^s}$$

as a Dirichlet series, then one may show that as $x\to\infty$,

$$\sum_{m\le x} b(m) \sim c\cdot x$$

for some constant $c>0$.

Another application of the Rankin–Selberg $GL(n)\times GL(n)$ convolution is to obtain a bound on the eigenvalues of a Maass form. This result was again first obtained in a much more general setting in (Jacquet and Shalika, 1981).

Proposition 12.1.9 *For $n\ge 2$, let f be a Maass form for $SL(n,\mathbb{Z})$ of type $(\nu_1,\ldots,\nu_{n-1})\in\mathbb{C}^{n-1}$ as in (12.1.1). Then*

$$\Re(\lambda_i(\nu)) \le \frac{1}{2}$$

for $i=1,2,\ldots,n-1$, with λ_i defined as in Theorem 10.8.6 and Remark 10.8.7.

Proof By Landau's lemma (Iwaniec and Kowalski, 2004), a Dirichlet series must be absolutely convergent up to its first pole. In the case of $L_{f\times f}(s)$ this implies that the Dirichlet series for $L_{f\times f}(s)$ converges absolutely for $\Re(s) > 1$. We also know that $\Lambda_{f\times f}(s)$, given in Theorem 12.1.4, has a meromorphic continuation to all $s \in \mathbb{C}$ with at most a simple pole at $s = 1$. The above remarks immediately imply that the Gamma factors in $\Lambda_{f\times f}(s)$ cannot have poles for $\Re(s) > 1$. One may then check, using Remark 10.8.7, that this implies Proposition 12.1.9. □

12.2 The $GL(n) \times GL(n+1)$ convolution

Fix $n \geq 2$, and let f, g be Maass forms of type $\nu_f \in \mathbb{C}^{n-1}$, $\nu_g \in \mathbb{C}^n$, for $SL(n, \mathbb{Z})$, $SL(n+1, \mathbb{Z})$, respectively, with Fourier expansions (see (9.1.2)):

$$f(z) = \sum_{\gamma \in U_{n-1}(\mathbb{Z})\backslash SL(n-1,\mathbb{Z})} \sum_{m_1=1}^{\infty} \cdots \sum_{m_{n-2}=1}^{\infty} \sum_{m_{n-1}\neq 0} \frac{A(m_1, \ldots, m_{n-1})}{\prod_{k=1}^{n-1} |m_k|^{k(n-k)/2}}$$
$$\times W_{\text{Jacquet}}\left(\begin{pmatrix} m_1 \cdots |m_{n-1}| & & & \\ & \ddots & & \\ & & m_1 & \\ & & & 1 \end{pmatrix} \cdot \begin{pmatrix} \gamma & \\ & 1 \end{pmatrix} z,\ \nu_f,\ \psi_{1,\ldots,1,\frac{m_{n-1}}{|m_{n-1}|}} \right), \tag{12.2.1}$$

$$g(z) = \sum_{\gamma \in U_n(\mathbb{Z})\backslash SL(n,\mathbb{Z})} \sum_{m_1=1}^{\infty} \cdots \sum_{m_{n-1}=1}^{\infty} \sum_{m_n\neq 0} \frac{B(m_1, \ldots, m_n)}{\prod_{k=1}^{n} |m_k|^{k(n+1-k)/2}}$$
$$\times W_{\text{Jacquet}}\left(\begin{pmatrix} m_1 \cdots |m_n| & & & \\ & \ddots & & \\ & & m_1 & \\ & & & 1 \end{pmatrix} \cdot \begin{pmatrix} \gamma & \\ & 1 \end{pmatrix} z,\ \nu_g,\ \psi_{1,\ldots,1,\frac{m_n}{|m_n|}} \right).$$

Definition 12.2.2 (Rankin–Selberg L-function) *For $n \geq 2$, let f, g be two Maass forms for $SL(n, \mathbb{Z})$, $SL(n+1, \mathbb{Z})$, respectively, as in (12.2.1). Let $s \in \mathbb{C}$. Then the Rankin–Selberg L-function, denoted $L_{f\times g}(s)$, is defined by*

$$L_{f\times g}(s) = \sum_{m_1=1}^{\infty} \cdots \sum_{m_n=1}^{\infty} \frac{A(m_2, \ldots, m_n) \cdot \overline{B(m_1, \ldots, m_n)}}{\left(m_1^n m_2^{n-1} \cdots m_n\right)^s},$$

which converges absolutely provided $\Re(s)$ is sufficiently large.

There are a number of significant differences between the $GL(n) \times GL(n+1)$ Rankin–Selberg convolution and the $GL(n) \times GL(n)$ convolution given in Section 12.1. For example, the Riemann zeta function does not appear in Definition 12.2.2 as it did in Definition 12.1.2. Even more surprising is the fact that the $GL(n) \times GL(n+1)$ convolution does not involve an Eisenstein series and the functional equation comes from the invariance of f, g under reflections of the Weyl group instead of the functional equation of an Eisenstein series. This explains the lack of the appearance of the Riemann zeta function in Definition 12.2.2. Another explanation comes from the theory of Schur polynomials and the applications to Hecke operators and Euler products. The convolution L-function $L_{f\times g}(s)$ has a simple Euler product whose explicit construction will be deferred to Section 12.3.

We now explain the $GL(n) \times GL(n+1)$ convolution. The key idea is to embed $GL(n)$ in $GL(n+1)$ into the upper left-hand corner and then integrate against a power of the determinant. Accordingly, let $z = xy \in \mathfrak{h}^n$ with

$$x = \begin{pmatrix} 1 & x_{n-1} & x_{1,3} & \cdots & x_{1,n} \\ & 1 & x_{n-2} & \cdots & x_{2,n} \\ & & \ddots & \ddots & \vdots \\ & & & 1 & x_1 \\ & & & & 1 \end{pmatrix}, \quad y = \begin{pmatrix} y_1 y_2 \cdots y_{n-1} & & & & \\ & y_1 y_2 \cdots y_{n-2} & & & \\ & & \ddots & & \\ & & & y_1 & \\ & & & & 1 \end{pmatrix},$$

as in Section 9.1, and consider

$$\left\langle f \cdot \bar{g},\ |\mathrm{Det}(*)|^{\bar{s}-\frac{1}{2}} \right\rangle = \int\limits_{SL(n,\mathbb{Z})\backslash\mathfrak{h}^n} f(z) \cdot \overline{g\left(\begin{pmatrix} z & \\ & 1 \end{pmatrix}\right)}\ |\mathrm{Det}(z)|^{s-\frac{1}{2}}\, d^*z.$$

The power $\mathrm{Det}(z)^{s-\frac{1}{2}}$, above, was chosen to make the final formulae as simple as possible. In view of the Fourier expansion of g given in (12.2.1), we may apply the unfolding trick to g and obtain

$$\begin{aligned} &\left\langle f \cdot \bar{g}, |\mathrm{Det}(*)|^{\bar{s}-\frac{1}{2}} \right\rangle \\ &= \sum_{m_1=1}^{\infty} \cdots \sum_{m_{n-1}=1}^{\infty} \sum_{m_n \neq 0} \frac{\overline{B(m_1,\ldots,m_n)}}{\prod\limits_{k=1}^{n} |m_k|^{k(n+1-k)/2}} \int\limits_{U_n(\mathbb{Z})\backslash\mathfrak{h}^n} f(z) \\ &\quad \times \bar{W}_{\mathrm{Jacquet}}\left(\begin{pmatrix} m_1 \cdots |m_n| & & & \\ & \ddots & & \\ & & m_1 & \\ & & & 1 \end{pmatrix}\begin{pmatrix} z & \\ & 1 \end{pmatrix}, \nu_g, \psi_{1,\ldots,1,\frac{m_n}{|m_n|}}\right) |\mathrm{Det}(z)|^{s-\frac{1}{2}} d^*z. \end{aligned}$$

Next, make the transformation

$$z \mapsto m_1^{-1} I_n z,$$

where I_n is the $n \times n$ identity matrix. Since $f(z)$ and d^*z are invariant under scalar multiplication, it follows that

$$\begin{aligned}
&\left\langle f \cdot \bar{g}, |\mathrm{Det}(*)|^{\bar{s}-\frac{1}{2}} \right\rangle \\
&= \sum_{m_1=1}^{\infty} \cdots \sum_{m_{n-1}=1}^{\infty} \sum_{m_n \neq 0} \frac{\overline{B(m_1, \ldots, m_n)}}{\prod_{k=1}^{n} |m_k|^{k(n+1-k)/2}} \int_{U_n(\mathbb{Z})\backslash \mathfrak{h}^n} f(z) \\
&\quad \times \overline{W_{\mathrm{Jacquet}}} \left(\left(\begin{pmatrix} m_2 \cdots m_n y_1 \cdots y_{n-1} & & & & \\ & m_2 \cdots m_{n-1} y_1 \cdots y_{n-2} & & & \\ & & \ddots & & \\ & & & m_2 y_1 & \\ & & & & 1 \end{pmatrix} \\ & & & & & 1 \end{pmatrix}, \nu_g, \psi_{1,\ldots,1} \right) \\
&\quad \times e^{-2\pi i (m_2 x_1 + m_3 x_2 + \cdots + m_n x_{n-1})} \, m_1^{-n(s-\frac{1}{2})} \, \left|\mathrm{Det}\right|(y)^{s-\frac{1}{2}} \, d^*z \\
&= \sum_{m_1=1}^{\infty} \cdots \sum_{m_{n-1}=1}^{\infty} \sum_{m_n \neq 0} \frac{A(m_2, \ldots, m_n)\, \overline{B(m_1, \ldots, m_n)}}{\left(m_1^n m_2^{n-1} \cdots |m_n|\right)^s} \\
&\quad \times \int_0^{\infty} \cdots \int_0^{\infty} W_{\mathrm{Jacquet}}\left(y, \nu_f, \psi_{1,\ldots,1}\right) \bar{W}_{\mathrm{Jacquet}}\left(\begin{pmatrix} y & \\ & 1 \end{pmatrix}, \nu_g, \psi_{1,\ldots,1} \right) \mathrm{Det}(y)^{s-\frac{1}{2}} d^*y,
\end{aligned} \tag{12.2.3}$$

because the above integral in d^*x picks off the $(m_2, m_3, \ldots, m_n)$th Fourier coefficient (Theorem 5.3.2) and $d^*y = \prod_{k=1}^{n-1} y_k^{-k(n-k)} dy_k/y_k$ as in (1.5.4).

Define

$$\Lambda_{f \times g}(s) := \left\langle f \cdot \bar{g}, \; |\mathrm{Det}(*)|^{\bar{s}-\frac{1}{2}} \right\rangle. \tag{12.2.4}$$

Theorem 12.2.5 (Functional equation) *Let f, g be Maass forms associated to $SL(n, \mathbb{Z}), SL(n+1, \mathbb{Z})$, respectively, with Fourier expansions given by (12.2.1). Then the Rankin–Selberg L-function, $L_{f \times g}(s)$, defined in Definition 12.2.2 has holomorphic continuation to all $s \in \mathbb{C}$ and satisfies the functional equation*

$$\Lambda_{f \times g}(s) = \Lambda_{\tilde{f} \times \tilde{g}}(1 - s),$$

where $\Lambda_{f \times g}(s)$ is defined in (12.2.4), (12.2.3), and $\tilde{f}, \tilde{g}$ denote the dual Maass forms as in Section 9.2.

Proof The functional equation is a consequence of the substitution

$$z \longrightarrow w \cdot {}^t(z^{-1}) \cdot w^{-1} := {}^\iota z,$$

where w is the long element as in Proposition 9.2.1. It follows from (9.2.2) that

$${}^\iota z = \begin{pmatrix} 1 & \delta x_1 & & & \\ & 1 & -x_2 & & * \\ & & \ddots & \ddots & \\ & & & 1 & -x_{n-1} \\ & & & & 1 \end{pmatrix} \begin{pmatrix} y_1 y_2 \cdots y_{n-1} & & & & \\ & y_2 y_3 \cdots y_{n-2} & & & \\ & & \ddots & & \\ & & & y_{n-1} & \\ & & & & 1 \end{pmatrix},$$

where $\delta = (-1)^{\lfloor n/2 \rfloor + 1}$. It follows from Proposition 9.2.1, and the fact that f, g are automorphic that

$$\tilde{f}({}^\iota z) = f(z), \qquad \tilde{g}\left(\begin{pmatrix} {}^\iota z & \\ & 1 \end{pmatrix}\right) = g\left(\begin{pmatrix} z & \\ & 1 \end{pmatrix}\right).$$

Consequently,

$$\begin{aligned} \Lambda_{f\times g}(s) &= \left\langle f \cdot \bar{g},\ |\mathrm{Det}(*)|^{\bar{s}-\frac{1}{2}} \right\rangle \\ &= \int\limits_{SL(n,\mathbb{Z})\backslash \mathfrak{h}^n} f(z) \cdot \overline{g\left(\begin{pmatrix} z & \\ & 1 \end{pmatrix}\right)}\ |\mathrm{Det}(z)|^{s-\frac{1}{2}}\, d^*z \\ &= \int\limits_{SL(n,\mathbb{Z})\backslash \mathfrak{h}^n} \tilde{f}({}^\iota z) \cdot \overline{\tilde{g}\left(\begin{pmatrix} {}^\iota z & \\ & 1 \end{pmatrix}\right)}\ |\mathrm{Det}(z)|^{s-\frac{1}{2}}\, d^*z. \end{aligned} \tag{12.2.6}$$

But d^*z is fixed under the transformation $z \to {}^t(z^{-1})$ while

$$|\mathrm{Det}({}^t(z^{-1}))|^{s-\frac{1}{2}} = |\mathrm{Det}(z)|^{\frac{1}{2}-s}. \tag{12.2.7}$$

It follows from the above remarks, after applying the substitution (12.2.7) to (12.2.6), that formally

$$\begin{aligned} \Lambda_{f\times g}(s) &= \int\limits_{SL(n,\mathbb{Z})\backslash \mathfrak{h}^n} \tilde{f}(wz) \cdot \overline{\tilde{g}\left(\begin{pmatrix} wz & \\ & 1 \end{pmatrix}\right)}\ |\mathrm{Det}(z)|^{\frac{1}{2}-s}\, d^*z \\ &= \int\limits_{SL(n,\mathbb{Z})\backslash \mathfrak{h}^n} \tilde{f}(z) \cdot \overline{\tilde{g}\left(\begin{pmatrix} z & \\ & 1 \end{pmatrix}\right)}\ |\mathrm{Det}(z)|^{\frac{1}{2}-s}\, d^*z \\ &= \Lambda_{\tilde{f}\times\tilde{g}}(1-s). \end{aligned}$$

The above formal proof can be made rigorous by breaking the integrals from 0 to ∞ into two pieces $[0, 1]$, $[1, \infty]$ and applying Riemann's method for obtaining the functional equation of the Riemann zeta function. We will not pursue this here. □

12.3 The $GL(n) \times GL(n')$ convolution with $n < n'$

Fix $2 \le n < n' - 1$. Let f, g be Maass forms of type $\nu_f \in \mathbb{C}^{n-1},\ \nu_g \in \mathbb{C}^{n'-1}$, for $SL(n,\mathbb{Z}),\ SL(n',\mathbb{Z})$, respectively, with Fourier expansions (see (9.1.2)):

$$f(z) = \sum_{\gamma \in U_{n-1}(\mathbb{Z})\backslash SL(n-1,\mathbb{Z})} \sum_{m_1=1}^{\infty} \cdots \sum_{m_{n-2}=1}^{\infty} \sum_{m_{n-1}\neq 0} \frac{A(m_1,\ldots,m_{n-1})}{\prod_{k=1}^{n-1} |m_k|^{k(n-k)/2}}$$
$$\times W_{\text{Jacquet}}\left(M \cdot \begin{pmatrix} \gamma & \\ & 1 \end{pmatrix} z,\ \nu_f,\ \psi_{1,\ldots,1,\frac{m_{n-1}}{|m_{n-1}|}}\right), \tag{12.3.1}$$

$$g(z) = \sum_{\gamma \in U_{n'-1}(\mathbb{Z})\backslash SL(n'-1,\mathbb{Z})} \sum_{m_1=1}^{\infty} \cdots \sum_{m_{n'-2}=1}^{\infty} \sum_{m_{n'-1}\neq 0} \frac{B(m_1,\ldots,m_{n'-1})}{\prod_{k=1}^{n'-1} |m_k|^{k(n'-k)/2}}$$
$$\times W_{\text{Jacquet}}\left(M \cdot \begin{pmatrix} \gamma & \\ & 1 \end{pmatrix} z,\ \nu_g,\ \psi_{1,\ldots,1,\frac{m_{n'-1}}{|m_{n'-1}|}}\right).$$

The $GL(n) \times GL(n+1)$ Rankin–Selberg method discussed in Section 12.2 does not naturally generalize to arbitrary $n' > n$. The best way to proceed in the more general situation is to apply a projection operator before taking the inner product. We follow (Bump, 1987), (Bump, to appear) and the exposition in Cogdell's chapter "On the analytic theory of L-functions for GL_n" in (Bernstein and Gelbart, 2003).

Definition 12.3.2 (Projection operator) *Fix integers $2 \le n < n' - 1$. We introduce the projection operator $\mathbb{P}_n^{n'}$ which acts on Maass forms for $SL(n',\mathbb{Z})$ and maps them to cuspidal automorphic functions for the parabolic subgroup*

$$P_{n,1}(\mathbb{Z}) = P_{n,1}(\mathbb{R}) \cap SL(n+1,\mathbb{Z})$$

where

$$P_{n,1}(\mathbb{R}) = \begin{pmatrix} GL(n,\mathbb{R}) & * \\ 0 & 1 \end{pmatrix} \subset GL(n+1,\mathbb{R}).$$

Let g be a Maass form for $SL(n', \mathbb{Z})$ as in (12.3.1). For $z \in P_{n,1}(\mathbb{R})$ we define

$$\mathbb{P}_n^{n'}(g)(z) := |\mathrm{Det}(z)|^{-(n'-n-1)/2} \int_0^1 \cdots \int_0^1 g\left(\left(\begin{array}{c|ccc} \boxed{z} & u_{1,n+2} & \cdots & u_{1,n'} \\ & \vdots & \cdots & \vdots \\ & u_{n+1,n+2} & \cdots & \vdots \\ & 1 & \ddots & \vdots \\ & & \ddots & u_{n'-1,n'} \\ & & & 1 \end{array}\right)\right)$$

$$\times\, e^{-2\pi i\left(u_{n+1,n+2}+u_{n+2,n+3}+\cdots+u_{n'-1,n'}\right)} \prod_{\substack{n+2\le j\le n' \\ 1\le i<j}} du_{i,j}.$$

It is clear that if $p \in P_{n,1}(\mathbb{Z})$, then $\mathbb{P}_n^{n'}(g)(pz) = \mathbb{P}_n^{n'}(g)(z)$.

Lemma 12.3.3 *Fix $2 \le n < n' - 1$, and let g be given by (12.3.1). Then for $z \in P_{n,1}(\mathbb{R})$, we have*

$$\mathbb{P}_n^{n'}(g)(z)$$

$$= |\mathrm{Det}(z)|^{-(n'-n-1)/2} \sum_{\gamma\in U_n(\mathbb{Z})\backslash SL(n,\mathbb{Z})} \sum_{m_{n'-n}=1}^{\infty} \cdots \sum_{m_{n'-1}=1}^{\infty} \frac{B(1,\ldots,1,m_{n'-n},\ldots,m_{n'-1})}{\prod\limits_{k=n'-n}^{n'-1} |m_k|^{k(n'-k)/2}}$$

$$\times\, W_{\mathrm{Jacquet}}\left(\begin{pmatrix} m_{n'-n}\cdots m_{n'-1} & & & & \\ & m_{n'-n}\cdots m_{n'-2} & & & \\ & & \ddots & & \\ & & & m_{n'-n} & \\ & & & & I_{n'-n} \end{pmatrix} \cdot \begin{pmatrix} \gamma z & \\ & I_{n'-n} \end{pmatrix}, \nu_g, \psi_{1,\ldots,1}\right),$$

where I_r denotes the $r \times r$ identity matrix.

Proof Let $z \in P_{n,1}(\mathbb{R})$. Since $\mathbb{P}_n^{n'}(g(z))$ is invariant under left multiplication by $P_{n,1}(\mathbb{Z})$ it follows from Section 5.3 that it has a Fourier expansion as in Theorem 5.3.2. After some computation, the proof follows as in the proof of Theorem 9.4.7. □

With these preliminaries out of the way, we may proceed to describe the Rankin–Selberg convolution. Let f, g be Maass forms for $SL(n,\mathbb{Z})$, $SL(n',\mathbb{Z})$, respectively as in (12.3.1). The requisite convolution is given by the inner product, $\langle f \cdot \mathbb{P}_n^{n'}(g), |\mathrm{Det}(*)|^{\bar{s}-\frac{1}{2}}\rangle$, taken over the fundamental domain

$SL(n,\mathbb{Z})\backslash GL(n,\mathbb{R})$. Lemma 12.3.3 allows us to unravel $\mathbb{P}_n^{n'}(g)$ in this inner product. It follows that

$$\left\langle f\cdot\overline{\mathbb{P}_n^{n'}(g)},\ |\mathrm{Det}(*)|^{\bar{s}-\frac{1}{2}}\right\rangle$$
$$= \int\limits_{U_n(\mathbb{Z})\backslash GL(n,\mathbb{R})} f(z) \sum_{m_{n'-n}=1}^{\infty}\cdots\sum_{m_{n'-1}=1}^{\infty} \frac{\overline{B(1,\ldots,1,m_{n'-n},\ldots,m_{n'-1})}}{\prod\limits_{k=n'-n}^{n'-1}|m_k|^{k(n'-k)/2}}$$
$$\times\, \overline{W}_{\text{Jacquet}}\left(\begin{pmatrix} m_{n'-n}\cdots m_{n'-1} & & & & \\ & m_{n'-n}\cdots m_{n'-2} & & & \\ & & \ddots & & \\ & & & m_{n'-n} & \\ & & & & I_{n'-n}\end{pmatrix}\cdot\begin{pmatrix} z & \\ & I_{n'-n}\end{pmatrix},\ \nu_g,\ \psi_{1,\ldots,1}\right)$$
$$\times\, |\mathrm{Det}(z)|^{s-\frac{n'-n}{2}}\, d^*z.$$

In the above integral, we may make the change of variables $z\to m_{n'-n}^{-1}I_n z$, noting that f and d^*z are invariant under this transformation. After this, the d^*x integral will pick off the $(m_{n'-n+1},\ldots,m_{n'-1})$ coefficient of f. It follows that

$$\left\langle f\cdot\overline{\mathbb{P}_n^{n'}(g)},\ |\mathrm{Det}(*)|^{\bar{s}-\frac{1}{2}}\right\rangle$$
$$= \sum_{m_{n'-n}=1}^{\infty}\cdots\sum_{m_{n'-1}=1}^{\infty} \frac{A(m_{n'-n+1},\ldots,m_{n'-1})\,\overline{B(1,\ldots,1,m_{n'-n},\ldots,m_{n'-1})}}{\left(m_{n'-n}^{n}m_{n'-n+1}^{n-1}\cdots m_{n'-1}\right)^{s}}$$
$$\times \int_0^\infty\cdots\int_0^\infty W_{\text{Jacquet}}(y,\nu_f,\psi_{1,\ldots,1})\bar{W}_{\text{Jacquet}}\left(\begin{pmatrix} y & \\ & I_{n'-n}\end{pmatrix},\ \nu_g,\ \psi_{1,\ldots,1}\right)$$
$$\times\, |\mathrm{Det}(z)|^{s-\frac{n'-n}{2}}\, d^*y.$$

By analogy with the Rankin–Selberg constructions given in Sections 12.1, 12.2, it is natural, after the above computations, to make the following definition.

Definition 12.3.4 (Rankin–Selberg L–function) *Fix $2\le n<n'$. Let f,g be two Maass forms for $SL(n,\mathbb{Z})$, $SL(n',\mathbb{Z})$, respectively, as in (12.3.1). Let $s\in\mathbb{C}$. Then the Rankin–Selberg L-function, denoted $L_{f\times g}(s)$, is defined by*

$$L_{f\times g}(s) = \sum_{m_1=1}^{\infty}\cdots\sum_{m_n=1}^{\infty} \frac{A(m_2,\ldots,m_n)\cdot\overline{B(m_1,\ldots,m_n,1,\ldots,1)}}{\left(m_1^n m_2^{n-2}\cdots m_n\right)^{s}},$$

which converges absolutely provided $\Re(s)$ is sufficiently large. Here, the coefficient $B(m_1,\ldots,m_n,\underbrace{1,\ldots,1}_{n'-n-1})$, has precisely $n'-n-1$ ones on the right-hand side.

Remark We may also take $n = 1$ in Definition 12.3.4 by letting f be the constant function. In this case the Rankin–Selberg L-function is just the Godement–Jacquet L-function as in Section 9.4. In fact, all constructions and proofs in this section will work when $n = 1$.

Theorem 12.3.5 (Euler product) *Fix $2 \le n < n'$. Let f, g be two Maass forms for $SL(n, \mathbb{Z})$, $SL(n', \mathbb{Z})$, respectively, as in (12.3.1) with Euler products*

$$L_f(s) = \sum_{m=1}^{\infty} \frac{A(m, 1, \ldots, 1)}{m^s} = \prod_p \prod_{i=1}^{n} (1 - \alpha_{p,i}\, p^{-s})^{-1},$$

$$L_g(s) = \sum_{m=1}^{\infty} \frac{B(m, 1, \ldots, 1)}{m^s} = \prod_p \prod_{i=1}^{n'} (1 - \beta_{p,i}\, p^{-s})^{-1},$$

then $L_{f\times g}$ will have an Euler product of the form:

$$L_{f\times g}(s) = \prod_p \prod_{i=1}^{n} \prod_{j=1}^{n'} (1 - \alpha_{p,i}\, \overline{\beta_{p,j}}\, p^{-s})^{-1}.$$

Proof The proof follows the ideas in Section 7.4, but requires a modification of Cauchy's identity (Proposition 7.4.20). For details, see (Bump, 1984, 1987, to appear). □

Finally, we consider the meromorphic continuation and functional equation.

Theorem 12.3.6 (Functional equation) *Fix $2 \le n < n'$. Let f, g be two Maass forms for $SL(n, \mathbb{Z})$, $SL(n', \mathbb{Z})$, respectively, as in (12.3.1), whose associated L-functions L_f, L_g satisfy the functional equations:*

$$\Lambda_f(s) := \prod_{i=1}^{n} \pi^{\frac{-s+\lambda_i(\nu_f)}{2}}\, \Gamma\left(\frac{s - \lambda_i(\nu_f)}{2}\right) L_f(s) = \Lambda_{\tilde{f}}(1 - s)$$

$$\Lambda_g(s) := \prod_{j=1}^{n'} \pi^{\frac{-s+\lambda_j(\nu_g)}{2}}\, \Gamma\left(\frac{s - \lambda_j(\nu_g)}{2}\right) L_g(s) = \Lambda_{\tilde{g}}(1 - s),$$

as in Remark 10.8.7 where $\tilde{f}$, $\tilde{g}$ are the dual Maass forms (see Section 9.2). Then the Rankin–Selberg L-function, $L_{f\times g}(s)$ (see Definition 12.3.4) has a holomorphic continuation to all $s \in \mathbb{C}$. Furthermore, $L_{f\times g}(s)$ satisfies the functional equation

$$\Lambda_{f\times g}(s) := \prod_{i=1}^{n} \prod_{j=1}^{n'} \pi^{\frac{-s+\lambda_i(\nu_f)+\overline{\lambda_j(\nu_g)}}{2}}\, \Gamma\left(\frac{s - \lambda_i(\nu_f) - \overline{\lambda_j(\nu_g)}}{2}\right) L_{f\times g}(s)$$
$$= \Lambda_{\tilde{f}\times\tilde{g}}(1 - s).$$

Remark Note that as in the remark after Theorem 12.1.4, the power of π in the above theorem simplifies to $\pi^{-nn's/2}$.

Proof This theorem has essentially been proved in the case $n' = n+1$ in Theorem 12.2.5, so we need only consider $n < n' - 1$.

Let $z \in \mathfrak{h}^n$. The functional equation is a consequence of the substitution

$$z \longrightarrow w \cdot {}^t(z^{-1}) \cdot w^{-1} := {}^{\iota}z,$$

where w is the long element as in Proposition 9.2.1. It follows from (9.2.2) that

$${}^{\iota}z = \begin{pmatrix} 1 & \delta x_1 & & & \\ & 1 & -x_2 & & * \\ & & \ddots & \ddots & \\ & & & 1 & -x_{n-1} \\ & & & & 1 \end{pmatrix} \begin{pmatrix} y_1 y_2 \cdots y_{n-1} & & & & \\ & y_2 y_3 \cdots y_{n-2} & & & \\ & & \ddots & & \\ & & & y_{n-1} & \\ & & & & 1 \end{pmatrix},$$

where $\delta = (-1)^{\lfloor n/2 \rfloor + 1}$. It follows from Proposition 9.2.1 and the fact that f, g are automorphic that

$$\tilde{f}({}^{\iota}z) = f(z),$$

and

$$\tilde{g}\left(\begin{pmatrix} {}^{\iota}z & \\ & I_{n'-n} \end{pmatrix}\right) = g\left(\begin{pmatrix} z & \\ & I_{n'-n} \end{pmatrix}\right).$$

Consequently,

$$\begin{aligned} \Lambda_{f\times g}(s) &= \left\langle f \cdot \overline{\mathbb{P}_n^{n'}(g)}, \quad |\mathrm{Det}(*)|^{\bar{s}-\frac{1}{2}} \right\rangle \\ &= \int\limits_{SL(n,\mathbb{Z})\backslash GL(n,\mathbb{R})} f(z) \cdot \overline{\mathbb{P}_n^{n'}(g)\left(\begin{pmatrix} z & \\ & I_{n'-n} \end{pmatrix}\right)} \, |\mathrm{Det}(z)|^{s-\frac{1}{2}} \, d^*z \\ &= \int\limits_{SL(n,\mathbb{Z})\backslash GL(n,\mathbb{R})} \tilde{f}({}^{\iota}z) \cdot \overline{\mathbb{P}_n^{n'}(\tilde{g})\left(\begin{pmatrix} {}^{\iota}z & \\ & I_{n'-n} \end{pmatrix}\right)} \, |\mathrm{Det}(z)|^{s-\frac{1}{2}} \, d^*z. \end{aligned} \tag{12.3.7}$$

But d^*z is fixed under the transformation $z \to {}^t(z^{-1})$ while

$$\left|\mathrm{Det}({}^t(z^{-1}))\right|^{s-\frac{1}{2}} = \left|\mathrm{Det}(z)\right|^{\frac{1}{2}-s}. \tag{12.3.8}$$

It follows from the above remarks, after applying the substitution (12.3.8) to (12.3.7), that formally

$$\begin{aligned}\Lambda_{f\times g}(s) &= \int\limits_{SL(n,\mathbb{Z})\backslash GL(n,\mathbb{R})} \tilde{f}(wz)\cdot \overline{\left(\iota\circ\mathbb{P}_n^{n'}\circ\iota\right)(\tilde{g})\left(\begin{pmatrix} z & \\ & 1\end{pmatrix}\right)}\,|\mathrm{Det}(z)|^{\frac{1}{2}-s}\,d^*z\\ &= \int\limits_{SL(n,\mathbb{Z})\backslash GL(n,\mathbb{R})} \tilde{f}(z)\cdot \overline{\left(\iota\circ\mathbb{P}_n^{n'}\circ\iota\right)(\tilde{g})\left(\begin{pmatrix} z & \\ & 1\end{pmatrix}\right)}\,|\mathrm{Det}(z)|^{\frac{1}{2}-s}\,d^*z\\ &= \tilde{\Lambda}_{\tilde{f}\times\tilde{g}}(1-s),\end{aligned}$$

where

$$\tilde{\Lambda}_{\tilde{f}\times\tilde{g}}(s) = \left\langle \tilde{f}\cdot\overline{\left(\iota\circ\mathbb{P}_n^{n'}\circ\iota\right)(\tilde{g})},\quad |\mathrm{Det}(*)|^{\bar{s}-\frac{1}{2}}\right\rangle.$$

The above formal proof can be made rigorous by breaking the integrals from 0 to ∞ into two pieces $[0, 1]$, $[1, \infty]$ and applying Riemann's method for obtaining the functional equation of the Riemann zeta function.

Once the form of the functional equation is obtained, the precise Gamma factors in the functional equation can be deduced by using the functional equation of the minimal parabolic Eisenstein series as a template as we did at the end of Section 12.1. We leave these computations to the reader. □

12.4 Generalized Ramanujan conjecture

Let f be a Maass form of type $\nu = (\nu_1, \ldots, \nu_{n-1}) \in \mathbb{C}^{n-1}$ for $SL(n, \mathbb{Z})$ with Fourier expansion as given in (9.1.2):

$$\begin{aligned}f(z) = \sum_{\gamma\in U_{n-1}(\mathbb{Z})\backslash SL(n-1,\mathbb{Z})} \sum_{m_1=1}^{\infty}\cdots\sum_{m_{n-2}=1}^{\infty}\sum_{m_{n-1}\neq 0} \frac{A(m_1,\ldots,m_{n-1})}{\prod\limits_{k=1}^{n-1}|m_k|^{k(n-k)/2}}\\ \times W_{\text{Jacquet}}\left(M\cdot\begin{pmatrix}\gamma & \\ & 1\end{pmatrix}z,\ \nu,\ \psi_{1,\ldots,1,\frac{m_{n-1}}{|m_{n-1}|}}\right),\end{aligned} \tag{12.4.1}$$

where

$$M = \begin{pmatrix} m_1\cdots m_{n-2}\cdot|m_{n-1}| & & & & \\ & \ddots & & & \\ & & m_1m_2 & & \\ & & & m_1 & \\ & & & & 1\end{pmatrix}, \qquad A(m_1,\ldots,m_{n-1}) \in \mathbb{C},$$

and

$$\psi_{1,\ldots,1,\epsilon}\left(\begin{pmatrix} 1 & u_{n-1} & & & * \\ & 1 & u_{n-2} & & \\ & & \ddots & \ddots & \\ & & & 1 & u_1 \\ & & & & 1 \end{pmatrix}\right) = e^{2\pi i\left(u_1+\cdots+u_{n-2}+\epsilon u_{n-1}\right)}.$$

We further assume that f is normalized so that $A(1,\ldots,1)=1$ and that f is an eigenfunction of the Hecke operators given in (9.3.5). It was shown in (9.4.2) that the Godement–Jacquet L-function

$$L_f(s)=\sum_{n=1}^{\infty}\frac{A(n,1,\ldots,1)}{n^s}$$

has an Euler product given by

$$\begin{aligned} L_f(s) &= \prod_p \Big(1-A(p,\ldots,1)p^{-s}+A(1,p,\ldots,1)p^{-2s}- \\ &\qquad \cdots+(-1)^{n-1}A(1,\ldots,p)p^{(1-n)s}+(-1)^n p^{-ns}\Big)^{-1} \\ &= \prod_p\prod_{i=1}^{n}(1-\alpha_{p,i}p^{-s})^{-1}, \end{aligned} \tag{12.4.2}$$

where $\alpha_{p,i}\in\mathbb{C}$ for $i=1,2,\ldots,n$.

Conjecture 12.4.3 (Ramanujan conjecture at finite primes) *For $n\ge 2$, let f be a Maass form for $SL(n,\mathbb{Z})$ as in (12.4.1). The generalized Ramanujan conjecture asserts that $\alpha_{p,i}$ given in (12.4.2) satisfy*

$$|\alpha_{p,i}|=1,$$

for all primes p and $i=1,\ldots,n$. Equivalently, for every rational prime p we have the bound

$$|A(p,1,\ldots,1)|\ \le\ n.$$

This conjecture has been proved for holomorphic modular forms on $GL(2)$ in (Deligne, 1974) (see Section 3.6) but is still a major unsolved problem for Maass forms. Deligne shows that every holomorphic modular form (Hecke eigenform) over $\mathbb{Q}$, say, is associated to an algebraic variety and that the pth Fourier coefficient can be interpreted in terms of the number of points on this variety over $\mathbb{F}_p$, the finite field of p elements. From this point of view, the Ramanujan conjecture is equivalent to the Riemann hypothesis for varieties over finite fields first conjectured in (Weil, 1949) and, in a stunning tour de force, finally proved in (Deligne, 1974). The problem with trying to generalize this approach to non-holomorphic automorphic forms is that there seem to be no visible connections between the theory of Maass forms and algebraic geometry.

There is yet another fundamental conjecture which was originally formulated by Selberg (1965) (see also Section 3.7). Classically this is known as the Selberg eigenvalue conjecture. For the cognoscenti, it is clear from the adelic point of view that the Selberg eigenvalue conjecture is really the generalized Ramanujan conjecture at the infinite prime. In the next section, we will give a more elementary explanation of why these two conjectures can be placed on an equal footing. Here is the generalized Selberg eigenvalue conjecture.

Conjecture 12.4.4 (Selberg eigenvalue conjecture) *For $n \geq 2$, let $f(z)$ be a Maass form of type $(\nu_1, \nu_2, \ldots, \nu_{n-1}) \in \mathbb{C}^{n-1}$ for $SL(n, \mathbb{Z})$ as in (12.4.1). Then*

$$\Re(\nu_i) = \frac{1}{n}, \qquad \Re(\lambda_i(\nu)) = 0,$$

for $i = 1, 2, \ldots, n-1$, with λ_i as defined in Theorem 10.8.6 and Remark 10.8.7.

Remark The first to observe that the classical Ramanujan conjecture concerning the Fourier coefficients of $\Delta(z)$ can be reformulated to a very general conjecture on $GL(n)$ appears to have been Satake (1966). The generalized Ramanujan conjecture has been established for automorphic cusp forms of $GL(n, F)$, of algebraic type satisfying a Galois conjugacy condition, where F is a complex multiplication field (see (Harris and Taylor, 2001)). The proof is a remarkable tour-de-force combining the Arthur–Selberg trace formula and the theory of Shimura varieties. The generalized Ramanujan conjecture is still unproven for Maass forms for $SL(n, \mathbb{Z})$ of the type considered in this book.

While the Selberg eigenvalue Conjecture 12.4.4 is not hard to prove for $SL(2, \mathbb{Z})$ (see Theorem 3.7.2) it is still an unsolved problem for congruence subgroups (see Conjecture 3.7.1); and, of course, Conjecture 12.4.4 can be easily generalized to higher level congruence subgroups of $SL(n, \mathbb{Z})$ with $n \geq 2$.

In (Kim and Sarnak, 2003), one may find the current world record for both the Ramanujan conjecture and Selberg eigenvalue conjecture for $GL(2)$ (this includes the case of congruence subgroups of higher level). The precise bounds obtained are

$$|\alpha_{p,1}|, \quad |\alpha_{p,2}| \;\leq\; p^{\frac{7}{64}},$$

and correspondingly,

$$\Re(\nu) \leq \frac{7}{64}.$$

In (Selberg, 1965) the bound $\le 1/4$ instead of $\le 7/64$ as in (Kim and Sarnak, 2003) was attained. Selberg's result was slightly improved to $< 1/4$ in (Gelbart and Jacquet, 1978). Further improvements were given in (Serre, 1981) (see also (Serre, 1977)), who obtained $|\alpha_{p,i}| \le p^{\frac{1}{5}}$, Shahidi (1988), $|\alpha_{p,i}| < p^{\frac{1}{5}}$, Duke and Iwaniec (1989), found a different proof of $|\alpha_{p,i}| \le 2p^{\frac{1}{5}}$, Bump, Duke, Hoffstein and Iwaniec (1992), $|\alpha_{p,i}| \le p^{\frac{5}{28}}$. In the case of $GL(n)$ with $n \ge 2$, Jacquet and Shalika (1981) obtained the bound $< 1/2$. This bound was proved for $SL(n, \mathbb{Z})$ in Propositions 12.1.6, 12.1.9. For the finite primes p, Serre (1981) suggested that one could do better by a clever use of the Rankin–Selberg L-function. This led to the bound

$$\le \frac{1}{2} - \frac{1}{dn^2+1} \tag{12.4.5}$$

for Maass forms on $GL(n)$ over a number field of degree d. In (Luo, Rudnick and Sarnak, 1995, 1999) the dependence on d was removed. We give an exposition of the Luo, Rudnick and Sarnak method (over $\mathbb{Q}$) in the next section.

12.5 The Luo–Rudnick–Sarnak bound for the generalized Ramanujan conjecture

As an application of the Rankin–Selberg method, we shall give a proof of the bound (12.4.5) obtained in (Luo, Rudnick and Sarnak, 1995, 1999). In order to simplify the exposition, we work over $\mathbb{Q}$, instead of a number field, so that $d = 1$. Let us restate the theorem.

Theorem 12.5.1 (Luo–Rudnick–Sarnak) *Fix $n \ge 2$. Let f be a Maass form of type $(\nu_1, \ldots, \nu_{n-1})$ for $SL(n, \mathbb{Z})$ which is an eigenfunction of the Hecke operators as in (12.4.1). Then for $\alpha_{p,i}$ as in (12.4.2), we have*

$$|\alpha_{p,i}| \le p^{\frac{1}{2}-\frac{1}{n^2+1}}, \qquad \Re(\lambda_i(\nu)) \le \frac{1}{2} - \frac{1}{n^2+1}$$

for all primes p, $1 \le i \le n$, and $1 \le j \le n-1$. Here λ_i is defined in Theorem 10.8.6 and Remark 10.8.7.

Remark In (Luo, Rudnick and Sarnak, 1999) the above bound was obtained for Maass forms on $GL(n)$ over an arbitrary number field.

Proof Recall the definition of $L_f(s)$ given in (12.4.2).

$$L_f(s) = \prod_{i=1}^{n} \prod_{p} (1 - \alpha_{p,i} p^{-s})^{-1}.$$

Now, let χ be a primitive Dirichlet character (mod q). Consider $f \otimes \chi$, which has associated L-function

$$L_{f\otimes\chi}(s) = \prod_p \prod_{i=1}^n (1 - \alpha_{p,i}\,\chi(p)\,p^{-s})^{-1}$$

We may now take the Rankin–Selberg convolution of f with $f \otimes \chi$. This leads to the Rankin–Selberg L-function

$$L_{(f\otimes\chi)\times f}(s) = \prod_p \prod_{i=1}^n \prod_{j=1}^n (1 - \alpha_{p,i}\overline{\alpha_{p,j}}\,\chi(p)\,p^{-s})^{-1}.$$

Further, as in the proof of Theorem 12.1.4, Remark 10.8.7, and the comments after it, the function $L_{(f\otimes\chi)\times f}(s)$ satisfies the functional equation

$$\begin{aligned}
&\Lambda_{(f\otimes\chi)\times f}(s)\\
&= \prod_{i=1}^n \prod_{j=1}^n \left(\frac{q}{\pi}\right)^{\frac{s+\mathfrak{a}_\chi-\lambda_i(\nu)-\overline{\lambda_j(\nu)}}{2}} \Gamma\left(\frac{s-\lambda_i(\nu)-\overline{\lambda_j(\nu)}+\mathfrak{a}_\chi}{2}\right) L_{(f\otimes\chi)\times f}(s)\\
&= \epsilon_\chi \cdot \Lambda_{(\tilde{f}\otimes\bar{\chi})\times\tilde{f}}(1-s), \qquad (12.5.2)
\end{aligned}$$

where

$$\epsilon_\chi = \left(\frac{i^{\mathfrak{a}_\chi}\sqrt{q}}{\tau(\chi)}\right)^{n^2}, \qquad \mathfrak{a}_\chi = \begin{cases} 0 & \text{if } \chi(-1) = 1,\\ 1 & \text{if } \chi(-1) = -1.\end{cases}$$

□

It follows from the methods used to prove Theorem 12.1.4 that $\Lambda_{(f\otimes\chi)\times f}(s)$ has a holomorphic continuation to all $s \in \mathbb{C}$ except for simple poles at $s = 1$ and $s = 0$, the latter simple poles can only occur if $\chi \equiv 1$ is the trivial character. Note that at finite level there can be finitely many χ for which this has a pole (of course, this impacts nothing). This result was also proved in much greater generality in the combination of papers: (Shahidi, 1981, 1985), (Jacquet and Shalika, 1981, 1990), (Jacquet, Piatetski-Shapiro and Shalika, 1983), (Moeglin and Waldspurger, 1989).

We also require, for any fixed prime p_0, the modified Rankin–Selberg L-function

$$L^{p_0}_{(f\otimes\chi)\times f}(s) = L_{(f\otimes\chi)\times f}(s) \cdot \prod_{i=1}^n \prod_{j=1}^n \left(1 - \alpha_{p_0,i}\overline{\alpha_{p_0,j}}\,\chi(p_0)\,p_0^{-s}\right), \qquad (12.5.3)$$

which is the same as $L_{(f\otimes\chi)\times f}(s)$ except that the Euler factor at p_0 has been removed.

The key idea in the proof of Theorem 12.5.1 is the following lemma.

Lemma 12.5.4 *Fix $n \geq 2$ and fix p_0 to be either 1 or a rational prime. Let f be a Maass form for $SL(n,\mathbb{Z})$ as above. Then for any real number $\beta > 1 - (2/(n^2+1))$, there exist infinitely many primitive Dirichlet characters χ such that $\chi(p_0) = 1$, $\chi(-1) = 1$, and*

$$L^{p_0}_{(f\otimes\chi)\times f}(\beta) \neq 0.$$

We defer the proof of this lemma until later and continue with the proof of Theorem 12.5.1.

It follows from (12.5.3) that

$$L_{(f\otimes\chi)\times f}(s) = L^{p_0}_{(f\otimes\chi)\times f}(s)\cdot\prod_{i=1}^{n}\prod_{j=1}^{n}\left(1-\alpha_{p_0,i}\,\overline{\alpha_{p_0,j}}\,\chi(p_0)\,p_0^{-s}\right)^{-1}.$$

Assume $\chi(-1) = 1$, so $\mathfrak{a}_\chi = 0$, and the Gamma factors in the functional equation take the form (12.5.5). Since $\Lambda_{(f\otimes\chi)\times f}(s)$ is holomorphic for χ non-trivial, we see that any pole of

$$\prod_{i=1}^{n}\prod_{j=1}^{n}\left(1-\alpha_{p_0,i}\,\overline{\alpha_{p_0,j}}\,\chi(p_0)\,p_0^{-s}\right)^{-1}$$

or

$$\prod_{i=1}^{n}\prod_{j=1}^{n}\Gamma\left(\frac{s-\lambda_i(\nu)-\overline{\lambda_j(\nu)}}{2}\right), \tag{12.5.5}$$

must be a zero of $L^{p_0}_{(f\otimes\chi)\times f}(s)$.

Assume that $\chi(p_0) = 1$ and for some $1 \leq i \leq n$ we have $|\alpha_{p_0,i}|^2 = p^\beta$ with $\beta > 1-(2/(n^2+1))$. Then $(1-|\alpha_{p_0,i}|^2 p_0^{-s})^{-1}$ has a pole at $s = \beta$. Similarly, assume that $\Re(\lambda_i(\nu)) = \beta$ for some $i = 1,\ldots,n-1$, and for some $\beta > 1 - \frac{2}{n^2+1}$. Then the Gamma factor

$$\prod_{i=1}^{n}\Gamma\left(\frac{s-2\Re(\lambda_i(\nu))}{2}\right)$$

would have a pole at $s = 2\beta$. It immediately follows, in both these cases, that $L^{p_0}_{(f\otimes\chi)\times f}(s)$ would have to have a zero in a region which contradicts Lemma 12.5.4. This proves Theorem 12.5.1. It also explains why the Ramanujan Conjecture 12.4.3 and the eigenvalue Conjecture 12.4.4 can be placed on an equal footing.

Remark The above proof yields a slightly better bound for the case of $GL(2)$ Maass forms when combined with the Gelbart–Jacquet lift. For example, if we start with a Maass form of type ν for $SL(2,\mathbb{Z})$ with L-function $\prod_p \prod_{i=1}^{2} (1-\alpha_{p,i}p^{-s})^{-1}$, then the Gelbart–Jacquet lift will yield a degree 3 Euler product as in Lemma 7.3.5 associated to a Maass form of type $(2\nu/3, 2\nu/3)$ for $SL(3,\mathbb{Z})$. When combined with the proof of Theorem 12.5.1, one gets the bounds

$$|\alpha_{p,i}| \le p^{\frac{1}{5}} \ (i=1,2), \qquad |\Re(\nu)| \le \frac{1}{5},$$

a result first obtained by Shahidi (1988) by quite different methods.

Proof of Lemma 12.5.4 We first prove Lemma 12.5.4 in the simpler case when $p_0 = 1$, as in (Luo, Rudnick and Sarnak, 1995). Afterwards, we sketch the proof in the more general case when p_0 is a prime, as in (Luo, Rudnick and Sarnak, 1999). □

With the notation of (12.5.3), for a primitive character $\chi \pmod q$, the functional equation (12.5.2) may be rewritten in the form.

$$L^{p_0}_{(f\otimes\chi)\times f}(s) = \epsilon_\chi \left(\frac{q}{\pi}\right)^{(n^2/2)-n^2 s} G_{p_0}(s)\, L^{p_0}_{(\bar{f}\otimes\bar{\chi})\times \bar{f}}(1-s), \tag{12.5.6}$$

where

$$G_{p_0}(s) = \frac{\prod\limits_{i=1}^{n}\prod\limits_{j=1}^{n} \Gamma\left(\frac{1-s-\lambda_i(\nu)-\overline{\lambda_j(\nu)}}{2}\right)\left(1-\alpha_{p_0,i}\overline{\alpha_{p_0,j}}\, p_0^{-(1-s)}\right)^{-1}}{\prod\limits_{i=1}^{n}\prod\limits_{j=1}^{n} \Gamma\left(\frac{s-\lambda_i(\nu)-\overline{\lambda_j(\nu)}}{2}\right)\left(1-\alpha_{p_0,i}\overline{\alpha_{p_0,j}}\, p_0^{-s}\right)^{-1}}.$$

It follows easily from the Euler product (Theorem 12.1.3) and the coefficient bound in Proposition 12.1.6 that for $\beta \ge 1$, we have $L_{(f\otimes\chi)\times f}(\beta) \ne 0$. This was proved in much greater generality in (Shahidi, 1981). Consequently, we may assume that $1-(2/(n^2+1)) < \beta < 1$. The key idea of the proof is to show that for $\epsilon > 0$, and Q sufficiently large that

$$\sum_{q\sim Q}\ \sum_{\chi \pmod q}^{*} L^{p_0}_{(f\otimes\chi)\times f}(\beta) \gg Q^{2-\epsilon}, \tag{12.5.7}$$

where $\sum^*$ means that the sum ranges over primitive characters $\chi \pmod q$ satisfying $\chi(p_0) = \chi(-1) = +1$, and $\sum\limits_{q\sim Q}$ means we sum over primes $Q \le q \le 2Q$. The lower bound (12.5.7) immediately implies Lemma 12.5.4.

To prove (12.5.7), we require an auxilliary compactly supported smooth function $h : [A,B] \to \mathbb{R}$, where $0 < A < B$, and $h(y) \ge 0$, and in addition

$$\int_0^\infty \frac{h(y)}{y}\, dy = 1.$$

Define

$$\tilde{h}(s) := \int_0^\infty h(y) y^s \frac{dy}{y},$$

to be the Mellin transform of h, and

$$h_1(y) := \frac{1}{2\pi i} \int_{\Re(s)=2} \tilde{h}(s) y^{-s} \frac{ds}{s}.$$

Then, by Mellin inversion,

$$h_1(y) = \int_y^\infty \frac{h(x)}{x}\, dx.$$

It follows that

$$0 \le h_1(y) \le 1; \qquad h_1(y) = 1; \qquad \text{if } 0 < y \le A,$$

while

$$h_1(y) = 0, \qquad \text{if } y \ge B.$$

Next, for $y > 0$, define

$$h_2(y) := \frac{1}{2\pi i} \int_{Re(s)=2} \tilde{h}(-s) G_{p_0}(-s+\beta) y^{-s} \frac{ds}{s}.$$

We shall forthwith show that h_2 satisfies the following bounds.

$$h_2(y) \ll_M y^{-M}, \text{ for } y \ge 1 \text{ and any positive integer } M, \tag{12.5.8}$$
$$h_2(y) \ll_\epsilon 1 + y^{1-\beta_0-\beta-\epsilon}, \text{ for } 0 < y \le 1 \text{ and any } \epsilon > 0, \tag{12.5.9}$$

where

$$\beta_0 = 2 \max_{1 \le i \le n} \Re(\lambda_i(\nu)),$$

and, we may recall that the $\lambda_i(\nu)$ occur in the Gamma factors of $G_{p_0}(s)$ as in (12.5.6). To prove (12.5.8) shift the line of integration to the right to the line $\Re(s) = M$. The result follows since $G_{p_0}(s)$ has at most polynomial growth in s in fixed vertical strips while $\tilde{h}(-s)$ has rapid decay. To prove (12.5.9), shift the line of integration to the left. We may assume $\beta_0 + \beta - 1 < 0$, otherwise, (12.5.9) is obvious. After shifting to the left, we will pick up the first simple pole at $\Re(s) = \beta_0 + \beta - 1$. The bound (12.5.9) follows immediately.

The next step in the proof is the derivation of an approximate functional equation for $L_{(f\otimes\chi)\times f}(s)$ of the type previously derived in Section 8.3. The asymmetric form of the functional equation (12.5.6) then leads to an asymmetric approximate functional equation.

Let

$$L_{(f\otimes\chi)\times f}(s) = \sum_{m=1}^{\infty} \frac{b(m)}{m^s}\chi(m).$$

It follows that for any $Y > 1$,

$$\begin{aligned}
\mathcal{I}(\beta, Y) &:= \frac{1}{2\pi i} \int\limits_{\Re(s)=2} \tilde{h}(s)\, L_{(f\otimes\chi)\times f}(s+\beta)\, Y^s\, \frac{ds}{s} \\
&= \sum_{m=1}^{\infty} \frac{b(m)\chi(m)}{m^\beta} \frac{1}{2\pi i} \int\limits_{\Re(s)=2} \tilde{h}(s) \left(\frac{Y}{m}\right)^s \frac{ds}{s} \\
&= \sum_{m=1}^{\infty} \frac{b(m)}{m^\beta} \chi(m) h_1\left(\frac{m}{Y}\right). \qquad (12.5.10)
\end{aligned}$$

On the other hand, we may shift the line of integration to the left, apply the functional equation (12.5.6), and then let $s \to -s$, to obtain

$$\mathcal{I}(\beta, Y) := L_{(f\otimes\chi)\times f}(\beta) + \frac{1}{2\pi i} \int\limits_{\Re(s)=-1} \tilde{h}(s)\, L_{(f\otimes\chi)\times f}(s+\beta)\, Y^s\, \frac{ds}{s}, \qquad (12.5.11)$$

where the integral on the right-hand side in (12.5.11) is equal to

$$\frac{\epsilon_\chi \left(\frac{q}{\pi}\right)^{\frac{n^2}{2}(1-2\beta)}}{2\pi i} \int\limits_{\Re(s)=1} \tilde{h}(-s) \left(\frac{\pi^{n^2} Y}{q^{n^2}}\right)^{-s} G_{p_0}(-s+\beta) L_{(\tilde{f}\otimes\bar{\chi})\times\tilde{f}}(1-\beta+s)\frac{ds}{s}$$

which is equal to

$$\epsilon_\chi \left(\frac{q}{\pi}\right)^{\frac{n^2}{2}(1-2\beta)} \sum_{m=1}^{\infty} \frac{\bar{b}(m)\bar{\chi}(m)}{m^{1-\beta}} h_2\left(mY\cdot\left(\frac{\pi}{q}\right)^{n^2}\right).$$

Combining (12.5.10), (12.5.11), and the above calculation, we obtain, for any $Y > 1$, the approximate functional equation

$$\begin{aligned}
L_{(f\otimes\chi)\times f}(\beta) &= \sum_{m=1}^{\infty} \frac{b(m)}{m^\beta}\chi(m)\cdot h_1\left(\frac{m}{Y}\right) \\
&\quad - \epsilon_\chi \left(\frac{q}{\pi}\right)^{\frac{n^2}{2}(1-2\beta)} \sum_{m=1}^{\infty} \frac{\bar{b}(m)}{m^{1-\beta}}\, \bar{\chi}(m)\cdot h_2\left(\frac{mY\pi^{n^2}}{q^{n^2}}\right).
\end{aligned}$$

This may be rewritten in the equivalent form

$$L_{(f\otimes\chi)\times f}(\beta) = \sum_{m=1}^{\infty} \frac{b(m)}{m^\beta}\chi(m)\cdot h_1\left(\frac{m}{Y}\right) - \frac{\overline{\tau(\chi)}^{n^2} q^{-n^2\beta}}{\pi^{\frac{n^2}{2}(1-\beta)}} \sum_{m=1}^{\infty} \frac{\bar{b}(m)}{m^{1-\beta}}\,\bar{\chi}(m)\cdot h_2\left(\frac{mY\pi^{n^2}}{q^{n^2}}\right). \tag{12.5.12}$$

The final step in the proof of Lemma 12.5.4 uses the approximate functional equation (12.5.12) to deduce (12.5.7). We shall restrict ourselves to prime q so that all non-trivial characters are automatically primitive. We require:

$$\sum_{\substack{\chi \pmod q\\ \chi\neq\chi_0,\ \chi(-1)=+1}} \chi(m) = \begin{cases} 0, & m\equiv 0 \pmod q\\ \frac{q-1}{2}-1, & m\equiv \pm 1 \pmod q\\ -1, & \text{otherwise.}\end{cases} \tag{12.5.13}$$

We then use (12.5.12) and (12.5.13) to estimate

$$\sum_{q\sim Q}\ \sum_{\substack{\chi \pmod q\\ \chi\neq\chi_0,\ \chi(-1)=+1}} L_{(f\otimes\chi)\times f}(\beta). \tag{12.5.14}$$

The contribution of the first sum on the right-hand side of (12.5.12) to the average (12.5.14) is

$$\sum_{q\sim Q}\ \sum_{\substack{\chi \pmod q\\ \chi\neq\chi_0,\ \chi(-1)=+1}}\ \sum_{m=1}^{\infty} \frac{b(m)}{m^\beta}\chi(m)\cdot h_1\left(\frac{m}{Y}\right) = \sum_{q\sim Q}\frac{q-1}{2}\sum_{m\equiv\pm 1 \pmod q} \frac{b(m)}{m^\beta}h_1\left(\frac{m}{Y}\right) - \sum_{q\sim Q}\sum_{(m,q)=1}\frac{b(m)}{m^\beta}h_1\left(\frac{m}{Y}\right). \tag{12.5.15}$$

We will show that the main contribution to (12.5.15) comes from the term $m=1$. All other terms will give a much smaller contribution. In fact, the contribution of the term $m=1$ is

$$\sum_{q\sim Q}\frac{q-1}{2}h_1\left(\frac{1}{Y}\right) = \sum_{q\sim Q}\frac{q-1}{2}.$$

The sum over $m\equiv 1 \pmod q$, $m\neq 1$ will contribute

$$\sum_{q\sim Q}\frac{q-1}{2}\sum_{d=1}^{\infty}\frac{b(1+dq)}{(1+dq)^\beta}h_1\left(\frac{1+dq}{Y}\right) \ll Q\sum_{m=1}^{\infty}\frac{b(m)m^\epsilon}{m^{\Re(\beta)}}\left|h_1\left(\frac{m}{Y}\right)\right|,$$

where we have used the fact that for $m \neq 1$, the number of different representations $m = 1 + dq = 1 + d'q'$ is $\mathcal{O}(m^{\epsilon})$. By the properties of the Rankin–Selberg convolution $L_{f\times f}(s)$, (see Section 12.1, and in particular, Remark 12.1.8) we see that $b(m) \geq 0$ of all $m = 1, 2, \ldots$, and

$$\sum_{m\leq x} b(m) \sim c_f x, \qquad (x \to \infty)$$

for some $c_f > 0$. It follows that $\sum\limits_{q\sim Q} \sum\limits_{m=1}^{\infty} \frac{b(m)m^{\epsilon}}{m^{\Re(\beta)}} |h_1(m/Y)| \ll QY^{1-\beta+\epsilon}$ and, therefore,

$$\sum_{q\sim Q} \frac{q-1}{2} \sum_{\substack{m\equiv\pm1 \pmod q\\ m\neq 1}} \frac{b(m)}{m^{\beta}} h_1\left(\frac{m}{Y}\right) \ll QY^{1-\beta+\epsilon}.$$

By the same type of computation, one also obtains a bound for the last sum on the right-hand side of (12.5.15):

$$\sum_{q\sim Q} \sum_{(m,q)=1} \frac{b(m)}{m^{\beta}} h_1\left(\frac{m}{Y}\right) \ll QY^{1-\beta+\epsilon}.$$

Finally, we consider the contribution of the second term on the right-hand side of (12.5.12). For this we make use of a deep bound of Deligne (1974) for hyper-Kloosterman sums. Define

$$K_n(r,q) := \sum_{x_1x_2\cdots x_n \equiv r \pmod q} e^{2\pi i\left(\frac{x_1+\cdots+x_n}{q}\right)},$$

to be the hyper-Kloosterman sum. Then Deligne proved that

$$K_n(r,q) \ll q^{(n-1)/2}. \qquad (12.5.16)$$

We shall use (12.5.16) to prove that for $m \in \mathbb{Z}$ and $q \nmid m$ that

$$\sum_{\substack{\chi \pmod q\\ \chi\neq\chi_0,\,\chi(-1)=+1}} \bar{\chi}(m)\tau(\chi)^{n^2} \ll q^{(n^2+1)/2}. \qquad (12.5.17)$$

Indeed, we have the identity

$$\sum_{\substack{\chi \pmod q\\ \chi\neq\chi_0,\,\chi(-1)=+1}} \bar{\chi}(m)\tau(\chi)^{n^2} = \frac{q-1}{2}\big(K_{n^2}(m,q) + K_{n^2}(-m,q)\big) - (-1)^{n^2},$$

from which (12.5.17) immediately follows.

We now return to the computation of the contribution of the second term on the right-hand side of (12.5.12). This contribution is

$$\sum_{q\sim Q}\ \sum_{\substack{\chi \pmod q\\ \chi\neq\chi_0,\chi(-1)=+1}} \frac{\overline{\tau(\chi)}^{n^2} q^{-n^2\beta}}{\pi^{(n^2/2)(1-\beta)}} \sum_{m=1}^{\infty} \frac{\bar{b}(m)}{m^{1-\beta}}\,\bar{\chi}(m)\cdot h_2\left(\frac{mY\pi^{n^2}}{q^{n^2}}\right)$$

$$\ll \sum_{q\sim Q} \frac{q^{\frac{n^2+1}{2}-n^2\beta}}{\pi^{\frac{n^2}{2}(1-\beta)}} \sum_{m=1}^{\infty} \frac{|b(m)|}{m^{1-\beta}}\cdot\left|h_2\left(\frac{mY\pi^{n^2}}{q^{n^2}}\right)\right|$$

$$\ll \sum_{q\sim Q} \frac{q^{\frac{n^2+1}{2}-n^2\beta}}{\pi^{\frac{n^2}{2}(1-\beta)}} \int_1^{\infty} \left|f_2\left(\frac{rY\pi^{n^2}}{q^{n^2}}\right)\right| r^{\beta}\,\frac{dr}{r}$$

$$\ll \sum_{q\sim Q} \frac{q^{\frac{n^2+1}{2}}}{\pi^{\frac{n^2}{2}(1+\beta)}\cdot Y^{\beta}}$$

$$\ll Q^{1+\frac{n^2+1}{2}}\cdot Y^{-\beta}.$$

Collecting together all the previous computations with the approximate functional equation (12.5.12) yields the asymptotic formula

$$\sum_{q\sim Q}\ \sum_{\substack{\chi \pmod q\\ \chi\neq\chi_0,\chi(-1)=+1}} L_{(f\otimes\chi)\times f}(\beta) = \sum_{q\sim Q}\frac{q-1}{2} + \mathcal{O}\left(QY^{1-\beta+\epsilon} + Q^{1+\frac{n^2+1}{2}}\cdot Y^{-\beta}\right).$$

Choosing $Y\sim Q^{(n^2+1)/2}$, we obtain Lemma 12.5.4 when $p_0=1$.

The case when p_0 is a prime is more difficult to deal with. In this situtation we replace (12.5.13) with

$$\sum_{\beta|q}\ \sideset{}{^*}\sum_{\substack{\chi \pmod \beta\\ \chi(p_0)=\chi(-1)=1}} \chi(m) = \begin{cases} = N_q & \text{if } m=1,\\ \geq 0 & \text{otherwise,}\end{cases} \tag{12.5.18}$$

where

$$N_q = \sum_{\beta|q}\ \sideset{}{^*}\sum_{\substack{\chi \pmod \beta\\ \chi(p_0)=\chi(-1)=1}} 1,$$

and Σ^* means that we sum over primitive characters only.

In this variation of the method, we do not assume that q is prime. If q_0 is the largest integer dividing q with the property that $(q_0, m)=1$, then

$$\sum_{\beta|q_0}\ \sideset{}{^*}\sum_{\substack{\chi \pmod \beta\\ \chi(p_0)=\chi(-1)=1}} \chi(m)\geq 0,$$

because the sum is going over all elements in a group.

The key point is that one needs to know that N_q is large for many values of q. More precisely, one requires that for every $\epsilon > 0$ and all Q sufficiently large

$$\sum_{Q<q<2Q} N_q \gg Q^{2-\epsilon}. \tag{12.5.19}$$

This may be proved using a construction of Rohrlich (1989) together with a variation of the Bombieri–Vinogradov theorem due to Murty and Murty (1987). With (12.5.18) and (12.5.19) in place, the proof of Lemma 12.5.4 proceeds as before. See (Luo, Rudnick and Sarnak, 1999) for precise details.

The case of a finite place and a number field is quite a bit more complicated. The reason is that the condition that χ be 1 at a finite place p, imposes a strong condition on the set of such χs. To get around this one resorts to the construction of special χs which have conductors which are highly divisible (Rohrlich, 1989), and in fact form a very sparse sequence. In this case in order to execute the averaging one needs to use the positivity of the coefficients of the Rankin–Selberg L-function $L(s, f \times \tilde{f})$. □

12.6 Strong multiplicity one theorem

As a final application of the Rankin–Selberg method, we give a proof of the strong multiplicity one theorem, originally due to Jacquet and Shalika (1981).

Theorem 12.6.1 (Strong multiplicty one) *Let f, g be two Maass forms for $SL(n, \mathbb{Z})$ as in (12.1.1) with Fourier coefficients*

$$A(m_1, \ldots, m_{n-1}), \qquad B(m_1, \ldots, m_{n-1}),$$

respectively, with $m_1 \geq 1, \quad \ldots \quad, m_{n-2} \geq 1, m_{n-1} \neq 0$. If

$$A(p, 1, \ldots, 1) = B(p, 1, \ldots, 1)$$

for all but finitely many primes p then $f = g$.

Proof If $f \neq g$, then the inner product $\langle f, g\rangle = 0$. By Theorem 12.1.4, the Rankin–Selberg L-function $L_{f\times g}(s)$ (see Definition 12.1.2) has a meromorphic continuation to all $s \in \mathbb{C}$ with at most a simple pole at $s = 1$ with residue proportional to $\langle f, g\rangle$. By our assumption that $f \neq g$, it follows that $L_{f\times g}(s)$ is entire and has no pole at $s = 1$. This is a contradiction because we are assuming that $A(p, 1, \ldots, 1) = B(p, 1, \ldots, 1)$ for all but finitely many primes p, which implies by Theorem 12.1.3 (up to a finite number of Euler factors) that $L_{f\times g}(s)$ is a Dirichlet series with positive coefficients, so it must have a pole. If it did not have a pole then by standard techniques in analytic number theory the sum $\sum_{p\le x} A(p,1,\ldots,1)\cdot\overline{B(p,1,\ldots,1)}$ would be small as $x\to\infty$. □

Theorem 12.6.1 can actually be proved in a much stronger form, i.e., it still holds provided $A(p,1,\ldots,1) = B(p,1,\ldots,1)$ for only finitely many primes p. By using the logarithmic derivative $L'_{f\times g}(s)/L_{f\times g}(s)$, Moreno (1985) obtained a general explicit bound on the number of primes required.

13

Langlands conjectures

About 25 years ago I was discussing analytic number theory with Jean-Pierre Serre. I distinctly recall how he went to the blackboard and wrote down the Euler product

$$\prod_p \prod_{i=1}^{n} (1 - \alpha_{p,i} p^{-s})^{-1}$$

corresponding to an automorphic form on $GL(n)$, and then pointed out that one of the most important problems in the theory of L-functions was to obtain the analytic properties of the higher kth symmetric power L-functions given by

$$\prod_p \prod_{1 \le i_1 \le i_2 \le \cdots \le i_k \le n} \left(1 - \left(\alpha_{p,i_1} \alpha_{p,i_2} \cdots \alpha_{p,i_k}\right) p^{-s}\right)^{-1}.$$

He then explained that if one knew that these L-functions (for $k = 1, 2, 3, \ldots$) were all holomorphic for $\Re(s) > 1$ then it would easily follow from Landau's lemma (a Dirichlet series converges absolutely up to its first pole, see (Iwaniec-Kowalski, 2004)) that

$$|\alpha_{p,i}| = 1$$

for all primes p and all $i = 1, 2, \ldots, n$. This, of course, is the famous generalized Ramanujan conjecture discussed in Section 12.4.

In January 1967, while at Princeton University, Langlands hand wrote a 17 page letter to André Weil. The letter outlines what are now commonly known as the "Langlands conjectures." Weil suggested that the letter be typed and it then circulated widely. It is now available at the Sunsite webpage:

http://www.sunsite.ubc.ca/DigitalMathArchive/Langlands/functoriality.html

The conjecture about the kth symmetric power L-functions shown to me by Serre is a special case of Langlands more general conjectures. One may also consider the kth exterior power L-function given by

$$\prod_p \prod_{i_1<i_2<\cdots<i_k} \left(1-\alpha_{p,i_1}\alpha_{p,i_2}\cdots\alpha_{p,i_k}p^{-s}\right)^{-1}.$$

Langlands predicts that the kth symmetric power L-function is associated to an automorphic form on $GL\left(\sum\limits_{1\le i_1\le i_2\le\cdots\le i_k\le n} 1\right)$ while the kth exterior power L-function is associated to an automorphic form on $GL\left(\sum\limits_{1<i_1<i_2<\cdots<i_k\le n} 1\right)$, i.e., it is on $GL(M)$ where M is simply given by the number of Euler factors in the Euler product. Although these conjectures are easy to state in terms of Euler products, it is very hard to get a grip on them from this point of view. The great insight of Langlands, in his letter to Weil, is to show that each L-function associated to an automorphic form on $GL(n)$, say, is also associated to a certain representation of an infinite dimensional Lie group, and by taking tensor powers or exterior powers of the representation one may validate the predictions.

Langlands came to his conjectures by carefully studying Eisenstein series and Artin L-functions. For example, we have shown in Theorem 10.8.6 that the L-functions associated to a minimal parabolic Eisenstein series for $SL(n,\mathbb{Z})$ are simply a product of shifted Riemann zeta functions. It is not hard to construct the higher symmetric and exterior powers which will again be products of other shifted Riemann zeta functions. One can then validate the Langlands conjectures by showing (with the method of templates, after Remark 10.8.7), that the higher symmetric and exterior products satisfy the expected functional equations.

Another compelling explanation for Langlands general conjectures, as explained by Langlands himself, is the striking analogy with Artin L-functions (Langlands, 1970). We will explore this analogy in the next two sections.

Langlands conjectures (see (Arthur, 2003), (Sarnak, to appear)) show that all automorphic forms should be encoded in the $GL(n)$ automorphic spectrum. It also follows from (Arthur, 1989, 2002) that the decomposition of a general group may be reduced to the study of $GL(n)$. So for the theory of L-functions, the group $GL(n)$ plays an especially important role.

The conjectures we have described up to now (symmetric and exterior powers of automorphic representations) are examples of the so-called "global Langlands functoriality conjectures over number fields."

Progress on the Langlands global functoriality conjectures over number fields has been slow, but with spectacular recent developments. Here is an account of the major developments up to the present.

- (Gelbart-Jacquet, 1978) *The symmetric square lift from $GL(2)$ to $GL(3)$ (see Section 7.3)* .
- (Ramakrishnan, 2002) *The tensor product lift: $GL(2) \times GL(2) \to GL(4)$.*
- (Kim and Shahidi, 2000, 2002) *Tensor product lift: $GL(2) \times GL(3) \to GL(6)$, the symmetric cube lift: $GL(2) \to GL(4)$.*
- (Kim, 2003), (Kim and Shahidi, 2000) *The exterior square lift: $GL(4) \to GL(6)$, symmetric fourth power lift: $GL(2) \to GL(5)$.*
- (Cogdell, Kim, Piatetski-Shapiro and Shahidi, 2001, 2004) *Lifting from split classical groups to $GL(n)$.*

The local Langlands conjectures (Carayol, 1992, 2000) have seen much greater advances. In (Laumon, Rapoport and Stuhler, 1993), the local Langlands conjecture was proved for local fields of prime characteristic. This was followed by Harris and Taylor (2001) who gave a proof for characteristic zero and then Henniart (2000) gave a simplified proof. In (Drinfeld, 1989) a proof of Langlands functoriality conjecture was obtained for $GL(2)$ over a function field. Finally Lafforgue (2002) established Langlands conjectures for $GL(n)$ (with $n \geq 2$) in the function field case.

Other references for Langlands conjectures include (Borel, 1979), (Bump, 1997), (Gelbart, 1984), (Arthur, 2003), (Bernstein and Gelbart, 2003), (Moreno, 2005).

13.1 Artin L-functions

Let K be an algebraic number field of finite degree (Galois extension) over another number field k with Galois group $G = \mathrm{Gal}(K/k)$. Let

$$\rho : G \to GL(V)$$

be a representation of G into a finite dimensional complex vector space V of dimension n. Then ρ is a homomorphism from G into the group $GL(V)$ of isomorphisms of V into itself. The group $GL(V)$ may be identified with $GL(n, \mathbb{C})$.

Example 13.1.1 (cubic field) Let $K = \mathbb{Q}(2^{\frac{1}{3}}, e^{2\pi i/3})$, and $k = \mathbb{Q}$. Then the Galois group $G = \mathrm{Gal}(K/\mathbb{Q})$ is S_3, the symmetric group of all permutations of three objects. A representation $\rho : G \to GL(3, \mathbb{R})$ can be explicitly given as follows. For $\alpha \in K$ and $g \in G$, we denote the action of g on α by α^g. Similarly, for $\alpha, \beta, \gamma \in K$ we define the binary operation $\circ$ by

$$\begin{pmatrix} \alpha \\ \beta \\ \gamma \end{pmatrix} \circ g := \begin{pmatrix} \alpha^g \\ \beta^g \\ \gamma^g \end{pmatrix}.$$

Now let

$$\alpha = 2^{\frac{1}{3}}, \qquad \beta = e^{2\pi i/3} \cdot 2^{\frac{1}{3}}, \qquad \gamma = e^{4\pi i/3} \cdot 2^{\frac{1}{3}}.$$

For each $g \in G$, we may define a matrix $\rho(g) \in GL(3, \mathbb{R})$ by the identity

$$\begin{pmatrix} \alpha \\ \beta \\ \gamma \end{pmatrix} \circ g = \rho(g) \cdot \begin{pmatrix} \alpha \\ \beta \\ \gamma \end{pmatrix}$$

where $\cdot$ denotes multiplication of a matrix by a vector. The matrices $\rho(g), \ g \in G$ are just the six permutation matrices

$$\begin{pmatrix} 1 & 0 & 0 \\ 0 & 1 & 0 \\ 0 & 0 & 1 \end{pmatrix}, \quad \begin{pmatrix} 0 & 1 & 0 \\ 0 & 0 & 1 \\ 1 & 0 & 0 \end{pmatrix}, \quad \begin{pmatrix} 0 & 0 & 1 \\ 1 & 0 & 0 \\ 0 & 1 & 0 \end{pmatrix}$$

$$\begin{pmatrix} 0 & 1 & 0 \\ 1 & 0 & 0 \\ 0 & 0 & 1 \end{pmatrix}, \quad \begin{pmatrix} 0 & 0 & 1 \\ 0 & 1 & 0 \\ 1 & 0 & 0 \end{pmatrix}, \quad \begin{pmatrix} 1 & 0 & 0 \\ 0 & 0 & 1 \\ 0 & 1 & 0 \end{pmatrix}.$$

Artin L-function – preliminary definition

Let K/k be a Galois extension with Galois group G. Let $\rho : G \to GL(V)$ be a representation of G as above. An Artin L-function is a meromorphic function of a complex variable s denoted $L(s, \rho, K/k)$ attached to this data. The precise definition gives a realization of $L(s, \rho, K/k)$ as an Euler product and requires the interplay between the representation ρ and prime numbers, which is determined by the Frobenius automorphism which we now discuss.

Definition 13.1.2 (Frobenius automorphism of a finite field extension) *Let $\mathbb{F}_q, \mathbb{F}_{q'}$ be finite fields of prime power orders q, q', respectively, where $q|q'$. The map $x \to x^q$ fixes $\mathbb{F}_q$ and permutes the elements of $\mathbb{F}_{q'}$ in such a way as to give an automorphism of $\mathbb{F}_{q'}$. This map is defined to be the Frobenius automorphism of $\mathbb{F}_{q'}/\mathbb{F}_q$. It is a particular element of* $\mathrm{Gal}(\mathbb{F}_{q'}/\mathbb{F}_q)$.

Let K/k be a Galois extension of number fields of degree n. Set $\mathcal{O}_K, \mathcal{O}_k$ to be the ring of integers of K, k, respectively. Let $\mathfrak{p}$ be a prime ideal of $\mathcal{O}_k$. Then the ideal $\mathfrak{p}\mathcal{O}_K$ factors into powers of prime ideals $\mathcal{P}_i$ $(i = 1, \ldots, r)$ of $\mathcal{O}_K$ as follows:

$$\mathfrak{p}\mathcal{O}_K = \mathcal{P}_1^{e_1} \cdots \mathcal{P}_r^{e_r}. \tag{13.1.3}$$

If we apply $g \in G$ to this, we get

$$\mathfrak{p}\mathcal{O}_K = \left(\mathcal{P}_1^g\right)^{e_1} \cdots \left(\mathcal{P}_r^g\right)^{e_r}. \tag{13.1.4}$$

By unique factorization, the two factorizations in (13.1.3), (13.1.4) must be the same. By varying $g \in G$ this implies that all the e_j must be equal to some integer $e \geq 1$. Therefore,

$$\mathfrak{p}\mathcal{O}_K = \mathcal{P}_1^e \cdots \mathcal{P}_r^e$$

and $n = efr$. We say $\mathfrak{p}$ is ramified if $e > 1$, otherwise, it is unramified.

Let $\mathcal{P}$ be one of the primes $\mathcal{P}_i$, $(i = 1, \ldots, r)$ which occur in the above factorization. We define the decomposition group $D_{\mathcal{P}}$, as the set

$$D_{\mathcal{P}} := \left\{ g \in G \;\middle|\; \mathcal{P}^g = \mathcal{P} \right\}.$$

The decomposition group tells us how $\mathfrak{p}\mathcal{O}_K$ splits. If we write G in terms of cosets of $D_{\mathcal{P}}$:

$$G = \bigcup_{i=1}^{\ell} D_{\mathcal{P}} \cdot g_i,$$

where $\ell = [G : D_{\mathcal{P}}]$ then the distinct conjugate divisors to $\mathcal{P}$ are just the divisors $\mathcal{P}^{g_i}$, $(1 \leq i \leq \ell)$. It follows that $\ell = r$ and the order of $D_{\mathcal{P}}$ is ef. For unramified primes this is just f.

Consider the residue fields $\mathbb{F}_{\mathcal{P}} := \mathcal{O}_K/\mathcal{P}$ and $\mathbb{F}_{\mathfrak{p}} := \mathcal{O}_k/\mathfrak{p}$, respectively. Then $\mathbb{F}_{\mathcal{P}}$, $\mathbb{F}_{\mathfrak{p}}$ are finite fields. It is not hard to see that the elements of the decomposition group $D_{\mathcal{P}}$ are automorphisms of $\mathbb{F}_{\mathcal{P}}/\mathbb{F}_{\mathfrak{p}}$. Indeed, if $g \in D_{\mathcal{P}}$ and

$$\alpha \equiv \beta \pmod{\mathcal{P}},$$

then $\alpha^g \equiv \beta^g \pmod{\mathcal{P}^g}$. But $\mathcal{P}^g = \mathcal{P}$ so that

$$\alpha^g \equiv \beta^g \pmod{\mathcal{P}}.$$

Consequently, elements of $D_{\mathcal{P}}$ take congruence classes (mod $\mathcal{P}$) to congruence classes (mod $\mathcal{P}$), which gives a homomorphism from $D_{\mathcal{P}}$ into $\mathrm{Gal}(\mathbb{F}_{\mathcal{P}}/\mathbb{F}_{\mathfrak{p}})$. Let

$$I_{\mathcal{P}} := \left\{ g \in G \;\middle|\; x^g \equiv x \pmod{\mathcal{P}},\ \forall x \in \mathcal{O}_K \right\},$$

denote the inertia group of $\mathcal{P}$. It turns out that the map

$$D_{\mathcal{P}}/I_{\mathcal{P}} \to \mathrm{Gal}(\mathbb{F}_{\mathcal{P}}/\mathbb{F}_{\mathfrak{p}})$$

is onto. It follows that there is an element of $D_{\mathcal{P}}$ which maps onto the Frobenius automorphism in $\mathrm{Gal}(\mathbb{F}_{\mathcal{P}}/\mathbb{F}_{\mathfrak{p}})$. Naturally, we shall call this element (denoted $\mathrm{Fr}_{\mathcal{P}}$) the Frobenius automorphism also. It is characterized by

$$x^{\mathrm{Fr}_{\mathcal{P}}} \equiv x^{N(\mathfrak{p})} \pmod{\mathcal{P}} \tag{13.1.5}$$

for all $x \in \mathcal{O}_K$. Here $N(\mathfrak{p})$ denotes the absolute norm of $\mathfrak{p}$ which is the cardinality of the finite field $\mathbb{F}_{\mathfrak{p}}$. Note that the Frobenius automorphism is well defined only modulo the inertia.

The prime $\mathfrak{p} \in \mathcal{O}_k$ is unramified in K/k if and only if $I_{\mathcal{P}}$ is trivial (contains only the identity element) for any prime $\mathcal{P}$ occurring in the factorization of $\mathfrak{p}\mathcal{O}_K$. Similarly, the prime $\mathfrak{p}\mathcal{O}_K$ factors into n prime ideals (splits completely) in K/k if and only if $D_{\mathcal{P}} = I_{\mathcal{P}} = \{1\}$.

Lemma 13.1.6 *Let K/k be a Galois extension of number fields with Galois group G. Let $g \in G$ and let $\mathcal{P}$ in $\mathcal{O}_K$ be a prime above the prime $\mathfrak{p}$ in $\mathcal{O}_k$. If $\mathrm{Fr}_{\mathcal{P}}$ is the Frobenius automorphism determined by (13.1.5) then*

$$\mathrm{Fr}_{\mathcal{P}^g} = g^{-1} \cdot \mathrm{Fr}_{\mathcal{P}} \cdot g.$$

If $D_{\mathcal{P}}$, $D_{\mathcal{P}^g}$ denote the decomposition groups of $\mathcal{P}$, $\mathcal{P}^g$, respectively, then

$$D_{\mathcal{P}^g} = g^{-1} \cdot D_{\mathcal{P}} \cdot g.$$

Proof If we apply $g \in G$ to (13.1.5), we otain

$$x^{\mathrm{Fr}_{\mathcal{P}} \cdot g} \equiv (x^g)^{N(\mathfrak{p})} \pmod{\mathcal{P}^g},$$

for all integers $x \in \mathcal{O}_K$. If we replace x by $x^{g^{-1}}$, it follows that

$$x^{g^{-1} \cdot \mathrm{Fr}_{\mathcal{P}} \cdot g} \equiv x^{N(\mathfrak{p})} \pmod{\mathcal{P}^g}.$$

□

Since the Galois group G is transitive on primes $\mathcal{P}$ lying over $\mathfrak{p}$, it follows that all the Frobenius elements $\mathrm{Fr}_{\mathcal{P}}$ are conjugate. Thus, attached to the prime $\mathfrak{p}$ of $\mathcal{O}_k$ is a conjugacy class of Frobenius elements in $\mathrm{Gal}(K/k)$. We are now ready to define the Artin L-function.

Definition 13.1.7 (Artin L-function) *Let K/k be a Galois extension of number fields with Galois group G. For a finite-dimensional complex vector space V, let $\rho : G \to GL(V)$ be a representation. Let $s \in \mathbb{C}$ with $\Re(s)$ sufficiently* *large. The Artin L-function (denoted $L(s, \rho, K/k)$) attached to this data is given by the Euler product*

$$\prod_{\mathfrak{p} \text{ unramified}} \mathrm{Det}\big(I - \rho\,(\mathrm{Fr}_{\mathcal{P}})\,N(\mathfrak{p})^{-s}\big)^{-1} \prod_{\mathfrak{p} \text{ ramified}} \mathrm{Det}\big(I - \rho\,(\mathrm{Fr}_{\mathcal{P}})\;\big|V^{I_{\mathcal{P}}}\,N(\mathfrak{p})^{-s}\big)^{-1},$$

where I is the identity matrix and $\mathcal{P}$ is any prime above $\mathfrak{p}$. Furthermore, for $\mathfrak{p}$ ramified, the quotient $D_{\mathcal{P}}/I_{\mathcal{P}}$ acts on the subspace $V^{I_{\mathcal{P}}}$ of V on which $I_{\mathcal{P}}$ acts trivially. The notation $\rho\,(\mathrm{Fr}_{\mathcal{P}})\;|V^{I_{\mathcal{P}}}$ means that the action of $\rho\,(\mathrm{Fr}_{\mathcal{P}})$ is restricted to $V^{I_{\mathcal{P}}}$.

Note The indicated determinant in Definition 13.1.7 is well defined (independent of the choice of $\mathcal{P}$ above $\mathfrak{p}$) by Lemma 13.1.6, since the determinant only depends on the conjugacy class of $\mathrm{Fr}_{\mathcal{P}}$. To see this note that for any $g \in G$, since the determinant is a multiplicative function and $\mathrm{Det}(g^{-1} \cdot g) = 1$,

$$\begin{aligned} \mathrm{Det}\big(I - \rho\,(\mathrm{Fr}_{\mathcal{P}})\,N(\mathfrak{p})^{-s}\big) &= \mathrm{Det}\big(g^{-1} \cdot (I - \rho(\mathrm{Fr}_{\mathcal{P}})N(\mathfrak{p})^{-s}) \cdot g\big) \\ &= \mathrm{Det}\big(I \;-\; (g^{-1} \cdot \rho(\mathrm{Fr}_{\mathcal{P}}) \cdot g)N(\mathfrak{p})^{-s}\big) \\ &= \mathrm{Det}\big(I - \rho(\mathrm{Fr}_{\mathcal{P}^g})N(\mathfrak{p})^{-s}\big). \end{aligned} \tag{13.1.8}$$

Here, the last equality follows from Lemma 13.1.6.

These L-functions were first introduced in (Artin, 1923, 1930) (see also (Roquette, 2000)). Artin made the following famous conjecture.

Conjecture 13.1.9 (Artin's conjecture) *Let K/k be a Galois extension of number fields with Galois group G. For a finite-dimensional complex vector space V, let $\rho : G \to GL(V)$ be a representation. If the representation ρ is irreducible and not the trivial representation, then $L(s, \rho, K/k)$ is an entire function of $s \in \mathbb{C}$.*

Artin himself proved Conjecture 13.1.9 when ρ is one-dimensional in (Artin, 1927). To quote from (Langlands, 1970):

> *Artin's method is to show that in spite of the differences in the definitions the function $L(s, \rho, K/F)$ attached to a one-dimensional ρ is equal to a Hecke L-function $L(s, \chi)$ where $\chi = \chi(\rho)$ is a character of $F^*\backslash I_F$. He employed all the available resources of class field theory, and went beyond them, for the equality of $L(s, \rho)$.*

It is not hard to show (see Heilbronn's article in (Cassels and Fröhlich, 1986)) that Artin's L-function $L(s, \rho, K/k)$ can be expressed as a product of rational powers of abelian L-functions of Hecke's type, where the abelian L-functions are associated to intermediate fields $k \subseteq \Omega \subseteq K$ with K/Ω abelian. A major advance on Artin's conjecture in the case when $\mathrm{Gal}(K/k)$ is not an abelian group was made by Brauer (1947) who proved that all irreducible representations ρ of the Galois group G can be expressed as $\mathbb{Z}$-linear combinations of induced representations of one-dimensional representations on subgroups of G. As a consequence, he showed that $L(s, \rho, K/k)$ can be expressed as a product of integer powers of abelian L-functions of Hecke's type. This proved that Artin's L-functions extended to meromorphic functions in the entire complex s plane and satisfied a functional equation. The problem was that Brauer could not exclude negative integral powers so that Artin's conjecture was still unproven.

Further advances on Artin's Conjecture 13.1.19 did not come until Langlands changed the entire landscape of research around this problem by making the striking conjecture that Artin's L-functions should be L-functions associated to Maass forms on $GL(n)$.

When $n = 2$ and the image of ρ in $PGL(2, \mathbb{C})$ is a solvable group, Artin's conjecture was solved in (Langlands, 1980). The ideas in this paper played a crucial role in Taylor and Wiles (1995) proof of Fermat's Last Theorem. In (Langlands, 1980) the conjecture is proved for tetrahedral and some octahedral representations and in (Tunnell, 1981) the results are extended to all octahedral representations. When $n = 2$ and the projective image is not solvable, the only possibility is that the projective image is isomorphic to the alternating group A_5. These representations are called icosahedral because A_5 is the symmetric group of the icosahedron. Joe Buhler's Harvard Ph.D. thesis (see (Buhler, 1978)) gave the first example where Artin's conjecture was proved for an icosahedral representation. The book (Frey, 1994) proves Artin's conjecture for seven icosahedral representations (none of which are twists of each other). In (Buzzard and Stein, 2002), the conjecture is proved for eight more examples. A further advance was made in (Buzzard, Dickinson, Shepherd-Barron and Taylor, 2001) who proved Artin's conjecture for an infinite class of icosahedral Galois representations which were disjoint from the previous examples.Very little is known for $n > 2$.

13.2 Langlands functoriality

The converse Theorem 3.15.3 of Hecke–Maass and the Gelbart–Jacquet symmetric square lift (Theorem 7.3.2) from $GL(2)$ to $GL(3)$ have been the principle motivation for writing this book. These results constitute one of the first important proofs of a special case of Langlands conjectures.

The Gelbart–Jacquet lift, or more accurately: *the Gelbart–Jacquet functorial image*, is an instance of Langlands functoriality. While it is beyond the scope of this book to give a definition of functoriality in the most general scenario, we shall attempt to motivate the definition and give a feeling for the important program that Langlands has created.

The key new idea introduced by Langlands in his 1967 letter to Weil, and also introduced independently in (Gelfand, Graev and Pyatetskii-Shapiro, 1990) is the notion of an **automorphic representation.** We have intensively studied Maass forms for $SL(n, \mathbb{Z})$. These are examples of automorphic forms. The leap to automorphic representation is a major advance in the subject with profound implications.It was first intensively researched, for the case of $GL(2)$, in (Jacquet and Langlands, 1970).

In the interests of notational simplicity and an attempt by the writer to explain in as simple a manner as possible the ideas behind the functoriality conjectures of Langlands, we shall restrict our discussion, for $n \geq 2$, to the group $SL(n, \mathbb{Z})$ which acts on $GL(n, \mathbb{R})$ by left matrix multiplication. An automorphic form is then a Maass form for $SL(n, \mathbb{Z})$ as in Definition 5.1.3, a Langlands Eisenstein series for $SL(n, \mathbb{Z})$ as in Chapter 10, or the residue of such an Eisenstein series. By Langlands spectral theorem (see (Langlands, 1966) and also Theorem 10.13.1 for the case of $GL(3)$) these automorphic forms generate the $\mathbb{C}$-vector space

$$\mathcal{V}_n := \mathcal{L}^2\big(SL(n, \mathbb{Z})\backslash GL(n, \mathbb{R})/O(n, \mathbb{R}) \cdot \mathbb{R}^\times\big).$$

We introduce the right regular representation which maps $g \in GL(n, \mathbb{R})$ to the endomorphism, $F \to \rho(g)F$ of $\mathcal{V}_n$, and is defined by

$$\rho : GL(n, \mathbb{R}) \to \mathrm{End}(\mathcal{V}_n),$$

where

$$(\rho(g)F)(z) := F(z \cdot g)$$

for all $F \in \mathcal{V}_n, \ z \in \mathfrak{h}^n = GL(n, \mathbb{R})/(O(n, \mathbb{R}) \cdot \mathbb{R}^\times)$, and all $g \in GL(n, \mathbb{R})$.

Remark This is a representation into the endomorphisms of an infinite dimensional vector space! The space of Maass forms (cusp forms) is invariant under this representation. It decomposes into an infinite direct sum of irreducible invariant subspaces. If π is the representation on one of these invariant subspaces then π is termed an automorphic cuspidal representation and corresponds to a Maass form. The L-function associated to π is then the L-function associated to the Maass form as in Definition 9.4.3, i.e., it is a Godement–Jacquet L-function.

At this point Langlands made a remarkable hypothesis which may be viewed as a naive form of his functoriality conjecture. He assumed:

- *that the properties of an automorphic representation mimic the properties of a Galois representation;*
- *that the properties of an L-function associated to an automorphic form (automorphic representation) mimic the properties of an Artin L-function.*

We shall illustrate these hypotheses with some simple examples. Let K/k be a Galois extension of number fields with Galois group G. For a finite-dimensional complex vector space V, let $\rho : G \to GL(V)$ be a representation, and consider the Artin L-function $L(s, \rho, K/k)$ given in Definition 13.1.7. If we let $g \in GL(n, \mathbb{R})$ and then use the first two identities in (13.1.8), it follows that, for $\mathfrak{p}$ unramified, we may diagonalize the matrix $\rho\,(\mathrm{Fr}_{\mathcal{P}})$, i.e.,

$$g^{-1} \cdot \rho\,(\mathrm{Fr}_{\mathcal{P}}) \cdot g = \begin{pmatrix} \lambda_{\mathfrak{p},1} & & \\ & \ddots & \\ & & \lambda_{\mathfrak{p},n} \end{pmatrix},$$

where $\lambda_{\mathfrak{p},1}, \lambda_{\mathfrak{p},2}, \ldots, \lambda_{\mathfrak{p},n}$ are the eigenvalues. A similar result holds when $\mathfrak{p}$ is ramified. The Euler product for the Artin L-function $L(s, \rho, K/k)$ then takes the form

$$L(s, \rho, K/k) = \prod_{\mathfrak{p}} \prod_{i=1}^{n} (1 - \lambda_{\mathfrak{p},i}\, N(\mathfrak{p})^{-s})^{-1}.$$

Now, it is possible to combine two Galois representations (of the type ρ above) and create a new Galois representation just as when we multiply two numbers to create a new number. In fact, there are many ways to do this. Such laws of composition have interesting images on the L-function side. One may think of the L-functions associated to Maass forms as the basic atoms which can be combined in various ways to form molecules, i.e., more complicated L-functions.

Let $\rho : G \to GL(V)$, $\rho' : G \to GL(V')$ be two Galois representations of the Galois group $G = \mathrm{Gal}(K/k)$ where V, V' are vector spaces (over $\mathbb{C}$) of dimensions n, n', respectively. Let

$$L(s, \rho, K/k) = \prod_{\mathfrak{p}} \prod_{i=1}^{n} (1 - \lambda_{\mathfrak{p},i}\, N(\mathfrak{p})^{-s})^{-1},$$

$$L(s, \rho', K/k) = \prod_{\mathfrak{p}} \prod_{j=1}^{n'} \left(1 - \lambda'_{\mathfrak{p},i}\, N(\mathfrak{p})^{-s}\right)^{-1},$$

be the corresponding Artin L-functions. An interesting way to combine ρ, ρ' is to form the tensor product $\rho \otimes \rho'$ which is a representation of G into $GL(V \otimes V')$. Every element of $V \otimes V'$ is a linear combination of terms of the form

$v \otimes v'$ with $v \in V,\ v' \in V'$. Then $\rho \otimes \rho'$ is defined by letting

$$(\rho \otimes \rho')(g)(v \otimes v') := \rho(g)(v) \otimes \rho(g)(v')$$

for all $g \in G, v \in V, v' \in V'$. Without loss of generality, we may assume that for some $g \in G$, the representations ρ, ρ' have been diagonalized so that $\rho(g), \rho'(g)$ correspond to diagonal matrices in $GL(n, \mathbb{C})$, $GL(n', \mathbb{C})$, respectively, where

$$\rho(g) = \begin{pmatrix} \lambda_1 & & \\ & \ddots & \\ & & \lambda_n \end{pmatrix}, \qquad \rho'(g) = \begin{pmatrix} \lambda'_1 & & \\ & \ddots & \\ & & \lambda'_{n'} \end{pmatrix}. \tag{13.2.1}$$

So, if $(e_1, \ldots, e_n)$ and $(e'_1, \ldots, e'_{n'})$ denote the standard bases for V, V', respectively, then it is clear that $\rho(g)$ maps $e_i \mapsto \lambda_i e_i$ while $\rho'(g)$ maps $e'_j \mapsto \lambda'_j e'_j$. It then follows that the tensor product of the two representations $(\rho \otimes \rho')(g)$ maps

$$e_i \otimes e'_j \mapsto \lambda_i \lambda'_j\, e_i \otimes e'_j \qquad (\text{for } 1 \le i \le n,\ 1 \le j \le n').$$

Consequently, we have shown that

$$L(s, \rho \otimes \rho', K/k) = \prod_{\mathfrak{p}} \prod_{i=1}^{n} \prod_{j=1}^{n'} (1 - \lambda_{\mathfrak{p},i} \lambda'_{\mathfrak{p},j} N(\mathfrak{p})^{-s})^{-1}.$$

But $L(s, \rho \otimes \rho', K/k)$ is another Artin L-function!It is also the Rankin–Selberg convolution as defined by the Euler product representation in Theorem 12.3.6. So if Langlands hypothesis (that automorphic representations mimic Galois representations) is correct it would have to follow that the Rankin–Selberg convolution of two automorphic representations is again automorphic. This is an important example of Langlands functoriality.

Another example comes from taking the symmetric product $V \vee V'$ of two vector spaces V, V' defined over $\mathbb{C}$. The vector space $V \vee V'$ then consists of all linear combinations of terms of the form $v \vee v'$ with $v \in V,\ v' \in V'$, and where $\vee$ satisfies the rules

$$v \vee v' = v' \vee v,$$
$$(a_1 v_1 + a_2 v_2) \vee v' = a_1 v_1 \vee v' + a_2 v_2 \vee v',$$

for all $v \in V, v' \in V'$, and $a_1, a_2 \in \mathbb{C}$. If $\rho : G \to GL(V),\ \rho' : G \to GL(V')$ are two Galois representations of $G = \mathrm{Gal}(K/k)$, then we may consider the symmetric product representation $\rho \vee \rho'$ which is a representation of G into $GL(V \vee V')$. Then $\rho \vee \rho'$ is defined by letting

$$(\rho \vee \rho')(g)(v \vee v') := \rho(g)(v) \vee \rho(g)(v')$$

for all $g \in G, v \in V, v' \in V'$. Let us consider the special case that $V = V'$ and $\rho = \rho'$ where V has the basis $e_1, \ldots, e_n$. We may view $V \vee V$ as a subspace of $V \otimes V$ with basis elements $e_i \otimes e_j$ where $1 \le i \le j \le n$. If we assume as before that $\rho(g), \rho'(g)$ correspond to diagonal matrices as in (13.2.1) then, in this case, we may consider the symmetric square representation $\rho \vee \rho$. It is easy to see that

$$L(s, \rho \vee \rho, K/k) = \prod_{\mathfrak{p}} \prod_{1 \le i \le j \le n} (1 - \lambda_{\mathfrak{p},i}\lambda_{\mathfrak{p},j} N(\mathfrak{p})^{-s})^{-1}.$$

Other examples of this type can be given by considering, for example, the exterior product of two vector spaces (see Section 5.6), and then forming the exterior product of two Galois representations or by taking higher symmetric or exterior powers.

Yet another type of interesting operation that can be done with representations is to consider induced representations. This corresponds to induction from a Galois subgroup (see (Bump, 1997). Langlands derived from this process his theory of base change (Langlands, 1980).

An even deeper theorem in Galois representations is Artin's reciprocity law (Artin, 1927) (see also the introductory article (Lenstra-Stevenhagen, 2000)) which generalized Gauss' law of quadratic reciprocity, and included all known reciprocity laws up to that time. Langlands formulated an even more general version of Artin reciprocity in the framework of automorphic representations.

Of course, Langlands did not stop here. This was the starting point. For a general connected reductive algebraic group G he introduced, in (Langlands, 1970), the dual group or what is now known as the L-group, ${}^LG$. He then formulated his now famous principle of functoriality which states that given any two connected reductive algebraic groups H, G and an L-homomorphism ${}^LH \to {}^LG$ then this should determine a transfer or lifting of automorphic representations of H to automorphic representations of G.

In fact, the modern theory of automorphic forms allows us to parameterize each automorphic representation of a reductive group such as H by a set of semisimple conjugacy classes $\{c_v\}$ in ${}^LH$, a complex Lie group, where v runs over almost all the places of the defining global field. The functoriality principle then roughly states that for each "L–homomorphism" $f\ {}^LH \to {}^LG$, the collection $\{f(c_v)\}$ defines, in fact, an automorphic representation of G, or more precisely a "packet" of automorphic representations of G.

List of symbols

$\langle\, , \,\rangle$

$\langle\, , \,\rangle_{\otimes^{\ell}(\mathbb{R}^n)}$

$\langle\, , \,\rangle_{\Lambda^{\ell}(\mathbb{R}^n)}$

$\lfloor\, , \,\rfloor$

$\otimes^{\ell}(\mathbb{R}^n)$

$\wedge$

$\vee$

$\otimes$

$A_{n_1,\ldots,n_r}$

$A(m_1, \ldots, m_{n-1})$

$\alpha_{p,i}$

α_p

B_n

$b_{i,j}$

$B(s_1, s_2)$

$C_G(\Gamma)$

C_+^{∞}

$\chi(n)$

D

$\mathcal{D}^n$

$\mathfrak{D}^n$

D_α

D_{ij}

$D_{\mathcal{P}}$

$d(n)$

$\mathrm{d}(z, z')$

d^*u

d^*z

Δ

Δ_1

Δ_2

$\Delta(z)$

Det

e_ψ

$E(z, s)$

$E^*(z, s)$

$E(z, s)$

$E(z, s, \chi)$

$E^*(z, s, \chi)$

$E_P(z, s)$

$E_P^*(z, s)$

$E_{P_{\min}}(z, s)$

$E_P(z, s, \phi)$

$\mathcal{E}_m(z, s)$

E_{ij}

ϵ_d

$\hat{f}$

$\tilde{f}$

$\mathbb{F}_q$

$F(z, h, q)$

$\mathrm{Fr}_{\mathcal{P}}$

$G_\nu(s)$

$\tilde{G}_\nu(s)$

$\gcd$

$\mathfrak{gl}(n, \mathbb{R})$

$GL(2, \mathbb{R})$

$GL(3, \mathbb{R})$

$GL(n, \mathbb{R})$

$GL(n, \mathbb{Z})$

$GL(V)$

G_w

$\mathcal{G}_w$

$\gamma_{\mathfrak{a}}$

Γ_∞

$\Gamma(N)$

$\Gamma_0(N)$

Γ_w

$h(d)$

$\mathfrak{h}$

$\mathfrak{h}^n$

$\heartsuit$

I_n

$I_{\mathcal{P}}$

I_r

I_s

$I_\nu(z)$

$I_\nu(z, P_{n_1,\ldots,n_r})$

K_s

$K_\nu(y)$

$k(u)$

$k(z)$

$K(z, w)$
$K(z, z')$
$L_{\mathfrak{b}}(s, \psi^n)$
$L_\chi(s)$
$L(s, \rho, K/k)$
$L_{E(*,w)}(s)$
$L_{E_v}(s)$
$L_f(s)$
$L_{\tilde{f}}(s)$
$L_{f\times g}(s)$
L_k
$\mathcal{L}_{k+1}$
$L(s, \vee^k)$
$\mathcal{L}^2(SL(2,\mathbb{Z})\backslash\mathfrak{h}^2)$
$\mathcal{L}^2_{\text{cusp}}(SL(2,\mathbb{Z})\backslash\mathfrak{h}^2)$
$\mathcal{L}^2_{\text{cont}}(SL(2,\mathbb{Z})\backslash\mathfrak{h}^2)$
$\mathcal{L}^2(SL(n,\mathbb{Z})\backslash\mathfrak{h}^n)$
λ_D
$\Lambda^n_b(s)$
$\Lambda_{E(*,w)}(s)$
$\Lambda_f(s)$
$\Lambda_{f\times g}(s)$
$\Lambda^\ell(\mathbb{R}^n)$
M
$M(n, \mathbb{R})$
$M_{n_1,\ldots,n_r}$
$M'_{n_1,\ldots,n_r}$
$M(g)$
$M_\lambda(g)$
$\mathfrak{m}_P$
n_i
N_i
$N_{n_1,\ldots,n_r}$
$O(2, \mathbb{R})$
$O(n, \mathbb{R})$
$\mathcal{O}_K,\ \mathcal{O}_k$
ω_i
$\Omega(P)$
$\Omega(P, P')$
$P_M(y, z')$
P_r
$\hat{P}_{n,r}$
$P_{n_1,\ldots,n_r}$
$\hat{\phi}_m(z)$
ψ_m
$\psi(u)$
$P(z, s, e_\psi)$
$\mathbb{Q}$
$\mathbb{Q}(\sqrt{-D})$
$\mathfrak{q}_w$
$\mathbb{R}$
$R_{r,\delta}$
$\mathcal{R}_{\Gamma,\Delta}$
$\rho(g)$
$\mathcal{S}$
$\mathcal{S}_n$
S_n
$S(n, c)$
$S_{k_1,k_2}(x_1, x_2, x_3)$
$S_k(x_1, \ldots, x_n)$
$S_w(\psi, \psi', c)$
$S_w(\psi_M, \psi_N, c)$
$SL(2, \mathbb{R})$
$SL(3, \mathbb{R})$
$SL(n, \mathbb{R})$
$SL(2, \mathbb{Z})$
$SL(3, \mathbb{Z})$
$SL(n, \mathbb{Z})$
$SO(n, \mathbb{R})$
$\mathfrak{S}_\lambda$
$\mathfrak{S}(T)$
σ_i
$\sigma_s(n)$
$\sigma_{\mathfrak{a}}$
$\Sigma_{a,b}$
$T(\mathbf{L})$
T_g
T_n
$\tau(\chi)$
$\theta(z)$
$\theta_\chi(z)$
U
$U(\mathbf{L})$
$U_n(\mathbb{R})$
$U_n(\mathbb{Z})$
U_w
$\tilde{U}_w$
$u(z, z')$
$\mu(n)$
$v_{j,k}$
V_n
$W(z)$
$W(z, \nu, \psi)$
$W_{\text{Jacquet}}(z, \nu, \psi)$
$W^*_{\text{Jacquet}}(z, \nu, \psi)$
w_0
w_n
W_n
$\mathbb{Z}$
Z_n
$Z(\psi, \psi', s)$
$\zeta(s)$

Appendix The **GL(n)pack** Manual

Kevin A. Broughan

A.1 Introduction

This appendix is the manual for a set of functions written to assist the reader to understand and apply the theorems on $GL(n, \mathbf{R})$ set out in the main part of the book. The software for the package is provided over the world wide web at

http://www.math.waikato.ac.nz/~kab

and is in the form of a standard Mathematica add-on package. To use the functions in the package you will need to have a version of Mathematica at level 4.0 or higher.

A.1.1 Installation

First connect to the website given in the paragraph above and click on the link for GL(n)pack listed under "Research" to get to the GL(n)pack home page. Instructions on downloading the files for the package will be given on the home page. If you have an earlier version of GL(n)pack first delete the file **gln.m**, the documentation **gln.pdf** and the validation program **glnval.nb**. The name of the file containing the package is **gln.m**. To install, if you have access to the file system for programs on your computer, place a copy of the file in the standard repository for Mathematica packages – this directory is called "Applications" on some systems. You can then type ≪**gln.m** and then press the Shift and Enter keys to load the package. You may need administrator or super-user status to complete this installation. Alternatively place the package file **gln.m** anywhere in your own file system where it is safe and accessible.

Instructions for Windows systems The package can be loaded by typing

```
SetDirectory["c:\your\directory\path"];
  <<gln.m;ResetDirectory[]<Shift/Enter>
```

or

```
Get["gln.m",Path->{"c:\your\directory\path"}]
  <Shift/Enter>
```

where the path-name in quotes should be replaced by the actual path-name of directories and subdirectories which specify where the package has been placed on a given computer.

Instructions for Unix/Linux systems These are the same as for Windows, but the path-name syntax style should be like **/usr/home/your/subdirectory**.

Instructions for Macintosh systems These are the same as for Windows, but the path-name syntax style should be like **HD:Users:ham:Documents:**.

All systems The package should load printing a message. The functions of GL(n)pack are then available to any Mathematica notebook you subsequently open.

A.1.2 About this manual

This appendix contains a list of all of the functions available in the package GL(n)pack followed by a manual entry for each function in alphabetical order. Many functions contain the transcript of an example and reference to the part of the text to which the function relates, as well as lists of related functions. Each function has both a Mathematica style long name and a 3 letter/digit abbreviated name. Either can be used, but the error messages and usage information are all in terms of the long names. To obtain information about bug fixes and updates to GL(n)pack consult the website for the package: given in A.I. above.

A.1.3 Assistance for users new to computers or Mathematica

On the GL(n)pack website (see above) there are links giving tutorial and other information for those people who want access to the package but are new or relatively new to computers. The manual entries assume familiarity with Mathematica, so some may require extra help. Alternatively sit down with someone familiar with Mathematica to see it at work.

There are many issues to do with computer algebra and mathematical software that will arise in any serious evaluation or use of Mathematica and GL(n)pack. A comment on one aspect: GL(n)pack function arguments are first evaluated and then checked for correct data type. If the user calls a function with an incorrect number of arguments or an argument of incorrect type, rather than issue a warning and proceeding to compute (default Mathematica style), GL(n)pack prints an error message, aborts the evaluation and returns the user to the top-level, no matter how deeply nested the function which makes the erroneous call happens to be placed. This is a tool for assisting users to debug programs which include calls to this package.

A.1.4 Mathematica functions

Each GL(n)pack function is a standard Mathematica function and so will work harmoniously with built-in Mathematica functions and user functions. Useful standard functions include those for defining functions (Module and Block for example) list and matrix manipulation and operations (“.” represents matrix multiplication), special functions (such as the BesselK), plotting functions and the linear algebra add-on package. Note however that a formatted matrix, returned by MatrixForm, is not recognized by Mathematica as a matrix. A matrix in Mathematica is just a list of lists of equal length.

A.1.5 The data type *CRE* (Canonical Rational Expression)

Many GL(n)pack functions take symbolic arguments which are either explicit integers or real or complex numbers (exact or floating point) or mathematical expressions which could evaluate to numbers. These expressions are expected to be in the class sometimes called “Canonical Rational Expressions” or *CRE*s. This class of expression is defined as follows: members are rational functions with numerical coefficients and with symbolic variables, any number of which may be replaced by function calls, or functions which are not evaluated (“noun forms”) or functions of any finite number of arguments each of which can be, recursively, a *CRE*. Some package functions will accept lists of *CRE*s or matrices with *CRE* elements. This should cover most user needs, but notice it excludes simple types like matrices with elements which are matrices. (The single exception is the GL(n)pack function MakeBlockMatrix, which takes as argument a matrix with matrix elements.) If a user is unsure regarding the data type of a mathematical expression, the GL(n)pack function **CreQ** can be used. See the manual entry.

A.1.6 The algorithms in this package

The reader who uses this package may notice that many functions appear to run almost instantaneously, for example MakeBlockMatrix or HeckeMultiplicativeSplit. Others however take considerable time to complete, minutes or hours rather than seconds. This is often because the underlying algorithm employed is exponential, or in at least one instance, more than exponential. Improvements in this completion time are of course possible: The GL(n)pack code is interpreted, so there may be speed-ups attainable using the Mathematica function Compile, even though it has a restricted domain of application. Existing algorithms could be replaced by faster algorithms. The existing algorithms could be re-implemented in a compile-load-and-go language such as C++ or Fortran, or an interactive language allowing for compilation such as Common Lisp. This latter would be the best choice, because execution speed for compiled code is quite comparable to that of the two former choices, but its range of data types is vast, certainly sufficient for all of the package. Functions which will slow significantly as the dimension increases include GetCasimirOperator, ApplyCasimirOperator, KloostermanSum, MPSymmetricPowerLFun, and SpecialWeylGroup.

A.1.7 Acknowledgements

Kevin Broughan had assistance from Columbia University and the University of Waikato and a number of very helpful individuals while writing the package. These included Ross Barnett, Mike Eastwood, Sol Friedberg, David Jabon (for the Smith Normal Form code), Jeff Mozzochi and Eric Stade, in addition to Dorian Goldfeld, who's detailed text and explicit approach made it possible. Their contribution is gratefully acknowledged.

A.2 Functions for GL(n)pack

ApplyCasimirOperator[m,expr,iwa] (aco): The operator acts on a CRE.
BruhatCVector[a] (bcv): The minors $(c_1, \ldots, c_{n-1})$.
BruhatForm[a] (bru): The four Bruhat factors of a symbolic matrix.
BlockMatrix[a,rows,cols] (blm): Extract a general sub-block of a matrix.
CartanForm[a] (car): The two Cartan factors of a numeric matrix.
ConstantMatrix[c,m,n] (com): Construct a constant matrix of given size.
CreQ[e] (crq): Check a Canonical Rational Expression.
DiagonalToMatrix[d] (d2m): Convert a list to a diagonal matrix.
EisensteinFourierCoefficient[z,s,n] (efc): The $GL(2)$ Fourier series nth term.
EisensteinSeriesTerm[z,s,ab] (est): The nth term of the series for $GL(2)$.
ElementaryMatrix[n,i,j,c] (elm): Construct a specified elementary matrix.
FunctionalEquation[vs, i] (feq): Generate the affine parameter maps.

GetCasimirOperator[m,n,"x","y","f"] (gco): The Casimir operators.
GetMatrixElement[a, i, j] (gme): Return a specified element.
GlnVersion[] (glv): Print the date of the current version.
HeckeCoefficientSum[m, ms, "x"](hcs): The right-hand side of the sum.
HeckeEigenvalues[m,n,"a"] (hev): The values of (λ_m) for $GL(n)$.
HeckeMultiplicativeSplit[m](hms): Prepare a Hecke Fourier coefficient.

HeckeOperator[n, z,"f"] (hop): of nth order for forms on $\mathfrak{h}^n$.
HeckePowerSum[e, es,"B"](hps): Exponents for the Hecke sum at any prime.
HermiteFormLower[a] (hfl): The lower left Hermite form.
HermiteFormUpper[a] (hfu): The upper left Hermite form.
IFun[ν,z] (ifn): The power function $I_\nu(z)$.
InsertMatrixElement[e,i,j,a] (ime): Insert an expression in a given matrix.
IwasawaForm([a] (iwf): The product of the Iwasawa factors of a matrix.
IwasawaXMatrix[w] (ixm): Get the x matrix from the Iwasawa form.
IwasawaXVariables[w] (ixv): Get the x variables from the Iwasawa form.
IwasawaYMatrix[z] (iym): Get the y matrix from the Iwasawa form.
IwasawaYVariables[z] (iyv): Get the y variables from the Iwasawa form.
IwasawaQ[z] (iwq): Test to see if a matrix is in Iwasawa form.

KloostermanBruhatCell[a,x,c,w,y] (kbc): Solve $a = x.c.w.y$ for x and y.
KloostermanCompatibility[t1,t2,c,w] (klc): Relations for a valid sum.
KloostermanSum[t1,t2,c,w] (kls): Compute an explicit Kloosterman sum.
LanglandsForm[p,d] (llf): The three matrices of the decomposition.
LanglandsIFun(g,d,s) [lif]: Summand for the Eisenstein series.
LeadingMatrixBlock[a,i,j] (lmb): Extract a leading sub-block of a matrix.
LongElement[n] (lel): Construct the matrix called the long element.
LowerTriangleToMatrix[l] (ltm): Construct a lower triangular matrix.

MakeBlockMatrix[mlist] (mbm): Construct a matrix from submatrices.
MakeMatrix["x",m,n] (mkm): Make a matrix with indexed elements.
MakeXMatrix[n,"x"] (mxm): Construct a symbolic unimodular matrix.
MakeXVariables[n,"x"] (mxv): Construct a list of Iwasawa x variables.
MakeYMatrix[n,"y"] (mym): Construct a symbolic diagonal matrix.
MakeYVariables[n,"y"] (myv): Construct Iwasawa y variables.
MakeZMatrix[n,"x","y"] (mzm): Construct a symbolic Iwasawa z matrix.
MakeZVariables[n,"x","y"] (mzv): A list of the x and y variables.
MatrixColumn[m,j] (mcl): Extract a column of a given matrix.
MatrixDiagonal[a] (mdl): Extract the diagonal of a matrix.
MatrixJoinHorizontal[a,b] (mjh): Join two matrices horizontally.
MatrixJoinVertical[a,b] (mjv): Join two matrices vertically.
MatrixLowerTriangle[a] (mlt): Extract the lower triangular elements.
MatrixRow[m,i] (mro): Extract a row of a matrix.
MatrixUpperTriangle[a] (mut): Extract the upper triangular elements.
ModularGenerators[n] (mog): Construct the generators for $SL(n, \mathbb{Z})$.
MPEisensteinGamma[s,v] (eig): Gamma factors for a parabolic series.
MPEisensteinLambdas[v] (eil): The $\lambda_i(v)$ shifts.
MPEisensteinSeries[s,v] (eis): Minimal parabolic Eisenstein series.
MPExteriorPowerGamma[s,v,k] (epg): Exterior power gamma factors.
MPExteriorPowerLFun[s,v,k] (epl): Minimal parabolic exterior power.
MPSymmetricPowerGamma[s,v,k] (spg): Symmetric power gamma.
MPSymmetricPowerLFun[s,v,k] (spf): Minimal parabolic symmetric power.
NColumns[a] (ncl): The column dimension of a matrix.
NRows[a] (nro): The row dimension of a matrix.

ParabolicQ[p,d] (paq): Test a matrix for membership in a given subgroup.
PluckerCoordinates[a] (plc): Compute the bottom row-based minors.
PluckerInverse[Ms] (pli): Compute a matrix with given minors.
PluckerRelations[n,v] (plr): Compute quadratic relations between minors.
RamanujanSum[n,c] (rsm): Evaluate the Ramanujan sum $s(n, c)$.
RemoveMatrixColumn[a,j] (rmc): Remove a matrix column.
RemoveMatrixRow[a,i] (rmr): Remove a row of a matrix non-destructively.
SchurPolynomial[k,x] (spl): The Schur multinomial $S_k(x_1, \ldots, x_n)$.
SmithElementaryDivisors[a] (sed): Smith form elementary divisors.
SmithForm[a] (smf): Compute the Smith form of an integer matrix.
SmithInvariantFactors[a] (sif): Smith form invariant factors.
SpecialWeylGroup[n] (swg): Weyl integer rotation group with det 1 elements.
SubscriptedForm[e] (suf): Print arrays with integer arguments as subscripts.
SwapMatrixColumns[a,i,j] (smc): Return a new matrix.

SwapMatrixRows[a,i,j] (smr): Return a new matrix with rows swapped.
TailingMatrixBlock[a,i,j] (tmb): Extract a tailing matrix block.
UpperTriangleToMatrix[u] (utm): Form an upper triangular matrix.

VolumeBall[r,n] (vbl): The volume of a ball in n-dimensions.
VolumeFormDiagonal["a",n] (vfd): The volume form for diagonal matrices.
VolumeFormGln["g",n] (vfg): The volume form for $GL(n)$.
VolumeFormHn["x","y",n] (vfh): Volume form for the upper half-plane.
VolumeFormUnimodular["x",n) (vfu): The form for the unimodular group.
VolumeHn[n] (vhn): The volume of the generalized upper half-plane.
VolumeSphere[r,n] (vsp): The volume of a sphere in n-dimensions.
Wedge[$\mathbf{f}_1, \cdots, \mathbf{f}_n$] (weg): The wedge product and the d operator.
WeylGenerator[n,i,j] (wge): Each matrix generator for the Weyl group.
WeylGroup[n] (wgr): Compute the Weyl group of permutation matrices.
Whittaker[z,v,psi] (wit): Compute the function W_{Jacquet} symbolically.
WhittakerGamma[v] (wig): Gamma factors for the Whittaker function.
WMatrix[n] (wmx): The long element matrix with determinant 1.
ZeroMatrix[m,n] (zmx): The zero matrix of given dimensions.

A.3 Function descriptions and examples

■ ApplyCasimirOperator (aco)

This function computes the Casimir operator acting on an arbitrary expression, or undefined function, of a matrix argument in Iwasawa form, and optionally other parameters, with respect to the Iwasawa variables, which must be specified by giving the matrix in Iwasawa form.

To simply compute the operator it is easier to use the function **GetCasimir-Operator**. Because this function uses symbolic differentiation, only functions or expressions which can be differentiated correctly and without error by Mathematica may be used as valid arguments. Note also that no check is made that the user has entered valid Mathematica variables. Since all arguments are evaluated it is good practice to use the function **Clear** to ensure arguments which should evaluate to themselves do so.

See Proposition 2.3.3, Example 2.3.4 and Proposition 2.3.5.

ApplyCasimirOperator[m, expr, iwa] ⟶ value

m is a positive integer with value **2** or more being the order of the operator,
expr is a Mathematica expression, normally in the Iwasawa matrix or variables and other parameters, which can be symbolically differentiated,
iwa is a numeric or symbolic matrix in Iwasawa form,
value is an expression or number being the result of applying the Casimir operator in the Iwasawa variables to the expression.

Example

```
In[1]:= z = MakeZMatrix[3, "x", "y"]

Out[1]= {{y[1] y[2], x[1, 2] y[1], x[1, 3]}, {0, y[1], x[2, 3]}, {0, 0, 1}}

In[2]:= ApplyCasimirOperator[2, Det[z.z], z]

Out[2]= 12 y[1]^4 y[2]^2

In[3]:= L = LongElement[3]

Out[3]= {{0, 0, 1}, {0, 1, 0}, {1, 0, 0}}

In[4]:= Simplify[ApplyCasimirOperator[3, a = IFun[{v1, v2}, L.z], z] / a]

Out[4]= 3 (2 v1^3 + 3 v1^2 v2 + v1 (-2 + 3 v2 - 3 v2^2) - 2 v2 (2 - 3 v2 + v2^2))
```

See also: GetCasimirOperator, MakeZMatrix, IwasawaForm, IwasawaQ.

■ BlockMatrix (blm)

This function returns a specified sub-block of a matrix. The entries of the sub-block must be contiguous.

Block matrices are used in a number of places, but most especially in Chapter 10 on Langlands Eisenstein series.

BlockMatrix[a, rows, columns] ⟶ b

a is a matrix of CREs,
rows is a list of two valid row indices for **a**, being the first and last sub-block rows,
columns is a list of two valid column indices for **a**, being the first and last sub-block columns,
b is the sub-block of **a** with the specified first and last row and column sub-block indices.

Example

```
In[168]:= A = {{4, 8, u, 1, 2}, {8, 7, 2, 0, 1}, {4, 5, 1, 2, 0}};
          B = {{1, 1, 1}, {1, 4, v}, {2, 2, 2}};

In[170]:= M = Transpose[A].B; MatrixForm[M]

Out[170]//MatrixForm=
```

$$\begin{pmatrix} 20 & 44 & 12+8\,v \\ 25 & 46 & 18+7\,v \\ 4+u & 10+u & 2+u+2\,v \\ 5 & 5 & 5 \\ 3 & 6 & 2+v \end{pmatrix}$$

```
In[171]:= MatrixForm[BlockMatrix[M, {1, 2}, {1, 3}]]

Out[171]//MatrixForm=
```

$$\begin{pmatrix} 20 & 44 & 12+8\,v \\ 25 & 46 & 18+7\,v \end{pmatrix}$$

See also: LeadingMatrixBlock, TailingMatrixBlock, MakeBlockMatrix, MakeMatrix.

■ BruhatCVector (bcv)

In the explicit Bruhat decomposition of a non-singular matrix **a**, the diagonal matrix c has a special form, each element being the ratio of absolute values of minor determinants (c_i) of the original matrix a with the element in the (i, i)th position being c_{n-i+1}/c_{n-i} for $2 \le i \le n-1$ with the (n, n)th element being $\mathbf{c_1}$ and the 1, 1th, $det(w)det(a)/c_{n-1}$. This function returns those c_i.

See Section 10.3 and Proposition 10.3.6.

BruhatCVector[a] ⟶ c

a is a non-singular $\mathbf{n \times n}$ square matrix of CREs,
c is a list of $\mathbf{n-1}$ CREs $\{\mathbf{c_1}, \ldots, \mathbf{c_{n-1}}\}$.

Example

```
In[17]:= LE3 = LongElement[3];
         BruhatCVector[{{1, 2, 3}, {4, 5, 7}, {7, 8, 9}}, LE3]

Out[18]= {7, 3}

In[19]:= LE4 = LongElement[4]; MatrixForm[LE4]

Out[19]//MatrixForm=
         ( 0 0 0 1 )
         ( 0 0 1 0 )
         ( 0 1 0 0 )
         ( 1 0 0 0 )

In[20]:= BruhatCVector[
          {{1, a, b, 3}, {0, 1, b, 1}, {a, 2, 0, b}, {1, 2, 3, 4}}, LE4]

Out[20]= {1, Abs[2 - 2 a], Abs[3 a + 2 b - 2 a b]}
```

See also: BruhatForm, LanglandsForm.

■ BruhatForm (bru)

This function finds the factors of a non-singular matrix, which may have entries which are polynomial, rational or algebraic expressions, so that the matrix can be expressed as the product of an upper triangular matrix with 1s on the leading diagonal (unipotent), a diagonal matrix, a permutation matrix (with a single 1 in each row and column), and a second unipotent matrix. When an additional constraint, namely, that the transpose of the permutation matrix times the second upper triangular matrix is lower triangular, then the factors are unique. This is the so-called Bruhat decomposition.

See Section 10.3 Proposition 10.3.6. The decomposition is used in Section 10.6 to derive the Fourier expansion of a minimal parabolic Eisenstein series.

BruhatForm[a] $\longrightarrow$ **{u₁,c,w,u₂}**

a is a non-singular square CRE matrix,
$\mathbf{u_1}$ is an upper triangular unipotent matrix,
c is a diagonal matrix,
w is a permutation matrix,
$\mathbf{u_2}$ is an upper triangular unipotent matrix.

Example

```
In[1]:=  B = {{x, y, z}, {u, v, 0}, {0, 2, 1}}

Out[1]=  {{x, y, z}, {u, v, 0}, {0, 2, 1}}

In[2]:=  b = BruhatForm[B];

In[3]:=  Map[MatrixForm, b]
```

Out[3]= $\left\{\begin{pmatrix} 1 & \frac{x}{u} & \frac{1}{2}\left(-\frac{vx}{u}+y\right) \\ 0 & 1 & 0 \\ 0 & 0 & 1 \end{pmatrix}, \begin{pmatrix} \frac{1}{2}\left(\frac{vx}{u}-y\right)+z & 0 & 0 \\ 0 & u & 0 \\ 0 & 0 & 2 \end{pmatrix}, \begin{pmatrix} 0 & 0 & 1 \\ 1 & 0 & 0 \\ 0 & 1 & 0 \end{pmatrix}, \begin{pmatrix} 1 & \frac{v}{u} & 0 \\ 0 & 1 & \frac{1}{2} \\ 0 & 0 & 1 \end{pmatrix}\right\}$

See also: CartanForm, HermiteForm, IwasawaForm, SmithForm, Langlands-Form.

■ CartanForm (car)

This function gives a form of the Cartan decomposition of a numeric real non-singular square matrix, namely the factorization $a = k \cdot exp(x)$ where k is orthogonal and x symmetric. It follows from this that the transpose of an invertible matrix satisfies an equation

$$^{t}a = k.a.k$$

for some orthogonal matrix k, where ^{t}a is the transpose of a. The function is restricted to numeric matrices because eigenvectors and eigenvalues are used.

CartanForm[a] $\longrightarrow$ **{k, s}**

a is a non-singular real numeric square matrix,
k is an orthogonal matrix,
s is the matrix exponential of a symmetric matrix.

Example

```
In[84]:= PrintCartan[g_] := Module[{ans, k, S},
           ans = N[Cartan[g]];
           Print[MatrixForm[First[ans]]];
           Print[MatrixForm[Part[ans, 2]]];
           Return[True]]

In[88]:= g = {{1, 2, 0}, {0, 2, 3}, {0, 0, 1}}

Out[88]= {{1, 2, 0}, {0, 2, 3}, {0, 0, 1}}

In[89]:= PrintCartan[g]

         ( -0.494923   -0.00367894  -0.868929  )
         (  0.112604    0.991288    -0.0683336 )
         (  0.86161    -0.131665    -0.490196  )

         ( -0.494923   -0.764638     1.19942   )
         ( -0.00367894  1.97522      2.8422    )
         ( -0.868929   -1.87453     -0.695197  )

Out[89]= True
```

See also: BruhatForm, IwasawaForm, LanglandsForm, HermiteFormLower, HermiteFormUpper, SmithForm.

■ ConstantMatrix (com)

This function constructs a constant matrix with specified element value.

This function can be used together with other functions to construct matrices.

ConstantMatrix[c, m, n] ⟶ a

c is a CRE, **m** is an integer with $\mathbf{m} \geq \mathbf{1}$,
n is an integer with $\mathbf{n} \geq \mathbf{1}$,
a is an **m** by **n** matrix having each element equal to **c**.

See also: ZeroMatrix, ElementaryMatrix, MatrixJoinHorizontal, MatrixJoinVertical.

■ CreQ (crq)

This function checks to see if its argument evaluates to a so-called Canonical Rational Expression (CRE), i.e. a number (real or complex, exact or floating point) or rational function in one or many variables with numerical coefficients, where any number of the variables can be replaced by function calls, including calls to undefined (so-called noun) functions of one or many arguments with arguments being canonical rational expressions. This is the data type expected by GL(n)pack functions.

See the introduction to the appendix.

CreQ[e] ⟶ P

e is a Mathematica expression,
P is **True** if **e** is a CRE and **False** otherwise.

Example

```
In[93]:= CreQ[{x, y}]

Out[93]= False

In[94]:= CreQ["x"]

Out[94]= False

In[95]:= CreQ[2 x + y / (x + 1 + Sin[x + y])]

Out[95]= True
```

See also: ParabolicQ, KloostermanSumQ.

■ DiagonalToMatrix (d2m)

This function takes a list and constructs a matrix with the list elements as the diagonal entries.

Diagonal matrices appear in many places, including in the Iwasawa and Bruhat decompositions.

DiagonalToMatrix[di] ⟶ a

di is a non-empty list of CREs,
a is a square matrix of size the length of **di**, with zeros in off-diagonal positions, and with the diagonal entries being the elements of **di** and in the same order.

See also: MatrixDiagonal.

■ EisensteinFourierCoefficient (efc)

This function returns the nth term of the Fourier expansion of an Eisenstein series for $GL(2)$, with an explicit integer specified for n.

See Section 3.1, especially Theorem 3.1.8.

EisensteinFourierCoefficient[z, s, n] ⟶ v

z is a CRE, **s** is a CRE,
n is an integer being the index of the **n**th coefficient,
v is a complex number or symbolic expression representing the **n**th Fourier term of the Eisenstein Fourier expansion for **GL(2)** with parameters **z** and **s**.

Example

```
In[1]:= EisensteinFourierCoefficient[z, s, 4]
```

Out[1]= $$\frac{2^{2s}\,(-1+2^{3-6s})\,e^{8 i \pi \mathrm{Re}[z]}\,\pi^{s}\,\mathrm{BesselK}[-\frac{1}{2}+s,\ 8\pi\,\mathrm{Im}[z]]\,\sqrt{\mathrm{Im}[z]}}{(-1+2^{1-2s})\,\mathrm{Gamma}[s]\,\mathrm{Zeta}[2\,s]}$$

```
In[2]:= EisensteinFourierCoefficient[z, s, 0]
```

Out[2]= $$\mathrm{Im}[z]^{s}+\frac{\sqrt{\pi}\,\mathrm{Gamma}[-\frac{1}{2}+s]\,\mathrm{Im}[z]^{1-s}\,\mathrm{Zeta}[-1+2\,s]}{\mathrm{Gamma}[s]\,\mathrm{Zeta}[2\,s]}$$

See also: EisensteinSeriesTerm, IFun, LanglandsIFun.

■ EisensteinSeriesTerm (est)

This function returns the term of the Eisenstein series $E(z, s)$ for $GL(2)$, namely the summand of:

$$E(z,s)=\frac{1}{2}\sum_{a,b\in\mathbb{Z},(a,b)=1}\frac{y^{s}}{|az+b|^{2s}}$$

with explicit values for the integers a, b.

See Definition 3.1.2.

EisensteinSeriesTerm[z, s, ab] ⟶ v

z is a CRE, **s** is a CRE,
ab is a list of two integers {**a**, **b**}, at least one of which must be non-zero,
v is a complex number or symbolic expression representing the term of the Eisenstein series for **GL(2)** with parameters **z**, **s**, **a**, **b**.

Example

```
In[1]:= EisensteinSeriesTerm[z, s, {3, 4}]
```

Out[1]= $\frac{1}{2}$ Abs[4 + 3 z]$^{-2 s}$ Im[z]^{s}

```
In[2]:= EisensteinSeriesTerm[z, s, {12, 16}]

Out[2]= 0
```

See also: EisensteinFourierCoefficient, LanglandsIFun.

■ ElementaryMatrix (elm)

This function returns a square matrix having 1s along the leading diagonal and with a given element in a specified off-diagonal position.

ElementaryMatrix[n, i, j, c] ⟶ e

n is a strictly positive integer being the size of the matrix,
i is a strictly positive integer being the row index of the off-diagonal entries,
j is a strictly positive integer with **i** $\neq$ **j**, representing the column index of the off-diagonal entries,
c is a CRE to be placed at the (**i**, **j**)th position,
[e] is an **n** by **n** matrix with 1s on the leading diagonal and all other elements zero, except in the (**i**, **j**)th position where it is **c**.

Example

```
In[182]:= MatrixForm[ElementaryMatrix[4, 3, 1, x]]
```

Out[182]//MatrixForm=

$$\begin{pmatrix} 1 & 0 & 0 & 0 \\ 0 & 1 & 0 & 0 \\ x & 0 & 1 & 0 \\ 0 & 0 & 0 & 1 \end{pmatrix}$$

```
In[183]:= {a1, a2, a3, a4}.%

Out[183]= {a1 + a3 x, a2, a3, a4}

In[184]:= %%.{{b1}, {b2}, {b3}, {b4}}

Out[184]= {{b1}, {b2}, {b3 + b1 x}, {b4}}
```

See also: SwapMatrixRows, SwapMatrixColumns.

■ FunctionalEquation (feq)

This function, for each index **i**, returns a list of affine combinations of its variables representing the ith functional equation for the Jacquet–Whittaker function of order $\mathbf{n} \geq \mathbf{2}$.

See Section 5.9, especially equations (5.9.5), (5.9.6) and Example 5.9.7.

FunctionalEquation[v, i] ⟶ vp

v is a list of CREs of length $\mathbf{n-1}$,
i is a strictly positive integer with $\mathbf{1 \leq i \leq n-1}$ being the index of the functional equation,
vp is a list of CREs representing the transformations required of the variables **vp** for the **i**th functional equation.

Example

```
In[96]:= Table[FunctionalEquation[{x1, x2, x3, x4}, i], {i, 1, 4}]
```

$$\text{Out[96]= } \left\{\left\{x1,\ x2,\ -\frac{1}{5}+x3+x4,\ \frac{2}{5}-x4\right\},\ \left\{x1,\ -\frac{1}{5}+x2+x3,\ \frac{2}{5}-x3,\ -\frac{1}{5}+x3+x4\right\},\ \left\{-\frac{1}{5}+x1+x2,\ \frac{2}{5}-x2,\ -\frac{1}{5}+x2+x3,\ x4\right\},\ \left\{\frac{2}{5}-x1,\ -\frac{1}{5}+x1+x2,\ x3,\ x4\right\}\right\}$$

See also: Whittaker, WhittakerGamma, WhittakerStar.

■ GetCasimirOperator (gco)

This function computes the Casimir operator acting on an arbitrary noun function and with respect to the Iwasawa variables. Note that this function makes an explicit brute-force evaluation of the operator, so is not fast, especially for $n \geq 3$.

See Proposition 2.3.3, Example 2.3.4 and Section 6.1.

GetCasimirOperator[m,n,"x","y","f"] ⟶ Operator

m is positive integer with value 2 or more being the order of the operator,
n is a positive integer with value 2 or more being the dimension of the Iwasawa form,
"x" is a string, being the name of the symbol such that the variables in the upper triangle of the matrix given by the Iwasawa decomposition are **x[i, j]**,
"y" is a string, being the name of the symbol such that the terms in the first $\mathbf{n-1}$ positions of the leading diagonal of the Iwasawa decomposition are

$$\mathbf{y[1] \cdots y[n-1],\ y[1] \cdots y[n-2],\ \ldots,\ y[1],}$$

"f" is a string being the name of a function of noun form (i.e. it should not be defined as an explicit Mathematica function or correspond to the name of an existing function) which will appear as partially differentiated by the computed Casimir operator, **Operator** is an expression in the variables

$$(\mathbf{x}[\mathbf{i},\mathbf{j}], \mathbf{1} \leq \mathbf{i} < \mathbf{j} \leq \mathbf{n}), (\mathbf{y}[\mathbf{i}], \mathbf{1} \leq \mathbf{i} \leq \mathbf{n}-\mathbf{1})$$

and the partial derivatives of the function with name **"f"** with respect to argument slots of **f** arranged in the order $(\mathbf{x}_{1,1}, \mathbf{x}_{1,2}, \ldots, \mathbf{y}_1, \ldots, \mathbf{y}_{n-1})$.

Example

```
In[8]:=  suf[GetCasimirOperator[3, 3, "x", "y", "f"]]
```

$$\text{Out[8]=}\ 3\, y_1 \big(-y_2\, f^{(0,0,0,1,1)}[x_{1,2}, x_{1,3}, x_{2,3}, y_1, y_2] + y_2^2\, f^{(0,0,0,1,2)}[x_{1,2}, x_{1,3}, x_{2,3}, y_1, y_2] + 2\, y_1\, f^{(0,0,0,2,0)}[x_{1,2}, x_{1,3}, x_{2,3}, y_1, y_2] - y_1\, y_2\, f^{(0,0,0,2,1)}[x_{1,2}, x_{1,3}, x_{2,3}, y_1, y_2] + 2\, y_1\, f^{(0,0,2,0,0)}[x_{1,2}, x_{1,3}, x_{2,3}, y_1, y_2] - y_1\, y_2\, f^{(0,0,2,0,1)}[x_{1,2}, x_{1,3}, x_{2,3}, y_1, y_2] + 4\, y_1\, x_{1,2}\, f^{(0,1,1,0,0)}[x_{1,2}, x_{1,3}, x_{2,3}, y_1, y_2] - 2\, y_1\, y_2\, x_{1,2}\, f^{(0,1,1,0,1)}[x_{1,2}, x_{1,3}, x_{2,3}, y_1, y_2] + 2\, y_1\, y_2^2\, f^{(0,2,0,0,0)}[x_{1,2}, x_{1,3}, x_{2,3}, y_1, y_2] + 2\, y_1\, x_{1,2}^2\, f^{(0,2,0,0,0)}[x_{1,2}, x_{1,3}, x_{2,3}, y_1, y_2] + y_1\, y_2^3\, f^{(0,2,0,0,1)}[x_{1,2}, x_{1,3}, x_{2,3}, y_1, y_2] - y_1\, y_2\, x_{1,2}^2\, f^{(0,2,0,0,1)}[x_{1,2}, x_{1,3}, x_{2,3}, y_1, y_2] - y_1^2\, y_2^2\, f^{(0,2,0,1,0)}[x_{1,2}, x_{1,3}, x_{2,3}, y_1, y_2] + 2\, y_1\, y_2^2\, f^{(1,1,1,0,0)}[x_{1,2}, x_{1,3}, x_{2,3}, y_1, y_2] + 2\, y_1\, y_2^2\, x_{1,2}\, f^{(1,2,0,0,0)}[x_{1,2}, x_{1,3}, x_{2,3}, y_1, y_2] + y_2^2\, f^{(2,0,0,1,0)}[x_{1,2}, x_{1,3}, x_{2,3}, y_1, y_2]\big)$$

See also: IwasawaForm, ApplyCasimirOperator.

■ GetMatrixElement (gme)

The specified element of a matrix is returned.

GetMatrixElement[a, i, j] ⟶ **e**

a is a matrix of CREs,
i is the row index of the element,
j is the column index of the element,
e is the **(i, j)**th element of **a**.

See also: MatrixColumn, MatrixRow, MatrixBlock.

■ GlnVersion (glv)

This function prints out the date of the version of GL(n)pack which is being used, followed by the version of Mathematica. It has no argument, but the brackets must be given.

GlnVersion[] ⟶ True

■ HeckeCoefficientSum (hcs)

This function takes a natural number m, a list of natural numbers $\{m_1, \ldots, m_{n-1}\}$ and a string for a function name and finds the terms in the sum right-hand side

$$\lambda_m A(m_1, \ldots, m_{n-1}) = \sum A(c_0 m_1/c_1, c_1 m_2/c_2, \ldots, c_{n-2} m_{n-1}/c_{n-1})$$

where the summation is over all (c_i) such that $\prod_{i=0}^{n-1} c_i = m$ and $c_i | m_i$, $1 \leq i \leq n-1$.

See Theorem 9.3.11 and equation (9.3.17).

HeckeCoefficientSum[m, ms, "A"] ⟶ s

m is a natural number (i.e. a strictly positive integer) representing the index of the eigenvalue $\lambda_{\mathbf{m}}$,

ms is a list of natural numbers representing the multi-index of the Fourier coefficient **A**,

"A" is a string giving the name of a noun function for the Fourier coefficient,

s is a term or sum of terms being the right-hand side of the expansion $\lambda_{\mathbf{m}} \mathbf{A}(\mathbf{m_1}, \ldots)$.

Example

```
In[99]:= HeckeCoefficientSum[6, {12, 4, 5}, "A"]

Out[99]= A[2, 24, 5] + A[4, 6, 10] + A[8, 12, 5] + A[18, 8, 5] + A[36, 2, 10] + A[72, 4, 5]
```

See also: HeckeOperator, SchurPolynomial, HeckePowerSum, HeckeEigenvalue.

■ HeckeEigenvalue (hev)

This function returns the value of the mth eigenvalue of the ring of HeckeOperators acting on square integrable automorphic forms $f(z)$ for $\mathfrak{h}^n$. Note that when the Euler product of a Maass form is known, the Fourier coefficients which appear in the expressions for the eigenvalues (the A in $\lambda_m = A(m, 1, \ldots, 1)$)

can be expressed in terms of Schur polynomials in the parameters which appear in the Euler product.

See Section 9.3, especially Theorem 9.3.11.

HeckeEigenvalue[m, n, a] ⟶ $\lambda_{\mathbf{m}}$

m is a natural number representing the index of the eigenvalue $\lambda_{\mathbf{m}}$,
n is a positive integer of size two or more being the dimension of **GL(n)**,
a is a string representing the name of a function of **n − 1** integers being the Fourier coefficients of a given Maass form which is an eigenfunction of all of the Hecke operators,
$\lambda_{\mathbf{m}}$ is an expression representing the **m**th Hecke eigenvalue.

Example

```
In[39]:= HeckeEigenvalue[2*3^4*5^2, 6, "a"]

Out[39]= a[2, 1, 1, 1, 1]
          (-a[1, 1, 1, 3, 1] + a[1, 3, 1, 1, 1]^2 + 2 a[1, 1, 3, 1, 1] a[3, 1, 1, 1, 1] -
            3 a[1, 3, 1, 1, 1] a[3, 1, 1, 1, 1]^2 + a[3, 1, 1, 1, 1]^4)
          (-a[1, 5, 1, 1, 1] + a[5, 1, 1, 1, 1]^2)
```

See also: HeckeOperator, HeckeMultiplicativeSplit, SchurPolynomial, Hecke-PowerSum.

■ HeckeMultiplicativeSplit (hms)

This function takes a list of natural numbers $\{m_1, \ldots, m_{n-1}\}$, finds the primes and their powers that divide any of the m_i, and returns a list of lists of those primes and their powers. The purpose of this function is the evaluation of the Hecke Fourier coeffients of a Maass form in terms of Schur polynomials when the Euler product coeffients of the form are known. If $p_1, \ldots, p_r$ are the primes and $k_{i,j}$ is the maximum power of p_i dividing m_j, then the Fourier coefficient

$$A(m_1, \ldots, m_{n-1}) = \prod_{i=1}^{r} A\big(p_i^{k_{i,1}}, \ldots, p_i^{k_{i,n-1}}\big).$$

See Theorem 9.3.11 and equation (7.4.14).

HeckeMultiplicativeSplit[m] ⟶ list

m is a list of natural numbers representing the multi-index of the Fourier coefficient,
list consists of sublists, each being a prime $\mathbf{p_i}$ and a list of **n − 1** powers of that prime $\mathbf{k_{i,j}}$.

Example

```
In[98]:= HeckeMultiplicativeSplit[{2*3^2*5^4, 2^3*3*5, 5}]

Out[98]= {{2, {1, 3, 0}}, {3, {2, 1, 0}}, {5, {4, 1, 1}}}
```

See also: HeckeOperator, SchurPolynomial, HeckeEigenvalue, HeckeCoefficientSum.

■ HeckeOperator (hop)

This function computes the nth order Hecke operator which acts on square integrable forms on $\mathfrak{h}^n$.

See Section 9.3, especially formula (9.3.5).

HeckeOperator[n, z, f] ⟶ $\mathbf{T_n(f(z))}$

n is a natural number being the order of the operator,
z is a square matrix of CREs of size **n**,
f is a string being the name of a function of a square matrix of size **n**,
$\mathbf{T_n(f(z))}$ is an expression representing the action of the **n**th Hecke operator on the matrix function **f(z)**.

Example

```
In[21]:= z = {{x, x^2, 2}, {1, x, x+1}, {0, 2, x}}

Out[21]= {{x, x², 2}, {1, x, 1+x}, {0, 2, x}}

In[22]:= HeckeOperator[2, z, "f"]

Out[22]= f[{{x, x², 2}, {1, x, 1+x}, {0, 4, 2 x}}] +
          f[{{x, x², 2}, {1, 2+x, 1+2 x}, {0, 4, 2 x}}] +
          f[{{x, x², 2}, {2, 2 x, 2 (1+x)}, {0, 2, x}}] +
          f[{{x, 2+x², 2+x}, {1, x, 1+x}, {0, 4, 2 x}}] +
          f[{{x, 2+x², 2+x}, {1, 2+x, 1+2 x}, {0, 4, 2 x}}] +
          f[{{2 x, 2 x², 4}, {1, x, 1+x}, {0, 2, x}}] +
          f[{{1+x, x+x², 3+x}, {2, 2 x, 2+2 x}, {0, 2, x}}]

In[26]:= f[z_] := Sum[z[[i, i]], {i, 1, Length[z]}]

In[28]:= HeckeOperator[12, z, "f"]

Out[28]= 2804 + 4688 x
```

See also: HeckeEigenvalue.

■ HeckePowerSum (hps)

This function takes a natural number m, a list of natural numbers $\{m_1, \ldots, m_{n-1}\}$ and a string for a function name and finds the powers of any fixed prime in the sum right-hand side

$$\lambda_m A(m_1, \ldots, m_{n-1}) = \sum A(c_0 m_1/c_1, c_1 m_2/c_2, \ldots, c_{n-2} m_{n-1}/c_{n-1})$$

where the summation is over all (c_i) such that $\prod_{i=0}^{n-1} c_i = m$ and $c_i | m_i$, $1 \le i \le n-1$ in case m and each of the m_i is a power of a fixed prime. The powers that appear in the expansion are the same for any prime. The purpose of this function is to simplify the study of the multiplicative properties of the Fourier coefficients.

See Theorem 9.3.11 and equation (9.3.17).

HeckePowerSum[a, as, "B"] ⟶ list

a is a non-negative integer, being the power **a** of any prime **p** such that $\mathbf{p^a}$ is the index of the eigenvalue $\lambda_{\mathbf{p^a}}$,

as is a list of non-negative integers representing the powers of a fixed prime which appear in the multi-index of a Fourier coefficient,

list consists of a sum of terms $\mathbf{B[b_{1,i}, \ldots, b_{n-1,i}]}$ such that the corresponding term in the Hecke sum would have a value $\mathbf{A(p^{b_{1,i}}, \ldots)}$.

Example

```
In[100]:=
        HeckePowerSum[2, {3, 4, 5}, "B"]

Out[100]=
        B[1, 6, 5] + B[2, 4, 6] + B[2, 5, 4] + B[3, 2, 7] +
         B[3, 3, 5] + B[3, 4, 3] + B[3, 5, 5] + B[4, 3, 6] + B[4, 4, 4] + B[5, 4, 5]
```

See also: HeckeOperator, SchurPolynomial, HeckeEigenvalue, HeckeCoefficientSplit,

■ HermiteFormLower (hfl)

This function computes the lower left Hermite form h of a non-singular integer matrix a, and a unimodular matrix l such that $a = lh$. This Hermite form is a lower triangular integer matrix with strictly positive elements on the diagonal of increasing size, and such that each element in the column below a diagonal entry is non-negative and less than the diagonal entry.

See Theorem 3.11.1.

HermiteFormLower[a] ⟶ {l, h}

a is a non-singular integer matrix,
l is a unimodular matrix,
h is a lower triangular integer matrix, being the Hermite form of **a**.

Example

```
In[4]:= m4 = {{5, 2, -4, 7}, {1, 6, 0, -3}, {1, 2, -2, 4}, {7, 1, 5, 6}};

In[8]:= {l, h} = HermiteFormLower[m4]

Out[8]= {{{-5, -2, -4, 7}, {1, 3, 0, -3}, {-3, -1, -2, 4}, {-4, -4, 5, 6}},
         {{241, 0, 0, 0}, {158, 4, 0, 0}, {35, 1, 1, 0}, {238, 2, 0, 1}}}

In[9]:= Map[MatrixForm, %]
```

Out[9]= $\left\{\begin{pmatrix} -5 & -2 & -4 & 7 \\ 1 & 3 & 0 & -3 \\ -3 & -1 & -2 & 4 \\ -4 & -4 & 5 & 6 \end{pmatrix}, \begin{pmatrix} 241 & 0 & 0 & 0 \\ 158 & 4 & 0 & 0 \\ 35 & 1 & 1 & 0 \\ 238 & 2 & 0 & 1 \end{pmatrix}\right\}$

```
In[10]:= l.h - m4

Out[10]= {{0, 0, 0, 0}, {0, 0, 0, 0}, {0, 0, 0, 0}, {0, 0, 0, 0}}
```

See also: HermiteFormUpper, SmithForm.

■ HermiteFormUpper (hfu)

This function computes the upper Hermite form h of a non-singular integer matrix a, and a unimodular matrix l such that $a = lh$. This Hermite form is an upper triangular integer matrix with strictly positive elements on the diagonal of increasing size, and such that each element in the column above a diagonal entry is non-negative and less than the diagonal entry.

See Theorem 3.11.1.

HermiteFormUpper[a] ⟶ {l, h}

a is a non-singular integer matrix,
l is a unimodular matrix,
h is an upper triangular integer matrix, being the Hermite form of **a**.

Example

```
In[107]:=
        m4 = {{5, 2, -4, 7}, {1, 6, 0, -3}, {1, 2, -2, 4}, {7, 1, 5, 6}};

In[108]:=
        MatrixForm[m4]

Out[108]//MatrixForm=
        ( 5  2  -4   7 )
        ( 1  6   0  -3 )
        ( 1  2  -2   4 )
        ( 7  1   5   6 )

In[109]:=
        {l, h} = HermiteFormUpper[m4];

In[110]:=
        Map[MatrixForm, %]

Out[110]=
        {( 5  2  -3  -2 )   ( 1  0  0  144 )
         ( 1  6  -3  -3 )   ( 0  1  1  262 )
         ( 1  2  -2  -1 ) , ( 0  0  2   91 )
         ( 7  1   2  -3 )   ( 0  0  0  482 )}

In[111]:=
        l.h - m4

Out[111]=
        {{0, 0, 0, 0}, {0, 0, 0, 0}, {0, 0, 0, 0}, {0, 0, 0, 0}}
```

See also: HermiteFormLower, SmithForm.

■ IFun (ifn)

This function returns the value

$$I_\nu(z) = \prod_{i=1}^{n-1} \prod_{j=1}^{n-1} y_i^{b_{i,j}\nu_j}$$

where $b_{i,j} = ij$ if $i + j \leq n$ and $(n-i)(n-j)$ if $i + j > n$. The $n \times n$ matrix z is real and non-singular, or has CRE elements which could evaluate to a real non-singular matrix. The variables y_i are those in the Iwasawa decomposition of z.

See Definition 2.4.1 and equation (5.1.1).

IFun[ν,z] $\longrightarrow$ **v**

ν is a list of $\mathbf{n-1}$ CREs,
z is an $\mathbf{n} \times \mathbf{n}$ non-singular matrix of CREs,
v is the product $\mathbf{I}_\nu(\mathbf{z})$.

Example

```
In[23]:= m = {{1, x[1, 2], x[1, 3]}, {0, 1, x[2, 3]}, {0, 0, 1}}.
            {{y[1] y[2], 0, 0}, {0, y[1], 0}, {0, 0, 1}};

In[22]:= MatrixForm[m]

Out[22]//MatrixForm=
```

$$\begin{pmatrix} y[1]\,y[2] & x[1,2]\,y[1] & x[1,3] \\ 0 & y[1] & x[2,3] \\ 0 & 0 & 1 \end{pmatrix}$$

```
In[16]:= IFun[{v1, v2}, m]
```

Out[16]= $y[1]^{v1+2\,v2}\,y[2]^{2\,v1+v2}$

See also: IwasawaForm, IwasawaYVariables.

■ InsertMatrixElement (ime)

An element is inserted into a matrix returning a new matrix and leaving the original unchanged.

InsertMatrixElement[e, i, j, a] ⟶ b

e is a CRE being the element to be inserted,
i is the row index of the position where the element is to be inserted,
j is the column index of the position where the element is to be inserted,
a is the original matrix of CREs,
b is a new matrix, being equal to **a** but with **e** in the (**i**, **j**)th position.

See also: DiagonalToMatrix, MatrixJoinHorizontal, MatrixJoinVertical.

■ IwasawaForm (iwf)

This function computes the Iwasawa form of a non-singular real matrix a. This consists of the product of an upper triangular unipotent matrix x and a diagonal matrix y with strictly positive diagonal entries such that, for some non-singular integer matrix u, real orthogonal matrix o and constant diagonal matrix δ, $a = u.x.y.o.\delta$. This function returns a single matrix $z = x.y$.

See Section 1.2.

IwasawaForm[a] ⟶ z

a is a non-singular square matrix of CREs,
z is an upper-triangular matrix with positive diagonal entries, being the Iwasawa form of a.

Example

The Iwasawa decomposition of the matrix

$$\begin{pmatrix} a & b \\ c & d \end{pmatrix}$$

is found.

```
In[8]:= g = {{a, b}, {c, d}};

In[9]:= IwasawaForm[g]
```

Out[9]= $\left\{\left\{\frac{b\,c-a\,d}{c^2+d^2},\ \frac{a\,c+b\,d}{c^2+d^2}\right\},\ \{0,\ 1\}\right\}$

```
In[12]:= g = {{y, x, x^2}, {2 y, 0, x}, {0, x, 1}};

In[13]:= MatrixForm[IwasawaForm[g]]
```

Out[13]//MatrixForm=

$$\begin{pmatrix} \frac{x\,(2+x-2\,x^2)\,y}{\sqrt{1+x^2}\,\sqrt{x^4+4\,y^2+4\,x^2\,y^2}} & \frac{\sqrt{\frac{x^4}{1+x^2}+4\,y^2}\,(-x^3+x^5+2\,y^2+2\,x^2\,y^2)}{\sqrt{1+x^2}\,(x^4+4\,y^2+4\,x^2\,y^2)} & \frac{2\,x^2}{1+x^2} \\ 0 & \frac{\sqrt{\frac{x^4}{1+x^2}+4\,y^2}}{\sqrt{1+x^2}} & \frac{x}{1+x^2} \\ 0 & 0 & 1 \end{pmatrix}$$

See also: IwasawaQ, MakeZMatrix, IwasawaXMatrix, IwasawaYMatrix, IwasawaXVariables, IwasawaYVariables.

■ IwasawaXMatrix (ixm)

This function returns the unipotent matrix x corresponding to the decomposition $z = x.y$ of a matrix z in Iwasawa form.

See Proposition 1.2.6 and Example 1.2.4

IwasawaXMatrix[w] ⟶ x

w is a square non-singular matrix of CREs which must be in Iwasawa form,
x an upper-triangular matrix with 1s on the diagonal and values $\mathbf{x_{i,j}}$ in each **(i, j)** position above the diagonal.

Example

In this example the x-matrix, x-variables, y-matrix and y-variables are extracted from a generic matrix in Iwasawa form.

```
In[23]:= m = {{1, x[1, 2], x[1, 3]}, {0, 1, x[2, 3]}, {0, 0, 1}}.
           {{y[1] y[2], 0, 0}, {0, y[1], 0}, {0, 0, 1}};

In[22]:= MatrixForm[m]

Out[22]//MatrixForm=
         ( y[1] y[2]  x[1, 2] y[1]  x[1, 3] )
         (     0          y[1]      x[2, 3] )
         (     0            0          1    )

In[17]:= MatrixForm[IwasawaXMatrix[m]]

Out[17]//MatrixForm=
         ( 1  x[1, 2]  x[1, 3] )
         ( 0     1     x[2, 3] )
         ( 0     0        1    )

In[18]:= IwasawaXVariables[m]

Out[18]= {x[1, 2], x[1, 3], x[2, 3]}

In[19]:= IwasawaYMatrix[m] // MatrixForm

Out[19]//MatrixForm=
         ( y[1] y[2]   0    0 )
         (     0     y[1]   0 )
         (     0       0    1 )

In[20]:= IwasawaYVariables[m]

Out[20]= {y[1], y[2]}
```

See also: IwasawaForm, IwasawaXMatrix, IwasawaYVariables, IwasawaY-Matrix.

■ IwasawaXVariables (ixv)

This function returns the x-variables from a matrix $z = x.y$ in Iwasawa form. These are the elements in the strict upper triangle of the matrix x in row order.

See Definition 1.2.3 and Proposition 1.2.6.

IwasawaXVariables[w] ⟶ l

w is a square non-singular matrix of CREs which must be in Iwasawa form,
l is a list of the form $\{\mathbf{x_{1,2}}, \ldots, \mathbf{x_{1,n}}, \mathbf{x_{2,3}}, \ldots, \mathbf{x_{n-1,n}}\}$.

See also: IwasawaForm, IwasawaXMatrix, IwasawaYVariables, IwasawaY-Matrix.

■ IwasawaYMatrix (iym)

This function returns the y-matrix from the decomposition $z = x.y$ of a matrix z in Iwasawa form.

See Definition 1.2.3 and Proposition 1.2.6.

IwasawaYMatrix[z] ⟶ y

z is a square non-singular matrix of CREs which must be in Iwasawa form, **y** a diagonal matrix where the **i**th diagonal slot has the value $\mathbf{y_1 \cdots y_{n-i}}$ for $\mathbf{1 \le i \le n-1}$ where the (**n**, **n**)th position has the value 1.

See also: IwasawaForm, IwasawaXMatrix, IwasawaYVariables, IwasawaX-Variables.

■ IwasawaYVariables (iyv)

This function returns a list of the y-variables from the Iwasawa decomposition of a matrix $z = x.y$.

See Definition 1.2.3 and Proposition 1.2.6.

IwasawaYVariables[z] ⟶ L

z is a square non-singular matrix of CREs which must be in Iwasawa form, **L** a list $\{\mathbf{y_1, \ldots, y_{n-1}}\}$ of the y-variables of the Iwasawa form.

See also: IwasawaForm, IwasawaYMatrix, IwasawaXVariables, IwasawaX-Matrix.

■ IwasawaQ (iwq)

This function tests a Mathematica form or expression to see whether it is a non-singular square matrix in Iwasawa form.

See Section 1.2.

IwasawaQ[z] ⟶ value

z is a Mathematica form,
value is **True** if **z** is a matrix of CREs in Iwasawa form, **False** otherwise.

See also: IwasawaForm, MakeZMatrix.

■ KloostermanBruhatCell (kbc)

This function takes an explicit permutation matrix w with all other arguments being symbolic. It returns rules which solve for x and y in the square matrix Bruhat decomposition equation $a = x.c.w.y$ assuming c is in "Friedberg form", x and y are unipotent and y satisfies ${}^t w.{}^t y.w$ is upper triangular. These rules are not unique.

See Chapter 11, especially Section 11.2. Also Lemma 10.6.3.

KloostermanBruhatCell[a,x,c,w,y] ⟶ rules

a is a symbol which will be used as the name of an **n** × **n** matrix,
x is a symbol which will be used as the name of a unipotent matrix,
c is a symbol which will be used as the name of an array **c[i]** representng a list of **n** − **1** non-zero integers specifying the diagonal of a matrix. (Note that the 1st element of the diagonal represents the term **det(w)/c[n − 1]**, the second **c[n − 1]/c[n − 2]** and so on down to the last **c[1]** as in the notation of (11.2.1).),
w is an **n** × **n** matrix which is zero except for a single 1 in each row and column, being an explicit element of the Weyl Group $\mathbf{W_n}$,
y is a symbol which will be used as the name of a unipotent matrix which satisfies $^t\mathbf{w}.^t\mathbf{y}.\mathbf{w}$ is upper triangular making the decomposition unique, given **a**,
rules is a list of rules of the form **x[i, j]** → **eij** or **y[i, j]** → **eij** where the **eij** are expressions in the **a[i, j]** and **c[i]**.

Example

```
In[29]:= w = WeylGroup[4][[19]]

Out[29]= {{0, 0, 0, 1}, {1, 0, 0, 0}, {0, 1, 0, 0}, {0, 0, 1, 0}}

In[30]:= suf[KloostermanBruhatCell[a, x, c, w, y]]
```

Out[30]= $\Big\{x_{1,2} \to \frac{c_2\, a_{1,1}}{c_3},\ x_{1,3} \to \frac{c_1\, a_{1,2}}{c_2},\ x_{1,4} \to \frac{a_{1,3}}{c_1},$
$x_{2,3} \to \frac{c_1\, a_{2,2}}{c_2},\ x_{2,4} \to \frac{a_{2,3}}{c_1},\ x_{3,4} \to \frac{a_{3,3}}{c_1},\ y_{1,2} \to 0,\ y_{1,3} \to 0,$
$y_{1,4} \to \frac{c_2\,(c_1\, a_{2,4} - a_{2,3}\, a_{4,4}) + c_1\, a_{2,2}\,(-c_1\, a_{3,4} + a_{3,3}\, a_{4,4})}{c_1\, c_3},$
$y_{2,3} \to 0,\ y_{2,4} \to \frac{c_1\, a_{3,4} - a_{3,3}\, a_{4,4}}{c_2},\ y_{3,4} \to \frac{a_{4,4}}{c_1}\Big\}$

See also: BruhatForm, BruhatCVector, KloostermanCompatibility, KloostermanSum.

■ KloostermanCompatibility (klc)

This function takes an explicit permutation matrix w, with remaining arguments symbolic, and returns a list of values, each element being a different type of constraint applicable to any valid Kloosterman sum based on w. The first element is a list of forms restricting the characters. The second is a set of divisibility relations restricting the values of the diagonal matrix c. And the third is the set of minor relations. A typical approach to forming Kloosterman sums would be to first run this function, determine a valid set or sets of parameters from the symbolic output, and then run **KloostermanSum** using explicit integer values of valid parameters.

See Chapter 11, Proposition 11.2.10, Lemma 10.6.3.

KloostermanCompatibility[t1,t2, c, w, v] ⟶ {characters, divisibilities, minors}

t1 is a symbol representing the character $\mathbf{e}^{2\pi i \sum_{i=1}^{n-1} \mathbf{t1[i,i+1]}}$ of $\mathbf{U_n}(\mathbb{R})$,
t2 is a symbol representing another character of $\mathbf{U_n}(\mathbb{R})$,
c is a symbol representing the diagonal of a matrix in Friedberg notation. (Note that the 1st element of the diagonal is the term $\mathbf{det(w)/c_{n-1}}$, the second $\mathbf{c_{n-1}/c_{n-2}}$ and so on down to the last $\mathbf{c_1}$ as in the notation of (11.2.1).),
w is an $\mathbf{n \times n}$ matrix which is zero except for a single 1 in each row and column, representing an explicit element of the Weyl Group $\mathbf{W_n}$,
v is a symbol representing the generic name of any bottom row-based minor,
characters is a list of expressions relating the elements of **t1**, **t2**, and the $\mathbf{c_i}$. Each expression must vanish if an explicit Kloosterman sum is to be valid,
divisibility is a list of lists, each sublist being of the form $\{\mathbf{c_i}, \mathbf{1}\}$ or $\{\mathbf{c_i}, \mathbf{c_j}\}$. The former means valid sums must have the $\mathbf{c_i} = \pm\mathbf{1}$. The latter means they must have $\mathbf{c_i | c_j}$,
minors is a list of rules of the form $\mathbf{v[\{j_1, j_2, \ldots, j_i\}] \to c_i}$ or $\mathbf{0}$, giving the constraints on minors.

Example

```
In[1861]:=
        MatrixForm[w = WeylGroup[4] [[17]]]

Out[1861]//MatrixForm=
        ( 0 0 1 0 )
        ( 0 0 0 1 )
        ( 1 0 0 0 )
        ( 0 1 0 0 )

In[1882]:=
        SubscriptedForm[klc[t1, t2, c, w, v]]

Out[1882]=
```

$$\left\{\left\{\frac{c_2\, t1_3}{c_1{}^2} - t2_1,\ \frac{c_2\, t1_1}{c_3{}^2} - t2_3\right\},\ \{\{c_1, c_2\}, \{c_3, c_2\}\},\ \{v[\{2\}] \to c_1,\ v[\{1\}] \to 0,\ v[\{1, 2\}] \to c_2,\ v[\{1, 2, 4\}] \to c_3,\ v[\{1, 2, 3\}] \to 0\}\right\}$$

See also: KloostermanBruhatCell, BruhatForm, BruhatCVector, KloostermanSum, PluckerRelations, PluckerCoordinates, PluckerInverse.

■ KloostermanSum (kls)

This function computes the generalized Kloosterman sum for $\mathbf{SL(n, \mathbb{Z})}$ for $\mathbf{n \ge 2}$ as given by Definition 11.2.2. When $\mathbf{n = 2}$ this coincides with the classical Kloosterman sum. More generally the sum is

$$S(\theta_1, \theta_2, c, w) := \sum_{\gamma = b_1 c w b_2} \theta_1(b_1)\theta_2(b_2)$$

where

$$\gamma \in U_n(\mathbb{Z})\backslash SL(n,\mathbb{Z}) \cap G_w / \Gamma_w$$

and $\Gamma_w = {}^t w.{}^t U_n(\mathbb{Z}).w \cap U_n(\mathbb{Z})$ and G_w is the Bruhat cell associated to the permutation matrix w. Since these sums are only well defined for some particular compatible values of the arguments the user is advised to first run **KloostermanCompatibility** with an explicit **w** to determine those values. Note that the complexity of the algorithm is $O(\prod_{1\le i\le n-1} |c_i|^n) = O(c^{n^2})$ where $c = \max|c_i|$.

See Chapter 11.

KloostermanSum[t1, t2, c, w] ⟶ value

t1 is a list of **n − 1** integers representing a character of $\mathbf{U_n}(\mathbb{R})$,
t2 is a list of **n − 1** integers representing another character of $\mathbf{U_n}(\mathbb{R})$,
c is a list of **n − 1** non-zero integers specifying the diagonal of a matrix. (Note that the 1st element of the matrix is $\mathbf{det(w)/c_{n-1}}$, the second $\mathbf{c_{n-1}/c_{n-2}}$ and so on down to the last $\mathbf{c_1}$ as in the notation of (11.2.1).),
w is an **n × n** matrix which is zero except for a single 1 in each row and column, representing an explicit element of the Weyl subroup $\mathbf{W_n}$ of **GL(n,** $\mathbb{R}$**)**,
value is a sum of complex exponentials being a Kloosterman sum when it is well defined.

Example

This $n = 4$ example shows how KloostermanCompatibility should be run after selecting a permutation matrix. Then KloostermanSum is called with compatible arguments.

```
In[37]:= w = WeylGroup[4][[17]]

Out[37]= {{0, 0, 1, 0}, {0, 0, 0, 1}, {1, 0, 0, 0}, {0, 1, 0, 0}}

In[31]:= suf[klc[t1, t2, c, w, v]]
```

Out[31]= $\left\{\left\{\frac{c_2\, tl_3}{c_1{}^2} - t2_1,\ \frac{c_2\, tl_1}{c_3{}^2} - t2_3\right\},\right.$
$\{\{c_1, c_2\}, \{c_3, c_2\}\}, \{v[\{2\}] \to c_1, v[\{1\}] \to 0,$
$\left. v[\{1, 2\}] \to c_2, v[\{1, 2, 4\}] \to c_3, v[\{1, 2, 3\}] \to 0\}\right\}$

```
In[32]:= kls[{3, 7, 12}, {4, 13, 1}, {3, 3, 3}, w]
```

Out[32]= $9\, e^{-\frac{2 i \pi}{3}} + 8\, e^{\frac{2 i \pi}{3}}$

Example

This first illustrates commutativity of the LongElement sums (c/f (Friedberg (1987), its Proposition 2.5)), then Proposition 11.4.1 and finally is given an example of a classical sum showing it is real.

In[21]:= **kls[{4, 13}, {6, 7}, {3, 3}, LongElement[3]]**

Out[21]= $4+3\,e^{-\frac{2 i \pi}{3}}+3\,e^{\frac{2 i \pi}{3}}$

In[22]:= **kls[{6, 7}, {4, 13}, {3, 3}, LongElement[3]]**

Out[22]= $4+3\,e^{-\frac{2 i \pi}{3}}+3\,e^{\frac{2 i \pi}{3}}$

In[23]:= **kls[{9, 7}, {1, 13}, {3, 3}, LongElement[3]]**

Out[23]= $4+3\,e^{-\frac{2 i \pi}{3}}+3\,e^{\frac{2 i \pi}{3}}$

In[40]:= **a = kls[{4, 13}, {6, 7}, {12, 31}, LongElement[3]]**

Out[40]=
$$2\,e^{-\frac{11 i \pi}{186}}+e^{\frac{11 i \pi}{186}}+e^{-\frac{17 i \pi}{186}}+2\,e^{\frac{17 i \pi}{186}}+2\,e^{-\frac{25 i \pi}{186}}+2\,e^{\frac{25 i \pi}{186}}+2\,e^{-\frac{29 i \pi}{186}}+2\,e^{\frac{29 i \pi}{186}}+$$
$$2\,e^{-\frac{i \pi}{6}}+e^{\frac{i \pi}{6}}+e^{-\frac{37 i \pi}{186}}+2\,e^{\frac{37 i \pi}{186}}+e^{-\frac{41 i \pi}{186}}+2\,e^{\frac{41 i \pi}{186}}+2\,e^{-\frac{47 i \pi}{186}}+e^{\frac{47 i \pi}{186}}+$$
$$e^{-\frac{53 i \pi}{186}}+2\,e^{\frac{53 i \pi}{186}}+2\,e^{-\frac{71 i \pi}{186}}+e^{\frac{71 i \pi}{186}}+2\,e^{-\frac{73 i \pi}{186}}+e^{\frac{73 i \pi}{186}}+e^{-\frac{77 i \pi}{186}}+2\,e^{\frac{77 i \pi}{186}}+$$
$$2\,e^{-\frac{79 i \pi}{186}}+e^{\frac{79 i \pi}{186}}+e^{-\frac{83 i \pi}{186}}+2\,e^{\frac{83 i \pi}{186}}+2\,e^{-\frac{91 i \pi}{186}}+e^{\frac{91 i \pi}{186}}+2\,e^{-\frac{95 i \pi}{186}}+e^{\frac{95 i \pi}{186}}+$$
$$2\,e^{-\frac{103 i \pi}{186}}+e^{\frac{103 i \pi}{186}}+2\,e^{-\frac{107 i \pi}{186}}+e^{\frac{107 i \pi}{186}}+e^{-\frac{109 i \pi}{186}}+2\,e^{\frac{109 i \pi}{186}}+e^{-\frac{113 i \pi}{186}}+$$
$$2\,e^{\frac{113 i \pi}{186}}+2\,e^{-\frac{115 i \pi}{186}}+e^{\frac{115 i \pi}{186}}+e^{-\frac{133 i \pi}{186}}+2\,e^{\frac{133 i \pi}{186}}+2\,e^{-\frac{139 i \pi}{186}}+e^{\frac{139 i \pi}{186}}+$$
$$e^{-\frac{145 i \pi}{186}}+2\,e^{\frac{145 i \pi}{186}}+e^{-\frac{149 i \pi}{186}}+2\,e^{\frac{149 i \pi}{186}}+e^{-\frac{5 i \pi}{6}}+2\,e^{\frac{5 i \pi}{6}}+2\,e^{-\frac{157 i \pi}{186}}+$$
$$2\,e^{\frac{157 i \pi}{186}}+2\,e^{-\frac{161 i \pi}{186}}+2\,e^{\frac{161 i \pi}{186}}+2\,e^{-\frac{169 i \pi}{186}}+e^{\frac{169 i \pi}{186}}+2\,e^{-\frac{175 i \pi}{186}}+e^{\frac{175 i \pi}{186}}$$

In[25]:= **b = kls[{35, 25}, {18, 38}, {12, 31}, LongElement[3]];**

In[26]:= **a - b**

Out[26]= 0

In[35]:= **s = KloostermanSum[{24}, {13}, {43}, LongElement[2]]**

Out[35]=
$$e^{-\frac{2 i \pi}{43}}+e^{\frac{2 i \pi}{43}}+2\,e^{-\frac{8 i \pi}{43}}+2\,e^{\frac{8 i \pi}{43}}+2\,e^{-\frac{10 i \pi}{43}}+2\,e^{\frac{10 i \pi}{43}}+2\,e^{-\frac{12 i \pi}{43}}+2\,e^{\frac{12 i \pi}{43}}+$$
$$2\,e^{-\frac{20 i \pi}{43}}+2\,e^{\frac{20 i \pi}{43}}+2\,e^{-\frac{24 i \pi}{43}}+2\,e^{\frac{24 i \pi}{43}}+2\,e^{-\frac{28 i \pi}{43}}+2\,e^{\frac{28 i \pi}{43}}+2\,e^{-\frac{30 i \pi}{43}}+$$
$$2\,e^{\frac{30 i \pi}{43}}+2\,e^{-\frac{32 i \pi}{43}}+2\,e^{\frac{32 i \pi}{43}}+2\,e^{-\frac{38 i \pi}{43}}+2\,e^{\frac{38 i \pi}{43}}+2\,e^{-\frac{42 i \pi}{43}}+2\,e^{\frac{42 i \pi}{43}}$$

In[36]:= **Im[ExpToTrig[s]]**

Out[36]= 0

See also: BruhatForm, BruhatCVector, KloostermanCompatibility, KloostermanBruhatCell, PluckerCoordinates, PluckerInverse, PluckerRelations.

■ LanglandsForm (llf)

This function returns a list of the three matrices of the Langlands decomposition of a square matrix in a parabolic subgroup specified by a partition of the matrix dimension.

See Section 10.2.

LanglandsForm[p, d] $\longrightarrow$ **{u, c, m}**

p is a square matrix of CREs,
d is a list of **r** positive integers of length at most **n** with sum **n**,
u is a unipotent block upper triangular matrix,
c is a diagonal matrix with **r** diagonal blocks each being a positive constant times the identity,
m is a block diagonal matrix with each diagonal block having determinant ± 1.

Example

```
In[1]:=  d = {2, 2}; a = {{1, 2, 3, 4}, {5, 6, 7, 8}, {0, 0, 1, 2}, {0, 0, 3, 4}}
         MatrixForm[a]
```

$$\begin{pmatrix} 1 & 2 & 3 & 4 \\ 5 & 6 & 7 & 8 \\ 0 & 0 & 1 & 2 \\ 0 & 0 & 3 & 4 \end{pmatrix}$$

```
In[19]:=  {u, c, m} = LanglandsForm[a, d];

In[20]:=  Map[MatrixForm, {u, c, m}]
```

Out[20]=
$$\left\{\begin{pmatrix} 1 & 0 & 0 & 1 \\ 0 & 1 & -2 & 3 \\ 0 & 0 & 1 & 0 \\ 0 & 0 & 0 & 1 \end{pmatrix}, \begin{pmatrix} 2 & 0 & 0 & 0 \\ 0 & 2 & 0 & 0 \\ 0 & 0 & \sqrt{2} & 0 \\ 0 & 0 & 0 & \sqrt{2} \end{pmatrix}, \begin{pmatrix} \frac{1}{2} & 1 & 0 & 0 \\ \frac{5}{2} & 3 & 0 & 0 \\ 0 & 0 & \frac{1}{\sqrt{2}} & \sqrt{2} \\ 0 & 0 & \frac{3}{\sqrt{2}} & 2\sqrt{2} \end{pmatrix}\right\}$$

```
In[21]:=  MatrixForm[u.c.m]
```

Out[21]//MatrixForm=
$$\begin{pmatrix} 1 & 2 & 3 & 4 \\ 5 & 6 & 7 & 8 \\ 0 & 0 & 1 & 2 \\ 0 & 0 & 3 & 4 \end{pmatrix}$$

See also: ParabolicQ, LanglandsIFun.

■ LanglandsIFun (lif)

This function computes the summand for Langlands' Eisenstein series with respect to a specified parabolic subgroup.

See Chapter 10, Definition 10.4.5.

LanglandsIFun[g, p, s] $\longrightarrow$ $\mathbf{I_s(g.z)}$

g is a non-singular matrix of CREs in the parabolic subgroup specified by the second argument,
p is a list of **r** positive integers representing a partition of the matrix dimension,
s is list of **r** CREs such that $\sum_{i=1}^{r} \mathbf{d_i s_i} = \mathbf{0}$,
$\mathbf{I_s(g.z)}$ is the summand for the Langlands Eisenstein series.

Example

```
In[20]:= g = {{1, x, 3}, {4, 5, x^2}, {0, 0, x + 1}};

In[21]:= MatrixForm[g]
```

Out[21]//MatrixForm=

$$\begin{pmatrix} 1 & x & 3 \\ 4 & 5 & x^2 \\ 0 & 0 & 1+x \end{pmatrix}$$

```
In[11]:= d = {2, 1};

In[23]:= LanglandsIFun[g, d, {1 + I, -2 - 2 I}]
```

Out[23]= $\text{Abs}[5 - 4\,x]^{1+i}\ \text{Abs}[1 + x]^{-2-2\,i}$

See also: LanglandsForm.

■ LeadingMatrixBlock (lmb)

This function extracts a leading matrix block of specified dimensions.

LeadingMatrixBlock[a, i, j] $\longrightarrow$ **b**

a is a matrix of CREs, **i** is a valid row index for **a**,
j is a valid column index for **a**,
b is the leading block of **a** with **i** rows and **j** columns.

See also: BlockMatrix, TailingMatrixBlock, GetMatrixElement.

■ LongElement (lel)

This function constructs the so-called long element of the group $GL(n,\mathbb{Z})$, a matrix with 1s along the reverse leading diagonal and 0s elsewhere.

See Chapter 5.

LongElement[n] $\longrightarrow$ **w**

n is a strictly positive integer,
w is an **n** by **n** matrix with 1s down the reversed leading diagonal and 0s elsewhere.

Example

```
In[267]:= MatrixForm[LongElement[4]]

Out[267]//MatrixForm=
```

$$\begin{pmatrix} 0 & 0 & 0 & 1 \\ 0 & 0 & 1 & 0 \\ 0 & 1 & 0 & 0 \\ 1 & 0 & 0 & 0 \end{pmatrix}$$

See also: WMatrix, ModularGenerators.

■ LowerTriangleToMatrix (ltm)

This function takes a list of lists of increasing length and forms a matrix with zeros in the upper triangle and the given lists constituting the rows of the lower triangle.

LowerTriangleToMatrix[l] $\longrightarrow$ **a**

l is a list of lists of CREs of strictly increasing length representing the elements of a lower triangular submatrix including the diagonal. The first has length 1 and each successive sublist has length 1 more than that preceding sublist,
a is a full matrix with 0 in each upper triangular position.

Example

```
In[185]:= MatrixForm[
            LowerTriangleToMatrix[
               {{a}, {b, c}, {d, e, f}}]]

Out[185]//MatrixForm=
```

$$\begin{pmatrix} a & 0 & 0 \\ b & c & 0 \\ d & e & f \end{pmatrix}$$

See also: UpperTriangleToMatrix.

■ MakeBlockMatrix (mbm)

This function takes a list of lists of matrices and creates a single matrix wherein the jth matrix element of the ith sublist constitutes the (i, j)th sub-block of this matrix. In order that this construction succeed, the original matrices must have compatible numbers of rows and columns, i.e. the matrices in each sublist must have the same number of rows for that sublist and for each j the jth matrix in each sublist must have the same number of columns. In spite of this restriction, the function is a tool for building matrices rapidly when they have a natural block structure.

MakeBlockMatrix[A] ⟶ B

A is a list of lists of equal length of matrix elements, each matrix having CRE elements,
B is a single matrix with sub-blocks being the individual matrices in **A**.

Examples

```
In[101]:=
        a11 = IdentityMatrix[2]; a12 = ZeroMatrix[2, 4];
        a21 = ConstantMatrix[x, 3, 2]; a22 = ConstantMatrix[1, 3, 4];
        MakeBlockMatrix[{{a22, a11}, {a21, a21}}]

        MakeBlockMatrix::arg3 : The submatrices must have compatible row and column numbers.

Out[103]=
        $Aborted

In[104]:=
        MakeBlockMatrix[{{a11, a12}, {a21, a22}}] // MatrixForm

Out[104]//MatrixForm=
```

$$\begin{pmatrix} 1 & 0 & 0 & 0 & 0 & 0 \\ 0 & 1 & 0 & 0 & 0 & 0 \\ x & x & 1 & 1 & 1 & 1 \\ x & x & 1 & 1 & 1 & 1 \\ x & x & 1 & 1 & 1 & 1 \end{pmatrix}$$

See also: ConstantMatrix, ZeroMatrix, LongElement, WeylGroup, SpecialWeylGroup.

■ MakeMatrix (mkm)

This function returns a symbolic matrix of given dimensions.

MakeMatrix["a", m, n] ⟶ A

"a" is a string being the name of the generic symbolic matrix element variable
a[i, j],
m is a strictly positive integer representing the number of rows of A,
n is a strictly positive integer representing the number of columns of A,
A is a symbolic matrix with **(i, j)**th entry **a[i, j]**.

Example

```
In[105]:=
        MakeMatrix["a", 5, 6] // MatrixForm

Out[105]//MatrixForm=
        ( a[1, 1]  a[1, 2]  a[1, 3]  a[1, 4]  a[1, 5]  a[1, 6] )
        ( a[2, 1]  a[2, 2]  a[2, 3]  a[2, 4]  a[2, 5]  a[2, 6] )
        ( a[3, 1]  a[3, 2]  a[3, 3]  a[3, 4]  a[3, 5]  a[3, 6] )
        ( a[4, 1]  a[4, 2]  a[4, 3]  a[4, 4]  a[4, 5]  a[4, 6] )
        ( a[5, 1]  a[5, 2]  a[5, 3]  a[5, 4]  a[5, 5]  a[5, 6] )
```

See also: MakeYMatrix, MakeZMatrix, MakeBlockMatrix, ZeroMatrix, ConstantMatrix, InsertMatrixElement, WeylGroup, ModularGenerators, LongElement, WMatrix, SpecialWeylGroup.

■ MakeXMatrix (mxm)

This function returns a symbolic upper triangular square matrix of given dimension with 1s on the leading diagonal, i.e. a unipotent matrix.

See Definition 1.2.3.

MakeXMatrix[n, "x"] ⟶ u

n is a strictly positive integer representing the size of the matrix,
"x" is a string being the name of the generic symbolic matrix element variable
x[i, j],
u is an upper-triangular symbolic matrix with 1s on the leading diagonal.

Example

```
In[8]:= MatrixForm[MakeXMatrix[4, "x"]]

Out[8]//MatrixForm=
 1  x[1, 2]  x[1, 3]  x[1, 4]
 0     1     x[2, 3]  x[2, 4]
 0     0        1     x[3, 4]
 0     0        0        1

In[9]:= MakeXVariables[4, "x"]

Out[9]= {x[1, 2], x[1, 3], x[1, 4], x[2, 3], x[2, 4], x[3, 4]}

In[11]:= MatrixForm[MakeYMatrix[4, "y"]]

Out[11]//MatrixForm=
 y[1] y[2] y[3]       0         0     0
       0         y[1] y[2]      0     0
       0              0       y[1]    0
       0              0         0     1

In[12]:= MakeYVariables[4, "y"]

Out[12]= {y[1], y[2], y[3]}

In[24]:= MatrixForm[MakeZMatrix[4, "x", "y"]]

Out[24]//MatrixForm=
 y[1] y[2] y[3]  x[1, 2] y[1] y[2]  x[1, 3] y[1]  x[1, 4]
       0             y[1] y[2]      x[2, 3] y[1]  x[2, 4]
       0                 0              y[1]      x[3, 4]
       0                 0                0          1

        MakeZVariables[3, "x", "y"]

Out[25]= {x[1, 2], x[1, 3], x[1, 4], x[2, 3], x[2, 4], x[3, 4], y[1], y[2], y[3]}
```

See also: MakeXVariables, MakeYMatrix, MakeYVariables, MakeZMatrix, MakeZVariables.

■ MakeXVariables (mxv)

This function returns a list of the x-variables which appear in the symbolic generic Iwasawa form of a square matrix of given dimension.

See Definition 1.2.3.

MakeXVariables[n, "x"] ⟶ l

n is a strictly positive integer representing the size of the matrix,
"x" is a string being the name of the generic list element variable **x[i, j]**,
l is a list of the **x**-variables in order of increasing row index.

See also: MakeXMatrix, MakeYMatrix, MakeYVariables, MakeZMatrix.

■ MakeYMatrix (mym)

This function returns a symbolic diagonal matrix of given dimension with values on the leading diagonal being the product of the y-variables of a matrix expressed in Iwasawa form.

See Definition 1.2.3 and the manual entry for MakeXMatrix.

MakeYMatrix[n, "y"] ⟶ d

n is a strictly positive integer representing the size of the matrix,
"y" is a string being the name of the generic symbolic matrix element variable **y[i]** such that the **j**th diagonal element is the product **y[1]y[2]** ··· **y[n − j]**,
d is a diagonal matrix.

See also: MakeXMatrix, MakeXVariables, MakeYVariables, MakeZMatrix.

MakeYVariables (myv)

This function returns a symbolic list of the $n - 1$ y-variables which would occur in the Iwasawa form of a matrix of size $n \times n$.

See Definition 1.2.3 and the manual entry for MakeXMatrix.

MakeYVariables[n, "y"] ⟶ l

n is a strictly positive integer representing the size of the matrix,
"y" is a string being the name of the generic variable **y[i]**,
l is a list of the form {**y[1]**, ..., **y[n − 1]**}.

See also: MakeXMatrix, MakeXVariables, MakeYMatrix, MakeZMatrix, MakeZVariables.

■ MakeZMatrix (mzm)

This function returns a symbolic upper triangular square matrix of given dimension being in generic Iwasawa form.

See Example 1.2.4 and the manual entry for MakeXMatrix.

MakeZMatrix[n, "x","y"] ⟶ u

n is a strictly positive integer representing the size of the matrix,
"x" is a string being the name of the generic symbolic Iwasawa x-variable **x[i, j]**,
"y" is a string being the name of the generic symbolic Iwasawa y-variable **y[i]**,
u is an upper-triangular symbolic matrix with **(i, j)**th element having the form **x[i, j]y[1]** ··· **y[n − j]**.

See also: MakeXMatrix, MakeXVariables, MakeYMatrix, MakeYVariables, MakeZVariables.

■ MakeZVariables (mzv)

This function returns a list of the variables which occur in the Iwasawa form for a matrix with generic symbolic entries and of given size.

See the manual entry for MakeXMatrix.

MakeZVariables[n, "x", "y"] ⟶ l

n is a strictly positive integer representing the size of the matrix,
"x" is a string being the name of the generic symbolic matrix element **x[i, j]** with **i** > **j**,
"y" is a string being the name of the generic symbolic matrix element **y[i]**,
l is a list of the Iwasawa variables with the **x**-variables first in order of increasing row index followed by the **y**-variables:

$$\{\mathbf{x[1,2]}, \ldots, \mathbf{x[1,n]}, \mathbf{x[2,3]}, \ldots, \mathbf{x[n-1,n]}, \mathbf{y[1]}, \ldots, \mathbf{y[n-1]}\}.$$

See also: MakeXMatrix, MakeXVariables, MakeYMatrix, MakeYVariables, MakeZMatrix.

■ MatrixColumn (mcl)

This function returns a given column of a matrix.

MatrixColumn[m, j] ⟶ c

m is a matrix of CREs,
j is a valid column index for **m**,
c is the **j**th column of **m** returned as a list.

See also: MatrixRow.

■ MatrixDiagonal (mdl)

This function extracts the diagonal of a matrix.

MatrixDiagonal[a] ⟶ d

a is a square matrix of CREs,
d is a list, being the diagonal entries of **a** in the same order.

See also: DiagonalToMatrix.

■ MatrixJoinHorizontal (mjh)

This function assembles a new matrix by placing one matrix to the right of another compatible matrix.

MatrixJoinHorizontal[a, b] ⟶ c

a is a matrix of CREs,
b is a matrix with the same number of rows as **a**,
c is a matrix with block decomposition **c** = [**a**|**b**].

Example

```
In[285]:= A = {{a, b, c}, {d, e, f}}; B = {{1, 1}, {2, 2}};

In[292]:= MatrixForm[MatrixJoinHorizontal[A, B]]

Out[292]//MatrixForm=
        ( a  b  c  1  1 )
        ( d  e  f  2  2 )
```

See also: MatrixJoinVertical.

■ MatrixJoinVertical (mjv)

This function assembles a new matrix by placing one matrix above another compatible matrix.

MatrixJoinVertical[a, b] ⟶ c

a is a matrix of CREs,
b is a matrix with the same number of columns as **a**,
c is a matrix with block decomposition having **a** above **b**.

See also: MatrixJoinHorizontal.

■ MatrixLowerTriangle (mlt)

This function extracts the elements in the lower triangle of a square matrix, including the diagonal, and returns them as a list of lists.

MatrixLowerTriangle[a] ⟶ t

a is a square matrix of CREs,
t is a list of lists where the **i**th element of the **j**th list represents the (**j**, **i**)th element of **a**.

See also: MatrixUpperTriangle, LowerTriangleToMatrix.

■ MatrixRow (mro)

This function returns a given row of a matrix.

MatrixRow[m, i] ⟶ r

m is a matrix of CREs,
i is a row index of **m**,
r is a list representing the **i**th row of **m**.

See also: MatrixColumn.

■ MatrixUpperTriangle (mut)

This function extracts the elements in the upper triangle, including the diagonal, of a square matrix and returns a list of lists of the elements from each row.

MatrixUpperTriangle[a] ⟶ t

a is a square matrix of CREs,
t is a list of lists of elements with the **i**th element of the **j**th list being the (**j**, **i** + **j** − **1**)th element of **a**.

Example

```
In[303]:= A = {{a, b, c}, {d, e, f}, {g, h, i}};

In[316]:= MatrixForm[A]

Out[316]//MatrixForm=
  a b c
  d e f
  g h i

In[320]:= MatrixUpperTriangle[A]

Out[320]= {{a, b, c}, {e, f}, {i}}
```

See also: MatrixLowerTriangle, UpperTriangleToMatrix.

■ ModularGenerators (mog)

This function returns a list of two matrix generators for the subgroup of the group of integer matrixes with determinant 1, i.e. generators of $SL(n, \mathbb{Z})$.

See Chapter 5, especially Section 5.9.

ModularGenerators[n] ⟶ g

n is a positive integer with **n** $\geq$ **2**,
g is a list of two **n** by **n** matrices which will generate **SL(n,** $\mathbb{Z}$**)**.

Example

```
In[174]:= Map[MatrixForm, ModularGenerator[6]]
```

$$\text{Out[174]=}\ \left\{\begin{pmatrix} 0 & 0 & 0 & 0 & 0 & -1 \\ 1 & 0 & 0 & 0 & 0 & 0 \\ 0 & 1 & 0 & 0 & 0 & 0 \\ 0 & 0 & 1 & 0 & 0 & 0 \\ 0 & 0 & 0 & 1 & 0 & 0 \\ 0 & 0 & 0 & 0 & 1 & 0 \end{pmatrix}, \begin{pmatrix} 1 & 1 & 0 & 0 & 0 & 0 \\ 0 & 1 & 0 & 0 & 0 & 0 \\ 0 & 0 & 1 & 0 & 0 & 0 \\ 0 & 0 & 0 & 1 & 0 & 0 \\ 0 & 0 & 0 & 0 & 1 & 0 \\ 0 & 0 & 0 & 0 & 0 & 1 \end{pmatrix}\right\}$$

See also: WeylGenerator, WeylGroup, SpecialWeylGroup, WMatrix, Long-Element.

■ MPEisensteinGamma (eig)

This function computes the gamma factors for the minimal parabolic Eisenstein series

$$G_{E_v}(s) = \pi^{-ns/2} \prod_{i=1}^{n} \Gamma\left(\frac{s-\lambda_i(v)}{2}\right).$$

See Chapter 10, Theorem 10.8.6.

MPEisensteinGamma[s,v] ⟶ G

s is a CRE representing a complex number,
v is a list of **n** − **1** CREs with **n** $\geq$ **2** representing complex parameters,
G is the gamma factor for the minimal parabolic Eisenstein series functional equation.

Example

```
In[5]:= suf[MPEisensteinGamma[s, {v[1], v[2]}]]
```

$$\text{Out[5]=}\ \pi^{-3\,s/2}\,\text{Gamma}\left[\frac{1}{2}\,(1+s-2\,v_1-v_2)\right] \text{Gamma}\left[\frac{1}{2}\,(s+v_1-v_2)\right]\,\text{Gamma}\left[\frac{1}{2}\,(-1+s+v_1+2\,v_2)\right]$$

See also: MPEisensteinLambdas, MPEisensteinSeries, MPExteriorPower-Gamma, MPExteriorPowerLFun, MPSymmetricPowerLFun, MPSymmetric-PowerGamma.

■ MPEisensteinLambdas (eil)

This function computes the functions $\lambda_i(v) : \mathbb{C}^{n-1} \to \mathbb{C}$ such that the L-function associated with the minimal parabolic Eisenstein series $L_{E_v}(s)$ is a product of shifted zeta values

$$L_{E_v}(s) = \prod_{i=1}^{n} \zeta(s - \lambda_i(v)).$$

See Chapter 10, (10.4.1) and Theorem 10.8.6.

MPEisensteinLambdas[v] ⟶ L

v is a list of **n − 1** CREs with **n ≥ 2** representing complex parameters,
L is a list of affine expressions in the elements of **v** representing the functions $\lambda_i(\mathbf{v})$.

Example

```
In[10]:=  suf[MPEisensteinLambdas[{v[1], v[2], v[3]}]]
```

Out[10]= $\left\{-\frac{3}{2} + 3\,v_1 + 2\,v_2 + v_3,\ -\frac{1}{2} - v_1 + 2\,v_2 + v_3,\ \frac{1}{2} - v_1 - 2\,v_2 + v_3,\ \frac{3}{2} - v_1 - 2\,v_2 - 3\,v_3\right\}$

See also: MPEisensteinSeries, MPEisensteinGamma, MPExteriorPower-Gamma, MPExteriorPowerLFun, MPSymmetricPowerLFun, MPSymmetric-PowerGamma.

■ MPEisensteinSeries (eis)

This function computes the L-function associated with the minimal parabolic Eisenstein series $E_v(z)$ as a product of shifted zeta values

$$L_{E_v}(s) = \prod_{i=1}^{n} \zeta(s - \lambda_i(v)).$$

See Chapter 10, (10.4.1) and Theorem 10.8.6.

MPEisensteinSeries[s,v] ⟶ Z

s is a CRE representing a complex number,
v is a list of **n − 1** CREs with **n ≥ 2** representing complex parameters,
Z is a product of **n** values of the Riemann zeta function at shifted arguments.

Example

```
In[4]:= suf[MPEisensteinSeries[s, {v[1], v[2]}]]

Out[4]= Zeta[1 + s - 2 v1 - v2] Zeta[s + v1 - v2] Zeta[-1 + s + v1 + 2 v2]
```

See also: MPEisensteinLambdas, MPEisensteinGamma, MPExteriorPowerGamma, MPExteriorPowerLFun, MPSymmetricPowerLFun, MPSymmetricPowerGamma.

■ MPExteriorPowerGamma (epg)

This function returns the gamma factors for the *k*th symmetric L-function associated with a minimal parabolic Eisenstein series.

See the introduction to Chapter 13.

MPExteriorPowerGamma[s,v,k] ⟶ G

s is a CRE representing a complex number,
v is a list of $\mathbf{n-1}$ CREs with $\mathbf{n \geq 2}$ representing complex parameters,
k is a natural number $\mathbf{k \geq 1}$ representing the order of the exterior power,
G is the gamma factor for the functional equation of the exterior power.

Example

```
In[9]:= MPExteriorPowerGamma[s, {v1, v2}, 2]
```

Out[9]= $\pi^{-3s/2}\,\text{Gamma}\left[\frac{1}{2}(1+s-v1-2\,v2)\right]\,\text{Gamma}\left[\frac{1}{2}(s-v1+v2)\right]\,\text{Gamma}\left[\frac{1}{2}(-1+s+2\,v1+v2)\right]$

See also: MPEisensteinLambdas, MPEisensteinSeries, MPEisensteinGamma, MPExteriorPowerLFun, MPSymmetricPowerLFun, MPSymmetricPowerGamma.

■ MPExteriorPowerLFun (epl)

This function returns the *k*th exterior power of the L-function of a minimal parabolic Eisenstein series as a product of zeta values.

See the introduction to Chapter 13.

This function can be used to show that exterior power L-functions satisfy a functional equation.

MPExteriorPowerLFun[s,v,k] ⟶ Z

s is a CRE representing a complex number,
v is a list of **n − 1** CREs with **n ≥ 2** representing complex parameters,
k is a natural number **k ≥ 1** representing the order of the exterior power,
Z is a product of Riemann zeta function values.

Example

```
In[7]:=  suf[MPExteriorPowerLFun[s, {v[1], v[2]}, 2]]

Out[7]=  Zeta[1 + s - v1 - 2 v2] Zeta[s - v1 + v2] Zeta[-1 + s + 2 v1 + v2]
```

See also: MPEisensteinLambdas, MPEisensteinSeries, MPEisensteinGamma, MPExteriorPowerGamma, MPSymmetricPowerLFun, MPSymmetricPowerGamma.

■ MPSymmetricPowerLFun (spf)

This function returns the *k*th symmetric power of the L-function of a minimal parabolic Eisenstein series as a product of zeta values.

See the introduction to Chapter 13.

This can be used to show that symmetric power L-functions satisfy a functional equation.

MPSymmetricPowerLFun[s,v,k] ⟶ Z

s is a CRE representing a complex number,
v is a list of **n − 1** CREs with **n ≥ 2** representing complex parameters,
k is a natural number **k ≥ 1** representing the order of the exterior power,
Z is a product of Riemann zeta function values.

Example

```
In[6]:=  suf[MPSymmetricPowerLFun[s, {v[1], v[2]}, 3]]

Out[6]=  Zeta[s] Zeta[1 + s - 3 v1] Zeta[-1 + s + 3 v1] Zeta[1 + s - 3 v2]
          Zeta[3 + s - 6 v1 - 3 v2] Zeta[2 + s - 3 v1 - 3 v2] Zeta[s + 3 v1 - 3 v2]
          Zeta[-1 + s + 3 v2] Zeta[-2 + s + 3 v1 + 3 v2] Zeta[-3 + s + 3 v1 + 6 v2]
```

See also: MPEisensteinLambdas, MPEisensteinSeries, MPEisensteinGamma, MPExteriorPowerGamma, MPExteriorPowerLFun, MPSymmetricPowerGamma.

■ MPSymmetricPowerGamma (spg)

This function returns the gamma factors for the kth symmetric L-function associated with a mimimal parabolic Eisenstein series.

See the introduction to Chapter 13.

MPSymmetricPowerGamma[s,v,k] $\longrightarrow$ **G**

s is a CRE representing a complex number,
v is a list of **n − 1** CREs with **n ≥ 2** representing complex parameters,
k is a natural number **k ≥ 1** representing the order of the exterior power,
G is the gamma factors for the **k**th symmetric power L-function.

Example

```
In[8]:=  suf[MPSymmetricPowerGamma[s, {v[1], v[2]}, 3]]
```

Out[8]=
$$\pi^{-5s}\,\mathrm{Gamma}\left[\frac{s}{2}\right]\mathrm{Gamma}\left[\frac{1}{2}(1+s-3v_1)\right]$$
$$\mathrm{Gamma}\left[\frac{1}{2}(-1+s+3v_1)\right]\mathrm{Gamma}\left[\frac{1}{2}(1+s-3v_2)\right]$$
$$\mathrm{Gamma}\left[\frac{1}{2}(3+s-6v_1-3v_2)\right]\mathrm{Gamma}\left[\frac{1}{2}(2+s-3v_1-3v_2)\right]$$
$$\mathrm{Gamma}\left[\frac{1}{2}(s+3v_1-3v_2)\right]\mathrm{Gamma}\left[\frac{1}{2}(-1+s+3v_2)\right]$$
$$\mathrm{Gamma}\left[\frac{1}{2}(-2+s+3v_1+3v_2)\right]\mathrm{Gamma}\left[\frac{1}{2}(-3+s+3v_1+6v_2)\right]$$

See also: MPEisensteinLambdas, MPEisensteinSeries, MPEisensteinGamma, MPExteriorPowerGamma, MPExteriorPowerLFun, MPSymmetricPowerLFun.

■ NColumns (ncl)

This function gives the number of columns of a matrix.

NColumns[a] $\longrightarrow$ **n**

a is a matrix of CREs,
n is the number of columns of **a**.

See also: NRows.

■ NRows (nro)

This function gives the number of rows of a matrix.

NRows[a] ⟶ m

a is a matrix of CREs,
m is the number of rows of **a**.

See also: NColumns.

■ ParabolicQ (paq)

This function tests a square matrix to see whether it is in a given parabolic subgroup as specified by a non-trivial partition of the matrix dimension.
See Chapter 10, especially Section 10.1.

ParabolicQ[a, d] ⟶ ans

a is an **n** × **n** square matrix with entries which are CREs,
d is list of at most **n** positive integers with sum **n**,
ans is **True** if **a** is in the specified subgroup and **False** otherwise.

Example

```
In[1]:= d = {2, 2}; a = {{1, 2, 3, 4}, {5, 6, 7, 8}, {0, 0, 1, 2}, {0, 0, 3, 4}}

Out[1]= {{1, 2, 3, 4}, {5, 6, 7, 8}, {0, 0, 1, 2}, {0, 0, 3, 4}}

In[2]:= MatrixForm[a]

Out[2]//MatrixForm=
```

$$\begin{pmatrix} 1 & 2 & 3 & 4 \\ 5 & 6 & 7 & 8 \\ 0 & 0 & 1 & 2 \\ 0 & 0 & 3 & 4 \end{pmatrix}$$

```
In[3]:= ParabolicQ[a, d]

Out[3]= True

In[57]:= a2 = {{1, 2, 3, 4}, {5, 6, 7, 8}, {0, 0, 1, 2}, {x, 0, 3, 4}}

Out[57]= {{1, 2, 3, 4}, {5, 6, 7, 8}, {0, 0, 1, 2}, {x, 0, 3, 4}}

In[58]:= ParabolicQ[a1, d]

Out[58]= False
```

See also: LanglandsForm, LanglandsIFun.

■ PluckerCoordinates (plc)

This function takes an $\mathbf{n} \times \mathbf{n}$ square matrix and returns a list of lists of the so-called Plücker coordinates, namely the values of all of the bottom $\mathbf{j} \times \mathbf{j}$ minors with $\mathbf{1} \le \mathbf{j} \le \mathbf{n} - \mathbf{1}$.

See Chapter 11, Section 11.3, Theorem 11.3.1.

PluckerCoordinates[a] ⟶ value

a is an $\mathbf{n} \times \mathbf{n}$ matrix of CREs,
value is a list of lists being the values of all of the $\mathbf{j} \times \mathbf{j}$ minor determinants with $\mathbf{1} \le \mathbf{j} \le \mathbf{n} - \mathbf{1}$ based on the bottom row and taking elements from the bottom $\mathbf{j}$ rows. The $\mathbf{j}$th sublist has the $\mathbf{j} \times \mathbf{j}$ minors in lexical order of the column indices.

Example

```
In[31]:= PluckerCoordinates[{{1, 2}, {4 x, 5 y}}]

Out[31]= {{4 x, 5 y}}

In[32]:= suf[PluckerCoordinates[MakeMatrix["x", 3, 3]]]
```

Out[32]= $\{\{x_{3,1}, x_{3,2}, x_{3,3}\},$
$\{-x_{2,2}\, x_{3,1} + x_{2,1}\, x_{3,2}, -x_{2,3}\, x_{3,1} + x_{2,1}\, x_{3,3}, -x_{2,3}\, x_{3,2} + x_{2,2}\, x_{3,3}\}\}$

```
In[33]:= m = {{12, 3, 4, -1, 7}, {3, 0, 2, 1, 0}, {4, 5, 6, 7, 0},
            {0, 2, 19, 3, 1}, {1, 2, 3, 4, 5}}; MatrixForm[m]

Out[33]//MatrixForm=
```

$$\begin{pmatrix} 12 & 3 & 4 & -1 & 7 \\ 3 & 0 & 2 & 1 & 0 \\ 4 & 5 & 6 & 7 & 0 \\ 0 & 2 & 19 & 3 & 1 \\ 1 & 2 & 3 & 4 & 5 \end{pmatrix}$$

```
In[34]:= plc[m]

Out[34]= {{1, 2, 3, 4, 5}, {-2, -19, -3, -1, -32, 2, 8, 67, 92, 11},
          {-45, 9, 37, 153, 374, 51, 99, 412, -1, -578},
          {360, 1310, 34, -1462, 414}}
```

See also: PluckerRelations, PluckerInverse, KloostermanSum.

■ PluckerInverse (pli)

This function takes a list of lists of integers, which could be the Plücker coordinates arising from a square matrix, and returns such a matrix having determinant 1. The matrix is not unique but **PluckerInverse** followed by

PluckerCoordinates gives the identity, provided the list of lists of integers is compatible, i.e. arises from some matrix.

See Section 11.3.

PluckerInverse[Ms] ⟶ a

Ms is a list of $\mathbf{n-1}$ sublists of integers $\mathbf{Ms} = \{\{\mathbf{M_1}, \ldots, \mathbf{M_n}\}, \{\mathbf{M_{12}}, \ldots\}, \ldots\}$, representing the Plücker coordinates of a matrix in lexical order,

a is an integer matrix having those Plücker coordinates or **False** in case they are incompatible.

Example

```
In[2162]:=
        g = ModularGenerators[4]; a = g[[1]]; b = g[[2]];
        m = a.a.a.b.b.b.b.a.b.b.b.a.b.b.a.a.a.b.a.a.a.b.b.b.a.a.b.a.a.a.a.a.a.b;
        MatrixForm[m]

Out[2163]//MatrixForm=
        ( 0  -1  -4  -10 )
        ( 0   0  -1   -3 )
        ( 0   0   0   -1 )
        ( 1   8  16   40 )

In[2164]:=
        Ms = plc[m]

Out[2164]=
        {{1, 8, 16, 40}, {0, 0, 1, 0, 8, 16}, {0, 0, 1, 8}}

In[2165]:=
        a = pli[Ms]; MatrixForm[a]

Out[2165]//MatrixForm=
        ( 0  -1   0   0 )
        ( 0   0  -1   0 )
        ( 0   0   0  -1 )
        ( 1   8  16  40 )

In[2166]:=
        plc[a]

Out[2166]=
        {{1, 8, 16, 40}, {0, 0, 1, 0, 8, 16}, {0, 0, 1, 8}}
```

See also: PluckerCoordinates, PluckerRelations, KloostermanSum, KloostermanCompatibility.

■ PluckerRelations (plr)

This function computes all the known quadratic relationships between the minors of a generic square $n \times n$ matrix known as the Plücker coordinates.

See Chapter 11.

In the case that $n = 2$ there are none and for $n = 3$ one. For **n > 3** the number grows dramatically. No claim is made that this function returns, for any given n, a complete set of independent relationships. By "complete" is meant sufficient to guarantee the coordinates arise from a member of $SL(n, \mathbb{Z})$.

PluckerRelations[n, v] ⟶ relations

n is a positive integer with $\mathbf{n \geq 2}$,

v is a symbol representing the generic name used for the Plücker coordinates so $\mathbf{v[\{i_1, \ldots, i_j\}]}$ is the matrix minor based on the last **j** rows and the columns indexed by $\mathbf{i_1, \ldots, i_j}$ with these indices in strictly increasing order,

relations is a list of quadratic expressions with coefficients $\pm\mathbf{1}$ in the $\mathbf{v[\{\cdots\}]}$s, which vanish whenever the values of the $\mathbf{v[\{\cdots\}]}$s come from the minors of an $\mathbf{n \times n}$ matrix.

Example

```
In[1]:= PluckerRelations[3, v]

Out[1]= {v[{3}] v[{1, 2}] - v[{2}] v[{1, 3}] + v[{1}] v[{2, 3}]}

In[2]:= PluckerRelations[4, v]

Out[2]= {v[{3}] v[{1, 2}] - v[{2}] v[{1, 3}] + v[{1}] v[{2, 3}],
         v[{4}] v[{1, 2}] - v[{2}] v[{1, 4}] + v[{1}] v[{2, 4}],
         v[{4}] v[{1, 3}] - v[{3}] v[{1, 4}] + v[{1}] v[{3, 4}],
         v[{4}] v[{2, 3}] - v[{3}] v[{2, 4}] + v[{2}] v[{3, 4}],
         v[{1, 4}] v[{2, 3}] - v[{1, 3}] v[{2, 4}] + v[{1, 2}] v[{3, 4}],
         v[{1, 4}] v[{1, 2, 3}] - v[{1, 3}] v[{1, 2, 4}] +
          v[{1, 2}] v[{1, 3, 4}], v[{4}] v[{1, 2, 3}] - v[{3}] v[{1, 2, 4}] +
          v[{2}] v[{1, 3, 4}] - v[{1}] v[{2, 3, 4}], v[{2, 4}] v[{1, 2, 3}] -
          v[{2, 3}] v[{1, 2, 4}] + v[{1, 2}] v[{2, 3, 4}],
         v[{3, 4}] v[{1, 2, 3}] - v[{2, 3}] v[{1, 3, 4}] +
          v[{1, 3}] v[{2, 3, 4}], v[{3, 4}] v[{1, 2, 4}] -
          v[{2, 4}] v[{1, 3, 4}] + v[{1, 4}] v[{2, 3, 4}]}

In[3]:= Map[Function[n, Length[PluckerRelations[n, v]]],
          {2, 3, 4, 5, 6, 7}]

Out[3]= {0, 1, 10, 47, 160, 458}
```

See also: PluckerCoordinates, PluckerInverse, KloostermanSum.

■ RamanujanSum (rsm)

This function computes the Ramanujan sum $s(n, c)$ for explicit natural number values of n, c, namely

$$s(n, c) = \sum_{r=1,(r,c)=1}^{c} e^{2\pi i(r/c)}.$$

See Definition 3.1.4 and Proposition 3.1.7.

RamanujanSum[n,c] ⟶ s

n is a strictly positive integer, **c** is a strictly positive integer,
s is an integer being the Ramanujan sum.

Example

```
In[27]:= Table[RamanujanSum[n, 1] - MoebiusMu[n], {n, 1, 10}]

Out[27]= {0, 0, 0, 0, 0, 0, 0, 0, 0, 0}

In[31]:= Table[RamanujanSum[n, 23 n] / EulerPhi[n], {n, 1, 20}]

Out[31]= {1, 1, 1, 1, 1, 1, 1, 1, 1, 1, 1, 1, 1, 1, 1, 1, 1, 1, 1, 1}
```

■ RemoveMatrixColumn (rmc)

A given row is removed from a matrix, creating a new matrix and leaving the original unchanged.

RemoveMatrixColumn[a, j] ⟶ b

a is a matrix of CREs,
j is a valid column index of **a**,
b is a matrix with all columns identical to **a** except the **j**th which is missing.

See also: RemoveMatrixRow.

■ RemoveMatrixRow (rmr)

A given row is removed from a matrix, leaving the original unchanged.

RemoveMatrixRow[a, i] ⟶ b

a is a matrix of CREs,
i is a valid row index of **a**,
b is a matrix with all rows identical to **a** except the **i**th which is missing.

See also: RemoveMatrixColumn.

SchurPolynomial (spl)

This function computes the Schur polynomial in n variables $x_1, \ldots, x_n$ with $n-1$ exponents $k_1, \ldots, k_{n-1}$, that is to say the ratio of the determinant of a matrix with (i, j)th element 1 for $i = n$ and $x_j^{k_1+\cdots+k_{i-1}+n-i}$ for $1 \leq i \leq n-1$, to the determinant of the matrix which is 1 for $i = n$ and x_j^{n-i} for $1 \leq i \leq n-1$.

See Section 7.4.

SchurPolynomial[x, k] ⟶ $\mathbf{S_k(x_1, \ldots, x_n)}$

x is list of **n** CREs,
k is a list of **n − 1** CREs,
$\mathbf{S_k(x_1, \ldots, x_n)}$ is the Schur polynomial.

Example

```
In[4]:= SchurPolynomial[{x, y, z}, {a, b}]
```

$$\text{Out[4]=}\ \frac{x^{2+a+b}\left(y^{1+a}-z^{1+a}\right)+y^{1+a}z^{1+a}\left(y^{1+b}-z^{1+b}\right)+x^{1+a}\left(-y^{2+a+b}+z^{2+a+b}\right)}{(x-y)\ (x-z)\ (y-z)}$$

```
In[5]:= SchurPolynomial[{1, x, x^2, x^3}, {2, 2, 2}]
```

$$\text{Out[5]=}\ x^8\left(1-x+x^2\right)^2\left(1+x+x^2\right)^5\left(1+x^3+x^6\right)$$

```
In[6]:= SchurPolynomial[{1, x^2, x, x^3}, {1, 2, 3}]
```

$$\text{Out[6]=}\ x^5\left(1+x+x^2\right)\left(1+2x+4x^2+7x^3+10x^4+13x^5+17x^6+19x^7+21x^8+22x^9+21x^{10}+19x^{11}+17x^{12}+13x^{13}+10x^{14}+7x^{15}+4x^{16}+2x^{17}+x^{18}\right)$$

See also: HeckeMultiplicativeSplit.

SmithElementaryDivisors (sed)

This function computes the elementary divisors of a non-singular $n \times n$ integer matrix a, i.e. for each j with $1 \leq j \leq n$, the gcd $d_j(a)$ of all of the $j \times j$ minor determinants. If s_j is the jth diagonal entry of the Smith form then $s_j = d_j(a)/d_{j-1}(a)$.

SmithElementaryDivisors[a] ⟶ l

a is a non-singular **n × n** integer matrix,
l is a list of the **n** Smith form elementary divisors of **a** in the order $\{\mathbf{d_1(a)}, \ldots, \mathbf{d_n(a)}\}$.

See also: SmithForm, SmithInvariantFactors, HermiteFormUpper, HermiteFormLower.

■ SmithForm (smf)

This function returns the Smith form diagonal matrix d of a square non-singular matrix a with integer entries. This matrix d satisfies $0 < d_{i,i}$ and $d_{i,i} \mid d_{i+1,i+1}$ for all $i \leq n-1$. It also returns unimodular matrixes l, r such that $a = l.d.r$.

See Theorem 3.11.2.

SmithForm[a] $\longrightarrow$ **{l, d, r}**

a is non-singular integer matrix,
l is a unimodular matrix,
d is a diagonal matrix, being the Smith Form of **a**, **r** is a unimodular matrix.

Example

The Smith form of a 4 by 4 matrix is computed and the result checked.

```
In[4]:=  m4 = {{5, 2, -4, 7}, {1, 6, 0, -3}, {1, 2, -2, 4}, {7, 1, 5, 6}};

         {l, s, r} = SmithForm[m4]

         {{{5, -4, 186, -23}, {1, 0, 0, 0}, {1, -2, 89, -11}, {7, 5, -178, 22}},
          {{1, 0, 0, 0}, {0, 1, 0, 0}, {0, 0, 1, 0}, {0, 0, 0, 964}},
          {{1, 6, 0, -3}, {0, -49, 1, 41}, {0, 118, 0, 1}, {0, 1, 0, 0}}}

         Map[MatrixForm, %]
```

$$\left\{\begin{pmatrix} 5 & -4 & 186 & -23 \\ 1 & 0 & 0 & 0 \\ 1 & -2 & 89 & -11 \\ 7 & 5 & -178 & 22 \end{pmatrix}, \begin{pmatrix} 1 & 0 & 0 & 0 \\ 0 & 1 & 0 & 0 \\ 0 & 0 & 1 & 0 \\ 0 & 0 & 0 & 964 \end{pmatrix}, \begin{pmatrix} 1 & 6 & 0 & -3 \\ 0 & -49 & 1 & 41 \\ 0 & 118 & 0 & 1 \\ 0 & 1 & 0 & 0 \end{pmatrix}\right\}$$

```
         l.s.r - m4

         {{0, 0, 0, 0}, {0, 0, 0, 0}, {0, 0, 0, 0}, {0, 0, 0, 0}}
```

See also: SmithElementaryDivisors, SmithInvariantFactors, HermiteForm-Lower, HermiteFormUpper.

■ SmithInvariantFactors (sif)

This function computes the invariant factors of the Smith form of a non-singular integer matrix a. These are all of the prime powers which appear in the diagonal entries of the Smith form of a.

SmithInvariantFactors[a] $\longrightarrow$ **l**

a is an $n \times n$ non-singular integer matrix,
l is a list of prime powers.

See also: SmithForm, SmithElementaryDivisors, HermiteFormUpper, HermiteFormLower.

■ SpecialWeylGroup (swg)

This function, for each natural number n, returns the group of $n \times n$ matrices with each entry being 0 or ± 1, and having determinant 1. There are $2^{n-1} n!$ such matrices.

See Sections 6.3 and 6.5.

SpecialWeylGroup[n] $\longrightarrow$ g

n is a natural number representing the matrix dimension,
g is a list of **n** $\times$ **n** matrices representing the Weyl group.

Example

```
In[3]:=  w = SpecialWeylGroup[3];

In[4]:=  Map[MatrixForm, w]
```

Out[4]=

$$\left\{ \begin{pmatrix} 1 & 0 & 0 \\ 0 & 1 & 0 \\ 0 & 0 & 1 \end{pmatrix}, \begin{pmatrix} 1 & 0 & 0 \\ 0 & 0 & 1 \\ 0 & -1 & 0 \end{pmatrix}, \begin{pmatrix} 1 & 0 & 0 \\ 0 & -1 & 0 \\ 0 & 0 & -1 \end{pmatrix}, \begin{pmatrix} 1 & 0 & 0 \\ 0 & 0 & -1 \\ 0 & 1 & 0 \end{pmatrix}, \right.$$
$$\begin{pmatrix} 0 & 1 & 0 \\ 1 & 0 & 0 \\ 0 & 0 & -1 \end{pmatrix}, \begin{pmatrix} 0 & 1 & 0 \\ 0 & 0 & 1 \\ 1 & 0 & 0 \end{pmatrix}, \begin{pmatrix} 0 & 1 & 0 \\ -1 & 0 & 0 \\ 0 & 0 & 1 \end{pmatrix}, \begin{pmatrix} 0 & 1 & 0 \\ 0 & 0 & -1 \\ -1 & 0 & 0 \end{pmatrix},$$
$$\begin{pmatrix} 0 & 0 & 1 \\ 1 & 0 & 0 \\ 0 & 1 & 0 \end{pmatrix}, \begin{pmatrix} 0 & 0 & 1 \\ 0 & 1 & 0 \\ -1 & 0 & 0 \end{pmatrix}, \begin{pmatrix} 0 & 0 & 1 \\ -1 & 0 & 0 \\ 0 & -1 & 0 \end{pmatrix}, \begin{pmatrix} 0 & 0 & 1 \\ 0 & -1 & 0 \\ 1 & 0 & 0 \end{pmatrix},$$
$$\begin{pmatrix} -1 & 0 & 0 \\ 0 & 1 & 0 \\ 0 & 0 & -1 \end{pmatrix}, \begin{pmatrix} -1 & 0 & 0 \\ 0 & 0 & 1 \\ 0 & 1 & 0 \end{pmatrix}, \begin{pmatrix} -1 & 0 & 0 \\ 0 & -1 & 0 \\ 0 & 0 & 1 \end{pmatrix}, \begin{pmatrix} -1 & 0 & 0 \\ 0 & 0 & -1 \\ 0 & -1 & 0 \end{pmatrix},$$
$$\begin{pmatrix} 0 & -1 & 0 \\ 1 & 0 & 0 \\ 0 & 0 & 1 \end{pmatrix}, \begin{pmatrix} 0 & -1 & 0 \\ 0 & 0 & 1 \\ -1 & 0 & 0 \end{pmatrix}, \begin{pmatrix} 0 & -1 & 0 \\ -1 & 0 & 0 \\ 0 & 0 & -1 \end{pmatrix}, \begin{pmatrix} 0 & -1 & 0 \\ 0 & 0 & -1 \\ 1 & 0 & 0 \end{pmatrix},$$
$$\left. \begin{pmatrix} 0 & 0 & -1 \\ 1 & 0 & 0 \\ 0 & -1 & 0 \end{pmatrix}, \begin{pmatrix} 0 & 0 & -1 \\ 0 & 1 & 0 \\ 1 & 0 & 0 \end{pmatrix}, \begin{pmatrix} 0 & 0 & -1 \\ -1 & 0 & 0 \\ 0 & 1 & 0 \end{pmatrix}, \begin{pmatrix} 0 & 0 & -1 \\ 0 & -1 & 0 \\ -1 & 0 & 0 \end{pmatrix} \right\}$$

```
In[5]:=  Map[Det, w]

Out[5]=  {1, 1, 1, 1, 1, 1, 1, 1, 1, 1, 1, 1, 1, 1, 1, 1, 1, 1, 1, 1, 1, 1, 1, 1}

In[7]:=  Length[SpecialWeylGroup[5]] - 2^4 5!

Out[7]=  0
```

See also: WeylGroup, WMatrix, WeylGenerator, ModularGenerators, LongElement.

■ SubscriptedForm (suf)

This function takes a Mathematica expression and prints it out in such a way that subexpressions of the form $\mathbf{x}[\mathbf{n}_1, \mathbf{n}_2, \ldots, \mathbf{n}_j]$, where the $\mathbf{n}_i$ are explicit integers, are printed in the style

$$\mathbf{x}_{\mathbf{n}_1,\mathbf{n}_2,\ldots,\mathbf{n}_j}.$$

The value of this function is for improving the look of expressions for inspection and should not be used otherwise. Compare the Mathematica function MatrixForm. Not all expressions can be subscripted using this function.

SubscriptedForm[e] ⟶ f

e is a Mathematica expression,
f is a subscripted rendition of the same expression.

Example

```
In[2395]:=
        a = MakeMatrix["x", 4, 4]

Out[2395]=
        {{x[1, 1], x[1, 2], x[1, 3], x[1, 4]}, {x[2, 1], x[2, 2], x[2, 3], x[2, 4]},
         {x[3, 1], x[3, 2], x[3, 3], x[3, 4]}, {x[4, 1], x[4, 2], x[4, 3], x[4, 4]}}

In[2396]:=
        SubscriptedForm[a]

Out[2396]=
```

$$\{\{x_{1,1}, x_{1,2}, x_{1,3}, x_{1,4}\}, \{x_{2,1}, x_{2,2}, x_{2,3}, x_{2,4}\}, \{x_{3,1}, x_{3,2}, x_{3,3}, x_{3,4}\}, \{x_{4,1}, x_{4,2}, x_{4,3}, x_{4,4}\}\}$$

```
In[2397]:=
        MatrixForm[%]

Out[2397]//MatrixForm=
```

$$\begin{pmatrix} x_{1,1} & x_{1,2} & x_{1,3} & x_{1,4} \\ x_{2,1} & x_{2,2} & x_{2,3} & x_{2,4} \\ x_{3,1} & x_{3,2} & x_{3,3} & x_{3,4} \\ x_{4,1} & x_{4,2} & x_{4,3} & x_{4,4} \end{pmatrix}$$

```
In[2398]:=
        b = x[0, 2, -4] + f[y[1] - 3 * y[2]] / (x[1, 2, 3] + y[3]);

In[2399]:=
        SubscriptedForm[b]

Out[2399]=
```

$$x_{0,2,-4} + \frac{f[y_1 - 3\,y_2]}{y_3 + x_{1,2,3}}$$

■ SwapMatrixColumns (smc)

Two columns of a matrix are exchanged creating a new matrix and leaving the original unchanged.

SwapMatrixColumns[a, i, j] ⟶ b

a is a matrix of CREs,
i is a valid column index for **a**,
j is a valid column index for **a**,
b is a matrix equal to **a** except the **i**th and **j**th columns have been exchanged.

See also: SwapMatrixRows, ElementaryMatrix.

■ SwapMatrixRows (smr)

Two rows of a matrix are exchanged creating a new matrix and leaving the original unchanged.

SwapMatrixRows[a, i, j] ⟶ b

a is a matrix of CREs,
i is a valid row index for **a**,
j is a valid row index for **a**,
b is a matrix equal to **a** except the **i**th and **j**th rows have been exchanged.

See also: SwapMatrixColumns, ElementaryMatrix.

■ TailingMatrixBlock (tmb)

This function returns a tailing matrix block of specified dimensions leaving the original matrix unchanged.

TailingMatrixBlock[a, i, j] ⟶ b

a is a matrix of CREs,
i is a positive integer less than the number of rows of **a**,
j is a positive integer less than the number of columns of **a**,
b is the tailing block of **a** with **i** rows and **j** columns.

See also: LeadingMatrixBlock, BlockMatrix.

■ UpperTriangleToMatrix (utm)

This function takes a list of lists of strictly decreasing length and forms a matrix with zeros in the lower triangle and with the given lists constituting the rows of the upper triangle. The length of the matrix is the length of the first sublist. The last sublist has length 1 and each successive sublist has length one less than the preceding sublist.

UpperTriangleToMatrix[u] ⟶ a

u is a list of lists of CREs of decreasing length representing the elements of an upper triangular submatrix including the diagonal,

a is a full matrix with 0 in each lower-triangular position.

See also: MatrixUpperTriangle, LowerTriangleToMatrix.

■ VolumeBall (vbl)

This function computes the volume of an n-dimensional ball with given radius.

VolumeBall[r, n] ⟶ Vol

r is a CRE representing the radius of the ball,
n is a positive integer, being the dimension of the ball,
Vol is the **n**-dimensional volume of the ball.

Example

```
In[342]:= {VolumeSphere[r, 3], VolumeSphere[r, 5]}
```

Out[342]= $\left\{4\,\pi\,r^2,\ \frac{8\,\pi^2\,r^4}{3}\right\}$

```
In[343]:= {VolumeBall[r, 3], VolumeBall[r, 5]}
```

Out[343]= $\left\{\frac{4\,\pi\,r^3}{3},\ \frac{8\,\pi^2\,r^5}{15}\right\}$

See also: VolumeSphere, VolumeHn.

■ VolumeFormDiagonal (vfd)

This function computes the differential volume form for the set of diagonal matrices

$$\bigwedge_{i=1}^{n} da_i,$$

where the product is the wedge product.

See Sections 1.4 and 1.5.

VolumeFormDiagonal["a", n] ⟶ Form

"a" is a string which will be the name of a one-dimensional array symbol,
n is a positive integer representing the dimension of the form,
Form is the diagonal volume form based on the variables **a[i]**.

See also: VolumeFormGln, VolumeFormHn, VolumeFormUnimodular.

■ VolumeFormGln (vfg)

This function computes the differential volume form for the matrix group $GL(n, \mathbb{R})$ using the wedge product.

See Sections 1.4 and 1.5 and Proposition 1.4.3.

VolumeFormGln["g", n] ⟶ Form

"g" is a string which will be the name of a two-dimensional array symbol,
n is a positive integer representing the dimension of the matrices,
Form is the diagonal volume form based on the variables **g[i, j]**.

Example

```
In[62]:= VolumeFormGln[g, 2]

            d[g[1, 1]] ∧ d[g[1, 2]] ∧ d[g[2, 1]] ∧ d[g[2, 2]]
Out[62]= ------------------------------------------------------
                  (-g[1, 2] g[2, 1] + g[1, 1] g[2, 2])^2
```

See also: VolumeFormHn, VolumeFormDiagonal, VolumeFormUnimodular.

■ VolumeFormHn (vfh)

This function computes the differential volume form for the generalized upper half-plane.

See Definition 1.2.3 and Proposition 1.5.3.

VolumeFormHn["x", "y", n] ⟶ Form

"x" is a string which will be the name of a two-dimensional array symbol,
"y" is a string which will be used as the name of a one-dimensional array symbol,
n is a positive integer representing the dimension of the matrices which appear in the Iwasawa decomposition,
Form is the volume form based on the variables **x[i, j]**, **y[j]**.

Example

```
In[60]:= VolumeFormHn["x", "y", 3]

            d[x[1, 2]] ∧ d[x[1, 3]] ∧ d[x[2, 3]] ∧ d[y[1]] ∧ d[y[2]]
Out[60]= -------------------------------------------------------------
                              y[1]^3 y[2]^3
```

See also: VolumeFormGln, VolumeFormDiagonal, VolumeFormUnimodular.

■ VolumeFormUnimodular (vfu)

This function computes the differential volume form for the group of unimodular matrices, i.e. real upper-triangular with 1s along the leading diagonal.

See Sections 1.4 and 1.5.

VolumeFormHn["x", n] ⟶ Form

"x" is a string being the name of an array symbol,
n is a positive integer representing the dimension of the matrices,
Form is the volume form based on the variables **x[i, j]**.

Example

```
In[57]:= VolumeFormUnimodular[x, 3]

Out[57]= d[x[1, 2]] ∧ d[x[1, 3]] ∧ d[x[2, 3]]
```

See also: VolumeFormHn, VolumeFormDiagonal, VolumeFormGln.

■ VolumeHn (vhn)

This function computes the volume of the generalized upper half-plane using the volume element **VolumeFormHn**.

See Example 1.5.2 and Proposition 1.5.3.

VolumeHn[n] ⟶ Vol

n is an integer with **n** $\geq$ **2** representing the order of the upper half-plane, being the size of the matrices appearing in the Iwasawa form,
Vol is a real number.

See also: VolumeBall, VolumeSphere.

■ VolumeSphere (vsp)

This function computes the n-dimensional volume of the sphere S^n in $\mathbb{R}^{n+1}$.

VolumeSphere[r, n] ⟶ Vol

r is a CRE being the radius of the sphere,
n is the dimension of the sphere,
Vol is the volume of the sphere computed using **n**-dimensional Lebesgue measure.

See also: VolumeBall, VolumeHn.

■ Wedge,d

This function computes the Wedge product of any finite number of functions or differential forms in an arbitrary number of dimensions. It works with the differential form operator **d**. Note that these functions have a different construction from others in GL(n)pack , and have limited error control. An alternative to the function **Wedge** is the infix operator which may be entered into Mathematica by typing a backslash, and open square bracket, the word "Wedge" and then a closing square bracket. It prints like circumflex, but is not the same. Note that wedge products of vectors are not currently supported.

See Sections 1.4, 1.5, 5.6, 5.7 and 5.8.

Wedge[$\mathbf{f_1}, \mathbf{f_2}, \ldots, \mathbf{f_n}$] $\longrightarrow$ value

$\mathbf{f_i}$ is an expression or a form,
value is the wedge product of the functions or forms $\mathbf{f_i}$.

Example

In this example the function Wedge is used in conjunction with the differential form generator function **d**. Note that symbols, such as **a**, can be declared to be constant explicitly by setting, **d[a] = 0**.

```
In[64]:= Wedge[a d[x] + b d[y], u d[x] + v d[y]]

Out[64]= -b u d[x] ∧ d[y] + a v d[x] ∧ d[y]

In[65]:= d[a] = 0; d[b] = 0; d[c] = 0;

In[68]:= (a d[x] + b d[y]) ∧ (Exp[a x] d[y])

Out[68]= a e^(a x) d[x] ∧ d[y]

In[69]:= d[%]

Out[69]= 0

In[71]:= d[a y d[x] + b d[y] + c x d[z]]

Out[71]= -a d[x] ∧ d[y] + c d[x] ∧ d[z]

In[72]:= d[x^4]

Out[72]= 4 x^3 d[x]
```

See also: VolumeFormGln, VolumeFormHn.

■ WeylGenerator (wge)

This function returns a set of matrix generators for the Weyl subgroup of the group of integer matrices with determiant ± 1, which consists of all matrices with exactly one ± 1 in each row and column. A single call returns a single generator.

See Chapter 6.
Also see the manual entry for SpecialWeylGroup.

WeylGenerator[n,i,j] ⟶ g

n is a positive integer with $\mathbf{n \geq 2}$,
i is a positive integer with $\mathbf{i \leq n}$,
j is a positive integer with $\mathbf{i \neq j \leq n}$,
g a matrix with 1s along the leading diagonal and zeros elsewhere, except in the (**i**, **j**)th position where the value is -1 and (**j**, **i**)th position where the value is 1 and where the corresponding (**i**, **i**)th and (**j**, **j**)th diagonal elements are 0.

Example

```
In[168]:= MatrixForm[WeylGenerator[5, 2, 3]]

Out[168]//MatrixForm=
```

$$\begin{pmatrix} 1 & 0 & 0 & 0 & 0 \\ 0 & 0 & -1 & 0 & 0 \\ 0 & 1 & 0 & 0 & 0 \\ 0 & 0 & 0 & 1 & 0 \\ 0 & 0 & 0 & 0 & 1 \end{pmatrix}$$

See also: ModularGenerators, LongElement.

■ WeylGroup (wgr)

This function returns, for each whole number *n*, a list of all of the Weyl group of *n* by *n* permutation matrices.

See the proof of Proposition 1.5.3.

WeylGroup[n] ⟶ $\{\mathbf{m_1}, \mathbf{m_2}, \ldots, \mathbf{m_k}\}$

n is a positive integer with $\mathbf{n \geq 1}$,
$\mathbf{m_j}$ is a matrix with a single 1 in each row and column.

Example

```
In[5]:= Map[MatrixForm, WeylGroup[3]]
```

Out[5]= $\left\{\begin{pmatrix} 1 & 0 & 0 \\ 0 & 1 & 0 \\ 0 & 0 & 1 \end{pmatrix}, \begin{pmatrix} 1 & 0 & 0 \\ 0 & 0 & 1 \\ 0 & 1 & 0 \end{pmatrix}, \begin{pmatrix} 0 & 1 & 0 \\ 1 & 0 & 0 \\ 0 & 0 & 1 \end{pmatrix}, \begin{pmatrix} 0 & 1 & 0 \\ 0 & 0 & 1 \\ 1 & 0 & 0 \end{pmatrix}, \begin{pmatrix} 0 & 0 & 1 \\ 1 & 0 & 0 \\ 0 & 1 & 0 \end{pmatrix}, \begin{pmatrix} 0 & 0 & 1 \\ 0 & 1 & 0 \\ 1 & 0 & 0 \end{pmatrix}\right\}$

See also: WeylGenerators, SpecialWeylGroup, LongElement.

■ Whittaker (wit)

This function computes a symbolic interated integral representatin of the generalized Jacquet Whittaker function W_{Jacquet} (also written W_J) of order n, for $n \geq 2$, as defined by Equation (5.5.1). See Proposition 3.4.6, Section 3.4, Equation (5.5.1) and Equation (5.5.5). The algorithm uses the recursive representation of the Whittaker function defined by Stade (1990, Theorem 2.1) related to that used in the book as follows. Let W_S and W_S^* be Stade's Whittaker and Whittaker starred functions respectively and let Γ represent the gamma factors for either form. Then

$$Q \cdot W_S^* = \Gamma \cdot W_J = W_J^* = W_S$$

where

$$Q = I_\nu(y) \prod_{j=1}^{n-1} y_j^{-\mu_j},$$

$$\mu_j = \sum_{k=1}^{n-j} r_{j,k},$$

$$r_{j,k} = \left(\sum_{i=k}^{k+j-1} \frac{n\nu_i}{2} \right) - \frac{j}{2}.$$

Whittaker[z, v, m, u] $\longrightarrow$ **{coef, char, gam, value}**

z is an $\mathbf{n} \times \mathbf{n}$ non-singular CRE matrix,
v is a list of $\mathbf{n-1}$ CREs,
m is a list of $\mathbf{n-1}$ CREs $(\mathbf{m_i})$ representing a character $\psi_{\mathbf{m}}(\mathbf{x}) = \mathbf{e}^{2\pi \mathbf{i}(\sum_{1 \leq i \leq n-1} \mathbf{m_i x_{i,i+1}})}$,
u is a symbol which will be used to form the dummy variables in the iterated integral,
coef is the coefficient $\mathbf{c}_{\nu,\mathbf{m}}$ defined in Proposition 5.5.2,
char is the value of the character $\psi_{\mathbf{m}}(\mathbf{x})$ for $\mathbf{z} = \mathbf{x.y}$ the Iwasawa form,
gam is a product of terms as returned by **WhittakerGamma**,
value is a symbolic expression being the value of the Whittaker function at **My** where $\mathbf{z} = \mathbf{x.y}$ with parameters **v** and character $\psi_{1,1,1,\ldots,1}$. In this expression the K-Bessel function at a complex argument and parameter $\mathbf{K}_\nu(\mathbf{z})$, is represented by the noun function $\mathbf{K}[\nu, \mathbf{z}]$.

Example

```
In[3]:= Whittaker[{{y, 0}, {0, 1}}, {v}, {m}, w]
```

$$\text{Out[3]= } \left\{\text{m Abs[m]}^{-3+\text{v}},\ 1,\ \pi^{-\text{v}}\,\text{Gamma[v]},\ \frac{2.\ \pi^{\text{v}}\sqrt{\text{v}}\ \text{K}[-\frac{1}{2}+\text{v},\ 2\,\pi\,\text{y Abs[m]}]}{\text{Gamma[v]}}\right\}$$

```
In[1]:= Whittaker[IdentityMatrix[6], {1, 1, 1, 1, 1}, {1, 1, 1, 1, 1}, u]
```

$$\text{Out[1]= } \Bigg\{1,\ 1,\ \frac{4998263614010039534235859662310294189453125}{8192\,\pi^{92}},$$

$$\Bigg(268435456\,\pi^{92}\int_0^\infty\int_0^\infty\int_0^\infty\int_0^\infty\Bigg(\int_0^\infty\int_0^\infty \frac{1}{\text{u[1] u[2]}}\Bigg(\text{K}\Big[\frac{19}{10},\ \frac{2\,\pi\,\text{u[1] u[4]}}{\text{u[2] u[5]}}\Big]$$

$$\text{K}\Big[\frac{57}{10},\ \frac{2\,\pi\sqrt{1+\frac{1}{\text{u[1]}^2}}\ \text{u[3]}}{\text{u[4]}}\Big]\ \text{K}\Big[\frac{57}{10},\ \frac{2\,\pi\sqrt{(1+\text{u[1]}^2)\left(1+\frac{1}{\text{u[2]}^2}\right)}\ \text{u[4]}}{\text{u[5]}}\Big]$$

$$\text{K}\Big[\frac{57}{10},\ \frac{2\,\pi\sqrt{1+\text{u[2]}^2}\ \text{u[5]}}{\text{u[6]}}\Big]\Bigg)\ \text{du[2] du[1]}\Bigg)\ \text{K}\Big[\frac{25}{2},$$

$$2\,\pi\sqrt{1+\frac{1}{\text{u[3]}^2}}\Big]\ \text{K}\Big[\frac{25}{2},\ 2\,\pi\sqrt{(1+\text{u[3]}^2)\left(1+\frac{1}{\text{u[4]}^2}\right)}\Big]$$

$$\text{K}\Big[\frac{25}{2},\ 2\,\pi\sqrt{(1+\text{u[4]}^2)\left(1+\frac{1}{\text{u[5]}^2}\right)}\Big]$$

$$\text{K}\Big[\frac{25}{2},\ 2\,\pi\sqrt{(1+\text{u[5]}^2)\left(1+\frac{1}{\text{u[6]}^2}\right)}\Big]$$

$$\text{K}\Big[\frac{25}{2},\ 2\,\pi\sqrt{1+\text{u[6]}^2}\Big]\ \text{du[6] du[5] du[4] du[3]}\Bigg)\Big/$$

$$4998263614010039534235859662310294189453125\Bigg\}$$

See also: WhittakerGamma.

■ WhittakerGamma (wig)

This function returns the gamma factors for the generalized Jacquet Whittaker function. See Definition 5.9.2. Note that although this definition differs from that in (Stade, 1990), the gamma factor that it represents is the same.

WhittakerGamma[v] ⟶ value

v is a list of $\mathbf{n-1}$ CREs which, if any are numerical, satisfy $\Re \mathbf{v_i} > \mathbf{1/n}$,
value is an expression, being the product of the gamma factors for the Whittaker function of order **n**.

See also: Whittaker.

■ WMatrix (wmx)

This function returns the so-called w-matrix, with $(-1)^{\lfloor n/2 \rfloor}$ in the $(1, n)$th position and 1 in every other reversed diagonal position, a member of $SL(n, \mathbb{Z})$.
See Section 5.5.

WMatrix[n] $\longrightarrow$ **w**

n is a strictly positive integer with $\mathbf{n} \geq \mathbf{2}$,
w is an **n** by **n** matrix with each element 0, except the $(\mathbf{1}, \mathbf{n})$th which is $(-\mathbf{1})^{\lfloor \mathbf{n}/\mathbf{2} \rfloor}$ and every $(\mathbf{i}, \mathbf{n} - \mathbf{i} + \mathbf{1})$th which is 1 for $\mathbf{2} \leq \mathbf{i} \leq \mathbf{n}$.

Example

```
In[2]:= Map[MatrixForm, Map[WMatrix, {2, 3, 4, 5, 6}]]
```

$$\text{Out[2]=} \left\{ \begin{pmatrix} 0 & -1 \\ 1 & 0 \end{pmatrix}, \begin{pmatrix} 0 & 0 & -1 \\ 0 & 1 & 0 \\ 1 & 0 & 0 \end{pmatrix}, \begin{pmatrix} 0 & 0 & 0 & 1 \\ 0 & 0 & 1 & 0 \\ 0 & 1 & 0 & 0 \\ 1 & 0 & 0 & 0 \end{pmatrix}, \begin{pmatrix} 0 & 0 & 0 & 0 & 1 \\ 0 & 0 & 0 & 1 & 0 \\ 0 & 0 & 1 & 0 & 0 \\ 0 & 1 & 0 & 0 & 0 \\ 1 & 0 & 0 & 0 & 0 \end{pmatrix}, \begin{pmatrix} 0 & 0 & 0 & 0 & 0 & -1 \\ 0 & 0 & 0 & 0 & 1 & 0 \\ 0 & 0 & 0 & 1 & 0 & 0 \\ 0 & 0 & 1 & 0 & 0 & 0 \\ 0 & 1 & 0 & 0 & 0 & 0 \\ 1 & 0 & 0 & 0 & 0 & 0 \end{pmatrix} \right\}$$

See also: LongElement.

■ ZeroMatrix (zmx)

This function returns a zero matrix of given dimensions.

ZeroMatrix[m,n] $\longrightarrow$ **Z**

m is a strictly positive integer representing the number of matrix rows,
n is a strictly positive integer representing the number of matrix columns,
Z is a zero matrix with **m** rows and **n** columns.

See also: ConstantMatrix.

References

Ahlfors L. (1966), *Complex Analysis*, Int. Series in Pure and Applied Math. New York, McGraw–Hill Book Company, pp. 133–7.

Arthur J. (1979), *Eisenstein series and the trace formula*, in Automorphic forms, representations and L-functions (Proc. Sympos. Pure Math., Oregon State University, Corvallis, OR, 1977), *Proc. Sympos. Pure Math.*, Vol. XXXIII, Providence, RI, Amer. Math. Soc., Part 1, pp. 253–74.

(1989), *Unipotent automorphic representations: conjectures, Orbites unipotentes et représentations, II, Astérisque* No. 171–2, 13–71.

(2002), *A note on the automorphic Langlands group*, Dedicated to Robert V. Moody, *Canad. Math. Bull.* **45** (4), 466–82.

(2003), *The principle of functoriality. Mathematical challenges of the 21st century*, (Los Angeles, CA, 2000). *Bull. Amer. Math. Soc. (N.S.)* **40** (1), 39–53.

Artin E. (1923), *Über eine neue Art von L-Reihen, Hamb. Math. Abh.* **3**, 89–108. (Collected papers, Edited by Serge Lang and John T. Tate. Reprint of the 1965 original. New York–Berlin, Springer-Verlag, (1982), pp. 105–24).

(1927), *Beweis des allgemeinen Reziprozitätsgesetzes, Hamb. Math. Abh.*, 353–63. (Collected papers, Edited by Serge Lang and John T. Tate. Reprint of the 1965 original. New York–Berlin, Springer-Verlag, (1982), pp. 131–41).

(1930), *Zur Theorie der L-Reihen mit allgemeinen Gruppencharakteren, Hamb. Math. Abh.* **8**, 292–306. (Collected papers, Edited by Serge Lang and John T. Tate. Reprint of the 1965 original. New York–Berlin, Springer-Verlag, (1982) pp. 165–79).

Arveson W. (2002), *A short course on spectral theory, Graduate Texts in Mathematics* **209**, New York, Springer-Verlag.

Atkin A. O. and Li W. C. (1978), *Twists of newforms and pseudo-eigenvalues of W-operators, Invent. Math.* **48** (3), 221–43.

Aupetit B. (1991), *A primer on spectral theory, Universitext.* New York, Springer-Verlag.

Baker Alan. (1971), *Imaginary quadratic fields with class number "2", Ann. Math.* (2) **94**, 139–52.

Baker Andrew. (2002) *Matrix groups, An introduction to Lie group theory, Springer Undergraduate Mathematics Series*, London, Springer-Verlag.

Banks W. (1997), *Twisted symmetric-square L-functions and the nonexistence of Siegel zeros on GL(3), Duke Math. J.* **87** (2), 343–53.

Bernstein J. (2002), *Meromorphic Continuation of Eisenstein Series*, IAS/Park City Lecture Notes, Park City, UT.

Bernstein J. and Gelbart S. (Editors) (2003), *An Introduction to the Langlands Program*, Boston, Birkhäuser.

Bombieri E., Friedlander J. B. and Iwaniec H. (1987), *Primes in arithmetic progressions to large moduli. II*, *Ann. Math.* **277** (3), 361–93.

Boothby W. M. (1986), *An introduction to differentiable manifolds and Riemannian geometry*, Second edition, *Pure and Applied Mathematics* **120**, Orlando, FL, Academic Press.

Borel A. (2001), *Essays in the History of Lie Groups and Algebraic Groups*, *History of Math.* **21**, Amer. Math. Soc., London Math. Soc., pp. 5–6.

(1966), *Automorphic forms*, in Algebraic Groups and Discontinuous Groups, *Proc. Sympos. Pure Math.* Vol. IX, Amer. Math. Soc., pp. 199–210.

(1979), *Automorphic L-functions. Automorphic forms, representations and L-functions*, (Proc. Sympos. Pure Math., Oregon State University, Corvallis, OR, 1977), Part 2, *Proc. Sympos. Pure Math.*, Vol. XXXIII, Providence, RI, Amer. Math. Soc., pp. 27–61.

Borel A. and Harish-Chandra (1962), *Arithmetic subgroups of algebraic groups*, *Ann. Math.* (2) **75**, 485–535.

Bourbaki N. (1998a), *Algebra I, Chapters 1–3*, Translated from the French. Reprint of the 1989 English translation, *Elements of Mathematics*, Berlin, Springer-Verlag.

(1998b), *Lie groups and Lie algebras, Chapters 1–3*, Translated from the French, Reprint of the 1989 English translation, *Elements of Mathematics*, Berlin, Springer-Verlag.

(2003), *Algebra II, Chapters 4–7*, Translated from the 1981 French edition by P. M. Cohn and J. Howie, Reprint of the 1990 English edition. *Elements of Mathematics*, Berlin, Springer-Verlag.

Brauer R. (1947), *On Artin's L-series with general group characters*, *Ann. Math.* (2) **48**, 502–14.

Brown W. (1988), *A second course in linear algebra*, A Wiley-Interscience Publication, New York, John Wiley.

Bruggeman R. W. (1978), *Fourier coefficients of cusp forms*, *Invent. Math.* **45** (1), 1–18.

Buhler J. (1978), *Icosahedral Galois representations*, *Lecture Notes in Mathematics* **654**, Berlin–New York, Springer-Verlag.

Bump D. (1984), *Automorphic Forms on* $GL(3,\mathbb{R})$, *Lecture Notes in Mathematics* **1083**, Springer-Verlag.

(1987), *The Rankin–Selberg method: A survey*, in *Number Theory, Trace Formulas and Discrete Groups*, Symposium in honor of Atle Selberg, Oslo, Norway Edited by K. E. Aubert, E. Bombieri, D. Goldfeld, Academic Press, pp. 49–109.

(1997), *Automorphic forms and representations*, *Cambridge Studies in Advanced Mathematics* **55**, Cambridge, Cambridge University Press.

(2004), *Lie Groups*, *Graduate Texts in Mathematics* **225**, Springer-Verlag.

(to appear), *The Rankin–Selberg method: An introduction and survey*.

Bump D., Duke W., Hoffstein J. and Iwaniec H. (1992), *An estimate for the Hecke eigenvalues of Maass forms*, *Int. Math. Res. Notices* **4**, 75–81.

Bump D. and Friedberg S. (1990), *The exterior square automorphic L-functions on* $GL(n)$, Festschrift in honor of I. I. Piatetski-Shapiro on the occasion of his sixtieth birthday, Part II (Ramat Aviv, 1989), *Israel Math. Conf. Proc.*, 3, Weizmann, Jerusalem, pp. 47–65.

Bump D., Friedberg S. and Goldfeld D. (1988), *Poincaré series and Kloosterman sums for* $SL(3,\mathbb{Z})$, *Acta Arith.* **50**, 31–89.

Bump D. and Ginzburg D. (1992), *The symmetric square L-functions on* $GL(r)$, *Ann. Math.* **136**, 137–205.

Bump D., Ginzburg D. and Hoffstein J. (1996), *The symmetric cube*, *Invent. Math.* **125** (3), 413–49.

Buzzard K., Dickinson M., Shepherd-Barron N. and Taylor R. (2001), *On icosahedral Artin representations*, *Duke Math. J.* **109** (2), 283–318.

Buzzard K. and Stein W. (2002), *A mod five approach to modularity of icosahedral Galois representations*, *Pacific J. Math.* **203** (2), 265–82.

Capelli A. (1890), *Sur les opérations dans la théorie des formes algébriques*, *Math. Annalen* **37**, 1–27.

Carayol H. (1992), *Variétés de Drinfeld compactes, d'aprés Laumon, Rapoport et Stuhler*, (French. French summary) [Compact Drinfeld varieties, after Laumon, Rapoport and Stuhler] Séminaire Bourbaki, Vol. 1991/92. *Astérisque* No. 206, Exp. No. 756, 5, 369–409.

(2000), *Preuve de la conjecture de Langlands locale pour* GL_n*: travaux de Harris-Taylor et Henniart*, (French, French summary) [Proof of the local Langlands conjecture for GL_n: works of Harris and Taylor and of Henniart] Séminaire Bourbaki, Vol. 1998/99. *Astérisque* No. 266, Exp. No. 857, 4, 191–243.

Casselman W. and Shalika J. (1980), *The unramified principal series of p-adic groups II: The Whittaker function*, *Compositio Math.* **41**, 207–31.

Cassels J. W. S. and Fröhlich A. (Eds.) (1986), *Algebraic number theory*, Proceedings of the instructional conference held at the University of Sussex, Brighton, September 1–17, 1965, Reprint of the 1967 original, London, Academic Press [Harcourt Brace Jovanovich, Publishers].

Chandrasekharan K. and Narasimhan R. (1962), *Functional equations with multiple Gamma factors and the average order of arithmetical functions*, *Ann. Math.* **76**, 93–136.

Cogdell J. W. and Piatetski-Shapiro I. I. (1990), *The arithmetic and spectral analysis of Poincaré series*, *Perspectives in Mathematics*, 13, Boston, MA, Academic Press.

(1994), *Converse theorems for* GL_n, *Publ. Math. IHES* **79**, 157–214.

(1999), *Converse theorems for* GL_n*, II*, *J. Reine Angew. Math.* **507**, 165–88.

(2001), *Converse theorems for* GL_n *and their applications to liftings*, *Cohomology of Arithmetic Groups, Automorphic Forms, and L-functions*, Mumbai 1998, Tata Inst. of Fund. Res., Narosa, pp. 1–34.

Cogdell J. W., Kim H. H., Piatetski-Shapiro I. I. and Shahidi F. (2001), *On lifting from classical groups to* GL_N, *Inst. Hautes Études Sci. Publ. Math.* **93**, 5–30.

(2004), *Functoriality for the classical groups*, *Inst. Hautes Études Sci. Publ. Math.* **99**, 163–233.

Cohen H. (1993). *A Course in Computational Algebraic Number Theory*, *Graduate Texts in Mathematics*, Springer-Verlag.

Conrey J. B. and Ghosh A. (1993), *On the Selberg class of Dirichlet series: small degrees*, *Duke Math. J.* **72**, 673–93.

Conway J. B. (1973), *Functions of One Complex Variable*, *Graduate Texts in Mathematics*, Springer-Verlag.

Curtis M. L. (1984), *Matrix groups*, Second edition, *Universitext*, New York, Springer-Verlag.

Dabrowski R. (1993), *Kloosterman sums for Chevalley groups*, *Trans. Amer. Math. Soc.* **337** (2), 757–69.

Dabrowski R. and Fisher B. (1997), *A stationary phase formula for exponential sums over $\mathbb{Z}/p^m\mathbb{Z}$ and applications to $GL(3)$-Kloosterman sums*, *Acta Arith.* **80** (1), 1–48.

Davenport H. (1974), *Multiplicative Number Theory*, Markham Publishing (1967), Second Edition (Revised by H. Montgomery) *Graduate Texts in Mathematics*, Springer-Verlag.

Deligne P. (1974), *La conjecture de Weil. I*, (French) *Inst. Hautes Études Sci. Publ. Math.* **43**, 273–307.

(1977), *Application de la formule des traces aux sommes trigonométriques*, in SGA4$\frac{1}{2}$, *Springer Lecture Notes* **569**, 168–232.

Deuring M. (1933), *Imaginäre quadratische Zahlkörper mit der Klassenzahl (1)*, *Math. Zeit.* **37**, 405–15.

Drinfeld V. G. (1989), *Cohomology of compactified moduli varieties of F-sheaves of rank* 2, (Russian) Zap. Nauchn. Sem. Leningrad. Otdel. Mat. Inst. Steklov. (LOMI) **162** (1987), Avtomorfn. Funkts. i Teor. Chisel. III, 107–158, 189; translation in *J. Soviet Math.* **46** (2), 1789–821.

Duke W. and Iwaniec H. (1989), *Estimates for coefficients of L-functions, I, Automorphic Forms and Analytic Number Theory* (Montreal, PQ, 1989), CRM, pp. 43–47.

Edelen D. G. B. (1985), *Applied exterior calculus*, A Wiley-Interscience Publication, New York, John Wiley.

Frey G. (editor) (1994), *On Artin's Conjecture for Odd 2-dimensional Representations*, Berlin, Springer-Verlag.

Friedberg S. (1987a), *Poincaré series, Kloosterman sums, trace formulas, and automorphic forms for $GL(n)$*, Séminaire de Théorie des Nombres, Paris 1985–86, *Progr. Math.* **71**, Boston, MA, Birkhäuser, 53–66.

(1987b), *A global approach to the Rankin–Selberg convolution for $GL(3,\mathbb{Z})$*, *Trans. Amer. Math. Soc.* **300**, 159–78.

(1987c), *Poincaré series for $GL(n)$: Fourier expansion, Kloosterman sums, and algebreo-geometric estimates*, *Math. Zeit.* **196** (2), 165–88.

Friedberg S. and Goldfeld D. (1993), *Mellin transforms of Whittaker functions*, *Bull. Soc. Math. France*, **121**, 91–107.

Garrett P. (2002), *Volume of $SL(n,\mathbb{Z})\backslash SL(n,\mathbb{R})$ and $Sp_n(\mathbb{Z})\backslash Sp_n(\mathbb{R})$*, preprint.

Gauss C. F. (1801), *Disquisitiones Arithmeticae*, Göttingen; English translation by A. Clarke, revised by W. Waterhouse, New Haven, Yale University Press, 1966; reprinted by Springer-Verlag, 1986.

Gelbart S. (1984), *An elementary introduction to the Langlands program*, *Bull. Amer. Math. Soc. (N.S.)* **10** (2), 177–219.

Gelbart S. and Jacquet H. (1978), *A relation between automorphic representations of $GL(2)$ and $GL(3)$*, *Ann. Sci. École Norm. Sup. 4e série* **11**, 471–552.

Gelbart S., Lapid E. and Sarnak P. (2004), *A new method for lower bounds of L-functions*, (English. English, French summary) *C. R. Math. Acad. Sci.* Paris **339** (2), 91–4.

Gelbart S. and Shahidi F. (1988), *Analytic properties of automorphic L-functions*, *Perspectives in Mathematics* **6**, Boston, MA, Academic Press.

Gelfand I. M., Graev M. I. and Pyatetskii-Shapiro I. I. (1990), *Representation theory and automorphic functions*, Translated from the Russian by K. A. Hirsch. Reprint of the 1969 edition. *Generalized Functions* **6**, Boston, MA, Academic Press.

Godement R. (1966), *The spectral decomposition of cusp-forms*, *Algebraic Groups and Discontinuous Subgroups* (Proc. Sympos. Pure Math., Boulder, CO, 1965), pp. 225–34 Amer. Math. Soc., Providence, RI.

Godement R. and Jacquet H. (1972), *Zeta functions of simple algebras*, *Lecture Notes in Mathematics*, Vol. 260. Berlin–New York, Springer-Verlag.

Goldfeld D. (1974), *A simple proof of Siegel's theorem*, *Proc. Nat. Acad. Sci. USA*. **71**, 1055.

(1976), *The class number of quadratic fields and the conjecture of Birch and Swinnerton-Dyer*, *Ann. Scuola Norm. Sup. Pisa* (4) **3**, 624–63.

(1985), *Gauss' class number problem for imaginary quadratic fields*, *Bull. Amer. Math. Soc.* **13**, 23–37.

(1987), *Kloosterman zeta functions for $GL(n,Z)$*, *Proceedings of the International Congress of Mathematicians*, Vol. 1, 2 (Berkeley, CA, 1986), Providence, RI, Amer. Math. Soc., pp. 417–24.

Goldfeld D., Hoffstein J., Lieman D. (1994), *An effective zero free region*, in the Appendix of *Coefficients of Maass forms and the Siegel zero*, *Ann. of Math.* **140**, 177–81.

(2004), *The Gauss class number problem for imaginary quadratic fields*, in Heegner Points and Rankin L-series, *MSRI Publications* **49**, 25–36.

Goldfeld D. and Satnak P. (1983), *Sums of Kloosterman sums*, *Invent. Math.* **71** (2), 243–50.

Gross B. and Zagier D. B. (1986), *Heegner points and derivatives of L-series*, *Invent. Math.* **84**, 225–320.

Halmos P. (1974), *Measure Theory*, *Graduate Texts in Mathematics* **18**, Springer-Verlag, 250–62.

Hamburger H. (1921), *Über die Riemannsche Funktionalgleichung der ζ–Funktion*, *Math. Zeit.* **10**, 240–54.

Harcos G. (2002), *Uniform approximate functional equations for principal L-functions*, *IMRN* **18**, 923–32.

Harish-Chandra (1959), *Automorphic forms on a semi-simple Lie group*, *Proc. Nat. Acad. Sci. U.S.A.* **45**, 570–3.

(1966), *Discrete series for semisimple Lie groups, II. Explicit determination of the characters*, *Acta Math.* **116**, 1–111.

(1968), *Automorphic forms on semisimple Lie groups*, Notes by J. G. M. Mars. *Lecture Notes in Mathematics*, No. 62, Berlin–New York, Springer-Verlag.

Harris M. and Taylor R. (2001), *The geometry and cohomology of some simple Shimura varieties*, With an appendix by Vladimir G. Berkovich, *Ann. Math.* **151**, Princeton, NJ, Princeton University Press.

Hecke E. (1920), *Eine neue Art von Zetafunktionen und ihre Beziehungen zur Verteilung der Primzahlen, Zweite Mitteilung*, *Math. Zeit.* **6**, 11–51.

(1936), *Über die Bestimmung Dirichletscher Reihen durch ihre Funktionalgleichung*, *Ann. Math.* **113**, 664–99.

(1937a), *Über Modulfunktionen und die Dirichletschen Reihen mit Eulerscher Produktententwicklung, I*, *Ann. Math.* **114**, 1–28.

(1937b), *Über Modulfunktionen und die Dirichletschen Reihen mit Eulerscher Produktententwicklung, II*, *Ann. Math.* **114**, 316–51.

Heegner K. (1952) *Diophantine Analysis und Modulfunktionen*, *Math. Zeit.* **56**, 227–53.

Heilbronn H. (1934), *On the class number in imaginary quadratic fields*, *Quarterly J. Math.* **5**, 150–60.

Hejhal D. (1976), *The Selberg Trace Formula for $PSL(2,\mathbb{R})$, Volume 1*, *Lecture Notes in Mathematics* **548**, Springer-Verlag.

(1983), *The Selberg Trace Formula for $PSL(2,\mathbb{R})$, Volume 2*, *Lecture Notes in Mathematics* **1001**, Springer-Verlag.

Henniart G. (2000), *Une preuve simple des conjectures de Langlands pour $GL(n)$ sur un corps p-adique*, (French) [A simple proof of the Langlands conjectures for $GL(n)$ over a p-adic field] *Invent. Math.* **139** (2), 439–55.

(2002), *Progrès récents en fonctorialité de Langlands*, (French) [Recent results on Langlands functoriality] Seminaire Bourbaki, Vol. 2000/2001. *Astérisque* No. 282 Exp. No. 890, ix, 301–22.

Hewitt E. and Ross K. A. (1979), *Abstract harmonic analysis, Vol. I. Structure of topological groups, integration theory, group representations*, Second edition, *Grundlehren der Mathematischen Wissenschaften* (Fundamental Principles of Mathematical Sciences), **115**, Berlin–New York, Springer-Verlag.

Hoffstein J. and Lockhart P. (1994), *Coefficients of Maass forms and the Siegel zero*, *Ann. Math.* **140**, 161–81.

Hoffstein J. and Murty M. R. (1989), *L-series of automorphic forms on $GL(3,\mathbb{R})$*, *Théorie des Nombres* (Quebec, PQ, 1987), Berlin, de Gruyter, pp. 398–408.

Hoffstein J. and Ramakrishnan D. (1995), *Siegel zeros and cusp forms*, *IMRN* **6**, 279–308.

Ivić A. (1995), *An approximate functional equation for a class of Dirichlet series*, *J. Anal.* **3**, 241–52.

Iwaniec H. (1990), *Small eigenvalues of Laplacian for $\Gamma_0(N)$*, *Acta Arith.* **56**, 65–82.

(1995), *Introduction to the Spectral Theory of Automorphic Forms*, Revista Matemática Iberoamericana.

Iwaniec H. and Kowalski E. (2004), *Analytic Number Theory*, Amer. Math. Soc. Colloq. Publications, Vol. 53.

Iwasawa K. (1949), *On some types of topological groups*, *Ann. Math.* **50**, 507–58.

Jacquet H. (1967), *Fonctions de Whittaker associes aux groupes de Chevalley*, (French) *Bull. Soc. Math. France* **95**, 243–309.

(1972), *Automorphic Forms on $GL(2)$, Part II*, *Lecture Notes in Mathematics* No. 278, Springer-Verlag.

(1981), *Dirichlet series for the group $GL(N)$*, in *Automorphic Forms, Representation Theory and Arithmetic*, Tata Inst. of Fund. Res., Springer-Verlag, pp. 155–64.

(1997), *Note on the analytic continuation of Eisenstein series*, An appendix to "Theoretical aspects of the trace formula for $GL(2)$" [in Representation theory and automorphic forms (Edinburgh, 1996), 355–405, Proc. Sympos. Pure Math., 61, Amer. Math. Soc., Providence, RI, 1997; MR1476505 (98k:11062)] by

A. W. Knapp. Representation theory and automorphic forms (Edinburgh, 1996), pp. 407–412, *Proc. Sympos. Pure Math.*, 61, Amer. Math. Soc., Providence, RI.

(2004a), *Integral Representation of Whittaker Functions*, Contributions to automorphic forms, geometry, and number theory, Baltimore, MD, Johns Hopkins University Press, 373–419.

(2004b), *Kloosterman identities over a quadratic extension*, Ann. Math., no. 2, 755–79.

Jacquet H. and Langlands R. P. (1970), *Automorphic forms on $GL(2)$, Lecture Notes in Mathematics*, Vol. 114. Berlin–New York, Springer-Verlag.

Jacquet H., Piatetski-Shapiro I. I. and Shalika J. (1979), *Automorphic forms on $GL(3)$, I, II, Ann. Math.* **109**, 169–258.

(1983), *Rankin–Selberg convolutions, Amer. J. Math.* **105**, 367–464.

Jacquet H. and Shalika J. (1976/77), *A non-vanishing theorem for zeta functions of GL_n, Invent. Math.* **38** (1), 1–16.

(1981), *On Euler products and the classification of automorphic representations, Amer. J. Math.* I: **103**, 499–588.

(1990), *Rankin–Selberg convolutions: Archimedean theory*, in *Festschrift in Honor of I. I. Piatetski-Shapiro*, Part I, Jerusalem, Weizmann Science Press, pp. 125–207.

Kaczorowski J. and Perelli A. (1999), *On the structure of the Selberg class, I: $0 \le d \le 1$, Acta Math.* **182**, 207–41.

(2002), *On the structure of the Selberg class, V: $1 < d < 5/3$, Invent. Math.* **150**, 485–516.

Kato S. (1978), *On an explicit formula for class-1 Whittaker functions on split reductive groups over p-adic fields*, preprint, University of Tokyo.

Kim H. (2003), *Functoriality for the exterior square of GL_4 and the symmetric fourth of GL_2, J. Amer. Math. Soc.* **16**, 139–83.

Kim H. and Sarnak P. (2003), Appendix 2 in *Functoriality for the exterior square of GL_4 and the symmetric fourth of GL_2, J. Amer. Math. Soc.* **16** (1), 139–83.

Kim H. and Shahidi F. (2000), *Functorial products for $GL_2 \times GL_3$ and functorial symmetric cube for GL_2, C. R. Acad. Sci. Paris SÉr. I Math.* **331** (8), 599–604.

(2002), *Functorial products for $GL_2 \times GL_3$ and the symmetric cube for GL_2, Ann. Math.* (2), **155** (3), 837–93.

Korkine A. and Zolotareff G. (1873), *Sur les formes quadratiques, Ann. Math.*, **6**, 366–89.

Kowalski E. (2003), *Elementary theory of L-functions, I, II*, in *An Introduction to the Langlands Program*, Edited by J. Bernstein and S. Gelbart, Birkhäuser, pp. 1–71.

Kuznecov N. V. (1980), *The Petersson conjecture for cusp forms of weight zero and the Linnik conjecture. Sums of Kloosterman sums*, (Russian) *Mat. Sb. (N.S.)* **111** (153) (3), 334–83, 479.

Lafforgue L. (2002), *Chtoucas de Drinfeld et correspondance de Langlands*, (French) [Drinfeld shtukas and Langlands correspondence] *Invent. Math.* **147** (1), 1–241.

Landau E. (1918), *Über die Klassenzahl imaginär-quadratischer Zahlkörper, Göttinger Nachtr.*, pp. 285–95.

(1935), *Bemurkungen zum Heilbronnschen Satz, Acta Arith.* **I**, 1–18.

Lang S. (1969), *Analysis II*, Mass. Addison-Wesley.

(2002), *Algebra*, Revised third edition, *Graduate Texts in Mathematics* **211**, New York, Springer-Verlag.

Langlands R. P. (1966), *Eisenstein series*, and *Volume of the fundamental domain for some arithmetical subgroups of Chevalley groups*, in *Proc. Symp. Pure Math.*, Vol. IX, Algebraic Groups and Discontinuous Subgroups, Amer. Math. Soc., pp. 235–57.

(1970), *Problems in the theory of automorphic forms*, in *Lecture Notes in Mathematics* **170**, Springer-Verlag.

(1971), *Euler Products*, Yale University Press.

(1976), *On the functional equations satisfied by Eisenstein series*, *Lecture Notes in Mathematics* **544**, Springer-Verlag.

(1980), *Base Change for* $GL(2)$, Princeton, NJ, Princeton University Press.

Larsen M. (1988), *Estimation of* $SL(3,\mathbb{Z})$ *Kloosterman sums*, in the Appendix to *Poincaré series and Kloosterman sums for* $SL(3,\mathbb{Z})$, *Acta Arith.* **50**, 86–9.

Laumon G., Rapoport M. and Stuhler U. (1993), $\mathcal{D}$*-elliptic sheaves and the Langlands correspondence*, *Invent. Math.* **113** (2), 217–338.

Lavrik A. F. (1966), *Funktional equations of Dirichlet L-functions*, *Soviet Math. Dokl.* **7**, 1471–3.

Lax P. D. and Phillips R. S. (1976), *Scattering Theory for Automorphic Functions*, *Ann. Math.*, No. 87. Princeton, NJ, Princeton University Press, pp. 7–11.

Lenstra H. W. and Stevenhagen P. (2000), *Artin reciprocity and Mersenne primes*, *Nieuw Arch. Wisk.* (5) **1**, 44–54.

Li W. C. (1975), *Newforms and functional equations*, *Ann. Math.* **212**, 285–315.

(1979), *L-series of Rankin type and their functional equations*, *Ann. Math.* **244** (2), 135–66.

Lieman D. B. (1993), *The* GL(3) *Rankin–Selberg convolution for functions not of rapid decay*, *Duke Math. J.* **69** (1), 219–42.

Lindenstrauss E. and Venkatesh A. (to appear), Existence and Weyl's law for spherical cusp forms.

Luo W., Rudnick Z. and Sarnak P. (1995), *On Selberg's eigenvalue conjecture*, *Geom. Funct. Anal.* **5** (2), 387–401.

(1999), *On the generalized Ramanujan conjecture for* $GL(n)$, Automorphic forms, automorphic representations, and arithmetic (Fort Worth, TX, 1996), *Proc. Sympos. Pure Math.* **66**, Part 2, Amer. Math. Soc., Providence, RI, 301–10.

Maass H. (1949), *Über eine neue Art von nichtanalytischen automorphen Funktionen und die Bestimmung Dirichletscher Reihen durch Funktionalgleichungen*, *Ann. Math.* **122**, 141–83.

(1964), *Lectures on Modular Functions of One Complex Variable*, Tata Inst. Fund. Research, Bombay.

Macdonald I. G. (1979), *Symmetric Functions and Hall Polynomials*, Oxford, Clarendon Press, Ch. 1, Sect. 4, pp. 32–4.

Manin J. I. and Pančiškin A. A. (1977), *Convolutions of Hecke series, and their values at lattice points*, (Russian) *Mat. Sb. (N.S.)* **104(146)** (4), 617–51.

Miller S. D. (2001), *On the existence and temperedness of cusp forms for* $SL_3(\mathbb{Z})$, *J. Reine Angew. Math.* **533**, 127–69.

Miller S. D. and Schmid W. (2004), *Summation formulas, from Poisson and Voronoi to the present*, in *Noncommutative Harmonic Analysis*, *Progr. Math.* **220**, Boston, MA, Birkhäuser, 419–40.

Moeglin C. and Waldspurger J. L. (1995), *Spectral Decomposition and Eisenstein Series*, Cambridge University Press.

(1989), *Le spectre résiduel de $GL(n)$*, *Ann. Sci. École Norm. Sup.* (4) **22**, 605–74.

Mordell L. J. (1934), *On the Riemann hypothesis and imaginary quadratic fields with a given class number*, *J. London Math. Soc.* **9**, 289–98.

Molteni G. (2002), *Upper and lower bounds at $s = 1$ for certain Dirichlet series with Euler product*, *Duke Math. J.* **111** (1), 133–58.

Moreno C. (1985), *Analytic proof of the strong multiplicity one theorem*, *Amer. J. Math.* **107** (1), 163–206.

(2005), *Advanced analytic number theory: L-functions*, *Mathematical Surveys and Monographs* **115**, Providence, RI, Amer. Math. Soc.

Motohashi Y. (1997), *Spectral theory of the Riemann zeta-function*, *Cambridge Tracts in Mathematics* **127**, Cambridge, Cambridge University Press.

Müller W. (2004), *Weyl's law for the cuspidal spectrum of SL_n*, *C. R. Math. Acad. Sci. Paris* **338** (5), 347–52.

Munkres J. R. (1991), *Analysis on manifolds*, *Advanced Book Program*, Addison-Wesley Publishing Company, Redwood City, CA.

Murty M. R. (1994), *Selberg's conjectures and Artin L-functions*, *Bull. Amer. Math. Soc. (N.S.)* **31** (1), 1–14.

Murty M. R. and Murty V. K. (1987), *A variant of the Bombieri–Vinogradov theorem*, Number theory (Montreal, Que., 1985), *CMS Conf. Proc.*, Vol. 7, Amer. Math. Soc. Providence, RI.

Neunhöffer H. (1973), *Über die analytische Fortsetzung von Poincaréreihen*, *S.-B. Heidelberger Akad. Wiss. Math.-Natur. Kl.*, pp. 33–90.

Osborne M. S. and Warner G. (1981), *The theory of Eisenstein systems*, *Pure and Applied Math.* **99**, Academic Press.

Philips R. S. and Sarnak P. (1985), On cusp forms for co-finite subgroups of PSL (2, R), *Invent. Math.* **80** (2), 339–64.

Phragmén E. and Lindelöf E. (1908), *Sur une extension d'un principe classique de l'analyse*, *Acta Math.* **31**, 381–406.

Piatetski-Shapiro I. I. (1975), *Euler subgroups. Lie groups and their representations*, (*Proc. Summer School, Bolyai Jnos Math. Soc.*, Budapest, 1971), New York, Halsted, pp. 597–620.

Ramakrishnan D. (2002), Modularity of solvable Artin representations of GO(4)-type, *Int. Math. Res. Not.*, Number 1, 1–54.

Ramakrishnan D. and Wang S. (2003), *On the exceptional zeros of Rankin–Selberg L-functions*, *Compositio Math.* **135** (2), 211–44.

Rankin R. (1939), *Contributions to the theory of Ramanujan's function $\tau(n)$ and similar arithmetic functions, I and II*, *Proc. Cambridge Phil. Soc.* **35**, 351–6, 357–73.

Richert H. E. (1957), *Über Dirichletreihen mit Funktionalgleichung*, *Publ. Inst. Math. Acad. Serbe Sci.* **11**, 73–124.

Riemann B. (1859), *Ueber die Anzahl der Primzahlen unter einer gegebener Grösse*, *Monatsberichte der Berliner Akad*, Werke (2nd ed.), 145–53.

Roelcke W. (1956), *Analytische Fortsetzung der Eisensteinreihen zu den parabolischen Spitzen von Grenzkreisgruppen erster Art*, *Ann. Math.* **132**, 121–9.

Rohrlich D. (1989), *Nonvanishing of L-functions for $GL(2)$*, *Invent. Math.* **97**, 383–401.

Roquette P. (2000), *On the history of Artin's L-functions and conductors. Seven letters from Artin to Hasse in the year 1930, Mitteilungen der Mathematischen Gesellschaft in Hamburg*, Band 19, pp. 5–50.

Sarnak P. (1990), *Some Applications of Modular Forms*, Cambridge University Press **99**.

(2004), *Nonvanishing of L-functions on $\Re(s) = 1$, Contributions to Automorphic forms, Geometry, and Number Theory*, Baltimore, MD, Johns Hopkins University Press, pp. 719–32.

(to appear), The generalized Ramanujan conjectures.

Satake I., *Spherical functions and Ramanujan conjecture*, (1966) *Algebraic Groups and Discontinuous Subgroups* (Proc. Sympos. Pure Math., Boulder, CO, 1965), Amer. Math. Soc., Providence, RI, 258–64.

Selberg A. (1940), *Bemerkungen über eine Dirichlesche reihe, die mit der theorie der modulformer nahe verbunden ist, Arch. Math. Naturvid.* **43**, 47–50.

(1956), *Harmonic analysis and discontinuous groups in weakly symmetric Riemannian spaces with applications to Dirichlet series, J. Indian Math. Soc.* **20**, 47–87.

(1960), *On discontinuous groups in higher dimensional symmetric spaces, in Contributions to Function Theory*, Tata Institute, Bombay, 147–64.

(1963), Discontinuous groups and harmonic analysis, *Proc. Int. Congr. Mathematicians (Stockholm*, 1962), pp. 177–89.

(1965), *On the estimation of Fourier coefficients of modular forms, Proc. Sympos. Pure Math.*, Vol. VIII, Amer. Math. Soc., Providence, RI, 1–15.

(1991), *Old and new conjectures and results about a class of Dirichlet series, Collected Papers (Vol. II)*, Springer-Verlag, 47–63.

Serre J. P. (1977), *Modular forms of weight one and Galois representations*, Algebraic number fields: L-functions and Galois properties (*Proc. Symp. Durham University*, 1975) Academic Press, 193–268.

(1981), *Letter to J. M. Deshouillers.*

Shahidi F. (1981), *On certain L-functions, Amer. J. Math.* **103**, 297–355.

(1985), *Local coefficients as Artin factors for real groups, Duke Math. J.* **52** (4), 973–1007.

(1988), *On the Ramanujan conjecture and finiteness of poles for certain L-functions, Ann. Math.* **127**, 547–84.

(1989), *Third symmetric power L-functions for $GL(2)$, Comp. Math.* **70**, 245–73.

(1990a), *On multiplicativity of local factors*, in Festschrift in honor of I. I. Piatetski-Shapiro, Part II, *Israel Math. Conf. Proc.* **3**, Jerusalem, Weizmann, 279–89.

(1990b), *A proof of Langlands conjecture on Plancherel measures; complementary series for p-adic groups, Ann. Math.* **132**, 273–330.

(1992), *Twisted endoscopy and reducibility of induced representations for p-adic groups, Duke Math. J.* (1) **66**, 1–41.

Shalika J. A. (1973), *On the multiplicity of the spectrum of the space of cusp forms of GL_n, Bull. Amer. Math. Soc.* **79**, 454–61.

(1974), *The multiplicity one theorem for GL(n), Ann. Math.* **100**, 171–93.

Shimura G. (1971), *Arithmetic Theory of Automorphic Functions*, Princeton University Press.

(1973), *On modular forms of half-integral weight, Ann. Math.* **97**, 440–81.

(1975), *On the holomorphy of certain Dirichlet series*, *Proc. London Math. Soc.* **31** (3), 79–98.
Shintani T. (1976), *On an explicit formula for class-1 "Whittaker Functions" on $GL(n)$ over p-adic fields*, *Proc. Japan Acad.* **52**.
Siegel C. L. (1935), *Über die Classenzahl quadratischer Zahlkörper*, *Acta Arith.* **I**, 83–6.
(1936), *The volume of the fundamental domain for some infinite groups*, *Trans. Amer. Math. Soc.* **30**, 209–18.
(1939), *Einheiten quadratischer Formen*, *Hamb. Abh. Math.* **13**, 209–39.
(1980), *Advanced Analytic Number Theory*, Tata Institute Fund. Research, Bombay.
Soundararajan K. (2004), Strong multiplicity one for the Selberg class, *Canad. Math. Bull.* **47** (3), 468–74.
Stade E. (1990), *On explicit integral formulas for $GL(n,\mathbb{R})$-Whittaker functions*, *Duke Math. J.* **60** (2), 313–362.
(2001), *Mellin transforms of* GL(n,$\mathbb{R}$) *Whittaker functions*, *Amer. J. Math.* **123** (1), 121–61.
(2002), *Archimedean L-factors on $GL(n) \times GL(n)$ and generalized Barnes integrals*, *Israel J. Math.* **127**, 201–19.
Stark H. (1967), *A complete determination of the complex quadratic fields of class number one*, *Mich. Math. J.* **14**, 1–27.
(1972), *A transcendence theorem for class number problems I, II*, *Ann. Math.* (2) **94** (1971), 153–73 and **96**, 174–209.
Stevens G. (1987), *Poincaré series on $GL(r)$ and Kloostermann sums*, *Ann. Math.* **277** (1), 25–51.
Tate J. (1968), *Fourier analysis in number fields and Hecke's zeta function*, Thesis, Princeton (1950), also appears in *Algebraic Number Theory*, edited by J. W. Cassels and A, Frohlich, New York, Academic Press, pp. 305–47.
Tatuzawa T. (1951), *On a theorem of Siegel*, *Japan J. Math.* **21**, 163–78.
Taylor R. and Wiles A. (1995), *Ring-theoretic properties of certain Hecke algebras*, *Ann. Math.* (2) **141** (3), 553–72.
Terras A. (1985), *Harmonic Analysis on Symmetric Spaces and Applications I*, Berlin, Springer–Verlag.
(1988), *Harmonic Analysis on Symmetric Spaces and Applications II*, Berlin, Springer–Verlag.
Titchmarsh E. C. (1986), *The Theory of the Riemann Zeta Function*, Second Edition revised by D. R. Heath-Brown, Oxford Science Publications, Oxford University Press.
Tunnell J. (1981), *Artin's conjecture for representations of octahedral type*, *Bull. Amer. Math. Soc. (N.S.)* **5** (2), 173–75.
Venkov A. B. (1983), *Spectral theory of automorphic functions*, A translation of *Trudy Mat. Inst. Steklov.* 153 (1981), Proc. Steklov Inst. Math. 1982, no. 4(153).
Vinogradov I. and Takhtadzhyan L. (1982), *Theory of Eisenstein series for the group $SL(3,\mathbb{R})$ and its application to a binary problem*, *J. Soviet Math.* **18**, 293–324.
Wallach N. (1988), *Real Reductive Groups, I*, Academic Press.
Watkins M. (2004), *Class numbers of imaginary quadratic fields*, *Math. Comp.* **73** (246), 907–38.
Weil A. (1946), *Sur quelques résultats de Siegel*, *Brasiliensis Math.* **1**, 21–39.

(1949), *Numbers of solutions of equations in finite fields*, *Bull. Amer. Math. Soc.* **55**, 497–508.

(1967), *Über die Bestimmung Dirichletscher Reihen durch Funktionalgleichungen*, *Ann. Math.* **168**, 149–56.

Weyl H. (1939), *Classical Groups, their Invariants and their Representations*, Princeton University Press, Ch. 7, sect. 6, 202–3.

Whittaker E. T. and Watson G. N. (1935), *A Course in Modern Analysis*, Cambridge University Press.

Ye Y. (2000), *A Kuznetsov formula for Kloosterman sums on GL_n*, *Ramanujan J.* **4** (4), 385–95.

(1998), *Exponential sums for $GL(n)$ and their applications to base change*, *J. Number Theory* **68** (1), 112–30.

Zagier D. (1982), *The Rankin–Selberg method for automorphic functions which are not of rapid decay*, *J. Fac. Sci. University of Tokyo Sect. IA Math.* **28**, 415–37.

(unpublished), *Unpublished notes on Kuznetsov's formula on $SL(2,\mathbb{Z})$*.

Index

Printed in the United States
by Baker & Taylor Publisher Services